MINISTRY OF AGRICULTURE, FISHERIES AND FOOD

GD1 (replaces Technical Bulletin 7)

PLANT NEMATOLOGY

Edited by
J. F. SOUTHEY
A.D.A.S. Plant Pathology Laboratory, Harpenden

LONDON
HER MAJESTY'S STATIONERY OFFICE

First published 1960 (as Technical Bulletin 7)
Third edition 1978

In the interests of factual reporting, occasional references in this publication to trade names and proprietary products may be inevitable. No endorsement of named products is intended, nor is any criticism implied of similar products which are not mentioned.

ISBN 0 11 240907 5*

Contents

Preface

As with earlier editions the primary aim of the third edition of this Technical Bulletin is to provide a practical survey of scientific information on the plant-parasitic nematodes of temperate regions and their control for the benefit of agricultural nematologists and zoologists, advisory entomologists, plant pathologists, crop protection workers and others who have to advise on plant nematode problems. It should also continue to be of service to research workers, lecturers and students in nematology and related fields. The work has been extensively revised, several of the chapters have been rewritten and numbers 2, 3, 9–11 and 18 are entirely new.

In keeping with the emphasis on the technical background rather than applied economic aspects, this latest edition excludes the special section on cyst-forming nematodes in agriculture that was a feature of the earlier editions. These aspects are now mostly treated in detail in M.A.F.F. Crop Pests Bulletins also published by H.M.S.O., e.g., Bulletin 162 *Sugar Beet Pests,* Bulletin 186 *Cereal Pests* and Bulletin 187 *Potato Pests* (in preparation). Nevertheless brief information on economic importance of the principal genera and species is included in the relevant chapters and the section on Control (now Part II) gives due prominence to cyst nematodes and other important pest groups.

I wish to thank all the other authors who have contributed to this edition for their co-operation throughout the time it has taken to assemble the material for the final text, figures and plates. Among the typing and other staff who helped with preparation of manuscripts and figures, Mrs. Janet Cowland of Nematology Department, Rothamsted Experimental Station, deserves special mention for hours of patient work preparing graphs and diagrams, checking references, and typing. Mr. J. R. Morrison of Plant Pathology Laboratory did photographic work for many of the text figures and some of the plates.

Grateful acknowledgment is due to the following for permission to reproduce published illustrations: North Holland Publishing Company (Associated Scientific Publishers B.V.), Amsterdam for Figs. 1.1 and 1.5; Commonwealth Institute of Helminthology (Director, Dr. Sheila Wilmott) for Fig. 1.7 from *Nematodes of Tropical Crops* ed. J. E. Peachey (1969) and for Figs. 4.4, 4.5, 5.5 (part), 5.9 (part), 8.5, 8.6, 9.1, 9.2, 9.3, 9.4, 9.5, 9.6, 9.8, 9.10 and 10.6, all from *CIH Descriptions of Plant-parasitic Nematodes*; the Board of Editors of the *Annals of Applied Biology* for Fig. 2.5; Edward Arnold (Publishers) Ltd. and Prof. H. R. Wallace for Fig. 2.7 from *The Biology of Plant Parasitic Nematodes* (1963); the same publisher and Prof. F. G. W. Jones for Figs. 2.10, 2.11 and 7.1 from *Pests of Field Crops* by F. G. W. & M. G. Jones (1974); the Editor of the *Proceedings of the Helminthological Society of Washington* for Figs. 4.1, 4.2 and 9.9; the Editor of the *Journal of Helminthology* for Fig. 4.6; the Editor-in-Chief of *Nematologica* for Figs. 5.3, 5.8 (part), 5.9 (parts—courtesy Dr. M. Susana N. de A. Santos and Dr. I. A. Elmiligy), 8.1 (parts), 8.2, 8.3 (part); Dr. A. G. Whitehead and The Zoological Society of London for Fig. 5.5 (parts) and 5.10; the McGraw-Hill Book Company for Figs. 6.3, 6.7, 12.12 and 12.14 from *Principles of Nematology* by the late Prof. G. Thorne; Dr. A. M. Shepherd and Commonwealth Institute of Helminthology for Fig. 7.2; Methuen & Company for Figs. 8.3 (part), 12.2, 12.4, 12.5, 12.6, 12.7 (parts), 12.8, 12.10, 12.11, 12.15, all from T. Goodey's *Soil and Freshwater Nematodes* revised by the late Dr. J. B. Goodey (1963), and 12.3 from the first edition of the same work (1951); Prof. J. Heyns for Fig. 9.7 from *A Guide to the Plant and Soil Nematodes of South Africa* (A. A. Balkema, 1971); Martinus Nijhoff (Publishers), The Hague for Fig. 12.13 from *A Monograph of the Nematode Genera* Dorylaimus *Dujardin,* Aporcelaimus *n.g.,* Dorylaimoides *n.g.* and Pungentus *n.g.* by the late Prof. G. Thorne & H. H. Swanger (1936, reprinted 1957). Authors' personal acknowledgments are made where necessary at the end of their chapters and credit for photographs is given in the legends to the Plates.

J. F. SOUTHEY

December, 1976

Part I

Biology, Morphology and Identification

CHAPTER I

Structure and Classification of Nematodes

D. J. HOOPER

Rothamsted Experimental Station, Harpenden

Introduction

Nematodes are a large ubiquitous group of invertebrates with representatives in almost every kind of environment. Many that live in soil and water are 'free-living', some feeding on bacteria, algae, or fungi, others preying on small animals including other nematodes. Some feed on higher plants and can damage crops; a few of these can transmit plant viruses. Although nematodes require moisture to be active, some stages of certain species can withstand dry conditions for long periods. About 5000 species of plant and soil nematodes are known and of these nearly 2000 belong to genera that contain plant parasites although the feeding habits of many species are unknown. There are also many marine nematodes some of which have very elaborate cephalic structures and cuticular ornamentation; Gerlach & Riemann (1973, 1974) list 3680 marine species excluding Dorylaimida. Finally, many other nematodes are parasitic in vertebrates, molluscs and arthropods; some have a life cycle that may include hosts from more than one of these groups and there may be a free-living and/or infective stage in soil or water.

Nematodes are unsegmented, usually circular in cross section, and basically bilaterally symmetrical although the anterior region is radially symmetric. The body is usually elongate (*nema*—Greek for thread), filiform, cylindrical to fusiform tapering at each end especially towards the tail; the females are often more adapted to parasitism than the males and may swell to become saccate or spherical. Adults of plant and soil nematodes range from 0·2 mm to just over 1 cm long but most are in the 0·5 to 1·5 mm range. Some marine forms are larger and some animal parasites are relatively enormous—the common intestinal parasite *Ascaris lumbricoides* reaches 40 cm and the giant kidney worm *Dioctophyme renale* can be over 1 m long; female *Placentonema gigantissimum* from whales may reach 8·5 m long and be up to 2·5 cm wide.

The basic body structure (Fig. 1.1) consists of a flexible body wall (cuticle, hypodermis and somatic muscles) surrounding a tube-like gut (oesophagus and intestine). The body cavity between the gut and the body wall is usually regarded as a pseudocoelom containing a pseudocoelomic fluid; it also sometimes contains fibrous tissue and large cells (pseudocoelomocytes). However, Zacheo *et al.*, (1975) conclude that the body cavity is a true coelom. The gonads are self-contained tubes and usually lie alongside the intestine in the pseudocoelom. Nematodes do not have a respiratory or circulatory system.

Nematodes have affinities with Rotifera, Gastrotricha, Kinorhyncha, Priapulida and Nematomorpha which, together with Nematoda, are treated by Hyman (1951) as separate Classes in the phylum Aschelminthes. Paramonov (1962) also ranks Nematoda as a Class and discusses their origin and relationships to allied groups with interesting observations on their ecological groupings and classification. De Coninck (1965) regards nematodes as a Class in the Nemathelminthes (= Aschelminthes). Chitwood & Chitwood (1950) reasoned that nematodes should be regarded as a separate Phylum and this is supported in a critical review by Maggenti (1971, 1976) and also by Coomans (1975). Whether or not Nematoda are regarded as a Phylum or Class, they are usually divided into two Classes/Subclasses, Adenophorea (= Aphasmidia) and Secernentea (= Phasmidia) although Goodey (1963), who regarded Nematoda as a Class, eschewed this binary division. Andrássy (1974, 1976) proposes a rather different classification in which the Class Nematoda (Phylum Nemathelminthes) is divided into three Subclasses. This system may be a better representation of the relationships between the major nematode groups but it is too early to know if it will be generally accepted.

In this account the Nematoda are regarded as a Class (Nematodea) and classified in outline to Suborders in Table 1, page 19.

Most plant-parasitic nematodes belong to the Tylenchida but a few, including those that transmit plant viruses, belong to Longidoridae Suborder Dorylaimina or Trichodoridae Suborder Diphtherophorina (see Chapter 10). Most of the free-living soil nematodes other than the plant parasites are Rhabditida but some belong to other groups (see Chapter 12).

For a comprehensive review of the Nematodea see Chitwood & Chitwood (1950) and Grassé (1965); Gerlach & Riemann (1973, 1974) and Hope & Murphy (1972) give detailed classifications of marine nematodes. Chitwood (1969) and Anderson, Chabaud & Willmott (1974) include information on the biology and/or classification of animal parasitic nematodes. *Helminthological Abstracts, Series B, Plant Nematology* covers most of the current literature on plant and soil nematodes.

General Structure of Nematodes

CUTICLE

The cuticle is a multilayered, flexible, exoskeleton which covers the whole body and also lines all the natural body openings. It usually has three layers: cortex or cortical, median or matrix, basal or fibre layers; these layers are often further subdivided in different nematode groups, eight layers have been identified in *Xiphinema index* (Roggen, Raski & Jones, 1967) and five in *Aphelenchus avenae* (Johnson, Van Gundy & Thomson, 1970b). The cuticle consists mainly of proteins resembling secreted collagens, although the outermost layer is lipid in nature which probably acts as a barrier to chemicals.

The cuticle is nearly always marked by transverse striations (forming annulations) but these are sometimes too fine to be seen with a light microscope; sometimes, as in *Criconemoides* spp., the annulations are conspicuous giving a segmented appearance. Longitudinal ridges occur on some nematodes and raised lateral alae (lateral field—Fig. 1.1 B & C) extend from the neck to the tail region. The lateral field varies from a simple single-line interruption of the annulation to a raised area which itself may have additional longitudinal lines (incisures); the latter are sometimes partially crossed by the annulations to give an 'areolated' pattern. In some males the lateral fields expand to form adcloacal caudal alae (bursa—Fig. 1.1C) which have a clasping function during copulation. Sometimes

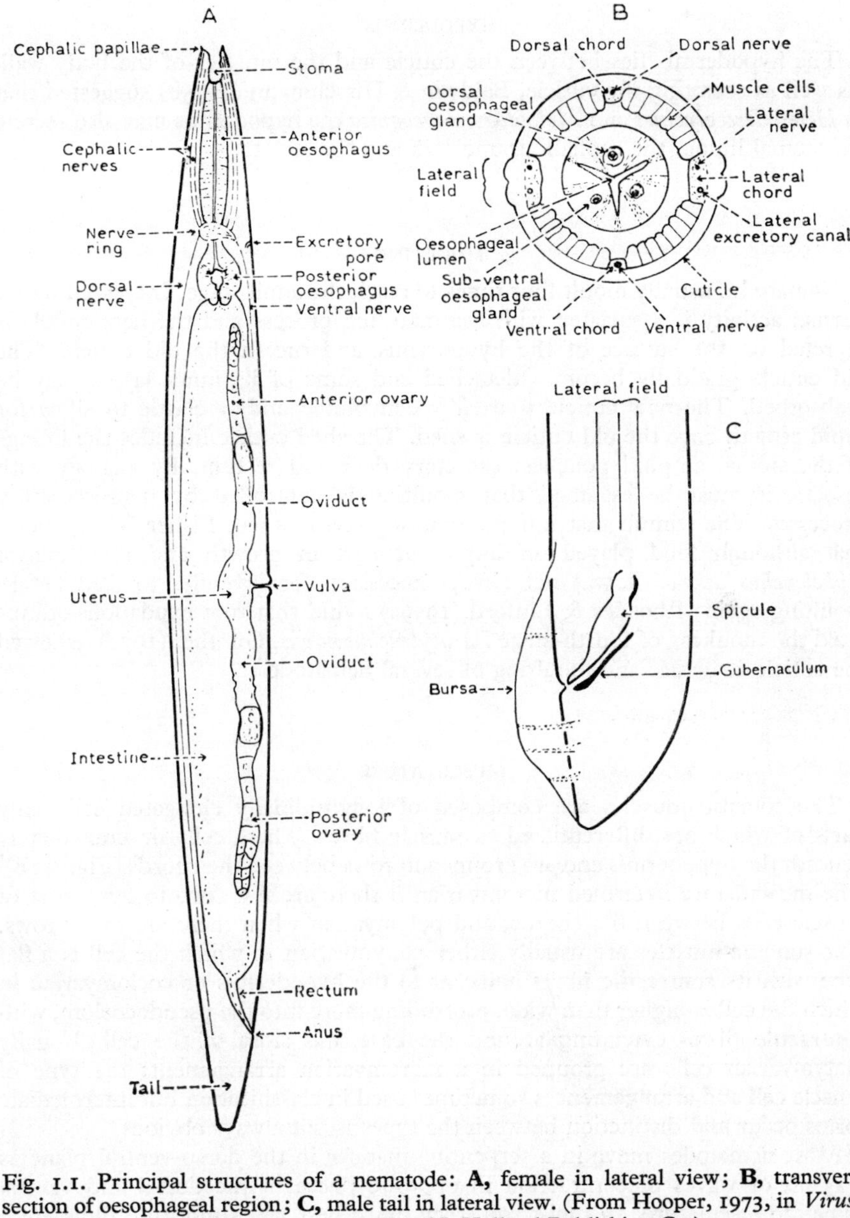

Fig. 1.1. Principal structures of a nematode: **A,** female in lateral view; **B,** transverse section of oesophageal region; **C,** male tail in lateral view. (From Hooper, 1973, in *Viruses and Invertebrates,* courtesy N. Holland Publishing Co.).

there are additional longitudinal ridges besides those of the lateral field. Bunonematid species (Rhabditida) are exceptional in having longitudinal ridges on one side of the body and rows of warts or tubercles on the other. Variations in cuticle ornamentation are of considerable taxonomic importance.

HYPODERMIS

The hypodermis lies between the cuticle and the muscles of the body wall. As well as secreting the cuticle, Baldwin & Hirschmann (1975b) suggested that in *Heterodera glycines* and *Meloidogyne incognita* the hypodermis may also secrete the stomal lining, the cephalic framework and the stylet.

MOULTING

Nematodes usually moult four times to reach the adult stage. Increased hypodermal activity is associated with the moulting process and the new cuticle is secreted on the surface of the hypodermis underneath the old cuticle. The old cuticle gradually becomes detached and some of its inner layers may be reabsorbed. The new cuticle is usually convoluted and/or elastic to allow for rapid growth once the old cuticle is shed. The shed cuticle includes the linings of the stoma, amphid pouches, excretory duct and rectum. By analogy with insects, it must be assumed that moulting is controlled by neurosecretory processes. The stimuli that initiate moulting seem varied. Fisher (1970) found that although food played an important part in growth and moulting of *Aphelenchus avenae* it was not always necessary for juveniles to feed before moulting again; Rhoades & Linford (1959a) found that root exudations stimulated the moulting of fourth-stage *Paratylenchus nanus*. Günther (1972) reviewed the cuticle structure and moulting of several nematodes.

MUSCULATURE

The somatic muscles are composed of longitudinally elongated cells only parts of which are differentiated as muscle fibres. These cells lie immediately beneath the hypodermis and are grouped in rows between the chords (Fig. 1.1B). The musculature is termed meromyarian if there are few (two to five) rows of muscle cells between the chords, and polymyarian when there are many rows. The somatic muscles are usually either platymyarian in which the cell is a flat type with its contractile fibres adjacent to the hypodermis or coelomyarian in which the cell is higher than wide, protruding more into the pseudocoelom, with contractile fibres extending around the base and sides of the cell. Usually platymyarian cells are grouped in a meromyarian arrangement; the type of muscle cell and arrangement is sometimes used in classification but intermediate forms occur and distinction between the types is not always obvious.

Most nematodes move in a serpentine manner in the dorso-ventral plane as a result of waves of contraction in opposite phase in the dorsal and ventral somatic musculature. In moist soil, they move best when moisture films are 2 to 5 μm thick at matric potentials less than about 200 cm water suction when their undulating motion exerts a force which propels them forward (see Chapter 2). A few plant-parasitic nematodes (e.g., *Aphelenchoides* spp.) move vigorously enough to swim when suspended in water and the vinegar eelworm (*Turbatrix aceti*) swims for most of its active life.

Nematodes also have specialized muscles such as those connected with the mouth parts, oesophagus, anus, vulva/vagina, spicules and cloaca, and those in the male tail that aid copulation.

NERVOUS SYSTEM

The most readily seen part of the nervous system is a commissure ('nerve ring'—Fig. 1.1A) that encircles the gut usually in the region of the oesophageal isthmus although Shepherd & Clark (1976) found that the nerve ring surrounded the oesophageal gland and anterior intestine in *Hexatylus viviparus*. Several nerves extend anteriorly from the nerve ring connecting with the amphids and papillae of the cephalic region. Posteriorly the body is served by proliferations from a main nerve in each of the four hypodermal chords, the one in the ventral chord being the most important. Ganglia occur throughout the body and, in the rectal/cloacal region, there are larger commissures and ganglia interconnecting the main body nerves. Small hypodermal commissures, usually in a ventro-lateral position, are visible on some nematodes especially in the Tylenchida; these include structures called cephalids, hemizonid, hemizonion and caudalid which are biconvex and refractive in lateral view. The hemizonid is the most conspicuous, extending through about three body annules just posterior to the nerve ring and close to the excretory pore. Its position relative to the excretory pore is often of taxonomic importance. The smaller hemizonion is one to three body widths posterior to the hemizonid and the caudalid is a similar inconspicuous structure just anterior to the anal region. The cephalids are about one body annule wide and encircle the neck. The anterior cephalids lie just behind the lip region and the posterior ones a few annules further back.

As indicated by Chitwood & Chitwood (1950) and De Coninck (1965), the basic head structure is hexaradiate with six lips. In nematodes, the head is often called the lip region and is usually regarded as being that part of the body from the oral aperture to the base of the head skeleton (cheilostom). A distinct head skeleton is not visible in some nematodes but the base of the head region is often indicated by a constriction or depression in the body contour. The head region may have fine transverse annulations, sometimes divided by longitudinal grooves, which although visible with a scanning electron microscope are sometimes too fine to be seen with a light microscope. 'Lips' often mean the sectors comprising the whole of the head region; however, the term sometimes applies only to the structures close to the oral aperture. The cephalic sensory organs comprise an inner and outer ring, each of six labial papillae, plus an outermost ring of four cephalic papillae (two subventral, two subdorsal) and a pair of lateral amphid openings (Fig. 1.2). In some Adenophoria this basic pattern can be seen in *en face* views with a light microscope although the lips may be fused. Scanning electron microscope studies by Sher & Bell (1975) of the lip pattern of some Tylenchida show it is very variable and, although the amphid openings are usually visible, the cephalic and labial papillae may not be. De Grisse, Lippens & Coomans (1974) noted that although the nerve arrangement for some Tylenchida follows the basic pattern, cephalic and outer labial nerves sometimes end within the cuticle without projecting papillae; the inner labial nerves sometimes open in the vestibule. Although cilia were formerly thought to be absent from nematodes, electron microscopy has shown that cilia are numerous in the amphids and that cilium-like structures are present in various papillae, especially cephalic ones.

Amphids occupy much space in the head region; they probably serve as chemoreceptors. The position and shape of the amphid aperture and associated pouch are important in systematics. In Tylenchid nematodes the apertures are labial and vary from rounded or oval pores to longitudinal or transverse slits or clefts. In Adenophoria the amphid apertures are post-labial and in some free-living, especially marine, forms they are characteristically shaped and may be ring-like or spiral. The amphid opening connects via a duct to a sensilla pouch; the sensilla consists of a number of cilia connected to the amphidial nerve. Baldwin & Hirschmann (1973, 1975a) described in detail the fine structure of the cephalic sense organs of *Meloidogyne* and *Heterodera* males. Occasionally a mucous-like secretion is emitted from the amphidial openings.

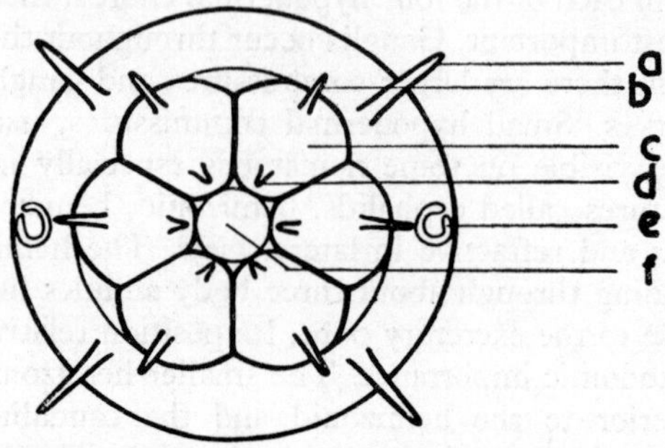

Fig. 1.2. Diagram of head of nematode, *en face* view: **a,** cephalic papilla; **b,** outer labial papilla; **c,** lip; **d,** inner labial papilla; **e,** amphid; **f,** stoma. (From Jones, 1965).

The phasmids are a pair of lateral sensory organs usually on the tail of Secernentian nematodes. They vary in size from minute innervated pores to plate-like structures (scutellae) or pouches but they are not readily visible in all species in this group. Like the amphids they also probably function as chemoreceptors.

EXCRETORY SYSTEM

The subdivision of Nematoda into Secernentia and Adenophoria is based on the type of excretory system (see Chitwood & Chitwood, 1950; Chitwood, 1958). However, this is not always helpful because the excretory system has not been observed in some groups and is difficult to interpret in others. An excretory duct opens to the exterior via an excretory pore which is usually mid-ventral near the nerve ring (Fig. 1.1A). An exception among plant-parasitic nematodes is *Tylenchulus semipenetrans,* where the pore is posterior and seems to have a secretory as well as excretory function (Maggenti, 1962). In Secernentia the excretory system is tubular and typically consists of two longitudinal canals each situated in the lateral chord (Fig. 1.1B). The canals are linked in the oesophageal region by a short transverse canal which connects via a median duct to the excretory pore, thus forming an 'H'. Paired subventral excretory glands may also be linked to the excretory duct as in the Rhabditoids. The anterior parts of the lateral canals may be absent, forming an inverted 'U' as in Cephaloboid species or only one lateral canal may be present giving an asymmetric system as in Tylenchoidea. Narang (1972) describes the ultrastructure and discusses the development of the excretory system of three Secernentian nematodes. In Adenophoria the system is thought more primitive and is basically a gland cell (the renette) connected by a duct of variable length to the excretory pore as in *Plectus* (see Maggenti, 1961a).

DIGESTIVE SYSTEM

The gut is an internal tube beginning at the oral opening and ending at the ventrally placed anus in juveniles and females and at the cloaca in males. A section at each end is lined with invaginated cuticle, presumably these portions are ectodermal in origin. The system is divided into three main sections: the stomodaeum (foregut), the mesenteron (intestine or midgut) and the proctodaeum (rectum or cloacal region). The cuticle-lined stomodaeum includes the stoma or buccal cavity (pharynx *sensu stricto*) and oesophagus (pharynx *sensu lato*); the oral aperture (cheilostom) is also sometimes included. The stomodaeum ends at its junction with the intestine where there is an oesophago-intestinal valve whose structure varies. This valve is prominent in some nematodes, and in Dorylaimid groups protrudes into the intestine as a heart-shaped organ and consequently is sometimes called a cardia. The structure of the stomodaeum is important in the classification of nematodes.

The buccal cavity varies from a simple tube to a wide chamber which may have small and/or large teeth, or may be armed with a hollow axial spear; sometimes a prominent head skeleton is also present. The general structure of this region indicates the method of feeding. Nematodes with a relatively simple tubular stoma (Rhabditids) are usually bacterial feeders, those with strongly developed teeth (e.g., Mononchids, see Chapter 12) are predators. Those known to feed on plant cells have a hollow axial spear resembling a hypodermic needle, except the Trichodorids which have an elongated, solid-tipped mural tooth. Plant-parasitic Tylenchida are thought to have evolved from Rhabditid forms, the head skeleton being formed by modification of the cheilostom and the hollow mouth spear by fusion of the sclerotized rhabdions (see Fig. 1.3). The different rhabdion components are most easily seen in Cephalobid nematodes. The Tylenchid spear has a ventral aperture close to its tip and because it is formed *in situ* from the wall of the stoma it is called a stomatostylet. It consists of three basic parts—an anterior cone, a middle shaft and three basal knobs (one dorsal and two subventral). There is much variation in the length and thickness of the stylet, and particularly in the size and shape of the basal knobs which may be large as in *Hoplolaimus* spp. or virtually absent as in *Aphelenchus* spp. (see Fig. 1.4).

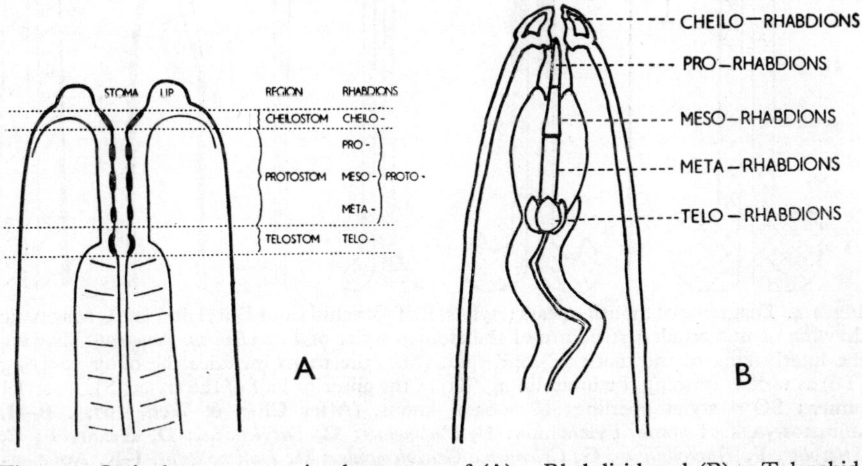

Fig. 1.3. Cuticular structures in the stomas of (A) a Rhabditid and (B) a Tylenchid. (From Jones, 1965).

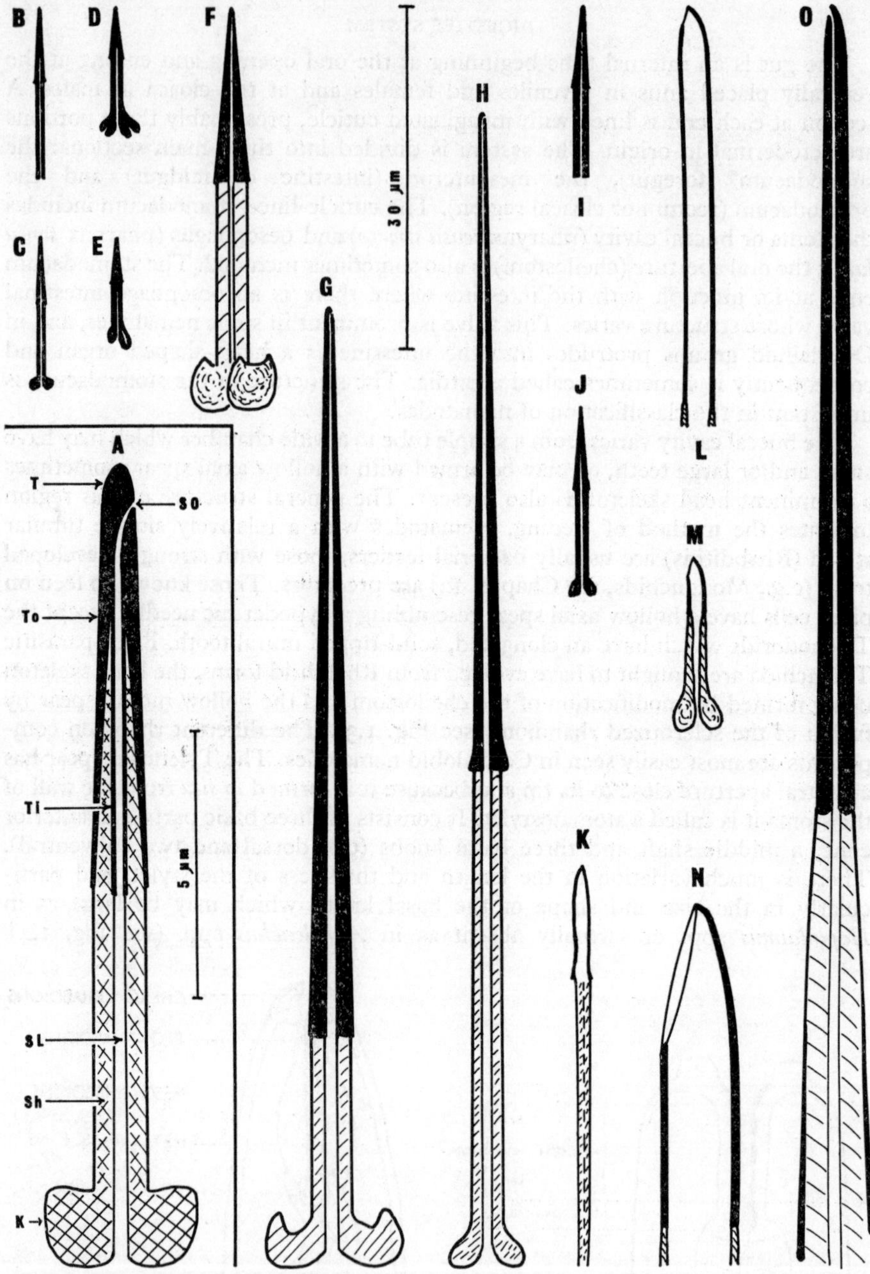

Fig. 1.4. Diagrams of mouth-spears (stylets) in Tylenchida and Dorylaimida. **A,** schematic drawing of the detailed structure of the stomatostylet of *Pratylenchus penetrans* showing the interlocking of the tooth (T) and shaft (Sh); the tooth includes the outer covering (To) as well as forming the inner lining (Ti) of the anterior half of the stylet; SL = stylet lumen; SO = stylet opening; K = basal knobs. (After Chen & Wen, 1972). **B–H,** stomatostylets of some Tylenchina: **B,** *Psilenchus*; **C,** *Ditylenchus*; **D,** *Hexatylus*; **E,** *Stictylus*; **F,** *Hoplolaimus*; **G,** *Criconema*/*Criconemoides*; **H,** *Dolichodorus*. **I–K,** Aphelenchina: **I,** *Aphelenchus*; **J,** *Aphelenchoides*; **K,** *Anomyctus*. **L–O,** odontostylets of some Dorylaimida: **L,** *Pungentus*; **M,** *Doryllium*; **N,** *Dorylaimus* (basic stylet type); **O,** *Longidorus*.

Muscles are attached to the base of the stylet (base of odontophore in Dorylaims) and extend forward to the base of the lip region, or to the head skeleton when present. When these muscles contract the stylet is thrust forward. Yuen (1967) and Chen & Wen (1972) studied the ultrastructure of the stomatostylet of *Ditylenchus dipsaci* and *Pratylenchus penetrans* respectively.

The axial spear of Dorylaimoid nematodes (Fig. 1.4) is thought to have evolved from a mural tooth and has a dorsal aperture; it is formed in the anterior part of the oesophagus and moves forward during moulting to replace the previous spear (see Coomans & De Coninck, 1963; also Chapter 10) at the same time the spear extension is reformed *in situ*. The Dorylaimoid spear is often called an odontostyle and its extension or basal part the odontophore (Fig. 1.5A & B). Care is necessary in interpreting the term spear or stylet in Dorylaims because the odontophore may or may not be included. Aboul-Eid (1969), Lopez-Abella *et al.*, (1967), Taylor *et al.*, (1970) and Wright (1965) show that the odontostyle has an eccentrically placed lumen with a dorsal slit throughout its length.

The Trichodorid spear is a ventrally curved, elongated, mural tooth (Fig. 1.5C). Its anterior tip is replaced during moulting and the replacement tip or tooth is formed in the buccal cavity and lies in a dorsal cleft in the basal part (onchiophore) of the spear. The whole spear is usually called an onchiostyle(t). Hirumi *et al.*, (1968), Bird (1971a) and Robertson & Taylor (1975) showed that the anterior tip of the onchiostyle is solid and that the pharyngeal lumen is ventral to it. Wyss (1971) described the feeding of *Trichodorus similis*, which establishes a 'feeding tube' through which plant cell contents are sucked into the pharyngeal lumen.

The oesophagus (pharynx *sensu lato*) is basically a pumping organ to transport food from the low pressure of the buccal cavity to the higher pressure of the intestine. In its simplest form, as in some Enoplid nematodes, the oesophagus is a cylinder with a lumen that is triradiate in section throughout at least part of its length. Usually much of the oesophagus is muscular with an internal cuticle-like layer and separated externally from the pseudocoelom by a basal lamella. Mononchids (see Figs. 12.16, 12.17) have a well developed cylindrical oesophagus. Oesophageal shape and musculature together with the position of valves, glands and gland-duct openings are important characters used in nematode systematics. In the commonly occurring free-living Rhabditid nematodes the muscular oesophagus is divided into procorpus, corpus or median oesophageal bulb, isthmus and prominent basal bulb having a complicated valve structure (Fig. 1.6). Doncaster (1962) described the form and function of the Rhabditid oesophagus. In *Panagrolaimus* and *Cephalobus* the median bulb is reduced or absent but the posterior bulb is similar to that of *Rhabditis* (Fig. 1.6), whereas in *Diplogaster* a valvular muscular median bulb is retained but the posterior bulb is glandular (Fig. 1.6). Many Tylenchida have a narrow procorpus, a valvular median bulb, isthmus, and basal oesophageal glands that may be retained in a pyriform bulb abutting onto the intestine (Fig. 1.7a) or may overlap the intestine (Fig. 1.7c). In Rhabditida and Tylenchida there are a dorsal and two subventral oesophageal glands situated within the wall of the posterior oesophagus; the dorsal gland duct is at or near the front of the oesophagus in Rhabditida and Tylenchoidea and the subventral gland ducts open into the base of the median bulb just behind the valve plates (Fig. 1.7a–e). In Aphelenchoidea, however, the dorsal gland duct also opens into the median bulb just in front of the valve plates (Fig. 1.7f). Yuen (1968) gave details of the

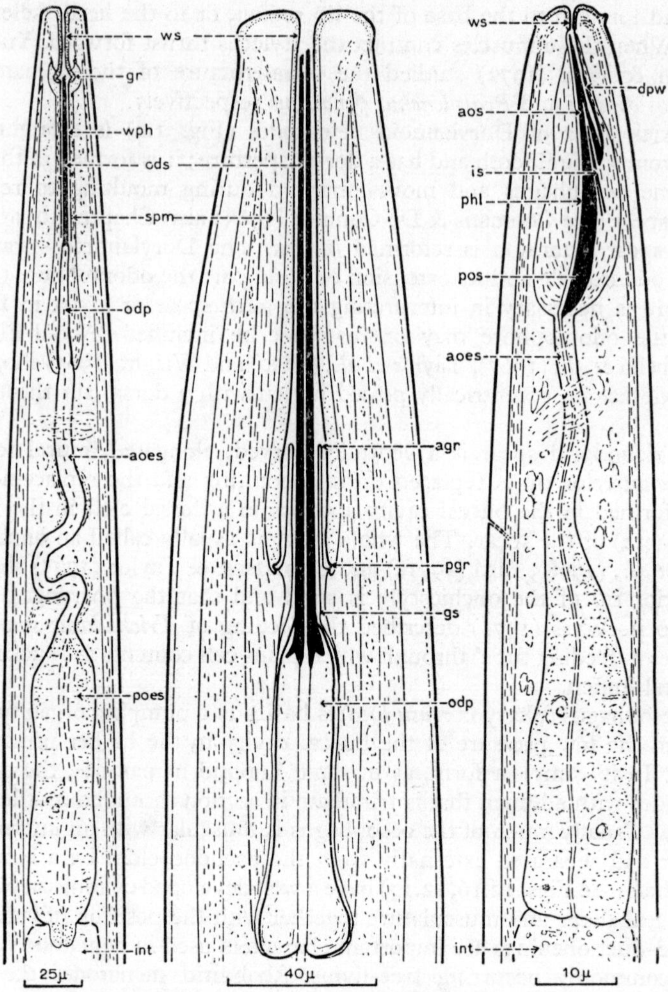

Fig. 1.5. **A,** *Longidorus elongatus*—anterior region; **B,** *Xiphinema diversicaudatum*—spear structure; **C,** *Paratrichodorus christiei*—anterior region.
agr = anterior guide ring; aoes = anterior oesophagus; aos = anterior outer spear; dpw = dorsal pharyngeal wall; gr = guide ring; gs = guiding sheath; int = intestine; is = inner spear; odp = odontophore; ods = odontostyle; phl = pharyngeal lumen; pgr = posterior guide ring; pos = posterior outer spear; poes = posterior oesophagus; spm = stylet protractor muscles; wph = wall of pharynx; ws = wall of stoma. (From Hooper, 1973, in *Viruses and Invertebrates,* courtesy N. Holland Publishing Co.).

ultrastructure of the oesophagus of *Ditylenchus dipsaci* and Doncaster (1966) described the feeding action of *D. destructor* and *D. myceliophagus*. Most Dorylaimida have a narrow, cylindrical, non-muscular anterior oesophagus with a circular lumen which expands into a posterior, muscular, cylindrical bulb with a triradiate lumen, the whole oesophagus having a bottle-shaped appearance (Fig. 1.5A). Basically there is a dorsal gland plus two pairs of subventral glands which, together with their duct openings, are within the basal bulb. The position of the glands and their duct openings is of taxonomic importance in Dorylaimida (see Coomans & Loof, 1970; Loof & Coomans, 1972). The Longidorids in this group, which include plant-virus vector species, have only one pair of subventral glands (see Chapter 10). The oesophago-intestinal valve (cardia) in some Dorylaims does not have an obvious lumen into the intestine but Seymour & Shepherd (1974) showed that, when closed, the valve in *Thornenema wickeni* is a convoluted path of closely apposed pairs of three-layered cell membranes which separate under pressure. In the Belondiroidea, order Dorylaimida, the basal portion of the oesophagus is surrounded by a sheath of muscles which are sometimes spirally arranged.

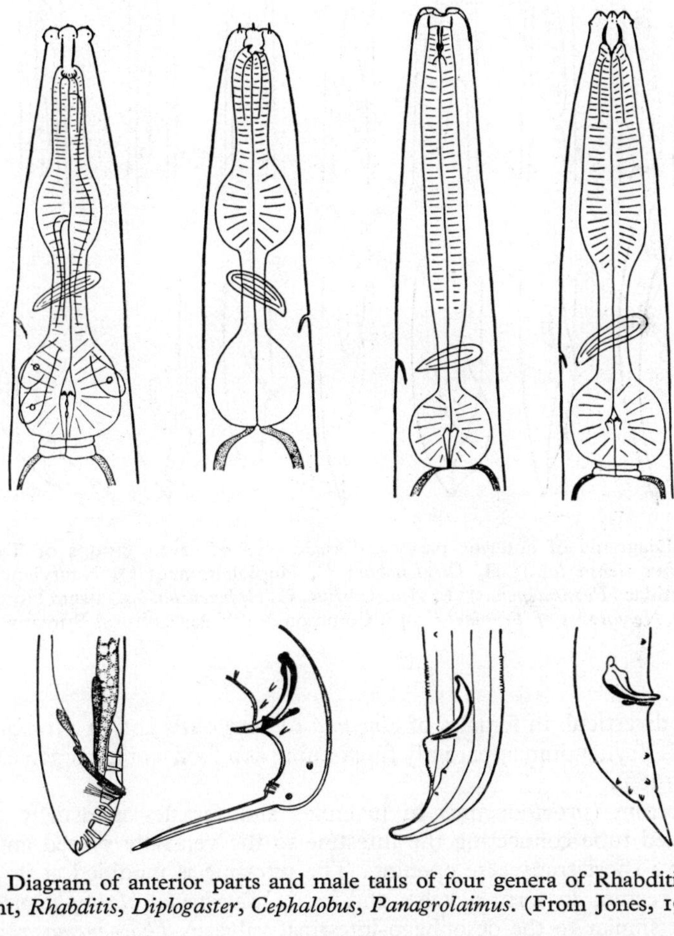

Fig. 1.6. Diagram of anterior parts and male tails of four genera of Rhabditida: left to right, *Rhabditis*, *Diplogaster*, *Cephalobus*, *Panagrolaimus*. (From Jones, 1965).

The intestine is a simple tube consisting of a single layer of epithelial cells. It has three regions: an anterior or ventricular, a mid-intestinal and a posterior pre-rectal, the last being most readily seen in some Dorylaims. The intestinal lumen is lined with microvilli which vary in number and shape in different species and according to their position along the intestine as noted by Hirumi & Hung (1969) for *Paratrichodorus christiei* and Shepherd & Clark (1976) for *Hexatylus viviparus*. Occasionally the intestine is modified. Wu (1968) noted

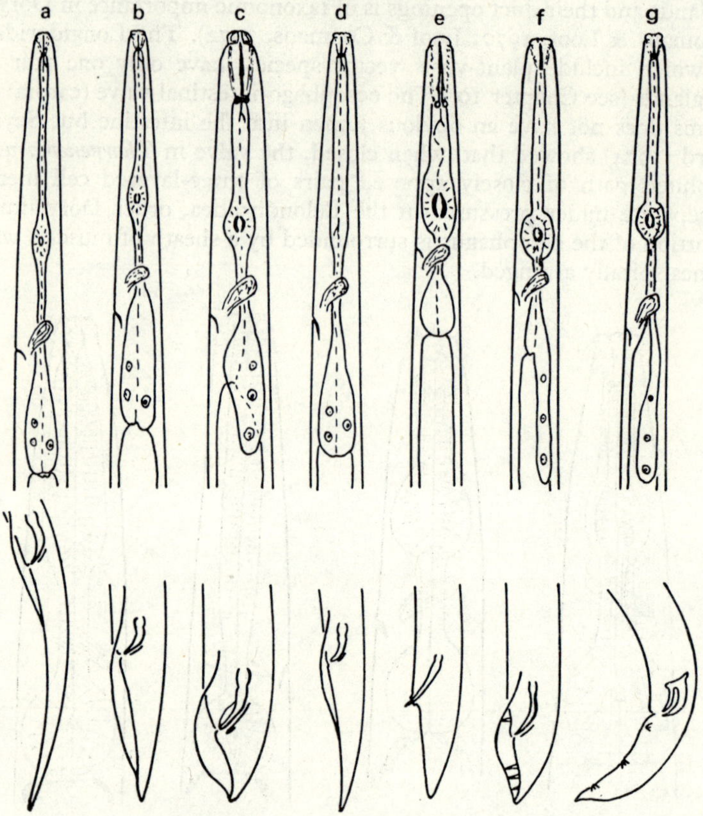

Fig. 1.7. Diagrams of anterior parts and male tails of seven groups of Tylenchida: **A,** *Tylenchus* (*sensu lato*); **B,** *Ditylenchus*; **C,** Hoplolaiminae; **D,** Neotylenchidae; **E,** Criconematidae (*Paratylenchus*); **F,** *Aphelenchus*; **G,** *Aphelenchoides*. (From Hooper, 1969, in *Nematodes of Tropical Crops*, Commonwealth Agricultural Bureaux).

intestinal diverticuli in females of *Anguina calamagrostis* and in *Meloidogyne* the intestine is a syncytium apparently functioning as a food storage organ (Maggenti & Allen, 1960).

The rectum (proctodaeum) in juveniles and females is usually a simple cuticle-lined tube connecting the intestine to the ventrally placed anus which is usually a short transverse opening. The intestine is modified at its junction with the rectum to form a valve and in *Aphelenchoides blastophthorus* this is somewhat similar to the oesophago-intestinal valve in *Thornenema wickeni* (see Seymour & Shepherd, 1974). Seymour & Doncaster (1973) describe defaecation of *A. blastophthorus*. Rectal glands may open into the rectum. Maggenti & Allen (1960) found that *Meloidogyne* females have six large unicellular glands which produce the gelatinous sac which envelops eggs ejected from the adjacent vulva. Although the anus and vulva are sometimes close to each other only exceptionally do they converge whereas in males the reproductive system always joins the rectum forming the cloaca which contains the spicules and other associated copulatory structures.

REPRODUCTIVE SYSTEM

Although bisexual reproduction is common in many nematodes, protandrous hermaphroditism and mitotic or meiotic parthenogenesis also occur. In some animal-parasitic groups bisexual and parthenogenetic generations alternate. The haploid chromosome number varies from 3 to 19 among various plant and soil nematodes but certain groups share a basic number, e.g., 9 in *Heterodera* spp. Triantaphyllou (1971) reviewed the genetics and cytology of plant and soil nematodes.

Females usually have a ventrally placed vulva in the mid-body region with paired reproductive tracts (didelphic) that are often opposed to each other (amphidelphic). A single reproductive tract (monodelphic) is usually directed anterior to the vulva (prodelphic) and the vulva is often then located more posteriorly. Occasionally the single reproductive tract is posterior to the vulva (opisthodelphic) and then the vulva may be located anteriorly (Fig. 1.8). When only one fully formed reproductive tract is present it may be opposed by a pre-/post-vulval sac which sometimes shows some differentiation of structure, especially in *Xiphinema* spp. (see Chapter 10). The reproductive tract consists of an ovary that is sometimes bent back upon the rest of the tract (reflexed) and an oviduct that is usually separated from the ovary by a constriction. The oviduct sometimes widens at the proximal end to form a spermatheca which may vary from round to elongate and may be continuous with, or set off from, the oviduct. The oviduct (or spermatheca) joins with the uterus which, at its distal end, is sometimes differentiated into large gland cells that may secrete part of the egg shell. The thicker walled proximal part of the uterus (ovejector in some nematodes) leads via a muscular vagina to the vulva. The vulva is often a transverse slit but in some species is pore-shaped or longitudinal. Sperm is sometimes stored in the uterus and post-vulval sac, especially if a well defined spermatheca is absent.

In plant and soil nematodes the ovary usually consists of a short apical germinal zone and an elongated growth zone; the oocytes are often in two or three rows distally but as they enlarge they tend to lie in tandem. In *Anguina* spp. and some animal-parasitic groups the oocytes are in multiple rows arranged radially about a central rachis.

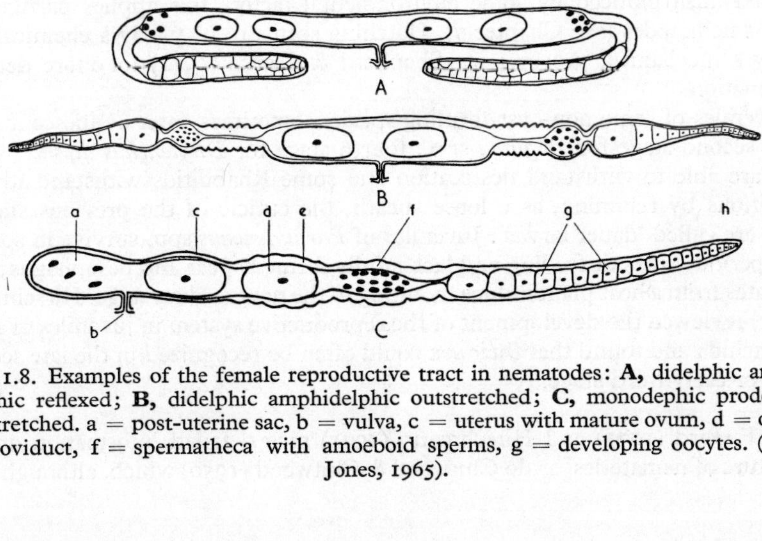

Fig. 1.8. Examples of the female reproductive tract in nematodes: **A,** didelphic amphidelphic reflexed; **B,** didelphic amphidelphic outstretched; **C,** monodephic prodelphic outstretched. a = post-uterine sac, b = vulva, c = uterus with mature ovum, d = ovum, e = oviduct, f = spermatheca with amoeboid sperms, g = developing oocytes. (From Jones, 1965).

Some females, especially old Rhabditids, lose the ability to deposit eggs which accumulate and develop inside the parent body. Eventually the juveniles hatch inside the body and kill the female—a phenomenon known as '*endotokia matricida*'.

The male reproductive system, especially of plant-parasitic Secernentia, usually consists of a single testis (monorchic), seminal vesicle and vas deferens. However, males of *Heterodera* and *Meloidogyne* spp. sometimes have paired testes (diorchic) directed anteriorly. Such males in *Meloidogyne* are thought to have developed from female juveniles by sex reversal caused by adverse environmental conditions (Triantaphyllou, 1960). Males of the Adenophoria, Dorylaimina, which include the plant-parasitic Longidoridae, have two opposed testes that join a single posteriorly directed vas deferens, whereas the Diphtherophorina, which include the plant-parasitic Trichodoridae, are monorchic. The vas deferens opens into the cloaca which has a ventrally placed opening. Males usually have a pair of copulatory spicules that lie in an invagination of the cloacal wall and there may also be associated sclerotized structures such as a gubernaculum or lateral accessory pieces which presumably guide the spicules (Figs. 1.1C and 10.1F—page 208). Spicule shape is often characteristic of particular nematode groups and consequently is of diagnostic importance.

EGGS AND JUVENILES ('LARVAE')

Eggs vary in shape: round, subspherical, kidney-shaped, or cylindrical with bluntly rounded ends. The surface is usually smooth although projections or spines occur on eggs of a few groups. The egg-shell consists of an external protein layer, a chitinous layer and an inner lipid layer. Eggs are often laid in an unsegmented condition. The cleavage pattern is similar in most nematodes but the time taken to complete embryological development and hatching varies from less than a day in *Rhabditis teres* to three or four weeks in Longidorid species and can be delayed even longer in *Heterodera* spp. The juvenile that hatches from the egg is often called a larva although, except for some Criconematids, this is a misnomer because the juvenile is similar to the young adult. There are four juvenile stages and, especially in Tylenchida, the first moult may occur inside the egg so that it is the second stage that emerges. Development and/or hatching may be delayed until induced by some environmental factor; this applies particularly to cyst nematodes (see Chapter 7). Hatching seems to be partly a chemical and partly a mechanical process; see Shepherd & Clarke (1971) for more detailed information.

Juveniles of some non-cyst-forming species sometimes enter a quiescent stage (e.g., second stage for *Anguina* spp., fourth stage for *Ditylenchus dipsaci*) when they are able to withstand desiccation and some Rhabditids withstand adverse conditions by retaining, as a loose sheath, the cuticle of the previous stage—these are called 'dauer larvae'. Juveniles of *Paratylenchus* spp. survive in soil for long periods without feeding and lack a fully formed spear and oesophagus; root exudates from a host plant induce moulting to the next feeding stage. Hirschmann (1971) reviewed the development of the reproductive system in juveniles of some Tylenchida and found that their sex could often be recognized in the late second stage or early third stage.

A. F. Bird (1971) and Hirschmann (1971) give detailed information on the structure of nematodes, as do Chitwood & Chitwood (1950) which, although now somewhat out of date, is still the main source of fundamental information on this group. Other useful publications are Lee & Atkinson (1976) on the physiology of nematodes, Croll (1976) on the organization of nematodes, Zuckerman, Mai & Rohde (1971), a two-volume treatise on various aspects of plant and soil nematodes, and Nicholas (1975) on the biology of free-living nematodes.

Handling, Mounting and Measuring Nematodes

These subjects are dealt with in detail in Southey (1970) and by Hooper (1969a) and Taylor (1971), consequently they are only briefly discussed here.

Before nematodes can be examined with a high-power microscope they must be immobilized. Usually they are killed by heat (about 65°C in water) or killed and fixed by adding an excess of hot fixative. Solutions of 5–10 per cent formalin (2 to 4 per cent formaldehyde) are frequently used for fixation and preservation of nematodes but body contents usually appear dark and granular. 'TAF' fixative (Courtney, Polley & Miller, 1955), which contains formalin (40 per cent formaldehyde) 7 ml, triethanolamine 2 ml and distilled water 91 ml, has the advantage that specimens retain their 'life-like' appearance in it for several hours. Seinhorst's (1966) method of killing and fixing with an excess of hot (100°C) formal acetic or formal propionic acid (F.A. or F.P. 4:1) gives good fixation of glands and gonads and enlarges nuclei which are more easily seen.

Many basic structures of a refractive nature, such as the mouth spear, oesophageal lumen, excretory pore and male genitalia, are best seen in freshly killed/fixed specimens. Unfortunately the reproductive tract and some other important structures are often obscured by the gut contents so it is necessary to clear the specimen. This can be done by transferring killed/fixed specimens to hot (65°–70°C) lactophenol and, when they have cleared sufficiently, they can be mounted in lactophenol to give a semi-permanent mount. Specimens can also be cleared by processing them to glycerol (Southey, 1970) but the procedure is more complicated and/or slower; however, most specimens for permanent slide collections are mounted in glycerol.

DIMENSIONS

Most nematode descriptions include certain dimensions and ratios; the de Man indices (modified) are frequently used:

L = total body length in mm or μm.
a = body length ÷ greatest body width.
b = body length ÷ distance from anterior end to junction of oesophagus and intestine.
b′ = body length ÷ distance from anterior end to posterior end of oesophageal glands—used when glands overlap the intestine.
c = body length ÷ tail length (anus or cloaca to tail terminus).
c′ = tail length ÷ body width at anus or cloaca.
V = distance of vulva from anterior end × 100 ÷ body length; small prefix and suffix numbers (e.g., $V = {}^{32}68^{10}$) indicate the proportions of body length from the vulva occupied by the anterior and posterior genital tracts respectively, also expressed as percentages of total body length.
T = distance from cloaca to anterior part of testis × 100 ÷ body length.
Length of stoma or spear in μm.

Other measurements besides those listed above are used for some groups and it is always worth while checking with previous publications to see which are necessary for diagnostic purposes rather than waste time working out values for all the above ratios/dimensions. It is important to remember that considerable intraspecific morphometric/morphological variation is known to occur (see reviews by Tarjan, 1967; Hooper, 1969b and Coomans, 1971) and great care should therefore be taken in assessing their significance. The usefulness of some of the above ratios/dimensions has been criticized (see Hooper, 1969) particularly 'a' and 'b' because their components often vary independently. Geraert (1968) noted that ratios only have value when the components are correlated so that the quotient remains more or less constant. Jones (1950) advocated the replacement of ratios by regression coefficients.

Some descriptions in older publications use the more elaborate 'Cobb formula' which is explained in Cobb (1920), Thorne (1961) and Southey (1970).

Many straight measurements may be made on suitably mounted (not flattened) specimens with a high-power microscope using an eyepiece scale that has been calibrated against a stage micrometer. To measure accurately curved whole specimens or structures it is necessary first to draw them. Most microscopes can be fitted with some form of drawing attachment so that the specimen and the drawing of it are visible at the same time. The curved drawing can then be measured with a map measurer or a piece of soft wire; the measurements obtained are calculated by comparison with a stage-micrometer scale drawn *at the same magnification* as the specimen.

Many of the distinguishing characters of nematodes are at the limits of resolution of the light microscope and consequently even the best instruments are barely adequate. Phase and, in particular, interference contrast (Smith or Nomarski) systems are often helpful in interpreting structures. Observations with light microscopes are being increasingly supplemented by scanning electron microscope and/or transmission electron microscope studies.

Classification

As indicated in the Introduction to this Chapter, there is disagreement regarding the ranking of the higher groups of the Nematodea; this also applies to classification at lower levels. This is partly because nematology is a relatively young science and classification systems are constantly having to be amended to accommodate new and often diverse species. Classification is also subjective and hence nematode systematists differ in their opinions as illustrated by the different classifications of Tylenchida/Tylenchoidea of Allen & Sher (1967), Golden (1971) and Siddiqi (1971). Although not rigidly adhered to, the Pearse system of nomenclature is often used in which the standard endings for the various higher categories frequently used in nematology are:

Phylum	—a
Class	—ea
Subclass	—ia
Order	—ida
Suborder	—ina
Superfamily	—oidea
Family	—idae
Subfamily	—inae

Table 1. *Classification of the Class NEMATODEA*

Subclass—ADENOPHORIA

Order	*Suborder*	*Bionomics*
Chromadorida	Chromadorina	
	Cyatholaimina	
Araeolaimida	Araeolaimina	
	Tripyloidina	
Monhysterida	Monhysterina	Aquatic or moist habitats, many marine; saprophagous, algal feeders and/or predatory
Desmodorida	Desmodorina	
	Draconematina	
Desmoscolecida		
Enoplida	Enoplina	
	Oncholaimina	
Dorylaimida	Dorylaimina	Saprophagous, predatory, algal feeders: few ectoparasitic on plant roots
	Diphtherophorina	
Mononchida	Mononchina	Mainly predatory
	Bathyodontina	
Isolaimida		
Trichosyringida	Trichosyringina	
Trichinellida	Trichinellina	Animal parasites, some groups require transport or intermediate hosts
Dioctophymatida	Dioctophymatina	

Subclass—SECERNENTIA

Order	*Suborder*	*Bionomics*
Tylenchida	Tylenchina	Fungivorous, plant parasitic, insect parasites; few predatory
	Aphelenchina	
Rhabditida	Rhabditina	Mainly saprophagous; few animal parasites
Strongylida	Strongylina	
	Trichostrongylina	
	Metastrongylina	Animal parasites, some requiring an intermediate host
Ascaridida	Ascaridina	
	Heterakina	
	Oxyurina	
Spirurida	Spirurina	Animal parasites requiring an intermediate host
	Camallanina	
	Filariina	

The International Code of Zoological Nomenclature gives rules and recommendations for the naming, description, and publication of taxa in the species (subspecies and species), genus (subgenus and genus) and family (subfamily, family and superfamily) groups. Jeffrey (1973) is a useful companion publication to the International Code and Mayr (1969) considers the broader aspects of systematic zoology. A classification of the Nematodea as a whole is given in Grassé (1965) and of soil and freshwater nematodes by Goodey (1963) although the latter is now somewhat out of date especially regarding the position of the plant-parasitic Trichodoridae. Goodey (1963) placed *Trichodorus* (now including *Paratrichodorus*—see Chapter 10) in the Diphtherophoroidea of the suborder Alaimina. However, Coomans & Loof (1970) raised Diphtherophoroidea to the suborder Diphtherophorina while the family Alaimidae is included by Gerlach & Riemann (1974) in the order Enoplida. This is probably a better arrangement but Diphtherophorina also have Enoplid affinities and further study might show that they would be better placed in the Enoplida. The following abbreviated classification of the Nematodea is concerned mainly with plant and soil forms and follows the scheme set out in Table I but the obligate animal parasitic orders are not considered further.

ABBREVIATED CLASSIFICATION OF THE NEMATODEA

Subclass ADENOPHORIA (=APHASMIDIA)

Excretory system without lateral canal(s); consists of one, rarely two, cells and a non-cuticularized (except in Plectidae) terminal duct. Amphid apertures usually post-labial and well developed; spiral or variants thereof to rarely slit- or pore-like. Phasmids and deirids absent. Caudal glands usually present but absent in DORYLAIMIDA and animal parasites. Somatic sensory organs setose or papilloid, usually present throughout the body length. Male usually with ventral preanal supplementary papillae; bursa usually absent, but present in *Paratrichodorus* spp.

Order CHROMADORIDA Amphid openings reniform to spiral. Cephalic sensory organs in three (6 + 6 + 4) or two (6 + 10) circlets. Oesophagus cylindrical with or without a terminal bulb. Stoma surrounded by oesophageal tissue and often containing a large dorsal tooth and/or smaller teeth. Cuticle often with strong punctations and/or other ornamentation, including setae, scattered over the body. Mainly marine.

Order ARAEOLAIMIDA Amphid openings often a single or double spiral, sometimes shepherd's-crook shape, rarely circular or pore-like. Cephalic sensory organs usually in three, sometimes in two circlets. Stoma usually a simple, funnel-shaped to elongate chamber, teeth often absent. Oesophagus with procorpus, isthmus and usually a terminal bulb. De Coninck (1965) divides the suborder ARAEOLAIMINA into five superfamilies: the Plectoidea is distinguished from the others in having a Rhabditid-like valve in the basal oesophageal bulb and the ovaries reflexed. Some Plectoid species are common in soil (see Chapter 12) but most of the other groups are mainly marine.

Order MONHYSTERIDA Amphid openings circular. Head with 4, 6 or 8 cephalic setae or multiples thereof. Stoma variable, often simple without prominent teeth. Oesophagus cylindrical and without a terminal bulb in the superfamily Monhysteroidea. Several genera occur in marine, freshwater and soil habitats of which the genus *Monhystera* is probably the commonest in soil (see Chapter 12).

Order DESMODORIDA Amphid openings variable: spiral, shepherd's crook, circular or slit-like but never reniform. Cephalic sensory organs usually in three circlets, often with additional subcephalic setae. Cuticle annulated but not punctate. Members of this order only reported from marine habitats. The suborder DRACONEMATINA is distinguished in having prominent subventral ambulatory setae.

Order DESMOSCOLECIDA Amphids vesiculate and usually in the region of four prominent cephalic setae. Cuticle plain, or ornamented with scales, warts, spines or concretion rings; the latter when present give these rather short (< 1·5 mm) relatively fat nematodes an unusual but characteristic segmented appearance. Almost entirely of marine habitat but a few non-marine species have occasionally been reported. Timm (1970), Decraemer (1974a, 1974b) and Freudenhammer (1975) have major publications on this order.

Order ENOPLIDA Amphids usually pocket-like with rather small transverse slit-like or oval apertures. Subventral oesophageal glands open anterior to the nerve ring in the stomatal region; oesophagus cylindrical to conoid, rarely multibulbar. Caudal glands usually present. Two suborders: (1) ENOPLINA in which the stoma is almost completely surrounded by oesophageal tissue; the stoma is usually weakly developed with rather small teeth if any; (2) ONCHOLAIMINA with only the posterior part of the stoma surrounded by oesophageal tissue; stoma usually well developed, strongly cuticularized, and armed with a dorsal and two subventral teeth. Although this order contains many marine species, some occur in moist soil, particularly those in the families Tripylidae and Alaimidae (suborder ENOPLINA)—see Chapter 12.

Order DORYLAIMIDA Amphid pouches and openings of variable shape but never spiral. Stoma with a mural tooth or teeth, or an axial stylet. Oesophagus with an anterior slender part and an expanded posterior portion which contains three or five gland nuclei; the gland duct openings are posterior to the nerve ring. Cephalic setae usually absent. Caudal glands absent.

Suborder DORYLAIMINA Amphids sublabial, pouch to inverted stirrup shape, their openings varying from a wide transverse slit to an indistinct pore. Anterior oesophagus a narrow tube encircled by a nerve ring; posterior part swollen and muscular. Stoma with an hollow axial stylet (odontostyle + odontophore) with an oblique dorsal aperture or with a ventral or ventrosublateral tooth. Replacement spear/tooth usually formed in anterior oesophagus of juveniles. Excretory duct absent or weakly developed. Pre-rectum present. Males with paired opposed testes, paired adanal supplementary papillae, and massive arcuate spicules with lateral accessory pieces; gubernaculum absent.

Superfamily Dorylaimoidea Definition as for DORYLAIMINA. According to Ferris (1971) this superfamily contains six families and includes forms with a hollow axial stylet or mural tooth. Also included is the family Longidoridae, characterized by an elongate odontostyle and odontophore and only one pair of subventral gland nuclei in the posterior oesophagus; further details of this plant-parasitic group are given in Chapter 10. Details of some other Dorylaims often found in soil are given in Chapter 12.

Superfamily Actinolaimoidea Walls of distal part of stoma heavily cuticularized, frequently bearing large teeth or denticles in addition to an hollow axial spear.

Superfamily Leptonchoidea Hollow axial spear with well developed basal extensions which may bear basal knobs or flanges. Basal expanded part of oesophagus relatively short, rarely as much as one third of the total oesophageal length. See also Chapter 12.

Superfamily Belondiroidea Basal portion of oesophagus surrounded by a thick sheath of muscles which often spiral. The anterior region usually tapers to a narrow lip region and the axial stylet is often shorter than the lip region width. Anterior branch of female reproductive tract is often rudimentary. See also Chapter 12.

Suborder DIPHTHEROPHORINA Body plump, usually less than 2 mm long; cuticle often appears thick and loose. Amphids pouch-shaped, post-labial, with ellipsoid apertures; separated from sensilla sac by a constriction. Pharynx (*sensu stricto*) with a complex spear (onchiostyle) composed of a dorsal tooth and supporting structures; replacement onchiostyle tip (onchium) formed in the pharynx. Oesophagus narrow anteriorly, swelling posteriorly to form a pyriform to elongate basal bulb that contains one dorsal and at least two subventral uninucleate glands. Excretory pore present. No distinct pre-rectum. Caudal glands and spinnerets absent. Females usually with paired opposed genital tracts, the ovaries being reflexed. Males with single testis, no ejaculatory muscles and no adanal pair of supplementary papillae. Spicules elongate, straight to somewhat curved ventrally; gubernaculum present, lateral accessory pieces absent. Each spicule surrounded by a capsule of suspensor muscles that replace the form and function of the typical Dorylaimoid protractor muscles.

Superfamily Diphtherophoroidea Spear a short complex structure, with basal knobs; guiding apparatus arch-like. Male supplements often reduced or absent. This superfamily contains the genera *Diphtherophora, Tylolaimophorus* and *Longibulbophora*; occasional species are found in soil but their feeding habits are unknown.

Superfamily Trichodoroidea Onchiostyle consists of an elongate dorsal tooth with an attenuated, ventrally arcuate extension without basal knobs. A simple guide ring surrounds the anterior third of the onchiostyle. Female genital tracts paired (except in one species) opposed with reflexed ovaries. Males with or without a bursa. Spicules almost straight to ventrally curved, with or without fine transverse striations. Tail not more than one anal body width long. For further details of the two genera (*Trichodorus* and *Paratrichodorus*) which are obligate plant parasites, see Chapter 10.

Order MONONCHIDA Amphid apertures sublabial, oval to small cup-like. Stoma usually wide, sometimes tubular, sclerotized with mural tooth or teeth.

Oesophagus a muscular cylinder with one dorsal and four ventro-sublateral oesophageal gland cells with their openings posterior to the nerve ring. Prerectum absent; caudal glands and terminal duct often present. Male with a single testis and single adanal supplement. Gubernaculum usually with elaborate lateral crura.

Suborder MONONCHINA Oesophagus of uniform thickness throughout its length. Dorsal gland nucleus far behind its oesophageal opening but anterior to the openings of the first subventral gland nuclei which are also far behind their openings. Stoma large and wide, not embedded in oesophageal tissue. Rhabdions heavily sclerotized with at least one large dorsal tooth.

Superfamily Mononchoidea Stoma tapers at the base and the oesophago-intestinal valve is non-tuberculate.

Superfamily Anatonchoidea Stoma with a broad flat base and oesophago-intestinal valve is tuberculate.

For further division and identification of this group of predacious nematodes that are common in soil see Chapter 12.

Suborder BATHYODONTINA Oesophagus of uniform thickness, dorsal gland nucleus far behind its oesophageal opening sometimes at the level of the first subventral gland nuclei which are also far behind their openings. Stoma rather narrow, embedded almost completely in oesophageal tissue, and composed of several sections. Rhabdion sclerotization variable but with only one distinct tooth of varying size located ventro-sublaterally. There are five genera in two superfamilies in this suborder (see Coomans & Loof, 1970; Jairajpuri, 1969 & 1971) but they are not often encountered in soil samples.

Order ISOLAIMIDA Body elongate (3–6 mm), cuticle annulated also with punctations in longitudinal and transverse rows. Six hollow tubes surround oral aperture and stoma; four or six small papillae also around oral aperture plus an outer ring of six large recessed papillae. Amphids apparently absent or represented by the dorso-lateral papillae in the outer ring. Stoma long, triradiate in section, with sclerotized walls that tend to be thicker at the anterior end. Oesophagus long, almost cylindrical, swelling slightly at the posterior end. Oesophago-intestinal valve prominent and elongate. Vulva transverse, about mid-body with paired opposed, reflexed, reproductive tracts. Male with paired outstretched testes; spicules stout, arcuate with short gubernaculum; pre-anal mid-ventral supplements and large caudal papillae also present. Tails of both sexes short, conoid, with bluntly rounded to pointed tip.

These large nematodes, represented by the single genus *Isolaimium*, are rare in most soils but are sometimes found in extractions made for Longidorid or other large nematodes. For further details see Timm (1969) who gives a key to the species.

Subclass SECERNENTIA (=PHASMIDIA)

Excretory system with lateral canal(s) sometimes with one or two subventral glands; terminal excretory duct cuticularized. Amphid apertures usually small, pore-like or small slits, usually labial in position. Phasmids and deirids usually present but sometimes difficult to see. Caudal glands absent. Somatic sensory organs usually absent except in posterior region of male. Males without pre-anal mid-ventral supplementary papillae.

Of the orders in this subclass, listed in Table 1, the plant parasitic species occur in the TYLENCHIDA and most of the free-living species found in soil are in the RHABDITIDA.

Order RHABDITIDA Stoma an open tube with its walls lined by rhabdions which may be separate or more or less fused. The various parts of a Rhabditid stoma are detailed in Fig. 1.3A. Amphids small and pocket-like, on the sides of the lateral lips or rarely behind the lips. Oesophagus consisting of a procorpus (with or without a corpus), isthmus and terminal bulb; valves present in corpus or terminal bulb or in both.

Superfamily Diplogasteroidea Oesophagus with median bulb containing crescentic valve plates; terminal bulb without valve plates (Fig. 1.6).

Family Diplogasteridae Stoma relatively short; oesophagus with procorpus and valvated median bulb. This family contains several genera having many species common in soil and decaying matter. See Chapter 12 for further details.

Family Cylindrocorporidae Stoma long; procorpus and corpus combined to form prominent cylindrical muscular bulb. Three genera comprising a few species recorded in this family. Associates of bark beetles and/or decaying matter, recorded only occasionally.

Superfamily Pseudodiplogasteroidoidea Median oesophageal bulb containing weak crescentic valve; terminal bulb also with valve plates. Only two monospecific genera recorded in this superfamily, both of them associates of wood-boring beetles.

Superfamily Rhabditoidea Anterior part of oesophagus cylindrical, sometimes with slight corporal swelling, but without valve plates. Terminal bulb with prominent, usually three, flap-like valve plates (Fig. 1.6A, C, D).

Family Rhabditidae With three or six lips or paired oral hooks. Protorhabdions fused, metastom formed of one dorsal and two subventral metarhabdions (called 'glottoid apparatus'). Female reproductive tracts usually paired and opposed; rarely single. Many species of several genera are free-living saprobes common in soil and decaying vegetable matter (see Chapter 12 for further details); some are parasites of arthropods, annelids, molluscs or vertebrates.

Family Brevibuccidae Stoma with indistinct cheilorhabdions; pro- and mesorhabdions fused; metarhabdions each with a tooth; telorhabdions obscure. Anterior part of oesophagus cylindrical; terminal bulb with Rhabditid valvular apparatus. Female with posterior vulva; single anterior genital tract with reflexed ovary. Male with prominent caudal papillae but no bursa. A monogeneric family, rarely reported, from decaying vegetable matter.

Family Panagrolaimidae Stoma with fused rhabdions forming a wide chamber that sometimes tapers at its base which is partly surrounded by the anterior end of the oesophagus. Anterior part of oesophagus cylindrical or with somewhat swollen, valveless, median bulb; isthmus narrow, basal bulb with Rhabditid valve (Fig. 1.6). Female with single anterior ovary which is reflexed posteriorly well down the body. Male tail without bursa but with typical arrangement of caudal papillae (Fig. 1.6). Representatives of this group are common in decaying matter and frass of wood-boring beetles. See also Chapter 12.

Family Myolaimidae Cheilo-, pro-, and mesorhabdions forming a short broad chamber inarching at the mouth. Dorsal segment of metastom with large hump-like tooth. Oesophagus and genital tract Panagrolaimoid. Male tail with bursa supported by seven or eight pairs of prominent papillae. Spicules and gubernaculum absent. A monogeneric family whose species occur where bacteria are plentiful, though they are not frequently reported.

Family Chambersiellidae Lip region with or without six tufted cephalic cirri; with a circlet of six prominent horn-shaped or hair-like setae, or large mammilliform papillae. Cheilorhabdions distinct; anterior part of stoma spacious, posterior part narrowing to oesophageal lumen and surrounded by oesophageal tissue. Females with one or two ovaries. Sanwal (1971) discusses this rarely recorded group and gives a key to the genera.

Family Cephalobidae Rhabdions of stoma distinctly jointed, rest of stoma narrower. Oesophagus cylindrical anteriorly, posterior bulb with Rhabditid-like valve. Lower metastomal rhabdion with small dorsal tooth. Oesophageal collar usually extending to base of cheilostom. Female genital tract prodelphic, reflexed posteriorly and often with an additional double flexure towards the end of the ovary. Male without bursa but with about eight pairs of caudal papillae arranged in a typical pattern. Some genera in this group have elaborate labial and/or cephalic probolae. See Chapter 12 for further details. Species are often found in soil and decaying plant tissues.

Superfamily Bunonematoidea Cuticle elaborately ornamented. Head mobile, retractile, asymmetrical with large bristles and lamellae. Oesophagus with procorpus, corpus, isthmus, and posterior bulb with Rhabditid-like valve plates.

Family Bunonematidae Hexagonal network of minute rods and/or raised warts or tubercles on right side; left side smooth but for five longitudinal ridges. Rather small nematodes ($<$ 0·5 mm long) of unusual appearance. Apparently bacterial feeders and found in moss, decaying vegetable matter, cow dung and rotting wood of bark-beetle tunnels. Six genera and many species have been described, see Andrássy (1971) for review of the superfamily.

Family Pterygorhabditidae Cuticle ornamented with transverse bars on both sides of the body. Head similar to Bunonematidae. Anterior part of oesophagus cylindrical. A monotypic family based on *Pterygorhabditis pakistanensis* from damp rice straw in Bangladesh (formerly East Pakistan).

Order TYLENCHIDA The classification of this order is dealt with in more detail than the others because it contains most of the plant-parasitic nematodes. An outline classification to subfamily is given in Table 2. This classification is a compromise between that of Golden (1971) and Siddiqi (1971) and includes some later additions and emendations. Space does not permit a full discussion but by following Siddiqi (1971) and leaving Nacobbinae in the Pratylenchidae and placing Rotylenchulidae at family rank in the Tylenchoidea, then Heteroderoidea *sensu* Golden (1971) remains as a distinct group; the females have two ovaries, are swollen and sedentary; the males have a characteristic short tail without a bursa. I consider that the superfamilies Tylenchulidoidea and Tylenchocriconematoidea of Raski & Siddiqi (1975) are overranked and unbalance the whole classification of the Tylenchida. Consequently they are accommodated within the Criconematoidea. Likewise, Atylenchoidea Golden, 1971 is returned to Tylenchoidea. In view of observations by Choi & Loof (1973) on the similarities between *Ditylenchus* and *Anguina* it does not seem appropriate to place these two genera in separate subfamilies and therefore Ditylenchinae Golden, 1971 is synonymized with Anguininae *sensu* Siddiqi (1971). For each category, the type superfamily, family, subfamily or genus is listed first.

Definition: Cuticle annulated, amphids small and pocket-like with openings on the lip region. Stoma with a protrusible, hollow, axial stylet (except in degenerate males and occasionally juveniles of some species). Stylet formed within the stoma (stomatostyle). Excretory system with a single lateral canal. Oesophagus usually consists of an anterior part (procorpus) which may or may not be amalgamated with a median bulb (metacorpus) which has valve plates (except in NEOTYLENCHOIDEA—p. 36); a narrow isthmus encircled by the nerve ring (except in some NEOTYLENCHIDAE—p. 7) and a glandular basal portion, without an obvious valvular apparatus, which may be a distinct bulb or lobed and overlapping the intestine. Caudal glands absent.

Suborder TYLENCHINA Dorsal oesophageal gland opening into lumen of oesophagus between stylet base and median bulb, often near base of stylet, or in that position if median bulb is absent.

Superfamily Tylenchoidea Small to moderate sized, usually vermiform nematodes; adults 0·5 to 4·0 mm long, often about 1 mm. Lip region distinct and body marked by fine but distinct transverse striations. Oesophagus with well defined, usually narrow procorpus distinct from the small to moderate sized valvated median bulb; isthmus long and narrow, well developed oesophageal glands enclosed in a distinct bulb or overlapping the intestine (Fig. 1.7a, b, c). Mostly vermiform (mature females of *Nacobbus* and *Rotylenchulus* swollen). Females monodelphic or didelphic. Males with a bursa, characteristically shaped spinules and gubernaculum (Fig. 1.7a–c); oesophagus rarely somewhat degenerate.

Family Tylenchidae Vermiform nematodes without marked sexual dimorphism apart from reproductive organs. Oesophageal glands usually within a basal bulb or with a slight overlap over the intestine (long overlap in *Pseudhalenchus*). Females usually monodelphic with post-equatorial vulva (except *Psilenchus* and *Leipotylenchus*). Males with a bursa. Head skeleton usually not or weakly sclerotized. Stylet with or without basal knobs and usually under 25 μm

Table 2 *Classification of the order TYLENCHIDA*

Suborder TYLENCHINA

Superfamily	*Family*	*Subfamily*
Tylenchoidea	Tylenchidae	Tylenchinae
		Dactylotylenchinae
		Leipotylenchinae
		Psilenchinae
		Pseudhalenchinae
		Anguininae
		Sychnotylenchinae
		Tylodorinae
		Antarctenchinae
	Atylenchidae	Atylenchinae
	Tylenchorhynchidae	Tylenchorhynchinae
		Trophurinae
	Dolichodoridae	Dolichodorinae
	Belonolaimidae	Belonolaiminae
		Telotylenchinae
	Pratylenchidae	Pratylenchinae
		Radopholinae
		Nacobbinae
	Hoplolaimidae	Hoplolaiminae
		Rotylenchinae
		Rotylenchoidinae
		Aphasmatylenchinae
		Rotylenchulinae
Heteroderoidea	Heteroderidae	Heteroderinae
		Ataloderinae
		Meloidoderinae
	Meloidogynidae	Meloidogyninae
Criconematoidea	Criconematidae	Criconematinae
		Hemicycliophorinae
	Paratylenchidae	Paratylenchinae
	Tylenchocriconematidae	Tylenchocriconematinae
	Tylenchulidae	Tylenchulinae
		Sphaeronematinae
Neotylenchoidea	Neotylenchidae	Neotylenchinae
	Paurodontidae	Paurodontinae
		Misticiinae
	Nothotylenchidae	Nothotylenchinae
		Boleodorinae
		Halenchinae
	Ecphyadophoridae	Ecphyadophorinae
	Sphaerulariidae	Insect parasites or associates—not classified further here.
	Allantonematidae	
	Contortylenchidae	

Suborder APHELENCHINA

Superfamily	*Family*	*Subfamily*
Aphelenchoidea	Aphelenchidae	Aphelenchinae
	Paraphelenchidae	Paraphelenchinae
	Aphelenchoididae	Aphelenchoidinae
		Rhadinaphelenchinae
		Anomyctinae
		Seinurinae
	Entaphelenchidae	Entaphelenchinae

long (about 100 μm long in *Tylodorus*). Lip region round, conoid or flattened but rarely offset (except in *Cephalenchus*).

Subfamily Tylenchinae Tail elongate-conoid to filiform. Lip region more or less continuous with body contour, rarely distinctly offset (*Cephalenchus*). Amphid openings indistinct on face of lip region. Deirids usually distinct, phasmids indistinct. Stylet relatively weak but usually with distinct basal knobs. Oesophageal glands in a distinct basal bulb; distinct oesophago-intestinal valve present. Vulva well posterior to mid-body, single anterior reproductive tract. Bursa small, adanal.

Genera: *Tylenchus* Bastian, 1865; *Filenchus* Andrássy, 1954 (Meyl, 1961); *Aglenchus* Andrássy, 1954 (Meyl, 1961); *Ottolenchus* Husain & Khan, 1967 (Wu, 1970); *Malenchus* Andrássy, 1968; *Cephalenchus* Goodey, 1962 (Geraert, 1968); *Pleurotylenchus* Szczygiel, 1969; *Lelenchus* Andrássy, 1954 (Meyl, 1961); *Miculenchus* Andrássy, 1959. A key to most of these genera is given in Chapter 9.

Subfamily Dactylotylenchinae Labial framework in form of finger-like sclerotized structure. Cuticle thick. Stylet with basal knobs. Oesophagus with distinct basal bulb. Female with posterior vulva, prodelphic reproductive tract. Male with adanal bursa. Tail long and slender. Type and only genus *Dactylotylenchus* Wu, 1968.

Subfamily Leipotylenchinae Amphids inconspicuous. Stylet without basal knobs. Oesophageal glands in a terminal bulb. Deirids present. Vulva median with paired, opposed reproductive tracts. Tail elongate. Bursa adanal. A monogeneric subfamily based on *Leipotylenchus* Sher, 1974.

Subfamily Psilenchinae Amphid openings distinct, slit-like, on the side of the lip region. Deirids and phasmids usually distinct. Stylet with or without basal knobs. Oesophageal glands in a distinct basal bulb. Vulva posterior with single anterior reproductive tract (except *Psilenchus*, vulva median with paired, opposed, reproductive tracts). Tail elongate, conoid to filiform, sometimes with clavate terminus. Male with small adanal bursa.

Genera: *Psilenchus* de Man, 1921; *Neopsilenchus* Thorne & Malek, 1968; *Basiria* Siddiqi, 1959; *Basiroides* Thorne & Malek, 1968; *Clavilenchus* Jairajpuri, 1966 (Thorne & Malek, 1968).

Subfamily Pseudhalenchinae Oesophageal glands overlapping intestine in an elongate lobe. Vulva in or near posterior fourth of body; single anterior reproductive tract. Tail elongate-conoid with rounded terminus. Bursa adanal to subterminal. Type and only genus *Pseudhalenchus* Tarjan, 1958.

Subfamily Anguininae Lip region low and flattened, labial skeleton very weak, stylet weak to moderate with distinct basal knobs. Cuticle with fine transverse striations. Median oesophageal bulb oval, not strongly developed, oesophageal glands usually well developed in a basal bulb or slightly overlapping the intestine. Oesophago-intestinal valve rudimentary or absent. Vulva posterior; single anterior reproductive tract with elongated axial spermatheca. Post-vulval sac usually present. Oocytes usually numerous in one or two rows and sometimes in multiple rows arranged about a rachis. Tail elongate to conoid, not filiform. Bursa subterminal.

Genera: *Anguina* Scopoli, 1777; *Paranguina* Kirjanova, 1955; *Cynipanguina* Maggenti, Hart & Paxman, 1974; *Ditylenchus* Filipjev, 1936; *Diptenchus* Khan, Chawla & Seshadri, 1969. See also Chapter 4.

Subfamily Sychnotylenchinae Similar to Anguininae but vulva more posterior (V = about 90). Female tail short, varying from bluntly rounded to pointed. Male with large bursa that surrounds the tail tip.

Genera: *Sychnotylenchus* Rühm, 1956; *Neoditylenchus* Meyl, 1961. Mainly associates of bark beetles.

Subfamily Tylodorinae Adults of moderate size, 1–2 mm long, Lip region continuous with body contour, four-lobed (lateral lips not distinct), with a lip cap but without annulation. Pore-like amphid openings distinct at lateral margins of lip cap. Head sclerotization absent. Stylet elongate (about 100 μm) with basal knobs. Deirids distinct. Oesophageal lumen convoluted in procorpus when stylet is retracted; procorpus short but distinctly set off from a well developed median bulb. Posterior oesophageal glands in a distinct pyriform basal bulb. Vulva posterior (V = 66) with single anterior reproductive tract and a post-vulval sac longer than the vulval body-width. Bursa adanal, fairly well developed. Tail of both sexes elongate to filiform.
Based on the monotypic genus *Tylodorus* Meagher, 1963 from Australia.

Subfamily Antarctenchinae Medium sized slender nematodes with long filiform tails in both sexes. Lip region annulated with moderate sclerotization; amphid apertures rather large, on face of lip region. Stylet of moderate size with basal knobs. Procorpus of oesophagus long, median bulb distinct, isthmus long ending in a distinct basal bulb. Female with subequatorial vulva and paired, opposed, outstretched reproductive tracts. Males with a distinct adanal bursa that does not incorporate the distinct post-anal phasmids. Based on a monotypic genus *Antarctenchus* Spaull, 1972 from Antarctic mosses.

Note—The **subfamily Tylodorinae** was placed in the family Dolichodoridae by Siddiqi, 1971 and subsequently the somewhat similar Antarctenchinae was also included. However, in this classification the narrower definition of Dolichodoridae *sensu* Golden (1971) is followed and hence these two subfamilies are moved to the Tylenchidae. They seem to represent a bridging group between the rest of the Tylenchidae and the Tylenchorhynchidae/Dolichodoridae.

Family Atylenchidae Lip region with four well developed setae; cephalic framework weak. Cuticle coarsely annulated with prominent protuberances arranged as 10 to 12 longitudinal ridges. Oesophagus with valvated median bulb; oesophageal glands in an elongated basal bulb. Vulva posterior with anterior reproductive tract and short post-vulval sac. Tail of both sexes elongate, filiform. Male with prominent adanal bursa (*Eutylenchus*) or without (*Atylenchus*).

Genera: *Atylenchus* Cobb, 1913; *Eutylenchus* Cobb, 1913. Sher, Corbett & Colbran (1966) redescribed these two rarely recorded genera.

Family Tylenchorhynchidae Medium sized, adults about 0·8 to 2 mm long, vermiform, without sexual dimorphism. Oesophageal glands in a distinct basal bulb. Lip region elevated, usually with some sclerotization. Stylet with basal knobs. Phasmids distinct and characteristically extend into bursal wings. Deirids indistinct. Vulva median, usually with paired, opposed reproductive tracts.

Subfamily Tylenchorhynchinae Lip region continuous with, or set off from, body contour. Amphid apertures on face of lip region. Cephalic sclerotization variable. Cuticular annulation variable, sometimes strong and sometimes with longitudinal striations. Lateral field with three to six incisures. Stylet well developed. Vulva median with paired opposed reproductive tracts. Phasmids distinct. Female tail two to three anal body-widths long, cylindrical to conoid with pointed or bluntly rounded terminus. Male with prominent bursa that envelops the tail (except *Triversus* in which bursa is subterminal).

Genera: *Tylenchorhynchus* Cobb, 1913; *Uliginotylenchus* Siddiqi, 1971; *Triversus* Sher, 1974; *Quinisulcius* Siddiqi, 1971; *Merlinius* Siddiqi, 1970; *Scutylenchus* Jairajpuri, 1971; *Sauertylenchus* Sher, 1974; *Geocenamus* Thorne & Malek, 1968; *Nagelus* Thorne & Malek, 1968. A key to separate *Tylenchorhynchus* and its closest relatives is given in Chapter 9.

Subfamily Trophurinae Cuticle usually weakly annulated but thickened or swollen on female tail terminus which is typically rounded. Stylet slender, short or elongate. Vulva median with paired, opposed, reproductive tracts or with anterior tract only plus a post-vulval sac. Male with prominent bursa that surrounds the tail tip.

Genera: *Trophurus* Loof, 1956; *Paratrophurus* Arias, 1970; *Macrotrophurus* Loof, 1958.

Family Dolichodoridae Medium to large, slender nematodes, 1–3 mm long. Oesophageal glands within a basal bulb. Lip region prominent with heavy sclerotization. Cuticle distinctly annulated, lateral incisures present. Stylet moderately to very long with distinct basal knobs. Vulva median with paired, opposed reproductive tracts. Female tail round to elongate-conoid. Male with distinct trilobed bursa.

Subfamily Dolichodorinae *sensu* Golden, 1971 Definition as for family.

Genera: *Dolichodorus* Cobb, 1914 (stylet long, > 60 μm, lip region often four-lobed; oesophagus with shortened procorpus containing coiled lumen of anterior oesophagus; female tail short, rounded or spicate); *Brachydorus* de Guiran & Germani, 1968 (stylet short; lip region continuous, with hexagonal framework; oesophageal procorpus long; female tail elongate-conoid).

Family Belonolaimidae Elongate forms. Little or no sexual dimorphism. Body with distinct annulation and lateral incisures. Lip region elevated. Stylet slender with basal knobs, of medium length or elongate. Oesophagus with well developed median bulb; oesophageal glands forming a lobe overlapping the intestine. Vulva median, reproductive tracts paired, opposed, outstretched. Female tail usually more than two anal body-widths long with distinct phasmids near the middle. Male with elongate, non-lobed bursa reaching to tail terminus.

Subfamily Belonolaiminae Lip region distinctly offset with two to four lobes; labial disc present. Cephalic framework sclerotized. Stylet long and slender, 60–150 μm (one species 30 μm). Oesophageal procorpus relatively short and swollen to accommodate coiled oesophageal lumen when stylet is retracted. Gubernaculum with titillae.

Genera: *Belonolaimus* Steiner, 1949; *Carphodorus* Colbran, 1965; *Morulaimus* Sauer, 1966. *Belonolaimus* spp. (sting nematodes) attack various crops in U.S.A.

Subfamily Telotylenchinae Lip region continuous with or offset from body contour but not usually lobed, labial disc absent. Stylet of moderate length (15–30 μm). Female tail cylindrical to elongate-conoid with rounded terminus.

Genera: *Telotylenchus* Siddiqi, 1960; *Histotylenchus* Siddiqi, 1971; *Trichotylenchus* Whitehead, 1959; *Telotylenchoides* Siddiqi, 1971.

Family Pratylenchidae Vermiform, except mature females of *Nacobbus*, with fine body annulation. Lip region relatively low, rounded to flattened anteriorly. Cephalic sclerotization distinct. Stylet stout, with distinct basal knobs, usually under 25 μm long (up to 45 μm in *Hirschmanniella* spp.). Oesophageal glands usually in a long lobe overlapping the intestine. Females with one or two reproductive tracts; tail usually at least two anal body-widths long, phasmids post-anal near centre of tail. Bursa terminal or subterminal. Marked sexual dimorphism in some forms.

Subfamily Pratylenchinae Oesophageal glands in a long lobe mainly overlapping intestine ventrally. Vulva posterior with single anterior reproductive tract and post-vulval sac, or median with paired opposed reproductive tracts. No marked sexual dimorphism.

Genera: *Pratylenchus* Filipjev, 1936; *Hirschmanniella* Luc & Goodey, 1963; *Zygotylenchus* Siddiqi, 1963. *Pratylenchus* spp. (root-lesion nematodes) are damaging root endoparasites; see Chapter 9.

Subfamily Radopholinae Oesophageal glands overlapping intestine mainly dorsally, usually in a long lobe. Female with single or paired reproductive tracts as in Pratylenchinae. Some sexual dimorphism varying from slight reduction of oesophageal glands in *Apratylenchoides* males to marked degeneration of oesophagus and differently shaped lip region in males of *Radopholus*.

Genera: *Radopholus* Thorne, 1949; *Radopholoides* de Guiran, 1967; *Apratylenchoides* Sher, 1973; *Pratylenchoides* Winslow, 1958; *Hoplotylus* s'Jacob, 1959. *Radopholus similis* (the burrowing nematode) is a damaging root endoparasite of bananas and citrus—see also Chapter 9.

Subfamily Nacobbinae Marked sexual dimorphism; mature female swollen, sedentary parasite in galls or swellings of plant roots. Vulva posterior, single anterior reproductive tract. Stylet stout, 20–25 μm long. Median oesophageal bulb in adults well developed with prominent valve plates; oesophageal glands in a long lobe overlapping the intestine dorsally. Tail of immature females just over an anal body-width long and bluntly rounded. Male tail enveloped by small bursa.

Monogeneric subfamily based on the genus *Nacobbus* Thorne & Allen, 1944, (the false root-knot nematode).

Family Hoplolaimidae Lip region elevated, conoid to hemispherical with moderate to strong cephalic sclerotization. Stylet strong, usually over 20 μm long, with distinct basal knobs. Oesophageal glands lobed, overlapping intestine. Phasmids, usually distinct, vary from pore-like to large (scutellae), variable in position. Female tail short, usually less than two anal body-widths long. Male tail usually with enveloping bursa. Some sexual dimorphism especially in lip-region shape in some groups.

Subfamily Hoplolaiminae Vermiform. Phasmids large (scutellae), variable in position in both sexes. Cephalic framework heavily sclerotized. Oesophageal glands overlap intestine mainly dorsally. Female tail less than two anal body-widths long; vulva about median with paired, opposed reproductive tracts. Bursa envelops male tail.

Genera: *Hoplolaimus* Daday, 1905; *Aorolaimus* Sher, 1963; *Scutellonema* Andrássy, 1958; *Peltamigratus* Sher, 1963. *Hoplolaimus* spp. (lance nematodes), and *Scutellonema* spp. are damaging root ecto/endoparasites especially in warmer soils.

Subfamily Rotylenchinae Vermiform. Phasmids pore-like, usually in anal region. Lip region and cephalic framework often somewhat weaker than in Hoplolaiminae. Oesophageal glands overlap intestine. Vulva median to post-median with paired, opposed, reproductive tracts. Bursa envelops male tail.

Genera: *Rotylenchus* Filipjev, 1936; *Helicotylenchus* Steiner, 1945; *Antarctylus* Sher, 1973. *Rotylenchus* and *Helicotylenchus* spp. (spiral nematodes) are root ecto/semi-endoparasites. See Chapter 9.

Subfamily Aphasmatylenchinae Vermiform. Phasmids absent. Lip region high and rounded with well developed sclerotization. Oesophageal glands with a narrow ventral overlap over the intestine. Vulva about median with paired, opposed, reproductive tracts. Male tail just over two anal body-widths long with an elongate enveloping bursa.

Genus: *Aphasmatylenchus* Sher, 1965. Recorded from Western Africa.

Subfamily Rotylenchoidinae Body cylindrical to slightly swollen. Phasmids pore-like, in anal region. Oesophageal glands overlapping intestine. Vulva posterior with single anterior reproductive tract and post-vulval sac. Female tail variable in shape and length but less than three anal body-widths long. Bursa enveloping tail where males known.

Genera: *Rotylenchoides* Whitehead, 1958 (body cylindrical, tail bluntly rounded and less than one anal body-width long); *Acontylus* Meagher, 1968 (mature female swollen somewhat in middle region; tail rounded, about one and a half anal body-widths long). Both genera rarely reported from West Africa and Australia respectively.

Subfamily Rotylenchulinae Sexual dimorphism pronounced; mature female swollen, kidney-shaped, vulva post-median with paired, opposed reproductive tracts. Lip region of immature females rounded to conoid, not offset, stylet less than 20 μm long. Oesophageal glands overlap intestine mainly laterally. Immature female tail conoid, two to four anal body-widths long. Male with weak stylet and somewhat degenerate oesophagus; tail with inconspicuous adanal bursa.

Genus: *Rotylenchulus* Linford & Oliveira, 1940. The females of *Rotylenchulus* spp. (reniform nematodes) are sedentary semi-endoparasites on the roots of a wide variety of plants especially in warmer soils. Dasgupta *et al.* (1968) review the genus and give a key to species.

Superfamily Heteroderoidea Marked sexual dimorphism. Oesophagus with distinct procorpus, well developed median bulb with prominent crescentic valve plates and oesophageal glands overlapping intestine mainly ventrally. *Mature female.* Swollen to subspherical, sedentary, sometimes forming a cyst. Endo/semi-endoparasites attached to plant roots sometimes causing and enclosed by root galls. Cuticle with or without annules; tail absent or very short. Vulva median to terminal; paired reproductive tracts. *Male.* Vermiform; tail very short, less than one and a half anal body-widths long, anus sometimes subterminal; large ventrally curved spicules and a small gubernaculum; bursa absent. Testis typically single but sometimes paired. Body often twists through at least 90° when killed by heat. Stylet prominent with distinct basal knobs, cephalic framework well developed. Cuticle distinctly annulated plus lateral incisures. *Second-stage juvenile.* Vermiform; the infective, migratory stage. Stylet moderate to strongly developed. Tail conical to roundly conical usually with a rounded terminus and a clear (hyaline) terminal area. Body annulation and lateral incisures distinct.

Family Heteroderidae *Female.* Tail absent, anus terminal or dorsally subterminal. Excretory pore posterior to level of valve in median oesophageal bulb. Basal plate divides head into six sectors, the two lateral ones being narrower than the others, Stylet more than 17 μm long. *Second-stage juvenile.* Basal plate subdivides head as in female. Stylet robust, more than 16 μm long; cephalic framework well developed. *Third-stage juvenile.* Swollen sedentary, with short blunt tail. Stylet present. *Fourth-stage juvenile.* Swollen, sedentary, with terminal anus. Stylet present. *Male.* With well developed stylet and cephalic framework. Basal plate subdivides head as in female. Testis single.

Subfamily Heteroderinae *Female.* Cyst formed after death. Cuticle non-annulated. Anus dorsal or on the inside of the dorsal vulva lip. Vulva terminal. *Second-stage juvenile.* Labial disc present but only seen clearly under a scanning electron microscope. Cuticle of labial region of even thickness. Oesophageal glands fill body width. *Male.* Up to 1·5 mm long. Labial disc present but with low profile or apparently absent. Tail present or absent.

Genera: *Heterodera* Schmidt, 1871; *Globodera* Skarbilovich, 1959 (Mulvey & Stone, 1976); *Sarisodera* Wouts & Sher, 1971; *Punctodera* Mulvey & Stone, 1976. See Chapter 6 for further details of some of these important plant-parasitic genera.

Subfamily Ataloderinae *Female.* No cyst formed. Cuticle non-annulated. Anus terminal; vulva terminal. *Second-stage juvenile.* Labial disc absent. Cuticle of labial region of even thickness. Oesophageal glands fill approximately half the body-width. *Male.* Up to 1·5 mm long. Labial disc present. Spicules more than 30 μm long.

Genera: *Atalodera* Wouts & Sher, 1971; *Sherodera* Wouts, 1974.

Subfamily Meloidoderinae *Female.* No cyst stage. Cuticle annulated. Anus terminal or dorsally subterminal. Vulva equatorial to terminal. *Second-stage juvenile.* Labial disc present; cuticle of labial region distinctly wider near basal plate than near labial disc. Oesophageal glands fill body width. *Male.* Less than 1 mm long. Labial disc present. Tail present; spicules less than 30 μm long.

Genera: *Meloidodera* Chitwood, Hannon & Esser, 1956; *Cryphodera* Colbran, 1966; *Zelandodera* Wouts, 1973.

Family Meloidogynidae *Mature female.* Swollen to subspherical. No cyst formed. Cuticle somewhat annulated especially in posterior region. Tail absent or very short. Anus and vulva subterminal to terminal. Excretory pore opposite or anterior to median oesophageal bulb. Basal plate divides head into six sectors, the two lateral sectors wider than the others. *Second-stage juvenile.* Basal plate subdivides head as in female. Stylet moderately to well developed. *Male.* With well developed stylet and cephalic framework. Amphid openings slit-like, partially covered by enlarged lateral lips (cheeks).

Subfamily Meloidogyninae Definition as for family.

Genera: *Meloidogyne* Goeldi, 1887; *Meloidoderella* Khan, 1972 or Khan & Husain, 1972; *Meloinema* Choi & Geraert, 1974; *Nacobbodera* Golden & Jensen, 1974.

Note—The genus *Meloidoderita* Pogosyan, 1966 was included by Wouts (1973) in Meloidogynidae but the female has a single reproductive tract. Pogosyan (1975) described bursate males with degenerate oesophagus and sharply pointed tails for the type species *M. kirjanovae.* This suggests that *Meloidoderita* would perhaps be better placed in the Criconematoidea near Sphaeronematinae.

Meloidogyne spp. (root-knot nematodes) damage many crops especially in warmer soils. See Chapter 5 for further details.

Superfamily Criconematoidea Both sexes generally small and vermiform but mature females in some groups may be swollen. Sexually dimorphic; males with degenerate oesophagus and stylet weakly formed or absent and body usually finely annulated; bursa present or absent. Female oesophagus Criconematoid, Fig. 1.7e, (i.e., the often short procorpus merges without a constriction into an enlarged median bulb (metacorpus) which has well developed crescentic valve plates. The isthmus is usually short and narrow and ends in a weak basal bulb containing the oesophageal glands). Vulva posterior with single anterior genital tract.

Family Criconematidae Both sexes usually vermiform but females often quite robust. Cuticle of females and juveniles with prominent annulation, sometimes retrorse, with or without scales, spines or serrations. Cuticular sheath present in females of Hemicycliophorinae [*sic*]. Female stylet well developed, elongated, the anterior portion being the longer. Males with or without an adanal bursa. Oesophageal procorpus very short and amalgamated with the metacorpus to give a large bulb with prominent crescentic valve plates. Isthmus short with weak basal bulb.

Subfamily Criconematinae Females usually short and stout, body annules typically less than 200, prominent, coarse, usually retrorse, with or without spines, scales or serrate markings. Stylet well developed with basal knobs usually concave anteriorly giving an anchor-shaped appearance. Males only, with distinct incisures on lateral fields but usually without a stylet; bursa present or absent, spicules straight or with slight ventral curvature.

Genera: *Criconema* Hofmänner & Menzel, 1914; *Bakernema* Wu, 1964; *Blandicephalanema* Mehta & Raski, 1971; *Criconemella* de Grisse & Loof, 1965; *Criconemoides* Taylor, 1936; *Crossonema* Mehta & Raski, 1971; *Discocriconemella* de Grisse & Loof, 1965; *Lobocriconema* de Grisse & Loof, 1965; *Macroposthonia* de Man, 1880; *Neolobocriconema* Mehta & Raski, 1971; *Nothocriconema* de Grisse & Loof, 1965; *Nothocriconemoides* Maas, Loof & de Grisse, 1971; *Pateracephalanema* Mehta & Raski, 1971; *Xenocriconemella* de Grisse & Loof, 1965.

Subfamily Hemicycliophorinae Female elongate with distinct non-retrorse body annules numbering about 200–400; body sometimes enclosed in a cuticular sheath (*Hemicycliophora*, *Hemicriconemoides*). Stylet elongate with spheroid basal knobs the outer margins of which usually slope backwards. Lateral fields may be present in both sexes. Female tail elongate-conoid, filiform or roundly conical. Male with or without adanal bursa, stylet usually absent; spicules almost straight or with marked, sickle-shaped, ventral curvature (*Hemicycliophora*).

Genera: *Hemicycliophora* de Man, 1921; *Caloosia* Siddiqi & Goodey, 1963; *Hemicriconemoides* Chitwood & Birchfield, 1957.

Family Paratylenchidae Generally small, vermiform, adults less than 0·5 mm long; mature females of some forms become obese and sedentary. Body annules fine, rounded, not retrorse. Female stylet variable, from 12 to 119 μm long, with rounded basal knobs. Lip region narrow, rounded to truncate. Labial sclerotization generally indistinct (except in *Paratylenchoides*). Vulva posterior with single anterior genital tract. Lateral fields usually present. Males with degenerate oesophagus; stylet weak or absent. Bursa usually absent or very small; spicule sheath often present; spicules with slight ventral curvature.

Subfamily Paratylenchinae Definition as for family.

Genera: *Paratylenchus* Micoletzky, 1922; *Cacopaurus* Thorne, 1943; *Gracilacus* Raski, 1962; *Paratylenchoides* Raski, 1973.

See Chapter 9 for further details of some of these root ectoparasitic nematodes within the Criconematidae.

Family Tylenchocriconematidae Both sexes vermiform. Marked sexual dimorphism. Male without stylet, oesophagus degenerate. Bursa well developed, extending to tip of tail. Male lip region asymmetrical, inclined ventrally. Female with low, truncate, lip region and distinct stylet with rounded basal knobs. Oesophageal procorpus of female enlarges gradually to large muscular metacorpus followed by a long slender isthmus and elongate basal bulb enclosing the oesophageal glands. Body annules fine; lateral field present.

Subfamily Tylenchocriconematinae Definition as for family. Based on the monotypic genus *Tylenchocriconema* Raski & Siddiqui, 1975.

Family Tylenchulidae Females elongate-saccate to subspherical with finely annulated cuticle. Stylet well developed in females and juveniles but not elongated. Oesophagus basically Criconematoid with narrow isthmus and distinct basal bulb but in some forms the procorpus is more elongate and less amalgamated with the typically large valvated median bulb. Males slender, oesophagus degenerate, stylet absent or reduced; bursa absent.

Subfamily Tylenchulinae Mature female obese to saccate. Excretory pore posterior to oesophagus, sometimes in posterior half of body. Males with or without weak stylet.

Genera: *Tylenchulus* Cobb, 1913; *Trophotylenchulus* Raski, 1957.

Subfamily Sphaeronematinae Mature females subspherical to obese. Cuticle rather thick, sometimes with distinct reticulate pattern. Excretory pore near nerve ring. Males without stylet.

Genera: *Sphaeronema* Raski & Sher, 1952; *Trophonema* Raski, 1957.

All of the Tylenchulidae are obligate, sedentary, root semi-endoparasites. *Tylenchulus semipenetrans* (the citrus nematode) is a serious pest causing 'slow decline' of citrus.

Superfamily Neotylenchoidea Median bulb of oesophagus absent or vestigial without valve plates (Fig. 1.7d) but a fusiform swelling may be present where the subventral glands open. Corpus, isthmus and enlarged glandular basal portion present; the latter may abut, overlap or protrude into the intestine. Both sexes of free-living forms vermiform. Females with posterior vulva and single anterior reproductive tract. Males usually with Tylenchoid-type spicules, bursa adanal, subterminal or terminal. Stylet distinct, with basal knobs, under 20 μm long. Cephalic framework weakly sclerotized, in six or eight sectors.

Family Neotylenchidae Cephalic framework in eight sectors. Oesophageal glands overlap intestine in a long lobe. Male bursa subterminal surrounding tail tip or absent.

Genera: *Hexatylus* Goodey, 1926 (syn. *Neotylenchus* Steiner, 1931); *Deladenus* Thorne, 1941; *Gymnotylenchus* Siddiqi, 1961; *Hadrodenus* Mulvey, 1969; *Scytaleum* Andrássy, 1961.

Some species of *Hexatylus* and *Deladenus* feed on fungi. *Hexatylus* spp. are sometimes associated with potato tuber rot. Some *Deladenus* species at times produce insect-parasitic females that have Allantonematid/Sphaerulariid characters and they are used in the biological control of some wood-boring wasps (*Sirex* spp.); see Bedding (1973 & 1974).

Family Paurodontidae Cephalic framework in six sectors. Basal oesophageal bulb with a stem-like extension into the anterior part of the intestine.

Subfamily Paurodontinae Excretory pore in vicinity of or posterior to the nerve ring. Tails of both sexes elongate, acute or subacute. Male bursa adanal or enveloping tail.

Genera: *Paurodontus* Thorne, 1941; *Bealius* Massey & Hinds, 1970; *Boleodoroides* Mathur, Khan & Prasad, 1966; *Paurodontella* Husain & Khan, 1968; *Paurodontoides* Jairajpuri & Siddiqi, 1969; *Stictylus* Thorne, 1941.

Subfamily Misticiinae Excretory pore anterior, not far from vicinity of stylet base. Tails of both sexes elongate-conoid. Male with large subterminla bursa.

Based on the monotypic genus *Misticius* Massey, 1967.

Family Nothotylenchidae Cephalic framework in six sectors. Basal oesophageal bulb pyriform to cylindrical, usually abutting intestine (long overlap in *Halenchus*).

Subfamily Nothotylenchinae Basal oesophageal gland pyriform to cylindrical abutting or slightly overlapping the intestine.

Genera: *Nothotylenchus* Thorne, 1941; *Anguillonema* Fuchs, 1938; *Dorsalla* Jairajpuri, 1966; *Luella* Massey, 1974; *Nothanguina* Whitehead, 1959; *Sakia* Khan, 1964; *Thada* Thorne, 1941. Many species of these genera are associates of bark beetles but *Nothanguina* spp. cause leaf and stem galling. See also Chapter 4.

Subfamily Boleodorinae Small nematodes, < 720 μm long, cuticle finely annulated, lateral field with four or six incisures. Lip region rounded, with or without slight oral depression, framework weakly developed; amphid openings more or less oval to oblique slits on the sides of the lateral lips. Stylet weakly developed with basal flanges rather than knobs. Oesophagus with weakly developed median bulb; posterior bulb abutting or slightly overlapping intestine. Tails of both sexes elongate-conoid; male with adanal bursa.

Genus: *Boleodorus* Thorne, 1941. Several species have been recorded from soils from around the world but their feeding habits are unknown. Geraert (1971) suggests that this subfamily should be placed next to Psilenchinae, Tylenchoidea, although it does not seem to have a valvated median bulb.

Subfamily Halenchinae Oesophageal glands form a long lobe overlapping the intestine. Excretory duct heavily sclerotized towards and at its outlet. Tails of both sexes elongate, convex-conoid, ending in a ventrally curved hook. Bursa rather narrow, adanal.

Genus: *Halenchus* Cobb, 1913. Species of this genus are marine and cause galls on some seaweeds.

Family Ecphyadophoridae Body slender. Cuticle with or without fine annulations. Lateral field present. Lip region high and asymmetrical or low and rounded. Cephalic sclerotization indistinct. Stylet delicate with basal knobs. Oesophagus a long narrow cylindrical tube without median swelling; oesophageal glands in a lobe overlapping the intestine. Male body narrows abruptly at the cloaca; prominent adanal bursa present. Tails of both sexes elongate-conoid to cylindrical with conoid terminus. Spicules straight or with slight ventral curvature; gubernaculum present or absent.

Genera: *Ecphyadophora* de Man, 1921; *Ecphyadophoroides* Corbett, 1964. Species of these two genera rarely reported from about plant roots.

The families Allantonematidae, Contortylenchidae and Sphaerulariidae, species of which are mainly parasites or associates of insects, are included by Massey (1974) in Neotylenchoidea. Nickle (1967) includes the first two within the single family Sphaerulariidae, whereas Poinar (1975) restricts this family to three genera but at the same time elevates it to superfamily rank. These three authors should be consulted for further information.

Suborder APHELENCHINA Dorsal oesophageal gland empties into lumen of oesophagus within the median bulb. Generally small vermiform nematodes about one mm long or less, although some insect-parasitic forms may be up to about 3 mm long.

Superfamily Aphelenchoidea Oesophagus consists of a narrow procorpus, a prominent oval to rectangular median bulb which often occupies most of the body diameter, a narrow isthmus and usually (except in Paraphelenchidae) an elongated lobe, containing the oesophageal glands, that overlaps the intestine mainly dorsally and laterally. The median bulb has distinct crescentic valve plates and the dorsal oesophageal gland duct opens into the oesophageal lumen just in front of the valve (an Aphelenchoid-type oesophagus, see Fig. 1.7f and g). Males usually without a cloacal bursa (except in *Aphelenchus* and some *Paraphelenchus*); sometimes a small terminal bursa present. Amphids inconspicuous, usually on the face of the lip region. Both sexes usually vermiform with no marked dimorphism except for sex organs. Cuticle marked by fine transverse annulations; lateral field usually present. Females with a posterior vulva and a single anterior genital tract. Cephalic sclerotization usually indistinct. Stylet weak to moderate, usually less than 20 μm long, with or without basal thickenings, seldom with conspicuous knobs.

Family Aphelenchidae Male with prominent bursa that envelops the tail and is supported by bursal rays (Fig. 1.7f); spicules slender, ventrally arcuate; gubernaculum present, V-shaped in dorsoventral view. Oesophageal glands overlap intestine dorsally and laterally, often in a long lobe. Lip region low, rounded to flattened anteriorly. Stylet without distinct basal thickenings. Females with relatively short blunt tail, $<$ two anal body widths long. Lateral field relatively wide with six to 14 incisures in the mid-body region.

Subfamily Aphelenchinae Definition as for family.

Genus: *Aphelenchus* Bastian, 1865. *Aphelenchus avenae* is a cosmopolitan, ubiquitous, fungivorous nematode common in soil and decaying plant tissue. See also Chapter 8.

Family Paraphelenchidae Oesophageal glands contained within a spathulate basal bulb that abuts the intestine. Lip region relatively low and usually flattened anteriorly. Stylet with or without basal thickenings. Bursa reduced or absent;

prominent caudal papillae present. Spicules slender, ventrally arcuate; gubernaculum linear to V-shaped in dorso-ventral view. Male and female tails short, conical, sometimes with terminal mucro. Lateral field with four to eight incisures.

Subfamily Paraphelenchinae Definition as for family.

Genus: *Paraphelenchus* Micoletzky, 1922 (Micoletzky, 1925). *Paraphelenchus* spp. seem to be fungivorous. See also Chapter 8.

Family Aphelenchoididae Oesophageal glands in a long lobe that overlaps the intestine mainly dorsally. Stylet with or without basal thickenings/knobs. Spicules usually rosethorn-shaped (= Aphelenchoid, Fig. 1.7g); gubernaculum usually absent. Large (*Aphelenchus*-type) bursa absent; subventral caudal papillae present; tail terminus sometimes with a small bursa.

Subfamily Aphelenchoidinae Lip region rounded to flattened anteriorly, usually relatively higher than in Aphelenchidae and often more distinctly offset from the body; six amalgamated lips present giving a hexaradiate pattern in face view. Annulation of lip region not usually visible with light microscope. Cephalic sclerotization usually indistinct except for refractive spear guide at the oral aperture. Stylet usually with basal thickenings. Female tail usually conical, usually less than five anal body-widths long, terminus conoid to blunt often with a terminal mucro; male tail similar, usually with two or three paris of small subventral papillae, sometimes with a small terminal bursa. Gubernaculum rarely present.

Note—This subfamily has many diverse genera so that a brief definitive diagnosis is not feasible; the group requires further study and probably subdivision. Genera: *Aphelenchoides* Fischer, 1894; *Aprutides* Scognamiglio, Talamé & s'Jacob, 1970; *Bursaphelenchus* Fuchs, 1937; *Cryptaphelenchus* Fuchs, 1937 (Rühm, 1954); *Cryptaphelenchoides* Goodey, 1960; *Ektaphelenchus* Fuchs, 1937 (Skrjabin *et al.*, 1954); *Huntaphelenchoides* Nickle, 1970; *Laimaphelenchus* Fuchs, 1937, *Megadorus* Goodey, 1960; *Papuaphelenchus* Andrássy, 1973; *Parasitaphelenchus* Fuchs, 1929; *Ruehmaphelenchus* Goodey, 1963; *Schistonchus* Cobb, 1927 (Fuchs, 1937); *Sheraphelenchus* Nickle, 1970; *Tylaphelenchus* Rühm, 1956.

Some species of *Aphelenchoides* are plant-parasitic and/or fungivorous (see Chapter 8). Species in most of the other genera listed are associates of insects, especially bark beetles, and many of these are probably fungivorous.

Subfamily Rhadinaphelenchinae Body extremely slender (a = about 100). Lip region high with sclerotized labial arches. Median oesophageal bulb about twice as long as wide with valve posterior to centre; oesophageal glands in long dorsal overlapping lobe. Vulva with wide overlapping flap; female tail elongated, almost cylindrical with narrow rounded terminus. Male tail curled ventrally upon itself when killed by heat; spicules small with prominent rostrum and elongated apex; tail terminus with small bursa.

Genus: *Rhadinaphelenchus* Goodey, 1960 based on the single species *R. cocophilus* which seems to be the cause of 'red-ring disease' of coconut palms. See also Chapter 8.

Subfamily Anomyctinae Cuticle more strongly annulated than in most other Aphelenchs. Lip region rounded, offset, with shallow, sclerotized, saucer-like frontal disc present. Oral opening surrounded by six small projections.

Stylet elongate (about 35 μm), anterior part unevenly sclerotized and about one third the length of the posterior part (extension) which is without basal knobs. Female tail conical with short terminal mucro. Male tail similar with large spicules and three pairs of subventral papillae; bursa and gubernaculum absent.

Genus: *Anomyctus* Allen, 1940. A monotypic group based on the rarely reported species *A. xenurus* from U.S.A. and England.

Subfamily Seinurinae Tails of both sexes elongate-filiform without a terminal mucro. Lip region high. Stylet slender but with relatively wide lumen, with or without small basal thickenings. Median oesophageal bulb elongate oval to rectangular with prominent posteriorly located valve plates. Male tail with two to four pairs of caudal papillae; spicules with elongated apex and prominent rostrum, gubernaculum present or absent.

Genera: *Seinura* Fuchs, 1931; *Paraseinura* Timm, 1960. Most species seem to be predacious.

Family Entaphelenchidae

Subfamily Entaphelenchinae Similar to Aphelenchoidinae but restricted to genera in which the species are obligate insect parasites.

Genera: *Entaphelenchus* Wachek, 1955; *Peraphelenchus* Wachek, 1955; *Praecocilenchus* Poinar, 1969; *Roveaphelenchus* Nickle, 1970.

KEY TO SUBFAMILIES OF PLANT AND SOIL TYLENCHIDA

This key serves as a guide for separating taxa down to subfamily level. However, it is stressed that there may be occasional exceptions that do not fit into this scheme and to try to include them would make it too unwieldy. Many of the differentiating characters are relative (i.e., head sclerotization distinct v. indistinct) and a beginner should refer to illustrations in other chapters of this book or preferably examine a selection of identified specimens to appreciate the differences concerned. Although of fundamental taxonomic importance, the nature of the overlap of oesophageal gland lobes over the intestine in some groups, especially Aphelenchoidea, is not readily seen and caution is advised where this character is used in the key.

1. Outlet of dorsal oesophageal gland duct in procorpus of oesophagus between base of stylet and median oesophageal bulb, often near base of stylet (Fig. 1.7a–e) . . . . **TYLENCHINA** 2
   Dorsal oesophageal gland duct opening within the median oesophageal bulb. (Females always with a single anterior genital tract and posteriorly placed vulva; males without a distinct bursa (except *Aphelenchus*); median oesophageal bulb usually large, oval to roundly rectangular and occupying nearly all the body diameter) (Fig. 1.7f & g) **APHELENCHINA** 43

2. Oesophagus without a median bulb; fusiform swelling sometimes present but valve-plates absent (Fig. 1.7d) . . . . . **Neotylenchoidea** 37
   Oesophagus with a swollen median bulb with refractive crescentic valve plates (Fig. 1.7a, b, c & e). 3

3\. Procorpus and median bulb usually amalgamated with prominent crescentic valve-plates (= Criconematoid, see Fig. 1.7e), or median bulb not distinctly offset from procorpus, basal bulb small and weakly defined. (Females, with posterior vulva and single anterior genital tract, usually vermiform but occasionally swollen and sessile; male oesophagus degenerate) **Criconematoidea** 32
Median bulb of oesophagus distinctly set off from procorpus (Fig. 1.7a, b, c) 4

4\. Mature females swollen, sessile, with paired genital tracts; male elongate vermiform, but with very short tail and well developed elongate spicules; bursa absent. . . . . **Heteroderoidea** 29
Both sexes vermiform and without marked sexual dimorphism (except *Rotylenchulus* and *Nacobbus*); male with distinct bursa **Tylenchoidea** 5

5\. Head with four distinct setae . . . . **Atylenchidae/inae**
Head without setae. 6

6\. Mature females swollen, saccate to kidney-shaped; immature females migratory in soil 7
Mature females usually vermiform, occasionally obese but not saccate 8

7\. Mature females subspherical to saccate, with single genital tract, embedded in root galls; males vermiform with well developed stylet and oesophagus and distinct small bursa enveloping the tail **Nacobbinae/Pratylenchidae**
Mature females kidney-shaped, with paired, opposed genital tracts; sedentary root parasites with most of body protruding from root; male with weaker spear and oesophagus and a small subterminal bursa **Rotylenchulinae/Hoplolaimidae**

8\. Head sclerotization distinct, head region elevated, stylet strong, $> 20\ \mu m$ long; male and female tail short, < 2 anal body-widths (a.b.w.) long; oesophageal gland lobes overlapping intestine; male with distinct bursa that envelops the tail . . . . **Hoplolaimidae** 26
[Young migratory females of *Nacobbus*—Pratylenchidae have somewhat similar head and tail shape but are readily distinguished by the posterior vulva that is within 2 a.b.w. of the anus.]
Female tail > 2 a.b.w. long 9

9\. Oesophageal glands overlapping oesophagus usually in a long lobe; if short overlap, head sclerotization distinct 10
Oesophageal glands abutting the intestine; if with short overlap then head sclerotization indistinct. 12

10\. Head region elevated, head sclerotization variable, body length moderate to long (up to 2 mm) with distinct cuticular annulation; stylet length variable sometimes up to 100 μm; female with cylindrical to elongate conical tail, vulva median with paired, opposed genital tracts; male with elongate enveloping bursa . . . . **Belonolaimidae** 24
Head region low, rounded, to flattened anteriorly; usually not more than 4 head annules 11

11. Female head with distinct sclerotization, stylet stout $< 45\ \mu$m, usually $< 25\ \mu$m; female tail 2–5 a.b.w. long, conoid with blunt to fine terminus; male with terminal or subterminal bursa
Pratylenchidae (except Nacobbinae) 25
Head without distinct sclerotization; female with single anterior genital tract and posterior vulva, tail elongate conoid with rounded terminus; male with adanal bursa . . . . . **Pseudhalenchinae/Tylenchidae**

12. Female with posterior vulva and single anterior genital tract; cuticle annulation usually indistinct . . . . . **Tylenchidae** (in part) 15
Female with median vulva, usually with paired, opposed genital tracts 13

13. Male with trilobed bursa, head of male and female heavily sclerotized; cuticle distinctly annulated; stylet moderate to elongate
Dolichodoridae/inae
Male bursa not trilobed but adanal, subterminal to terminal . . . 14

14. Female tail 2–6 a.b.w. long, cylindrical to conoid with pointed to bluntly rounded terminus; bursa prominent, enveloping male tail or almost so; phasmids distinct on mid-tail of female and on bursal wings of male
Tylenchorhynchidae 23
Female tail elongate to filiform . . . . **Tylenchidae** (in part) 15

15. Female with vulva about median with paired, opposed genital tracts 16
Female with posterior vulva, single anterior genital tract 18

16. Head with moderate sclerotization; amphid apertures on face of head region; stylet with distinct basal knobs; both sexes with elongate cylindrical to filiform tails; male with adanal bursa . . . . **Antarctylenchinae**
Head without distinct sclerotization; stylet without basal knobs; tail elongate conoid; bursa adanal. 17

17. Head region flattened anteriorly, amphid apertures inconspicuous
Leipotylenchinae
Distinct oval to slit-like amphid apertures on side of head
Psilenchinae (in part—*Psilenchus*)

18. Stylet elongate, about 100 μm, with large basal knobs; cuticle distinctly annulated; tails of both sexes filiform, bursa adanal . . **Tylodorinae**
Stylet short, $< 25\ \mu$m; body annulation usually inconspicuous . . 19

19. Body cuticle thickened, head with finger-like sclerotization; tails of both sexes elongate and slender; bursa adanal. . . **Dactylotylenchinae**
Body cuticle normal, head sclerotization indistinct 20

20. Amphid apertures oval to slit-like, on side of head; stylet with basal knobs; deirids and phasmids distinct; tails of both sexes elongate-conoid to filiform
Psilenchinae (in part)
Amphid apertures indistinct, usually on face of head region . . . 21

21. Tail elongate conoid to filiform; bursa adanal, basal oesophageal bulb small but distinct, pyriform, abutting intestine; distinct oesophago-intestinal valve present . . . . **Tylenchinae**
Tail conoid to bluntly rounded; basal oesophageal gland bulb well developed, sometimes slightly overlapping the intestine; oesophago-intestinal valve inconspicuous; head region usually low and flattened anteriorly 22

22. Female tail conical to short-conical, rarely elongated; male with elongate but subterminal bursa . . . . **Anguininae**
Female tail short, usually bluntly rounded; V = about 90; male with prominent bursa that envelops tail terminus . . **Sychnotylenchinae**

23. Cuticle distinctly to strongly annulated, typically not thickened or swollen at tail terminus; amphid apertures on face of head region; basal oesophageal bulb well formed, abutting intestine; paired, opposed genital tracts always present in female . . . . **Tylenchorhynchinae**
Cuticle weakly annulated and thickened or swollen at tail terminus; basal oesophageal bulb rather weak and misshapen; vulva median with single anterior genital tract or if genital tracts paired, then stylet > 90 μm long **Trophurinae**

24. Head region prominent, distinctly offset, with labial disc; stylet long and slender 60–150 μm (one species 30 μm); oesophageal procorpus usually short and somewhat swollen . . . . **Belonolaiminae**
Head region smaller, offset or continuous with body contour, labial disc absent; stylet 15–50 μm; oesophageal procorpus not swollen or shortened **Telotylenchinae**

25. Oesophageal glands overlapping intestine in a long lobe mainly ventrally; no marked sexual dimorphism . . . . **Pratylenchinae**
Oesophageal glands overlapping intestine dorsally, usually in a long lobe (short in *Pratylenchoides*); some sexual dimorphism, anterior end of male modified and/or degenerate . . . . **Radopholinae**

26. Vulva about median with paired, opposed genital tracts 27
Vulva in posterior quarter of body with single anterior genital tract **Rotylenchoidinae**

27. Phasmids absent; oesophageal gland lobe overlaps intestine ventrally; male tail 2 a.b.w. long with elongate bursa. . . . **Aphasmatylenchinae**
Phasmids present, oesophageal glands overlap intestine dorsally and laterally, or ventrally and laterally 28

28. Phasmids large (scutellae), variable in position on body; oesophageal glands overlap intestine mainly dorsally . . . . . **Hoplolaiminae**
Phasmids small, pore-like, in anal region . . . . **Rotylenchinae**

29. Mature female spherical to subspherical, cuticle does not form cyst; excretory pore opposite or anterior to median oesophageal bulb; males with prominent lateral lips (cheeks) . . . . **Meloidogynidae/inae**
Mature female swollen to subspherical; anus terminal or dorsally subterminal; excretory pore posterior to level of valve in median oesophageal bulb; male without prominent lateral 'cheeks' . . **Heteroderidae** 30

30. Female cuticle annulated, does not form cyst; cuticle of labial region of second-stage juvenile widest at level of basal plate . . **Meloidoderinae**
Female cuticle not annulated; cuticle of labial region of second-stage juvenile of even thickness 31

31. Female not forming a cyst; males with prominent labial disc; second-stage juveniles with narrow oesophageal glands occupying < half body-width **Atalodorinae**

Female forming a cyst; males with indistinct labial disc; oesophageal glands occupy whole body-width in second-stage juvenile **Heteroderinae**

32. Cuticular annulation of female very coarse, cuticular sheath present in some forms; oesophageal isthmus short with weak basal bulb **Criconematidae** 33

Cuticular annulation of female fine; oesophageal isthmus usually long 34

33. Females usually short and stout (a = < 20), without a cuticular sheath; stylet knobs with forward pointing outer margins (anchor-shaped) **Criconematinae**

Females vermiform, sometimes with a cuticular sheath; stylet knobs rounded often sloping backwards . . . . . **Hemicycliophorinae**

34. Mature female slender (a = > 40) with long oesophageal isthmus; both sexes with elongate-conoid tail; male with prominent, elongate, enveloping bursa. . . . . **Tylenchocriconematidae/inae**

Mature female stout to very swollen (a = < 40); male without bursa 35

35. Mature female elongate-saccate, sometimes coiled, to subspherical, stylet < 20 μm . . . . **Tylenchulidae** 36

36. Excretory pore posterior to oesophagus, sometimes in posterior half of body; mature female elongate-saccate . . . . . **Tylenchulinae**

Excretory pore in vicinity of nerve ring; mature female subspherical with thick cuticle, or elongate saccate and coiled . . . **Sphaeronematinae**

37. Adults very slender (a = > 70), male body narrows abruptly at cloaca **Ecphyadophoridae/inae**

Adults vermiform (a usually < 70), males not narrowing abruptly at cloaca 38

38. Cephalic framework in eight sectors, oesophageal glands in long lobe overlapping intestine . . . . **Neotylenchidae/inae**

Cephalic framework in six sectors 39

39. Basal oesophageal bulb with stem-like extension into the anterior intestine **Paurodontidae** 40

Basal oesophageal bulb abutting or overlapping intestine, without stem-like extension . . . . **Nothotylenchidae** 41

40. Excretory pore in vicinity of or posterior to nerve ring **Paurodontinae**

Excretory pore anterior, close to vicinity of stylet base. . **Misticiinae**

41. Basal oesophageal bulb pyriform to cylindrical, abutting or slightly overlapping intestine 42

Basal oesophageal bulb in long overlapping lobe; excretory duct heavily cuticularized; tail elongate with terminal hook . . . . **Halenchinae**

42. Amphid openings oval to oblique slits on outer margins of head **Boleodorinae**

Amphid openings indistinct, on the face of a usually somewhat flattened head . . . . **Nothotylenchinae**

43. Male with elongate spicules and distinct gubernaculum (Fig. 1.7f); females with short (< 3 a.b.w.) tails 44

Male with rose-thorn or irregular shaped spicules (Fig. 1.7g), gubernaculum usually absent or indistinct; female tail usually > 2 a.b.w. long **Aphelenchoididae** 45

44. Male with prominent bursa that envelops tail and is supported by bursal 'ribs' (Fig. 1.7f); lateral field with 6–14 incisures; oesophageal glands overlap intestine dorsally/laterally; stylet without basal thickenings; female tail short and usually bluntly rounded . . . **Aphelenchidae/inae**

Bursa absent or greatly reduced, prominent caudal papillae present, tails of both sexes short, conical, with or without mucro; oesophageal glands in basal bulb abutting intestine . . . . . **Paraphelenchidae/inae**

45. Body extremely slender (a = about 100), head region with sclerotized arches; male tail with terminal bursal flap; oesophageal glands in long dorsal overlapping lobe . . . . **Rhadinaphelenchinae**

Body not extremely slender (a = < 60) 46

46. Head region rounded, offset, with sclerotized saucer-like frontal disc; stylet elongate (35 μm) without basal knobs . . . . **Anomyctinae**

Head region without sclerotized frontal disc. 47

47. Male and female with elongate filiform tails without a terminal mucro, male without a bursa; median oesophageal bulb elongate-oval to rectangular with posteriorly placed valve plates; stylet with or without basal thickenings **Seinurinae**

Tails usually conical, usually < 5 a.b.w. long, not filiform, often with terminal mucro; adanal bursa absent, terminal bursa sometimes present; stylet usually with basal thickenings; lateral field with 2–6 incisures **Aphelenchoidinae**

CHAPTER 2

The Soil–Plant Environment

F. G. W. JONES
Rothamsted Experimental Station, Harpenden

Some plant feeding nematodes live entirely in the soil. Other, endoparasitic species live mainly within plants, but spend periods of varying duration in the soil when they are dependent on soil conditions while host finding, feeding and mating. Yet other nematodes, which feed for long periods with their stylets deeply inserted into host tissue or with the head and fore part of the body embedded, obtain some of the benefits of endoparasitism. They are able to continue feeding and to maintain their water balance when movement and feeding would otherwise be halted by relatively slight soil-moisture deficits. Unlike concealed endoparasites, however, they get no protection from enemies operating in the soil spaces.

The Soil Environment

To plough depth, ploughing and cultivating ensures that the main constituents of arable soils are uniformly mixed (viz. sand, silt, clay and old organic matter) but soil structure (i.e., the arrangement of the above constituents into aggregates) varies with depth according to recent treatments. However, heavy rainfall may cause the redistribution of particles and that of recently applied fertilizers, and the distribution of fresh organic matter is not random. Essentially there are two soil layers, the cultivated topsoil and the subsoil beneath it which is influenced little by cultivations and may itself be layered. Between topsoil and subsoil a smear-layer plough pan may develop, which, depending on its porosity, may be difficult for gases, water, roots and nematodes to penetrate.

THE PORE SYSTEM

Most British agricultural soils are maintained at pH 7 or thereabouts by regular liming. Consequently pH and the chemistry of the parent materials affect nematodes little, except perhaps for elements like copper to which *Trichodorus* spp. appear to be sensitive (Cooper, 1971). Consequently, for most purposes the soil can be regarded as a system of pore spaces ramifying within an inert framework. Nematodes within the pore space are influenced by its morphology, its volume, the amount of water it contains, the surface tension forces binding water to the matrix and the oxygen concentration within it. Under their own power nematodes move over short distances only and their relative immobility is accentuated by the constraints of the soil pore system.

Most vermiform nematodes range from 300 to 2000 μm long and from 15 to 60 μm diameter. Some nematodes are much longer reaching 10 to 12 mm. Exceptionally slender nematodes may be no more than 5 μm diameter and exceptionally stout ones as much as 150 μm diameter (Jones, Larbey & Parrott,

1969). Unlike roots, nematodes cannot exert much force and therefore are limited to the existing pore system. The fraction they can occupy is determined by the cross sectional diameter of the pore necks through which they can just pass. Usable space is mainly in pores of somewhat larger diameter, greater than say 30 to 60 μm diameter because, at the narrower end of the range, usage is restricted by tortuosity and by blind ends. The usability of large pores (voids) has not been studied but possibly longer nematodes can traverse larger spaces better than short ones can for the latter are forced to migrate around the walls. For them the morphology of the wall surface and its physical properties, which influence the thickness of water films, could be important factors affecting movement.

Studies in graded sands show that nematodes move fastest when the particle size is about one third of the body length, i.e., when the particles fit into the wave pattern adopted by nematodes when moving in water free from obstructions (Wallace, 1963). This suggests that nematodes would move best through mono-particulate soils with spheroidal particles having diameters about a third of their body length, or over surfaces with closely spaced, rounded projections of that order of size. A few podzols and dune soils may approximate to this ideal but most soils do not. The size of the particles ranges widely, many are angular and some plate-like. The spongy mesophylls of leaves and the surface of leaves and stems of some plants approach the ideal but the internal surfaces of soil voids must rarely do so.

MEASUREMENT OF TOTAL PORE AND USABLE PORE SPACE

The porosity of a soil P is given by $P = 1 - D_b.D_p^{-1}$ where D_b is the bulk density of an undisturbed soil sample (mass, less that of contained water, divided by its apparent volume) and D_p is the particle density (mass, after drying, divided by actual volume). For mineral soils containing little organic matter and not recently ploughed from grass, the particle density is approximately constant at 2·65. It follows that for these soils porosity is proportional to bulk density. The total pore space can also be measured from the moisture characteristic (Fig. 2.1) which relates the volume of water retained to the suction pressure applied. At saturation when these forces are zero the pores are filled with water. After oven drying at 105°C, which is equivalent to pF 7 (pF is log suction in cm water), the pores are empty and the volume of water lost is the volume of the pore space, which can be expressed as a fraction (P) of the volume of the undisturbed sample or as a percentage (P per cent) of the same.

In coarse, structureless sandy soils virtually all the pore space is in usable macropores between the grains (30 to 60 per cent or more by volume depending on packing). In other soils as much as half, though usually less, may be within aggregates and therefore unusable by nematodes.

The total pore space can be partitioned into that usable and unusable by a nematode of a given diameter, from the equation $h = 3000d^{-1}$ (Wallace, 1963) where h is the suction pressure in cm water and d is the nematode diameter in μm. Thus the pores through which a nematode of diam. 30 μm can just pass are emptied at a suction of 100 cm and the porosity in pores equal to or greater than this diameter (Ca(0·1)) (Ca = air capacity; Reeve, Smith & Thomasson, 1973) is given by $Ca(0{\cdot}1) = \theta_{(0)} - \theta_{(0{\cdot}1)}$ where $\theta_{(0)}$ is the water content by volume at saturation and $\theta_{(0{\cdot}1)}$ at a suction of 0·1 bar (100 cm water). In the artificial soil depicted in Fig. 2.1 this would include almost all the soil pore

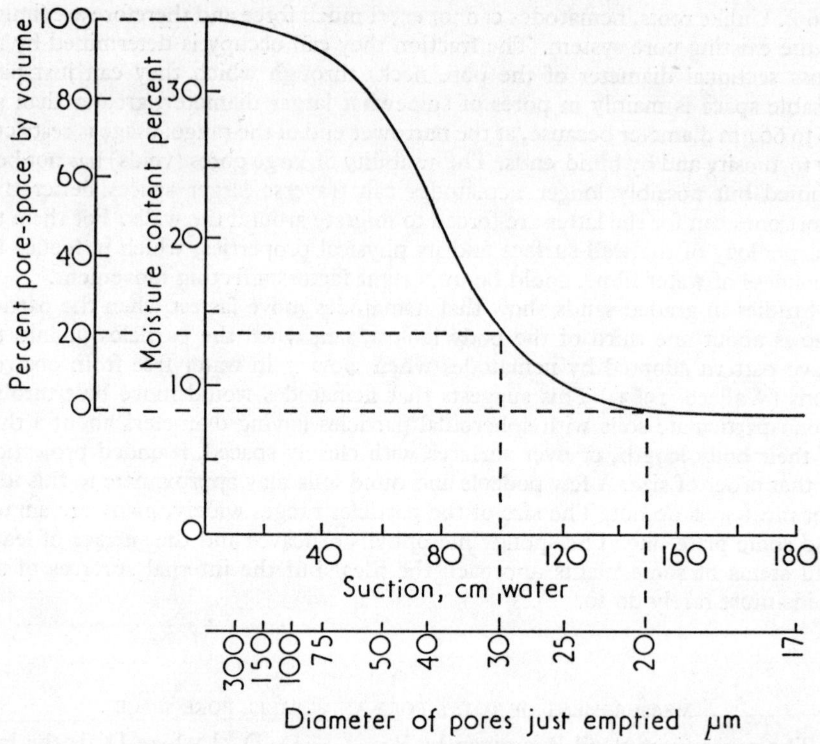

Fig. 2.1. The moisture characteristic of a coarse graded sand which shows the relationship between water retained and suction in cm water. Additional scales show the percentage pore space and the diameter of pores just emptied at a given suction. For this system almost all the pore space is coarse (> 20 μm diam.) which is emptied at 150 cm suction. At 100 cm suction pore space greater than 30 μm diam. is empty and accounts for 80 per cent of the total. (Partly after Wallace, 1958a).

space, and that usable by a nematode of 60 μm diameter and defined by Ca (0·05) would include almost half the soil porosity. In real soils the moisture characteristic is much flatter and is ill-defined at small suctions. In them the difference between Ca(0·05) and Ca(0·1) amounts only to one or two per cent of the pore space (Jones, 1975a).

Measurements of bulk density and of moisture tension assume that the pore system in the soil sample is undisturbed during collection and measurement. Measurements of moisture retention are suspect because the structure changes during wetting and drying due to the swelling and shrinking of the clay fraction and to the partial collapse of aggregates, the amount of collapse depending on the soil strength (Williams, 1971). The amount of space in very large pores is difficult to measure because dimensions of the sample are not appreciably greater than the suction in cm of water to be applied.

Although exceptionally slender nematodes can pass through microvoids (5–30 μm diam.), the pores usable by most nematodes are classified as mesovoids (30–75 μm diam.) and macrovoids (> 75 μm diam.) (Brewer, 1964). Most soil spaces with these dimensions (macropores) lie between soil aggregates and their volume varies greatly with cultivation according to the degree of packing. The volume in smaller spaces (micropores) within aggregates is relatively more constant and depends on the organic matter content, properties and past history of a particular soil. It follows that bulk density, the easiest measure of porosity in mineral soils, is not an adequate indicator of space usable by nematodes when comparing soils from different textural classes. Nevertheless it is useful when similar soils belonging to the same textural class are being compared (Jones & Thomasson, 1976).

TEMPERATURE

In soils the pattern of temperature accumulation is similar from year to year and place to place. Moreover, in arable fields, differences due to season and locality are moderated by the changes in planting date they enforce. For spring-sown crops especially, this tends to ensure that, regardless of season and locality, the development of the host and that of the nematodes feeding upon it are initiated and continue under temperature conditions that vary rather narrowly. However, significant differences between fields arise, because of aspect, soil composition and water content. In districts free from early frosts, e.g., in Britain, on the Ayrshire coast, in parts of Cornwall and in Pembrokeshire, where early planting of potatoes is practised, a temperature differential may exist between host-root development and nematode attack. In these circumstances roots may be able to grow at temperatures too low for rapid nematode activity and this situation may persist until mid-April when the spring temperature rise becomes pronounced. For similar reasons autumn-sown crops may make winter growth during periods when nematodes are less active. However, these cool moist periods probably favour nematodes like *Ditylenchus dipsaci* which is active in the soil surface. Such species are inactivated more by the drying and warming of the soil surface in spring and summer.

In British soils, except at or near the surface, the highest temperatures reached during summer rarely rise far above the optimum for most native plant nematodes but, during the winter months, soil temperatures often sink below the minimum bringing development and activity to a standstill. Fig. 2.2 shows smoothed temperature curves at depths of 10, 30, 50 and 100 cm beneath grass for a typical year and Fig. 2.3 the change of the monthly mean temperature with depth, also under grass. From March to August there is a net heat gain at the soil surface which is propagated downwards to the layers beneath. In September and February there is a net loss from the surface. During this period heat is transferred upwards. On bare soil, when the weather is clear and sunny, the temperature of the ultimate layer that intercepts solar radiation is raised considerably but its value is difficult to measure accurately (Fig. 2.4). Fig. 2.4 also shows that the variability of soil temperature is much smaller than that of air temperature. Fig. 2.5 contains typical temperature traces from a clay-with-flints soil under short grass in April, May and June, 5, 10 and 20 cm deep. A vegetation cover lessens diurnal temperature fluctuations in the soil below. Water, which is a large sink for heat, and peat, which is a poor conductor, both tend to damp diurnal oscillations which are greatest in dry mineral soils. In most reasonably moist soils and in those with a vegetation cover, diurnal temperature changes are slight below 5 to 10 cm and can often be neglected.

After entering their host roots as juveniles and beginning to feed, root endo-parasitic nematodes such as the females of *Heterodera* spp. are able to obtain an

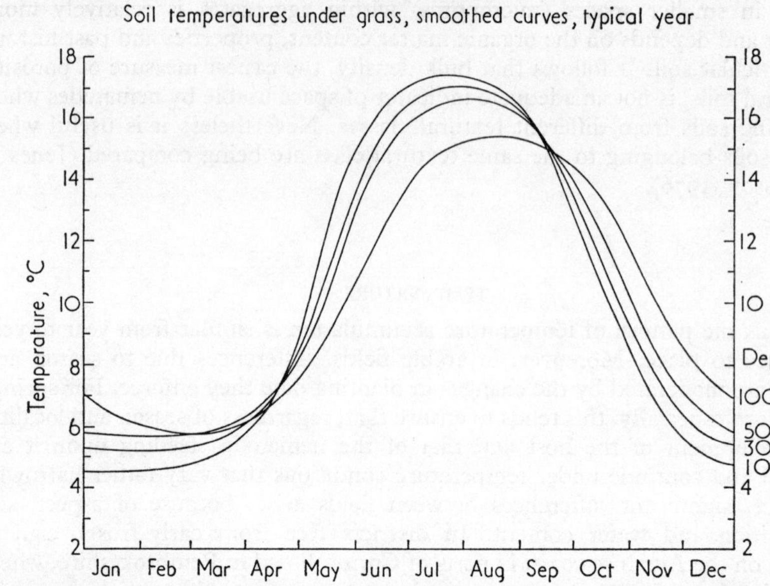

Fig. 2.2. Smoothed soil temperatures for a typical year, under grass, 10, 30, 50 and 100 cm deep. Note the rapid rise in May, the inversion in April after which heat transfer is downwards, and that in September after which transfer is upwards. (From data supplied by J. A. Currie, Rothamsted).

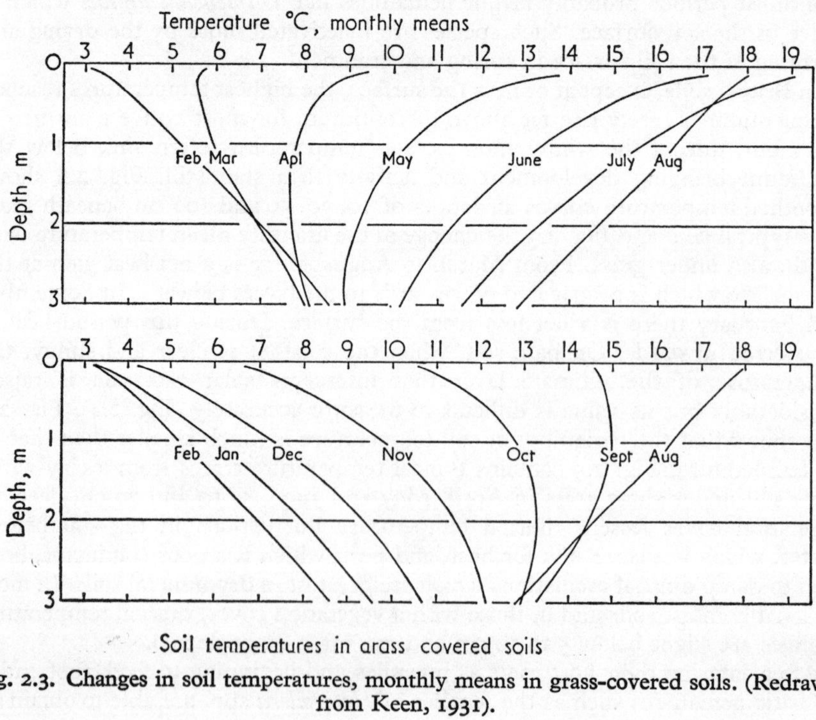

Fig. 2.3. Changes in soil temperatures, monthly means in grass-covered soils. (Redrawn from Keen, 1931).

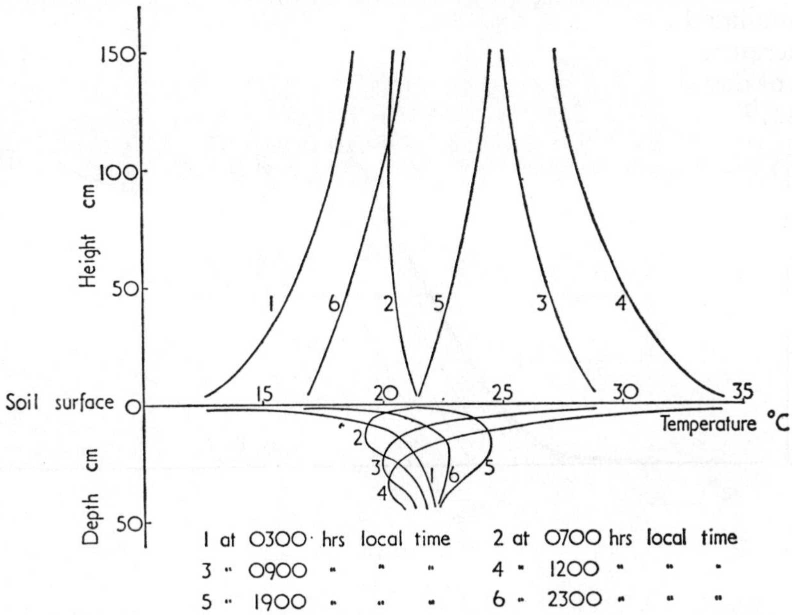

Fig. 2.4. Comparative variability of air and soil temperatures. Daily variation at Seabrook, New Jersey, U.S.A. during clear weather in spring. (After Geiger, 1965).

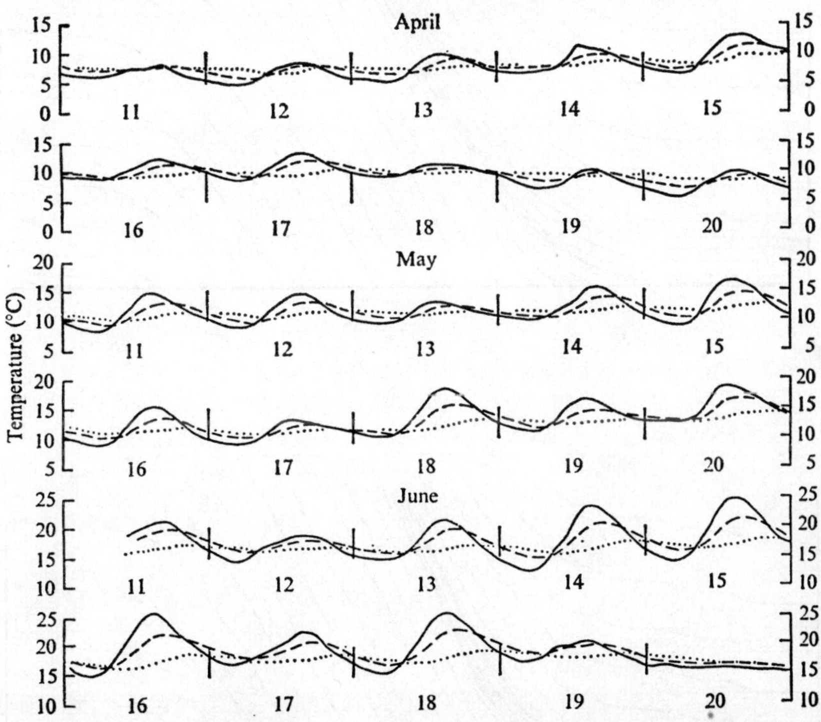

Fig. 2.5. Typical temperature traces. Under short grass, April, May and June, 1973. Depths: ——— 5 cm; - - - 10 cm; …… 20 cm. (From Jones, 1975, *Ann. appl. Biol.* **79**, 129; data from J. A. Currie, Rothamsted).

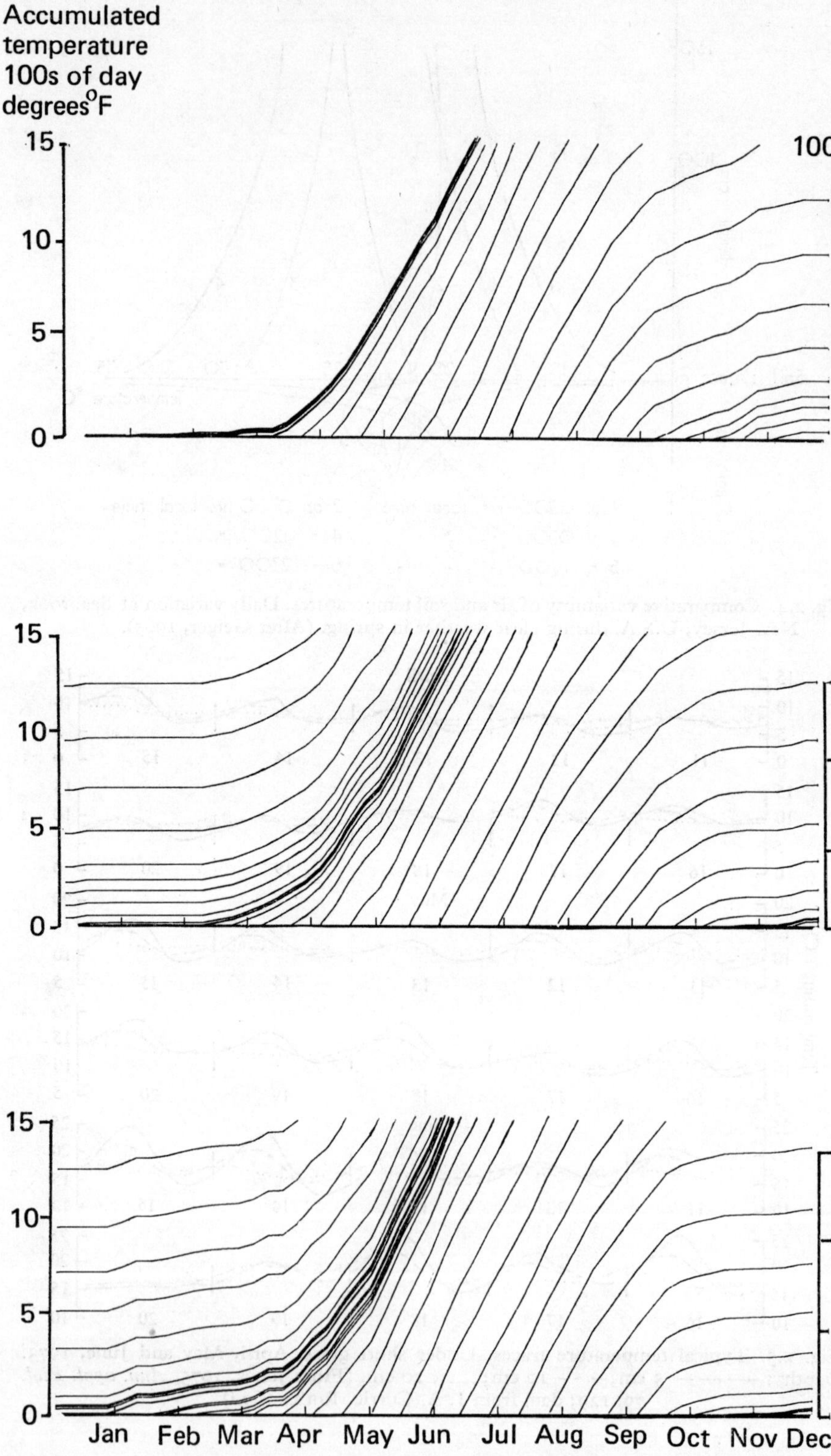
Accumulated temperature 100s of day degrees°F
100s°C
15
10
5
0
7.5
5
2.5
0
Jan
Feb
Mar
Apr
May
Jun
Jul
Aug
Sep
Oct
Nov
Dec

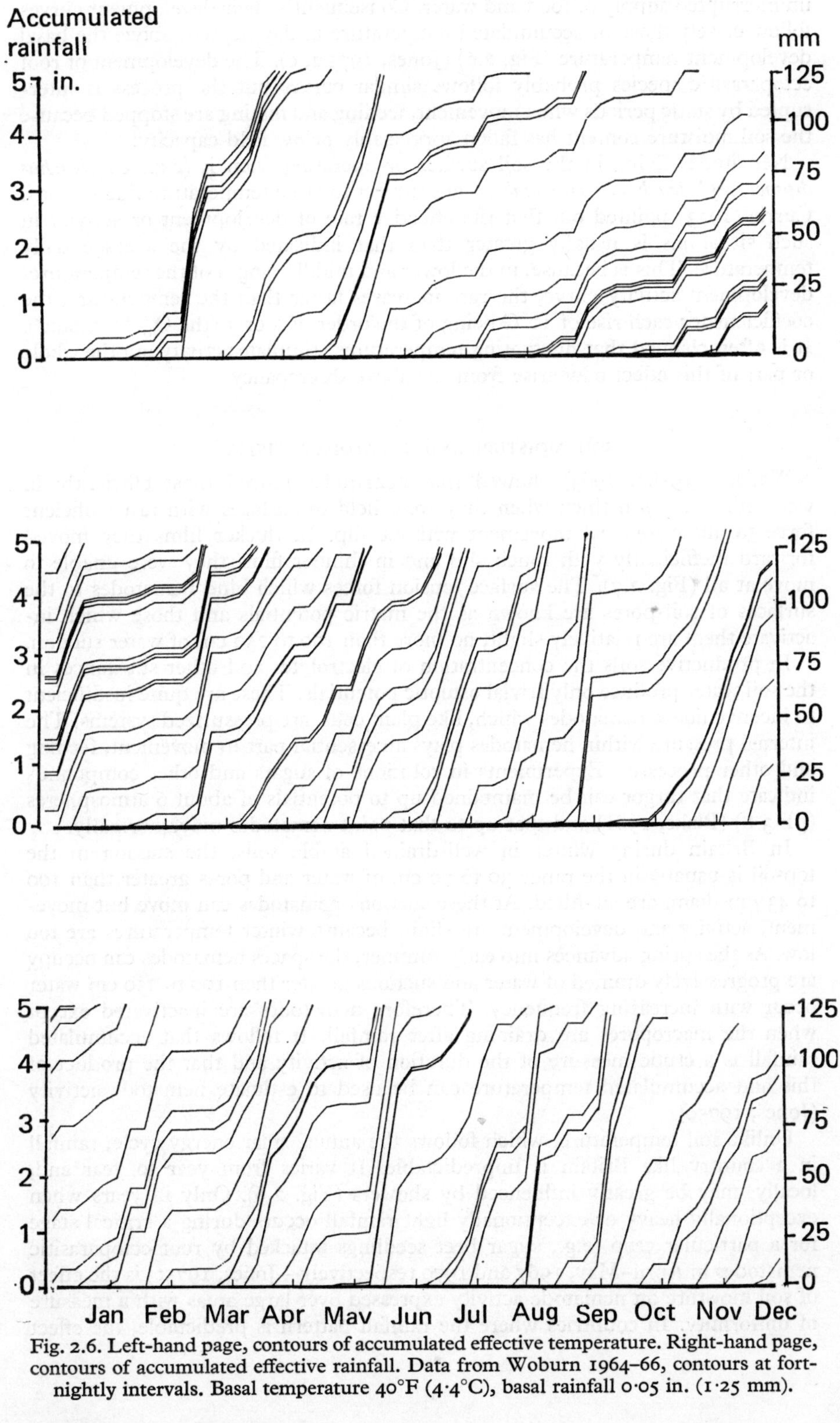

Fig. 2.6. Left-hand page, contours of accumulated effective temperature. Right-hand page, contours of accumulated effective rainfall. Data from Woburn 1964–66, contours at fortnightly intervals. Basal temperature 40°F (4·4°C), basal rainfall 0·05 in. (1·25 mm).

uninterrupted supply of food and water. Consequently their development curves follow closely those of accumulated temperature in day degrees above the basal development temperature (Fig. 2.6) (Jones, 1975 a, c). The development of root ectoparasitic species probably follows similar curves but the process is interrupted by static periods when movement, feeding and mating are stopped because the soil moisture content has fallen appreciably below field capacity.

Nematodes living in the soil surface or operating from it (e.g., *Ditylenchus dipsaci, Aphelenchoides ritzemabosi*) are subject to large temperature fluctuations. Currie (1973) pointed out that the effective rate of development or activity in such situations is usually greater than that indicated by the average daily temperature. This is because, in the lower and middle ranges of the temperature/development/activity curve, the rate increases faster than the temperature, the coefficient for each rise of 10°C being of the order of 2 or 3 (the Q_{10} 'constant'). It is often claimed that fluctuating temperatures stimulate activity but the whole or part of this effect may arise from the above discrepancy.

SOIL MOISTURE AND NEMATODE ACTIVITY

Wallace (1958a, 1963) showed that nematodes moved most efficiently in water films 2–5 μm thick when they were held to surfaces with just sufficient force to allow forward movement without slip. In thicker films they moved forward inefficiently with much slip and in thinner films they were unable to move at all (Fig. 2.7). The surface tension forces which bind nematodes to the surfaces of soil pores are known as the matric potentials and those which inactivate them are relatively slight, no more than 100 to 250 cm of water suction.

In productive soils the concentration of electrolytes and other substances in the soil water produce only trivial osmotic potentials. These are quite insufficient to inconvenience nematodes which, like plant cells, are pressurized systems. The internal pressure within nematodes plays an essential part in movement, feeding and other processes. Experiments in solutions of sugars and other compounds indicate that turgor can be maintained up to potentials of about 6 atmospheres (pF 3·8) (Blake, 1961) and that up to that point nematodes move normally.

In Britain during winter in well-drained arable soils, the suction in the topsoil is usually in the range 30 to 70 cm of water and pores greater than 100 to 43 μm diam. are air-filled. At these suctions nematodes can move but movement, activity and development are slight because winter temperatures are too low. As the spring advances into early summer, the spaces nematodes can occupy are progressively drained of water and suctions greater than 100 to 250 cm water occur with increasing frequency. Therefore nematodes are inactivated except when the macropores are draining after rainfall. It follows that accumulated rainfall is a crude measure of the duration of activity and that the product of this and accumulated temperature can be used to estimate nematode activity (Jones, 1975c).

Unlike soil temperature, which follows the annual solar energy cycle, rainfall in a country like Britain is unpredictable. It varies from year to year and, locally, may be greatly influenced by showers (Fig. 2.6). Only in years when exceptionally heavy or exceptionally light rainfall occurs during a critical stage for a particular crop (e.g., sugar beet seedlings attacked by root ectoparasitic nematodes in April–May, 1967 and 1970 respectively—Jones, 1975c) is the effect of soil moisture on nematode activity expressed over large areas with a measure of uniformity. In countries where the rainfall pattern is predictable, the effect

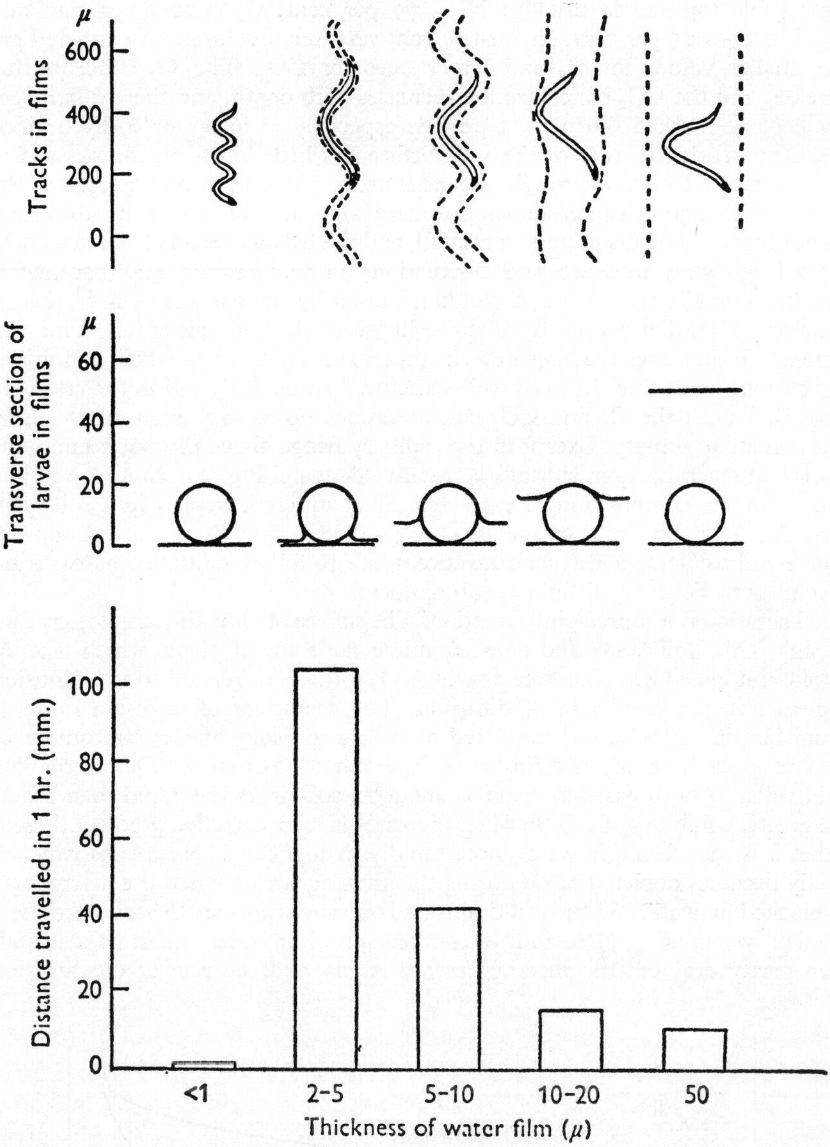

Fig. 2.7. Relationship between thickness of water films and movement of nematodes on a glass plate. (From Wallace, 1963).

of soil moisture should also be predictable provided it is adequate to keep macropores topped up with moisture without causing flooding or wrecking soil structure.

AERATION

Since nematodes live in macropores, our concern is with their aeration. The atmosphere above the soil surface consists on average of: $N_2 = 79$ per cent, $O_2 = 21$ per cent, $CO_2 = 0{\cdot}03$ per cent. The composition of large samples of air from arable top soil is usually: $N_2 = 79$ per cent, $O_2 = 20{\cdot}3$ per cent and $CO_2 = 0{\cdot}15$–$0{\cdot}65$ per cent. Estimates from very small volumes of extracted gas give smaller values for O_2 and larger ones for CO_2. The O_2 concentration decreases and the CO_2 concentration increases with depth, and these differences may become marked during wet periods especially in heavy or badly drained soils. Apparently crusting of the soil surface has little effect on the passage of gases because of cracks through and beneath it. More than 90 per cent of the gaseous exchange between the atmosphere and the soil air is by diffusion, 6–8 per cent of O_2 gains ingress in rainfall, and the balance results from mass flow induced by wind pressure and fluctuations in temperature and barometric pressure. Local compaction and ponding caused by the passage of implements, irrigation or rainfall results from the collapse of the soil aggregates. Fine soil fractions fill the pores creating a dense impervious layer below which conditions may become anaerobic. In most well-structured arable soils and in the cracks in dense clay soils, the O_2 and CO_2 concentrations down to plough depth rarely limit nematode activity. Except in the capillary fringe above the water table, the aeration of macropores in subsoils is usually adequate. Fig. 2.8 shows the annual changes in the composition of extracted air at different depths in a soil from North America and Fig. 2.9 seasonal changes in per cent CO_2 in a British topsoil. From April to October the concentration tends to follow rainfall, whereas from November to February it follows soil temperature.

Soil aeration is a source-sink situation. The source of O_2 is the atmosphere and the sink is the soil fauna and flora including the roots of plants which take up oxygen and emit CO_2. The rate at which O_2 diffuses is related to the diffusion gradient and the coefficient of diffusion. The coefficient of diffusion in air is diminished in soil where it is related to the morphology and water content of the pore space. The rate of diffusion of O_2 is about $1{\cdot}25$ that of CO_2 and the rate of diffusion of both gases in water is about 10 000 times less rapid than in air. The greater solubility of CO_2 in water increases its concentration gradient therein so that it is transferred in water more rapidly than is O_2. This explains why soil rapidly becomes depleted of O_2 during the growing season when the macropores are blocked by water. As the soil drains and air spaces appear, the soil becomes a complex system of air-filled and water-filled spaces, in series and in parallel and, when partly drained, the macropores are empty and the rate of O_2 diffusion

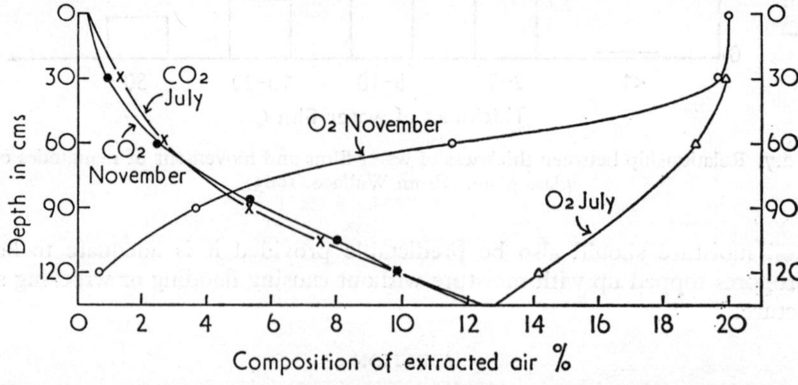

Fig. 2.8. Composition of the soil atmosphere, O_2 and CO_2, with depth in November and July. The concentration of N_2 remains approximately constant. (Redrawn from Baver, Gardner & Gardner, 1972).

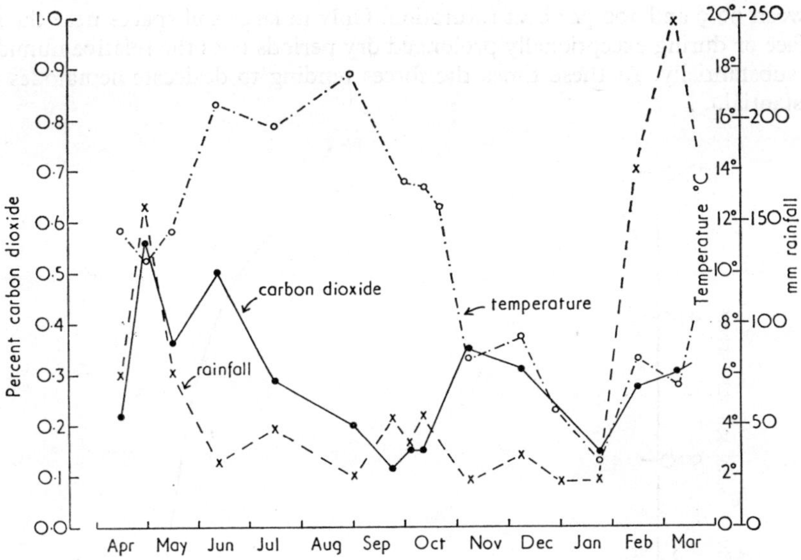

Fig. 2.9. Monthly changes in CO_2 content of soil atmosphere compared with soil temperature and rainfall. (Redrawn from Baver, Gardner & Gardner, 1972).

increases to about one fortieth of that in air. The effects of water in the pore system are such that a water path of 1 mm in length decreases the rate of passage of oxygen by about ten times.

The measurement of oxygen diffusion rates at particular points in soil by platinum electrodes is empirical and depends on several factors (O_2 concentration, diffusion coefficient, thickness of moisture films, length of diffusion path and electrode radius). To be effective the whole surface of the electrode must be covered by a moisture film. This ruptures in dry, coarse-textured soils (Baver, Gardner & Gardner, 1972). Nevertheless, Van Gundy *et al.* (1962) found that oxygen electrodes were satisfactory in studying the activity and survival of nematodes in soil. All the nematodes in their samples were killed by zero oxygen concentrations for 14 days but some survived concentrations as small as 2 per cent better than others. As would be expected, the oxygen diffusion rates in moist soils were less than in drier ones. The O_2 concentration applied to the soil surface was not critical but that at the air/water film interface within soil was important. Baxter & Blake (1969) avoided difficulties with oxygen electrodes by using a diffusion column to obtain a range of oxygen concentrations the effect of which they studied on the hatch of *Meloidogyne* eggs in graded sands at several moisture suctions. Suctions from pF 0 to 3·4 (0–2500 cm water) had no effect on oxygen requirements. Embryo development, hatch and movement were equally dependent on O_2. Metabolic activity was slight at near zero O_2 concentrations and increased asymptotically to a maximum at 10–18 per cent O_2. They concluded that a favourable environment was one where moist soil drained rapidly to allow oxygen concentrations to reach 10 per cent or more.

As already explained, the suction forces on soil moisture are related to the moisture retained by the moisture-characteristic curve. The relative humidity of the air soil space is also related to both as in Fig. 2.10 from which it can be seen that in soils which are suitable for nematodes the relative humidity lies between 99·9 and 100 per cent saturation. Only in large soil spaces near the soil surface or during exceptionally prolonged dry periods does the relative humidity fall substantially. At these times the forces tending to desiccate nematodes are substantial.

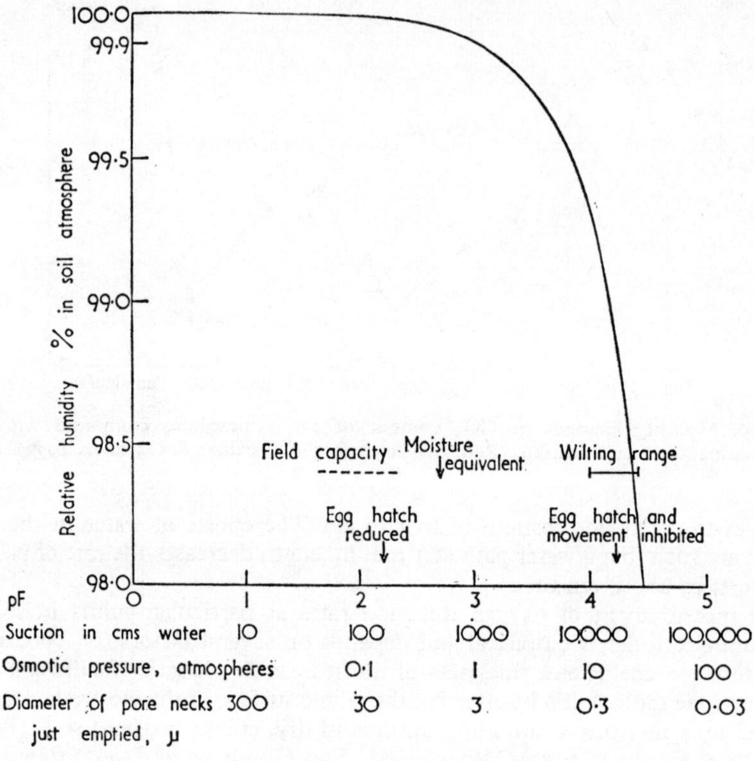

Fig. 2.10. The relationship between the relative humidity of the soil atmosphere and log suction pressure (pF) in cm water. The equivalent osmotic pressure (atmospheres) and the size of pore necks just emptied (μm) are also shown. (From Jones & Jones, 1974).

Much of the work on soil assumes that it is a homogeneous medium. This is not so. Its structure varies greatly from point to point and so do the micro-climatic conditions to which individual nematodes and colonies of nematodes are subjected. In moist, rapidly respiring soils the O_2 concentration in local, water-saturated pockets probably approaches zero while air-filled spaces nearby contain large concentrations (Greenwood, 1968). Nematodes near an active oxygen sink must compete with it. Moreover, the oxygen must move from the air space into the soil water and thence to the nematode. Except immediately after rain when the smaller macropores are filled, nematodes and soil particles are surrounded by thin water films. At suctions greater than 300 cm water, films around soil particles are thought to be less than 0·1 μm thick (Kemper, 1960) and, at the optimum for nematode movement, in the range 2–5 μm thick, which is a fraction of the thickness of the nematode. The O_2 concentration difference across water films so thin, although greater than that in air, is apparently not a serious obstacle to gaseous exchange.

ENEMIES

Enemies of nematodes are listed in Table 1 (See also Chapter 13). Tardigrades, predatory nematodes, sporozoans, amoebae, some ciliates, bacteria and the hyphae and motile zoospores of parasitic fungi can use the same sized pore spaces as plant nematodes. The movement of motile stages is probably governed by the soil moisture conditions as is nematode movement, i.e., they are immobilized at suctions that immobilize nematodes in the range 100–250 cm water. Fungal hyphae may be able to operate at somewhat greater suctions. Enemies with diameters larger than nematodes, especially the predatory arthropods, are confined to large pore spaces. In coarse sandy soils nematodes may avoid larger enemies by retreating into small pores but, in clay soils with dense peds, plant roots, nematodes and their enemies are confined to the same major cracks and crevices (Jones, 1974).

Table 1

Enemies and Competitors of Plant Nematodes
(Esser & Sobers, 1964; Sayre, 1971)

Predators	*Parasites*	*Competitors*
Predacious fungi	Endoparasitic fungi	Organisms
Amoebae	? Viruses	injurious
Ciliates	Rickettsias	to roots
Tardigrades	Bacteria	
Turbellarians	Sporozoans	Fungi
Nematodes		Bacteria
Enchytraeids		Other nematodes
Collembolans		
Mites		
? Insects		

EFFECTS OF PLANTS ON THE SOIL ENVIRONMENT

The introduction of plants complicates the soil environment greatly. Some of the complications are illustrated in Fig. 2.11. The layer immediately around the plant root contains a complex of substances derived primarily from exudations from roots and the contents of ruptured or degraded cells. This leads to a growth of bacteria, fungi and other organisms the activity of which adds further substances. Nematodes, when numerous, also add their excretory products and sometimes there is a complex interaction between these and micro-organisms as in the formation of the subcrystalline-layer of some cyst nematodes (Brown *et al.*, 1971). The layer of soil around roots about 1 to 2 mm thick is known as the rhizosphere and has been much studied (Rovira & Davy, 1974) in connection with its effect on fungal pathogens. Little is known about its effect on nematodes.

Nematode parasites of roots, especially the more highly adapted ones, respond to substances (phytomones) given out by roots, e.g., hatching factors for cyst-nematode eggs. Nematodes also respond to the sex attractants (pheromones) of their own and related species (Green & Plumb, 1970). To operate in soil spaces and interlock the activity of host and parasite, and the sexes of the parasite when this is essential, these substances must have properties different from the

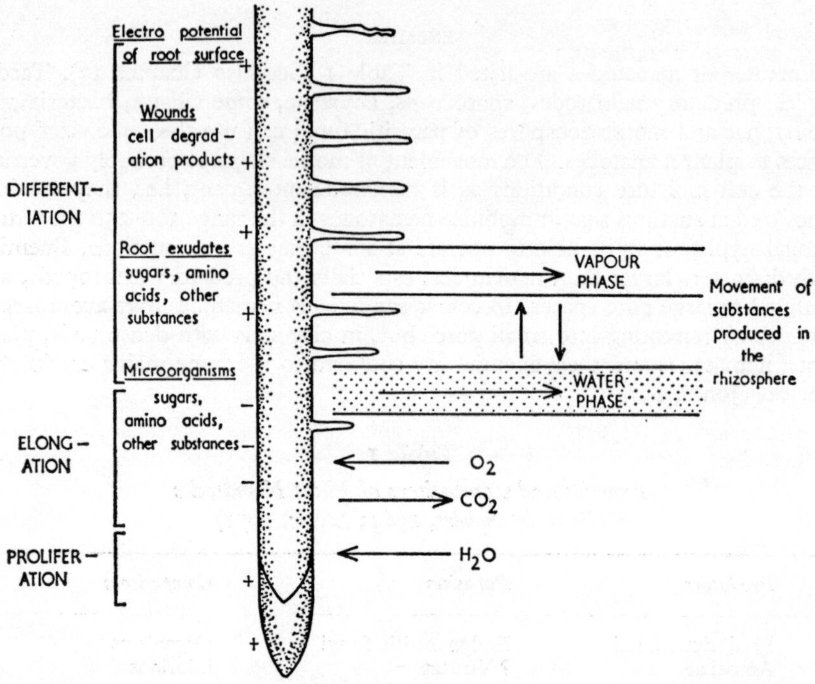

Fig. 2.11. Some factors affecting host finding by plant-feeding nematodes in soil. Excretions of nematodes may be added to the substances in the rhizosphere. (From Jones, 1960; Jones & Jones, 1974).

phytomones and pheromones operating in the aerial environment. They cannot be chemically very active or they would bind to the walls of the soil matrix. Those studied to date are water soluble, polar with many OH groups, of moderate to high molecular weight and slightly or non-volatile. These properties make them difficult to isolate, purify and characterize.

The roots of actively growing plants extract water from the soil around them which causes local shrinkage of soils containing an appreciable clay fraction and creates the so-called soil-root gap. The dimensions of this gap are ill-defined but probably it helps to produce an elongate macropore more or less continuous along the path of the root. Such a gap would aid aeration and assist the movement of nematodes along the length of the root and facilitate the finding of suitable feeding and penetration sites. Hydraulic flow to the root creates a gradient of suction. However, when soils are moist enough for nematode movement suctions are small and the gradient is negligible (Newman, 1974).

Roots emit CO_2 which is known to be a general attractant for nematodes. Micro-organisms active in the rhizosphere must increase the concentration locally, especially around actively metabolizing root tips, but there must be other locally intense sources of CO_2 competing for the attention of nematodes.

In recently tilled soils, the crop is absent. After planting and germination, the root system extends and ramifies. The nematode-root interaction unfolds in a time sequence related to root extension, which itself may be modified if many harmful nematodes are present. Also, as roots extend, nematodes scattered throughout the soil accumulate on the root system and those attracted to root tips, for example, move through the soil as the tips grow. The whole is a complex dynamic system difficult to visualize and analyse.

Host-Parasite Relationships

Regardless of species and organs, the environment within plant tissues has certain common features. It consists of pressurized cells surrounded by walls of varying thickness and composition. The cells may or may not have air spaces between them and groups of cells are crossed or bounded by fibrous tissue and by conducting tissue of two types, one (the xylem) carries water, salts and other substances in solution upwards towards the leaves and another (the phloem) carries metablic products downwards towards the roots. There is also a centripetal flow inwards to the central conducting tissue in roots, and a centrifugal flow outwards in other plant organs. In addition, there is evidence of a movement of metabolic products through non-conducting tissue, albeit slowly. Unless subjected to excessive water stresses, plant tissues retain their turgor. The system contains an abundance of food available to nematodes as cell sap which they extract via the mouth stylet. Cell turgor aids nematode feeding which is passive when the pressure in the cell exceeds that in the nematode appreciably.

Some nematodes are almost fully facultative, that is, they can feed on a great range of plant hosts with which they have only casual relationships. *Aphelenchoides ritzemabosi* is such a nematode so also are species of *Trichodorus*. The first is able to swim and to climb plants when they are moist. It treats the spaces in the folded leaves of buds and in the mesophyll of leaves as it would soil spaces and feeds much as does *Trichodorus* on the epidermal cells of roots. *A. ritzemabosi* is unable to penetrate cells, and it cannot separate cells in compact plant tissues or pass through conducting tissue or fibres. The saliva which such nematodes inject into cells tends soon to induce harmful tissue changes from which the cell may die but the nematode feeds only briefly and has moved elsewhere before the cell dies. *Pratylenchus* spp. are endoparasites causing cavities within plant roots which ultimately are invaded by weakly pathogenic fungi and bacteria. Some species of *Pratylenchus* cause disorganization of tissue surrounding the lesion, possibly as a result of toxins in their saliva or excretory products. Such species are less well attuned to their hosts than those not producing toxins. The behaviour of *Ditylenchus dipsaci* is somewhat similar although it tends to cause swellings rather than lesions and, typically, modifies invaded host tissues by inducing the dissolution of the middle lamellae of cell walls by enzyme activity, producing spaces between cells that must increase the nematode's mobility within the tissue. Dissolution of middle lamellae by action of pectinase enzymes seems characteristic of *D. dipsaci* (and probably *D. destructor*) in 'good' hosts, in contrast to *Aphelenchoides* spp. which presumably do not produce (or induce) these enzymes. Again, it is to the nematode's advantage not to cause excessive necrosis of the cells on which it feeds.

For nematodes that insert their stylets deeply or feed for long periods at one point from one or a few cells, the possession of non-toxic saliva is essential. Several species with immobile or nearly immobile females, e.g., *Anguina*, *Rotylenchulus*, *Heterodera*, *Meloidogyne*, go further and induce tissue changes to their advantage. In *Anguina* these changes take the form of characteristic galls. In the other genera the changes take the form of groups of cells modified to transport nutrients from the conducting tissue to the nematode. The type of transfer-cell system induced varies with the nematode and, in minor details, with the host parasitized. *Rotylenchulus* produces a sheet of cells in the pericycle just beneath the root endodermis. The feeding cell, against which the head is fixed by cell wall intrusions into cuticular folds, and into which the stylet is thrust, is a multinucleate syncytium produced by the breakdown of cell walls; its modified neighbours are enlarged (giant) cells. A feeding tube develops in the feeding cell. *Meloidogyne* spp. induce multinucleate giant cells without cell-wall breakdown. In most parasitized plant tissues the stimulus from the nematode induces cell division in surrounding tissues which may possibly serve to protect the embedded females and egg masses. *Heterodera* spp. induce multinucleate syncytia from enlarged cells by cell wall dissolution (Razak, 1975; Jones & Northcote, 1972a, 1972b). In such nematode-plant associations, the induction of galls and of long-lasting transfer systems can only succeed if the nematode saliva is compatible with that of the contents of the host cell into which it is injected (Jones, 1975b).

The development of such complex nematode-plant interrelationships is evidently the result of the co-evolution of host and parasite which has led to an increasing dependence of the parasite on its host. A facultative nematode requires food from its host adequate in quantity and suitable in quality only. Whether its saliva is compatible with the contents of the host cell is largely immaterial. Nevertheless the host can influence the nutrition of its parasite by producing toxins or substances that repel, deter or otherwise make the plant unattractive, or those that diminish its reproductive success. These resistance mechanisms are probably controlled by a galaxy of major and minor genes which, if numerous, are difficult for a parasite to circumvent. The hosts of specialized parasites may do the same and also may produce substances that interfere with gall or transfer cell formation. Usually this results from acquiring single major genes (H), that interact with corresponding major gene (P) products in the saliva of the parasite. A hypersensitive reaction results. However, the situation is not permanent, for a simple mutation which deletes the major gene (P) from the parasite results in recessive individuals (pp) and their saliva is compatible with the host cell sap. In the co-evolution of Solanaceae and round-cyst nematodes, and in that of cereals and cereal cyst-nematodes, several of these complementary gene systems seem to have developed (see Anon., 1972b; Sidhu, 1975; Jones, 1975b).

In the soil-plant system, modifying the first to the disadvantage of the plant parasitic nematode is difficult. Indeed most systems of cultivation devised to favour the plant also favour nematodes. Modifying the plant is easier by changing the species or variety to one resistant or more tolerant, or by arranging the planting sequence to avoid growing a susceptible crop on heavily infested land (Jones, 1974).

CHAPTER 3

Interactions of Nematodes with other Pathogens

R. S. PITCHER

East Malling Research Station, Kent

Introduction

The ability of nematodes to interact with other pathogens has been increasingly recognized by plant pathologists, especially since the 1950s. Most of the literature relates to interactions with fungi, bacteria or viruses already known as plant pathogens in their own right. However, recent work in the U.S.A. has shown that some nematodes, particularly *Meloidogyne* spp., may incite damage by soil fungi which are not normally associated with plant disease. The bulk of the literature describes interactions with fungi, dating from a record of synergism between root-knot nematodes and a *Fusarium* sp. causing a wilt disease of cotton (Atkinson, 1892). In contrast, less than 20 major papers have appeared on interactions of nematodes with bacteria since Hunger (1901) found that nematodes incited tomato wilt. Virus transmission by a nematode was not reported until 1958 (Hewitt, Raski & Goheen, 1958), but the subsequent literature already greatly exceeds that on bacterial interactions and proportionately more time has been spent on investigations of the mechanisms of virus transmission. Little more will be said in this chapter about virus transmission, which is fully covered in Chapter 11. Nor is it proposed to cover the small amount of literature, mostly of recent date, on interactions, competitive or otherwise, between two or more nematode species parasitizing a common host plant. It is not intended to present an extensive literature review, but to deal only with the broad principles of the more important papers on fungi and bacteria, especially those from northern temperate zones. Recent reviews on interactions cover fungi (Powell, 1963; Dittmann, 1963; Bergeson, 1972), bacteria (Pitcher, 1963), fungi and bacteria (Pitcher, 1961; Mountain, 1965) and viruses (Raski & Hewitt, 1963; Cadman, 1963; Proeseler, 1964; Taylor, 1972), while Pitcher (1965) and Weischer (1968a) cover all three types of pathogen.

All the interactions considered here have three major components in common, namely the plant host, the nematode and the 'other' pathogen. The most familiar type of interaction is that in which the pathogen is able to parasitize a host which it would otherwise be incapable of attacking, or to make a more damaging attack than it could have done in the absence of the nematode. This, however, is only part of the picture; in a few cases the nematode's activities decrease the damage done by the pathogen and, probably more often than is usually realized, the nematode is affected by the pathogen, either to its benefit or to its detriment.

The *schema* (Fig. 3.1) indicates the principles underlying three-component interactions. Whereas the host is attacked directly by the nematode and the pathogen, any interaction between them is usually indirect, acting *via* the changes each brings about in the substrate on which they both depend, namely the plant host. Occasionally the nematode and the pathogen may interact directly, as when a mycophagous nematode actually feeds on the hyphae of a fungal parasite.

Nematodes are well-fitted to interact with other plant pathogens, especially when both are soil-borne. All truly parasitic nematodes must pierce (i.e., wound) the epidermis in order to feed, while the endoparasites follow this by entering the plant tissue bodily, often with fungal spores or bacterial cells

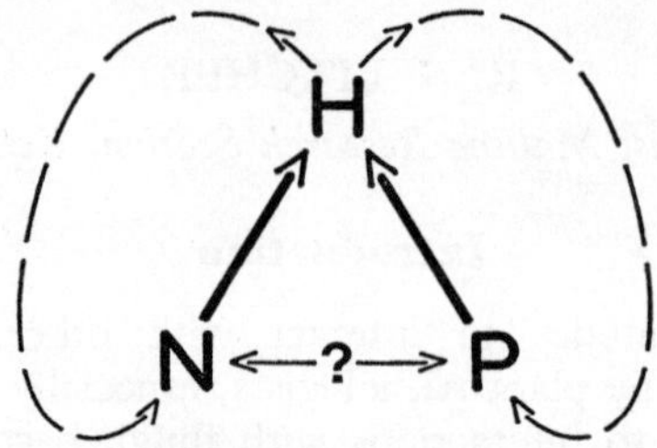

Fig. 3.1. *Schema* of the interactions of the host plant (H), nematode (N) and pathogen (P). Continuous lines indicate a direct interaction and broken lines an indirect interaction, via the host plant.

adhering to their cuticle. In the course of feeding, all inject salivary enzymes, whose prime function is to modify cellular tissue, either to aid immediate ingestion or to prepare a long-term feeding site. It is during the immediate feeding stage that all known viruses are either injected into or sucked out of the host cells. The moist environment necessary for nematode activity is also one which favours the development of many fungi and bacteria, while the soil, which is so often the moisture source, is well known to be a vast reservoir of micro-organisms, including the propagules of many plant pathogens.

Examples of Interactions

Tables 1 and 2 show selected examples of papers giving experimental proof of interactions of nematodes with fungi or bacteria, respectively. Some examples of negative experimental results are included to illustrate the specificity of the positive results, which constitute the majority. All refer to cultivars sensitive to or tolerant of the other pathogen.

Several points emerge clearly from a consideration of Tables 1 and 2. The most important is that in 12 out of 20 listed cases of positive interaction the nematode is believed to act only as an assistant to the pathogen, propagules of which are already close to a vulnerable organ of the host and able to invade unaided. It is the extent, speed or severity of their invasion which is increased by the activities of the nematode. In only three cases does the nematode play an essential role, i.e., without the nematode, the pathogen cannot become established. The various mechanisms involved are discussed on pages 65 *et seq.* The marked ability of *Meloidogyne* spp. to assist a variety of pathogens, especially *Fusarium* spp., indicates that the unique host-parasite relationships of this nematode genus confer a markedly greater advantage than does the simpler feeding of most other genera. The smaller size of Table 2 reflects the well-known fact that, amongst plant-parasitic species, fungi greatly outnumber bacteria, but it may also indicate the greater difficulties inherent in working with bacteria and a virtual neglect of all but well-known plant-parasitic bacteria.

Table 3 summarizes the salient features of the known interactions. It emphasizes the versatility of the fungi, which allows them to participate in all but the fully systemic infections. This versatility is due in large measure to the wide range of adaptation found amongst fungi, often in the form of specialized organs and/or enzymes which aid their entry into plant tissue, coupled with a degree of mobility conferred by efficient spore dispersal and the development of exploratory hyphae. In contrast, bacteria are less well-equipped, being less mobile and more wound-dependent. For infection, viruses are virtually completely dependent on the wound and transport provided by the nematode vector although, once established within the plant, they alone are capable of systemic infection of the host tissues.

Plant Reaction to Nematode Attack

Probably no two nematode species produce an identical plant response to parasitism. Even plant reactions which appear outwardly very similar may differ biochemically. Recent reviews (Seinhorst, 1961; Krusberg, 1963; Wallace, 1963; Mountain, 1965; Dropkin, 1969a) emphasize the understandably wide range of reactions which arise from the enormous number of possible combinations of plant species or cultivars and nematodes, and the varied physiology of both. Table 4 is a brief summary of the main types of plant response.

With the possible exception of plant stimulation by low nematode populations, representative examples of all the types of reaction listed have been studied under aseptic conditions. As Mountain (1960) and others have said, unless a reaction is confirmed under monoxenic conditions (Dougherty, 1953), when the host and parasite are the only organisms present, one cannot be certain that it represents the true nature of the host-parasite relationship.

Mechanisms of Plant Reactions to Nematode Feeding

As mentioned in the Introduction, all the reactions listed in Table 4 follow the mechanical rupture of some of the plant cells by the nematode stylet and the injection of salivary enzymes. The resulting mechanical wound may be simply a micro-puncture in an outer cell wall, possibly readily repairable. In the case of an endoparasite this is followed by inter- or intracellular penetration, sometimes of many cells, usually occurring in the cortex, but occasionally reaching vascular tissue. Biochemical modification of cell tissue also varies widely. In example 2, Table 4, interference with the vital processes within the cell may be so slight that even protoplasmic streaming (cyclosis) is not interrupted (Dropkin, 1969a). In most cases, however, a cell or cells are killed or they may undergo major changes, providing a long-term substrate for a sedentary nematode (Dropkin, 1972). In every case, the cell contents are modified biochemically, if only temporarily. This is not surprising, as it is unlikely that nematodes would have evolved their well-developed salivary glands if their substrates could be ingested without modification. Krusberg (1963) and others have demonstrated in nematodes a variety of enzymes capable of modifying plant components, e.g., cellulases, pectinases, proteases etc. The role of growth-regulatory substances has

Table 1

Nematode-fungal Interactions

Part of plant diseased	Fungi	Nematodes	Role of nematode	References
Aerial meristem	Dilophospora	Anguina	Essential	Atanasoff, 1925
Hypocotyl (damping-off)	Rhizoctonia	Meloidogyne	Assists	Taylor & Wylie, 1959
Leaves, stems, flowers	Alternaria	Meloidogyne	Assists	Powell & Batten, 1969
	Leptosphaeria	Aphelenchoides	Deters	Nonaka, 1959
Root (cortical rot)	Cylindrocarpon	Pratylenchus	Assists	Slootweg, 1956
	Rhizoctonia	Globodera*	Assists	Dunn & Hughes, 1964
	Aphanomyces	Tylenchorhynchus	Assists	Haglund & King, 1961
	Macrophomina	Pratylenchus	None	Norton, 1958
	Pyrenochaeta	Globodera	None	James, 1968
	Pythium	Aphelenchus	Deters	Rhoades & Linford, 1959b
	Pythium	Meloidogyne*	Assists	Melendéz & Powell, 1969

Root (vascular wilt)	Fusarium	Meloidogyne	Assists	Many
		Heterodera	Assists	Ross, 1965
		Rotylenchus	Assists	Labruyère *et al.*, 1959
		Tylenchorhynchus	Assists	Holdeman, 1956
		Pratylenchus*	Assists	Seinhorst & Kuniyasu, 1971
		Meloidogyne	None	Binder & Hutchinson, 1959
		Helicotylenchus	None	Schindler *et al.*, 1961
		Rotylenchus	None	
		Heterodera*	None	Jorgenson, 1970
		Trichodorus	None	Martin *et al.*, 1956
		Tylenchorhynchus	None	
		Helicotylenchus	None	
		Ditylenchus	Deters	Schindler & Stewart, 1956
	Phytophthora	Meloidogyne	Assists	Powell & Nusbaum, 1960
		Pratylenchus	None	Graham, 1958
		Rotylenchus	None	
	Verticillium	Globodera	Assists	Corbett & Hide, 1971
		Pratylenchus	Assists	McKeen & Mountain, 1960
		Meloidogyne	Assists	Khoury & Alcorn, 1973

*Nematode development inhibited by fungus.

Table 2

Nematode-bacterial Interactions

Part of plant diseased	Bacteria	Nematodes	Role of nematode	References
Aerial meristem	Corynebacterium	Anguina* Aphelenchoides	Essential	Gupta & Swarup, 1972 Pitcher & Crosse, 1958
Leaves	Xanthomonas	Aphelenchoides	Assists	Riedel & Larsen, 1974
Crown buds (soft rot)	Erwinia	Ditylenchus	Assists	Metcalfe, 1940
Crown buds (vascular wilt)	Corynebacterium	Ditylenchus	Assists	Hawn, 1963
Root (galls)	Agrobacterium	Meloidogyne	Assists	Griffin *et al.*, 1968
Root (vascular wilt)	Pseudomonas	Meloidogyne* Helicotylenchus	Assists	Johnson & Powell, 1969 Stewart & Schindler, 1956 Libman *et al.*, 1964
		Ditylenchus	Deters	Stewart & Schindler, 1956
		Xiphinema Tylenchorhynchus	None	Stewart & Schindler, 1956 Lucas & Krusberg, 1956

*Nematode development inhibited by bacterium.

Table 3

Major Types of Nematode-pathogen Interaction

Frequency*	Site of interaction	Examples	Pathogen types involved		
			Fungi	Bacteria	Viruses
x	Aerial meristem (shoot or flower)	Anguina/Dilophospora Aphelenchoides/Corynebacterium	+	+	—
x	Leaves	Meloidogyne/Alternaria	+	—	—
x	Hypocotyl or stem	Meloidogyne/Rhizoctonia	+	—	—
xx	Feeder root cortex	Pratylenchus/Cylindrocarpon	+	—	—
xxx	Vascular tissues (usually *via* root)	/Fusarium Meloidogyne/Verticilium /Pseudomonas	+	+	—
xx	Systemic (*via* root)	Xiphinema/nepo viruses	—	—	+

*Indicates the relative frequency of interaction in each site; *x* = rare, *xxx* = common.
+ = Some cases known, — = no cases known.

Table 4

Major Types of Plant Reaction to Nematode Attack

Reaction	Examples
Stimulation—may result from a light attack	Heterodera, Meloidogyne, Xiphinema
No visible reaction—indicates well-adjusted parasitism, but some tissue modification must occur	Pratylenchus minyus, Tylenchorhynchus
Necrosis—often of parenchyma (e.g. root cortex)	Pratylenchus penetrans
Hyperplasia—increase in cell numbers (e.g. gall tissue)	Meloidogyne, Ditylenchus, Xiphinema
Hypertrophy—increase in cell size (e.g. giant cells, syncytia)	Meloidogyne, Heterodera

also received attention and some have been identified in nematodes or nematode-infested plants, but their origin and role are still little understood (Viglierchio, 1971). Profound chemical changes in host tissue have, however, been noted after nematode attack, both in mineral content (Bergeson, 1966) and in organic content, e.g., large increases in amino-acids associated with *Heterodera* attack (Doney, Fife & Whitney, 1970) and of amino-acids and proteins in the case of *Meloidogyne* spp. (Owens & Specht, 1966). Such changes may well provide a metabolite 'sink' at or near the site of nematode attack which is much more favourable for the establishment of pathogenic fungi or bacteria than is healthy tissue. Biochemical reactions to parasitism are not confined to nematode attack, but it is not clear whether plant reaction to all forms of parasitism is fairly standardized or whether the reaction differs according to the parasite involved. It is, however, quite clear that nematodes, by virtue of their superior mobility and invasive powers, are often able to make the initial attack, provoking these biochemical changes in advance of the establishment of other pathogens. Experiments involving manipulation of the timing of inoculation of *Meloidogyne* and *Fusarium* have shown that the fungus becomes established most readily and that the resulting wilt disease is most severe when nematode inoculation precedes that of the fungus by a period of three to four weeks (Melendéz & Powell, 1967). The above considerations, together with the differing reactions of closely similar tissues to the same nematode (e.g., necrosis in one tobacco variety parasitized by *Pratylenchus minyus*, but not in another (Mountain, 1954, 1961)) or of the same tissue to closely related nematodes (e.g., *Pratylenchus penetrans* causes necrosis in peach, while *P. minyus* does not (Mountain & Patrick, 1959)), suggest that biochemical reactions between host and nematode are more important than mechanical injury. Moreover, biochemical changes are almost universal in nematode-plant relationships and may, alone, cause necrosis in the less well-adjusted relationships (Pitcher, Patrick & Mountain, 1960).

The Role of the Nematode in Interactions

It is helpful to make clear-cut distinctions between the different circumstances in which nematodes can interact with other pathogens, but it must be stressed that the nematode often plays more than one role in any given interaction.

PATHOGEN REQUIRES INTRODUCTION INTO HEALTHY HOST TISSUE

Here the nematode acts as a vector, carrying inoculum internally. The transmission of nepo- or tobra-viruses (Chapter 11) provides the only obvious example of this role, in which nematodes closely parallel the better-known arthropod vectors of other viruses.

PATHOGEN REQUIRES ACCESS TO SUSCEPTIBLE HOST TISSUE

Again the nematode acts as a vector, but the inoculum is carried externally. Typical examples are those in which only the host's apical shoot meristem is susceptible to the pathogen. Most apical meristems are closely ensheathed by bud scales, leaf sheaths, overlapping stipules etc., which makes them quite inacessible to non-motile pathogens such as most fungal spores or bacterial cells. The vector role of the nematode is usually limited to carrying the pathogen only a few centimetres, from soil to plant, or from the outer plant tissue to the meristem. A good example is the infection of the developing wheat inflorescence with the bacterium *Corynebacterium tritici*, carried by larval *Anguina tritici* which are seeking developing ovules in which to mature. The larvae will have come from the ear-cockles produced by the previous generation of nematodes and are contaminated with the bacteria, which cause yellow ear-rot disease (Gupta & Swarup, 1972). Without the active participation of the invading nematodes, the bacteria cannot reach the meristem and the plant remains healthy. A similar situation arises with the same nematode and the fungus *Dilophospora alopecuri*, whose spores carry bristle-like appendages which help them to adhere to the nematode (Atanasoff, 1925), while *Corynebacterium fascians* is unable to reach the meristem of the strawberry plant without the aid of *Aphelenchoides* spp. (Pitcher & Crosse, 1958).

PATHOGEN REQUIRES WOUND IN HEALTHY HOST TISSUE

This was the original, simple concept, once thought likely to explain all nematode-pathogen interactions, but it now seems probable that a wound, *per se*, is of less importance than was at first believed. Doubts have arisen partly because artificial wounds have often failed to allow the establishment of fungal pathogens (Jenkins & Coursen, 1957; Porter & Powell, 1967; Oyekan & Mitchell, 1971). However, occasionally wounds do seem to favour fungi (Inagaki & Powell, 1969; Polychronopoulos, Houston & Lownsbery, 1969) and, more frequently, bacteria (Lucas, Sasser & Kelman, 1955; Stewart & Schindler, 1956;

Libman, Leach & Adams, 1964). The greater benefit derived by bacteria may reflect their inferior adaptation to epidermal penetration, but some bacteria and most fungi are quite capable of invading root tissue in the absence of artificial wounding. It seems therefore, that, except in the case of the virus vectors, the provision of a simple wound is not the most important aid which nematodes afford to other plant pathogens.

PATHOGEN REQUIRES A NECROTIC WOUND IN HOST TISSUE

Some pathogens, especially facultative parasites, may be unable to establish themselves in healthy tissue, even if wounded. The availability of a small pocket of necrotic cells, as in a root cortex damaged by *Pratylenchus* spp., may, however, provide a micro- 'food-base' (Garrett, 1956) on which they can become established and from which they are able to invade healthy tissue. Probably some of the imperfectly understood root rots (e.g., of peach (Mountain & Patrick, 1959) or of strawberry (Morgan & McAllan, 1962)) begin in this way. Some highly specialized fungal pathogens may also benefit from necrotic infection courts. Mountain & McKeen (1965) found that the vascular parasite *Verticillium dahliae* showed 'a distinct affinity' for the lesions caused by *Pratylenchus penetrans* in eggplant roots, and Conroy, Green & Ferris (1972) found comparable lesions on tomato to be more important than any general physiological change.

PATHOGEN BENEFITS FROM MODIFICATION OF THE HOST SUBSTRATE

As mentioned on page 65, nematode feeding must always be accompanied by some modification of the host tissue, although this may be slight, as in the case of many ectoparasites, or extensive and complex, as in the case of sedentary endoparasites. Host tissue modification could help a pathogen in either or both of the following ways:

(1) By providing a more favourable substrate for the establishment of the pathogen. Such effects may be localized at or near the site of nematode attack, or the nematode injury may cause systemic changes in the host physiology, so that sites remote from the nematodes may also be rendered more favourable to the pathogen. In some cases these favourable effects may even extend outside the host, e.g., into the soil.

(2) By the disruption or destruction of some natural defence mechanism against the pathogen, e.g., by altering cell permeability. In practice, it is often hard to distinguish between these two effects.

Localized substrate changes. The first detailed study of such a case was that of Powell & Nusbaum (1960) who showed that *Phytophthora parasitica* var. *nicotianae* was able to enter the roots of both susceptible and resistant tobacco cultivars, but that fungal growth was halted in the cortex of resistant cultivars. The entry tracks of *Meloidogyne incognita acrita* larvae had little effect on this resistant reaction, but if gall formation had begun, hyphae adjacent to the gall grew much more luxuriantly in the outer hyperplastic parenchyma and invaded the giant cells, eventually destroying their contents. From this point of vantage the hyphae moved rapidly into the vascular tissue, even in resistant cultivars. This clear demonstration of the advantage conferred on the fungus by substrate modification, as distinct from the simple penetration of epidermis and cortex, marked a turning point in conceptual thought on nematode-pathogen interactions. Later work by Melendéz & Powell (1967) on *M. incognita* and *Fusarium oxysporum* f. sp. *nicotianae* gave similar results on tobacco cultivars either susceptible or resistant to the fungus. Moreover they showed the nematode to be a much more effective incitant when it was inoculated three to four weeks in advance of the fungus, by which time gall-tissue was well developed. In general, other workers supported these findings, although Perry (1963), studying another *Meloidogyne-Fusarium* interaction and making no nematode inoculations in advance of the fungus, favoured a generalized change in host physiology rather than a localized advantage conferred by gall tissue as the major predisposing factor. Bacteria, also, seem able to exploit *Meliodogyne*-modified tissue as a more favourable substrate than healthy tissue (Johnson & Powell, 1969).

In view of the close similarities between the giant cells induced by *Meloidogyne* and the syncytia of *Heterodera* and *Globodera* spp. it is surprising that so few cases of fungal interaction with the latter genera have been recorded. The first was that of *H. glycines* as an incitant of *Fusarium oxysporum* on soybean (Ross, 1965), on which *Meloidogyne incognita* was a less effective incitant. *G. rostochiensis* was later shown to interact positively with *Rhizoctonia solani* and *Colletotrichum atramentarium* on tomato (Dunn & Hughes, 1964) and with *Verticillium dahliae* on potato (Corbett & Hide, 1971). However, in the apparent absence of any published histopathological studies, it is difficult to compare the modes of action of cyst nematodes and *Meloidogyne* as incitants. Differences in incitant ability may well be associated with obvious differences in the host-parasite relationships of these two groups, e.g., the absence of hyperplastic gall-tissue in cyst nematodes or differences between syncytial and giant cell formation (Jones & Northcote, 1972b; Endo, 1971). These, in turn probably derive from differences in nematode biochemistry, e.g., a wider range of proteolytic enzymes in *Meloidogyne* (Myuge, 1961). As Dropkin (1969a) points out, research is now at a stage in which new techniques, such as electron microscopy (Paulson & Webster, 1970; Jones & Northcote, 1972a, b) are being used to learn more of the complex, but fascinating details of the host-parasite relationships of nematodes. Such knowledge should, in turn, allow a better understanding of nematode-pathogen interactions and their dependence on host substrate changes.

Translocatable substrate changes. The first clear evidence of a translocatable factor favouring the establishment of a fungal pathogen was presented by Bowman & Bloom (1966), who showed that split-rooted tomatoes with a polygenic resistance to *Fusarium oxysporum* f. sp. *lycopersici* wilted when the fungus was added to one half root system and *Meloidogyne incognita* to the other. Such plants were as badly affected as those with entire root systems inoculated with both pathogens. Their findings, therefore, tend to agree with those of Perry (1963), mentioned above, that galling is not the only, nor necessarily the most important factor in *Meloidogyne-Fusarium* interactions. Further indication of a translocatable factor, in a *Pratylenchus penetrans-Verticillium dahliae* interaction, was provided by Faulkner, Bolander & Skotland (1970) using mint stem pieces with two isolated root systems. Here also, fungus inoculation of one root system caused more wilting when nematodes were established on the other, the severity of the effect being proportional to the number of nematodes present. In this case the provision of localized nematode infection courts was clearly not the only way in which *Pratylenchus* damage encouraged fungal invasion.

Pathogen-favouring root exudates. Attention has only recently turned to the possibility that physiological changes in the host could affect fungal pathogens before entry. Bergeson, Van Gundy & Thomason (1970) found many more propagules of *Fusarium oxysporum* f. sp. *lycopersici* in the rhizosphere soil of tomatoes infested with *Meloidogyne javanica* than in corresponding nematode-free plants. Golden & Van Gundy (1972) using C^{14} tracer techniques, showed that *Meloidogyne incognita*-infested tomato roots began to leak labelled electrolyte five days after nematode inoculation. This leakage, which reached a maximum three to five weeks later, was accompanied by increased growth of *Rhizoctonia solani* and *Thielaviopsis basicola,* even when they were separated from the galled roots by a semi-permeable membrane. Such exudation may well prove to be more important than hitherto realized and might, for instance, encourage germination of dormant fungal spores or those inhibited by soil fungistasis (Dobbs & Hinson, 1953).

NEMATODE INTERACTS WITH 'NON-PATHOGENS'

There is a small but growing body of evidence that nematodes may interact with organisms not generally recognized even as facultative plant parasites. Most of the available information concerns the genus *Meloidogyne* and probably reflects its ability to initiate the extensive and complex changes in the host tissues which make this genus pre-eminent as an incitant of known plant pathogens. Fungi shown to damage tobacco roots only in the presence of *M. incognita* include *Trichoderma harzianum* (Melendéz & Powell, 1969). This fungus and species of *Curvularia, Aspergillus* and *Penicillium* were reported on similarly by Powell, Melendéz & Batten (1971), together with *Botrytis cinerea,* normally considered a pathogen only of the aerial parts of plants. *Trichoderma viride* may interact with *Pratylenchus penetrans,* but not to the direct disadvantage of the plant (Edmunds & Mai, 1966). (See page 75). Fungi, including *Trichoderma* sp. were found to be less common than bacteria on tomato seedlings co-inoculated with *Meloidogyne incognita* and a soil extract (Mayol & Bergeson, 1970). A comparison of the growth of such plants and others inoculated aseptically with *M. incognita* led these authors to conclude that 'secondary microbial invaders of galled roots were the major cause of plant growth reduction', although they were unable to judge the relative importance of fungi and bacteria in this context. Their work and that of Powell and his co-workers emphasize the potential of the long-neglected and possibly crucial role played by ill-defined but widespread soil organisms in the etiology of root disease under field conditions.

HOST CULTIVAR IS RESISTANT TO THE PATHOGEN, ACTING ALONE

The observations which first drew attention to nematode-pathogen interactions were of dramatic failures of certain disease-resistant cultivars in the presence of *Meloidogyne* spp. (Hunger, 1901; Sasser, Lucas & Powers, 1955; Jenkins & Coursen, 1957). However, the predisposing effects of nematode attack are by no means limited to resistance-breaking (see also page 111). In susceptible cultivars there is usually an acceleration or intensification of infection with and damage by the other pathogen. Present evidence suggests that the pattern of resistance-breaking is often similar, as the pathogen can usually enter the resistant host, but fails to progress as well as it would in a susceptible cultivar. Possibly the nematode supplies something which is lacking in the resistant cultivar, or alters cell permeability. Thus, the reactions of disease-susceptible and resistant cultivars may not be as distinct as is sometimes thought and resistance-breaking may be merely a striking and disturbing variant of the response of susceptible or tolerant cultivars.

NEMATODE DETERS THE OTHER PATHOGEN

The few cases of deterrence mentioned in the literature probably represent only part of the picture, as the absence of disease attracts less attention than its presence. Most of the available evidence comes from inoculation experiments; those involving fungi often disclosed direct feeding by a mycophagous nematode on the hyphae of the pathogen (Schindler & Stewart, 1956; Rhoades & Linford, 1959b; Barker, 1964). Such 'grazing' is unlikely to account for the deterrence of bacteria (Stewart & Schindler, 1956), which are too large to pass up the lumen of the stylet and probably difficult to puncture. Other possible mechanisms include the transmission of a bacteriophage, as yet recorded only in a saprozoic nematode (Chantanao & Jensen, 1969), or a generalized increase in respiratory rate and associated enzymes, to which Nonaka (1959) attributed the deterrence of a fungus by *Aphelenchoides besseyi.*

NEMATODE IS AFFECTED BY THE OTHER PATHOGEN

Experimenters have been, in general, slow to recognize that the nematode may derive either benefit or disadvantage from co-existence with other plant pathogens. Adverse effects on the nematode are not unexpected, as two obligate parasites are often sharing, and thereby competing for, the same host substrate. The nematode, usually the initiator of the interaction, seems particularly vulnerable, not only from impaired nutrition, but also from harmful by-products of the other pathogen (Christie & Perry, 1959). Sometimes the effects are more specific, e.g., *Pyrenochaeta lycopersici* inhibits the hatching of *Globodera* eggs on tomato (James, 1966); the same fungus may also inhibit syncytium formation (Roy, 1968). Equally or more damaging are direct attacks on fully formed giant cells induced by *Meloidogyne*, e.g., by *Fusarium* (Melendéz & Powell, 1967), *Phytophthora* (Powell & Nusbaum, 1960), or bacteria (Johnson & Powell, 1969). *Fusarium* may also deter the multiplication of *Pratylenchus penetrans* (Seinhorst & Kuniyasu, 1971) while *Corynebacterium tritici* can reduce the number and size of ear-cockles produced by *Anguina tritici* on wheat inflorescences (Gupta & Swarup, 1972). It is also possible for nematodes to be adversely affected by micro-organisms in the soil, before any parasitism has taken place. Hollis & Rodriguez-Kabana (1966) identified fatty acids produced by *Clostridium* spp. as a factor responsible for the death of *Tylenchorhynchus martini* in flooded soils; on a longer-term basis hydrogen sulphide can also kill (Rodriguez-Kabana, Jordan & Hollis, 1965). Finally, virus infection of a plant can depress the numbers of nematodes able to subsist thereon (Weischer, 1969a); there is, however, no clear evidence that nematode-transmitted viruses adversely affect their vectors.

Benefit to the nematode has also been noted, arising in several ways. The fungus *Trichoderma viride,* not usually considered to be a plant parasite, makes the roots of alfalfa more attractive to *Pratylenchus penetrans,* even when present in small amounts, whereas the vascular parasite *Fusarium oxysporum* was similarly effective only when present in large amounts (Edmunds & Mai, 1966). In the latter case nematode attraction to infected roots was believed to be associated with an increased output of carbon dioxide (Edmunds & Mai, 1967). Most workers agree that *Pratylenchus* spp. multiply more readily on roots infected by *Verticillium dahliae* than on healthy roots (Mountain & McKeen, 1962 & 1965; Faulkner & Skotland, 1965), while, on wilt tolerant cultivars of soybean, *Fusarium* favoured *Heterodera glycines* (Ross, 1965) and, on peas, *Rotylenchus uniformis* (Seinhorst & Kuniyasu, 1971). There are reports that tobacco rattle virus infection on the host plant favours the vector *Trichodorus allius* (Ayala & Allen, 1966), while *Meloidogyne*, a non-vector, benefited from other viruses (Bird, 1969; Khurana, Goswami & Raychaudhuri, 1970). The influence of pathogen on nematode is often a dynamic relationship, changing as the interaction progresses, which may account for some conflict between the findings of different authors.

Methods Commonly used in Studies of the Interactions of Nematodes with Fungi or Bacteria

When field studies indicate the probability of an interaction involving nematodes, the next step is to determine whether or not the two (or more) suspect organisms are consistently associated with the disease problem in the field. If such observations are affirmative one or more of the following techniques, which are listed in ascending order of standard of proof, may be used to establish the etiology of the field disease.

(1) In some soil-borne diseases a narrow-spectrum nematicide, e.g., ethylene dibromide, dichloropropene, or dibromochloropropane, thought not to have important side-effects on soil fungi or bacteria, may be used to control the nematode. Reduction or elimination of the disease when the nematode is controlled may be regarded as circumstantial evidence that the nematode is inciting the disease. However, no nematicide is truly selective, so the possibility remains that the disease reduction is due, at least in part, to a direct control of the other pathogen by the chemical which was applied to the soil. A positive result from this type of approach therefore requires confirmation by one or more of the following techniques.

(2) Unsterile nematode inoculum, preferably monospecific, may be added to a field soil known to be infected with the other pathogen but free from the suspect nematode. A positive result at this stage affords supporting evidence that the nematode is an incitant.

(3) The most frequently used technique is the 'four-treatment' experiment, usually conducted in open pots of partially sterilized soil, free from plant-parasitic nematodes. The following treatments are applied to test plants growing in such soil.

(i) Nematode only (monospecific).
(ii) Fungus or bacterium only (pure culture) with or without mechanical wounding.
(iii) Nematode plus fungus or bacterium.
(iv) No treatment.

Such experiments are relatively easy to conduct and to evaluate. They are most informative when a clear-cut disease symptom, such as wilt, can readily be seen, but disease records should be supplemented with those of weight, plant size etc. If no unequivocal disease symptom is expected, it is often possible to rate root systems for decay etc. on a points system, but in some cases plant weight and/or size is the only parameter available. In the latter case the combined treatment (iii) of a true interaction should show a growth-depressing effect greater than the sum of those of the two components acting separately (i.e., (i) plus (ii)).

Various refinements may be applied to four-treatment experiments to improve their standard of proof or to increase their yield of information. They do, however, make the experimenter's task more complex and time-consuming.

(a) Aseptic techniques, applied to all three components (i.e., nematode, pathogen and host) and to the growth medium, can give the most reliable evidence that an interaction exists between the nematode and another pathogen and can prove that this is independent of all other culturable organisms. This has not often been done (Edmunds & Mai, 1967; Melendéz & Powell, 1970; Pitcher & Crosse, 1958), but the techniques to achieve it exist (Zuckerman, 1969) and further work on these lines would be advantageous. Even surface sterilization of nematode inoculum is an advance, freeing it from micro-organisms with which it is commonly contaminated in the field.

(b) Variations in inoculation timing—in which the nematode is added before, with or after the other pathogen—usually give the clearest indication whether or not prior substrate modification by the nematode is an important predisposing factor. (See page 72).

(c) Split-root techniques—in which the nematode and the pathogen are inoculated to separate halves of a root system which has been divided at ground level and planted in separate containers. Such tests give the clearest indication whether any pathogen-favouring substrate changes brought about by nematode feeding are localized near the damaged cells or are translocated throughout the root system. (See page 73).

(d) Varied inoculum dosage—of either nematode or pathogen—will often give a much better picture of the mechanisms involved than a single dosage of either component, which is often chosen arbitrarily. Such experiments also lend themselves to mathematical analysis and to the determination of threshold inocula, which greatly strengthen experimenters' conclusions (Seinhorst & Kuniyasu, 1971; McKeen & Mountain, 1960).

CHAPTER 4

Ditylenchus, *Anguina* and related genera

D. J. HOOPER
Rothamsted Experimental Station, Harpenden
& J. F. SOUTHEY
Plant Pathology Laboratory, Harpenden

The genera *Ditylenchus* and *Anguina* are placed in the subfamily Anguininae Paramonov, 1962 *sensu* Siddiqi, 1971. Although Golden (1971) placed *Ditylenchus* in a separate new subfamily, Ditylenchinae, this does not seem feasible because, as Choi & Loof (1974) indicate, it is difficult to find consistent anatomical characters to separate *Anguina* from the closely related *Ditylenchus*. Further detailed morphological study is required on these genera but in the meantime the composition of the genus *Anguina* follows that given by Choi & Loof (1974) except that *Anguillula radicicola* Greeff, 1872 is returned to *Ditylenchus*.

The systematic position of these two genera is: Secernentia: Tylenchida: Tylenchoidea: Tylenchidae: Anguininae; type genus *Anguina* Scopoli, 1777. See Chapter 1 for classification and definitions of these groups.

DITYLENCHUS

Genus *Ditylenchus* Filipjev, 1936

Anguininae. Both sexes vermiform, female rarely obese. Cuticle marked by fine transverse striations; lateral fields with four to six incisures. Deirids inconspicuous, phasmids rarely visible. Lip region relatively low and flattened, continuous with or slightly offset from body contour, with very fine annulations that are not easily seen with a light microscope. Cephalic sclerotization weak; stylet weak to moderate, usually less than 15 μm long, with distinct basal knobs. Basal bulb of oesophagus clavate or variously expanded sometimes overlapping the intestine for a short distance. Vulva usually in the posterior quarter of the body; genital tract single, anterior, outstretched, ovary sometimes partially reflexed; oocytes usually arranged in one or two rows except in the germinal zone; not arranged about a rachis. Bursa enveloping a quarter to three-quarters the tail length but not enclosing the tail terminus. Spicules typically Tylenchoid (Fig. 4.1g); small gubernaculum present.

Some *Ditylenchus* species are important parasites of higher plants, some feed on higher plants and/or fungi while the feeding habits of others are unknown.

Type species: *Ditylenchus dipsaci* (Kühn, 1857) Filipjev, 1936.
This species has many junior synonyms—see Baker (1962), Goodey (1963), Hooper (1972).

Other species:

D. angustus (Butler, 1913) Filipjev, 1936
D. askenasyi (Bütschli, 1873) Goodey, 1951
D. ausafi Husain & Khan, 1967
D. bacillifer (Micoletzky, 1922) Filipjev, 1936
D. beljaevae Karimova, 1957
D. brenani (Goodey, 1945) Goodey, 1951
D. brevicauda (Micoletzky, 1925) Filipjev, 1936
D. caudatus Thorne & Malek, 1968
D. clarus Thorne & Malek, 1968
D. convallariae Sturhan & Friedman, 1965
D. cyperi Husain & Khan, 1967
D. deiridus Thorne & Malek, 1968
D. destructor Thorne, 1945
D. dipsacoideus (Andrássy, 1952) Andrássy, 1956
D. drepanocercus Goodey, 1953
D. emus Khan, Chawla & Prasad, 1969
D. equalis Heyns, 1964
D. geraerti Paramonov, 1970
D. humuli Skarbilovich, 1972
D. inobservabilis Kir'yanova, 1938
D. intermedius (de Man, 1880) Filipjev, 1936
D. karakalpakensis Erzhanov, 1964
D. medicaginis Wasilewska, 1965
D. melongena Bhatnagar & Kadyan, 1969
D. microdens Thorne & Malek, 1968
D. minutus Husain & Khan, 1967
D. mirus Siddiqi, 1963
D. misellus Andrássy, 1958
D. myceliophagus J. B. Goodey, 1958
D. nanus Siddiqi, 1963
D. nortoni (Elmiligy, 1971) Bello & Geraert, 1972
D. obesus Thorne & Malek, 1968
D. paragracilis (Micoletzky, 1922) Sher, 1970
D. procerus (Bally & Reydon, 1931) Filipjev, 1936
D. pumulis Karimova, 1957
D. radicicolus (Greeff, 1872) Filipjev, 1936
D. sapari Atakhanov, 1958
D. sibiricus German, 1969
D. taleolus (Kir'yanova, 1938) Kir'yanova, 1961
D. tausaghyzatus (Kir'yanova, 1938) Kir'yanova 1961
D. tenuidens Gritsenko, 1971
D. triformis Hirschmann & Sasser, 1955
D. tulaganovi Karimova, 1957
D. valveus Thorne & Malek, 1968
D. virtudesae Tobar Jiménez, 1964

Meyl (1961) includes the following as *species inquirendae*:

D. darbouxi (Cotte, 1912) Filipjev, 1936
D. eurycephalus (de Man, 1921) Filipjev, 1936
D. sycobius (Cotte, 1920) Filipjev, 1936

Geraert & Kheiri (1970) suggest that *Pseudhalenchus anchilisposomus* Tarjan, 1958 is somewhat similar to *D. destructor* and should probably be transferred to *Ditylenchus*.

Many of the above listed species are inadequately described and their taxonomic status is questionable and/or their biological role is unknown.

The most important plant-parasitic species are: *D. dipsaci*—the stem nematode; *D. destructor*—the potato tuber nematode; *D. myceliophagus*—which destroys mushroom mycelium; *D. angustus*—which causes 'Ufra' disease of rice. The first three are common in Europe and of these, *D. dipsaci* is one of the most damaging plant-parasitic nematodes especially in temperate regions.

Ditylenchus dipsaci (STEM NEMATODE—Fig. 4.1)

Measurements: (ex teasel, *Dipsacus fullonum* L., type host; after Thorne, 1945)

Females: L = 1·0–1·3 mm; a = 36–40; b = 6·5–7·1; c = 14–18;
$V = {}^{60-70}80^{7}$. Eggs about 2·5 vulval-body-widths long.

Males: L = 1·0–1·3 mm; a = 37–41; b = 6·5–7·3; c = 12–15; T = 65–72.

'Giant Race' (ex broad bean, *Vicia faba* L.; after Goodey, 1941).—

Females: L = 1·73–2·23 mm; a = 50–64; b = 7–12; c = 15·8–20·0;
V = 76–84.

Males: L = 1·51–1·93 mm; a = 58–74; b = 6–8; c = 14·6–19·1.

Note—There is considerable variation in the measurements of adults depending on their race and/or their host (see Barraclough & Blackith, 1962; Blake, 1962b; Goodey, 1941; Metlitzky, 1969). D. J. Hooper (unpublished) has found that the size ranges of adults of the oat and giant races from field beans (*V. faba minor*) overlap.

Description

Female: Body almost straight when killed by heat. Cuticle marked by transverse striae about 1 μm apart; lateral fields with four incisures occupying $\frac{1}{6}$ to $\frac{1}{8}$ of body width. Lip region low, unstriated, slightly flattened, only slightly set off from the body contour. Head skeleton moderately developed, spear about 10–12 μm long with distinct basal knobs. Procorpus of oesophagus cylindrical, narrowing slightly as it joins the fusiform median bulb. Isthmus narrow, surrounded by nerve ring where it begins to expand into a clavate posterior oesophageal bulb that abuts or slightly overlaps the intestine (Fig. 4·1A & E). Tail conoid, four to five anal-body-widths long with a sharply pointed terminus (Fig. 4.1H). Vulva distinct, anterior genital tract outstretched, ovary with oocytes usually in a single, occasionally double row that sometimes reaches the oesophageal region. Post-vulval sac present extending about half-way to the anus.

Male: Anterior region similar to female. Body almost straight when killed by heat. Tail similar to female with sharply pointed terminus, bursa present and extends about three-quarters of the tail length (Fig. 4.1F). Spicules curled ventrally and expanded anteriorly (Fig. 4.1G); gubernaculum short and simple.

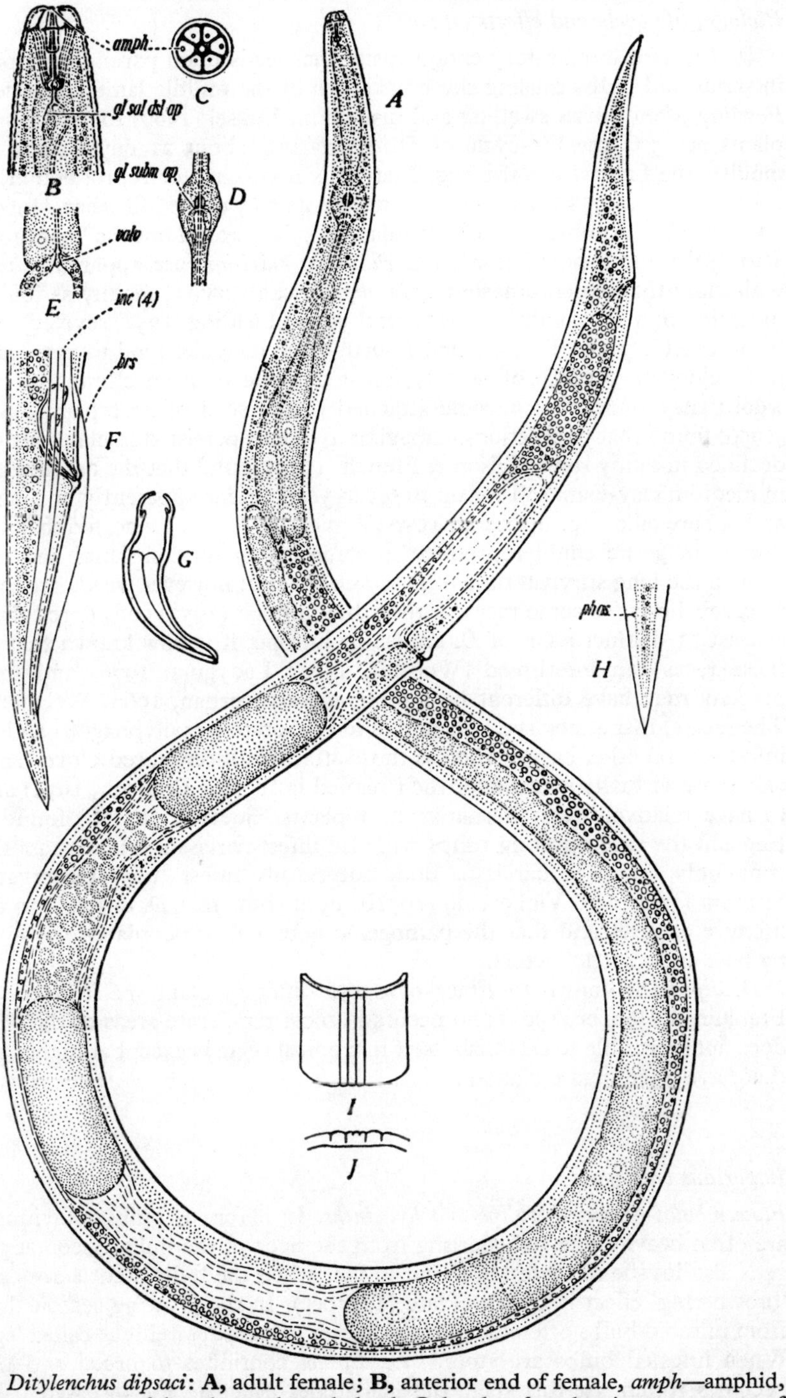

Fig. 4.1. *Ditylenchus dipsaci*: **A,** adult female; **B,** anterior end of female, *amph*—amphid, *gl sal dsl ap*—aperture of dorsal oesophageal gland; **C,** head *en face* showing arrangment of amphids and 4 labial papillae, *amph*—amphid; **D,** dorsal view of median oesophageal bulb, *gl sbm ap*—apertures of submedian oesophageal glands; **E,** junction of intestine and oesophageal lumen; **F,** posterior portion of male; **G,** spiculum; **H,** terminus, *phas*—phasmid (much exaggerated); **I,** section of cuticle at mid-body showing 4 incisures; **J,** cross section of lateral field (note the two minute incisures in the central element). (From Thorne, 1949).

Biology, life cycle and effects on host

D. dipsaci is a migratory endoparasite that feeds upon parenchymatous tissue in stems and bulbs causing the breakdown of the middle lamellae of cell walls. Feeding often causes swelling and distortion. Yuksel (1960) found that in onion plants at 15°C the life-cycle of *D. dipsaci* was about 21 days; there are four moults, the first within the egg. Mating is necessary for reproduction, females lay from 207 to 498 eggs each and live for 45 to 73 days. Sturhan (1969) found $2n = 24$ chromosomes in most populations of *D. dipsaci* but the 'giant race' from *Vicia faba* and a population from *Plantago maritima* are apparently tetraploid with more than 50 chromosomes. *D. dipsaci* can survive, mainly as fourth-stage juveniles, in a dry condition for several years; Fielding (1951) revived specimens from teasel kept dry for 23 years. Fourth-stage juveniles tend to aggregate on or just below the surface of heavily infested tissue to form clumps of 'eelworm wool'; they sometimes become attached to the seed of host plants. Seinhorst (1956) found that populations damaging to onions persisted in clay soils but soon declined in sandy soils. Wilson & French (1975) found that the oat race persisted in medium clay-loam soil for up to seven years under apparently non-host crops and in bare fallow for up to four years. Viglierchio (1971b) reported that a population from garlic could reproduce to some extent on soil fungi and this may explain the long survival time of *D. dipsaci* in soil; however weeds are important reservoir hosts for some races. Although Seinhorst (1957) considered there were at least 11 distinct races of *D. dipsaci* in Europe, it is now known that some of these races can interbreed (Webster, 1964; Ladygina, 1976) and that their progeny may have different host preferences (Sturhan, 1966; Webster, 1967). The race(s) that attack rye, oats and onions seem to be polyphagous and can also infest several other crops, whereas those attacking lucerne, red clover and white clover are virtually specific for their named hosts (see Eriksson, 1974) and seem to have relatively few alternative host plants. Southey (1957b) found that in England the race infesting tulips will also infest narcissus whereas another race commonly found in narcissus does not readily infest tulip. Observations by Sturhan (1969) and Viglierchio (1971b) both show that *D. dipsaci* is a complex mixture of races and that the pathogenic potential of populations is influenced by host and genetic factors.

D. dipsaci is known to attack over 450 different plant species (see Goodey, Franklin & Hooper, 1965) and occurs in most temperate areas of the world but does not seem able to establish itself in tropical regions except at higher altitudes that have a temperate climate.

Important crop hosts

Flower bulbs—narcissi, tulips and hyacinths. In narcissus bulbs individual scales are often heavily infested, starting from the neck of the bulb. Secondary pathogens also invade the infested tissue so that when the bulb is cut across a typical 'brown-ring' effect is visible (Plate II upper). In the growing season the leaves from infested bulbs often have small, lighter coloured, swellings called 'spickels'. When infested bulbs are stored, *D. dipsaci* continues to breed and the bulb becomes rotten. At this stage the nematodes evacuate the diseased tissue and often form clumps of eelworm wool usually at the base of the bulb. Pieces of eelworm wool attached to small particles of bulb tissue can easily infest other bulb stocks especially if they contaminate handling machinery. Infested hyacinth

Item held for:

P1563865

Classmark:	SB 998.N3
Barcode:	79018831
Date:	August 05, 2016

Holds self-collection

If you have reserved this book you will be able to issue it using the self-issue machine in this area.

Collecting your hold

Please note that there may be more than one copy of a book on the Hold Shelves. Holds are arranged by classification order. Please make sure that you are taking the **correct copy** by checking that the number above under 'item held for' **matches your** University number.

Need Help?

Please **leave the item** on the shelf and ask for help at the Helpdesk or from a Library Assistant.

Please note

Removal of any items from this area, without self-issuing or Library staff authorisation, breaches Library rules.

bulbs have a similar appearance to those of narcissi. Infested tulip bulbs do not produce distinct brown-ring symptoms; infestation usually starts from the base of the bulb, infested tissue appears as brown or greyish patches and the whole bulb eventually rots. Flower stems from infested bulbs usually have whitish or purplish streaks that increase in size as the flower develops. On infested stems the flower often bends and the affected parts of the flower stay green. *D. dipsaci* is generally spread in lightly infested bulb stocks. Narcissus and hyacinth bulbs can be disinfested by hot-water treatment when they are dormant but tulips are sensitive to this treatment and dormant bulbs are usually soaked in a solution of thionazin. See Chapters 14 & 15.

Onions and garlic. These are very susceptible to stem eelworm. Seedlings quickly become infested and the invaded tissue becomes swollen and distorted causing characteristic 'bloat' symptoms (Plate I upper). Nematodes continue to multiply in infested onion bulbs during storage. When lightly infested bulbs are replanted for seed, *D. dipsaci* is carried upwards with the scape and infests the flowers so that juveniles become attached to and are dispersed by the seed. Control is achieved mainly by fumigating seed with methyl bromide to ensure that it is free from *D. dipsaci*; soil treatment is possible but expensive. See Chapter 14.

Oats and rye. In infested susceptible cultivars the leaf bases of the main stem and tillers are attacked, the nematodes multiply rapidly and the leaf bases swell to give a swollen-based plant—a condition known as 'tulip-root' (Plate I lower). Cool moist conditions favour invasion and therefore winter-sown cultivars tend to become more heavily infested than spring-sown ones. Heavily infested plants are usually stunted and often fail to produce panicles. Most oat cultivars are susceptible to *D. dipsaci* but the winter cvs. Maris Quest and Peniarth and the spring cv. Manod are resistant (see Chapter 16). Although these cultivars can be invaded by *D. dipsaci* the nematodes tend to become isolated in necrotic lesions and ultimately decrease in number (Goodey & Hooper, 1962). Rye may also be infested by *D. dipsaci* and the symptoms are similar to those on oats. Infestation of this crop is uncommon in Britain but very heavy losses occur in continental Europe where rye is more extensively grown. The European rye cv. Heertvelder is resistant but it is outyielded by other cultivars on non-infested land. Wheat and barley are generally resistant to *D. dipsaci* (Goodey & Hooper, 1962) but seedlings could be damaged by invading nematodes if planted on heavily infested land.

Field beans (Vicia faba). Both the race that commonly attacks oats and the 'giant race' infest winter and spring cultivars of field beans and sometimes infested seed is produced. Seed-borne infestations usually have little effect on the vigour of the first bean crop (Dierks & Klewitz, 1962; Hooper, 1971) but they are important in distributing *D. dipsaci* to non-infested land. Unless the land is heavily infested at planting, e.g., after an infested oat or bean crop, the oat race has little effect on plant growth but the infested stem tissue has a characteristic red-brown coloration and a little infested seed may be produced. The 'giant race' causes much more damage, the stems are often more discoloured and distorted, and the infestation frequently reaches the pod-bearing region so that more infested seed is produced. Also seed yield may be much decreased. Although probably introduced from the Mediterranean region, the giant race seems to persist in British soils for several years (Hooper, 1976). Infested seed can be fumigated with methyl bromide; a large dosage is required to ensure eradication but germination of the seed may then be affected (Powell, 1974).

Teasel. This plant, which is the type host of *D. dipsaci*, is still grown commercially in parts of South-west England. The seed heads are used to produce a fine nap on tweed cloth. Infested plants are stunted and distorted, the leaf bases and midribs are swollen—a condition called 'cabbagy'. Heavily infested plants may not produce seed heads but in lightly infested plants, set out to produce a crop of teasel heads in the second year, the nematodes infest the flower head and a soft malformed seed-head is produced. *D. dipsaci* is readily dispersed on the seed and associated plant debris but can be eradicated by fumigating with methyl bromide. See Chapter 14.

Clover and lucerne. Red clover, white clover and lucerne each seem to be attacked by their own specific race of *D. dipsaci* and rarely by races from other hosts. Infested plants are stunted, and stems and leaf petioles are often swollen. Flower heads become infested and seed contaminated. Severe field infestations often lead to thin or bare patches in the crop, especially in the second and third year. Some resistant cultivars of each of these legumes are known and to some extent have been bred for local situations (see Chapter 16). Infested seed is an important source of infestation but *D. dipsaci* is readily eliminated by fumigation with methyl bromide (see Chapter 14).

Strawberry. Several races of *D. dipsaci* can infest strawberry plants causing distortion and swelling of the leaves and petioles. The main crown may be killed and several weak secondary crowns produced. Runners from infested plants can also be infested but the nematodes can be eradicated by hot-water treatment (Chapter 15). Infested patches can be controlled by treating the soil with a suitable fumigant. See Chapter 14.

Other crops sometimes infested, especially by the oat race, include carrots, sugar beet, mangel, rape, peas, parsnip and rhubarb. In The Netherlands potatoes are sometimes attacked but serious damage has been reported only once in Britain (Newton & Duthoit, 1954). *D. dipsaci* has caused serious damage to maize in France and Germany (Caubel, 1973; Hirling, 1974) and tobacco is sometimes attacked (Cairaschi, 1954; Grujičić, 1974). *D. dipsaci* can also be a troublesome pest of hydrangea (*Hydrangea macrophylla*) and phlox (*Phlox paniculata* or *P. drummondii*). More information on *D. dipsaci* as a pest of many crops is given in Thorne (1961), Webster (1972), Hooper (1972) and in advisory leaflets issued by the Ministry of Agriculture, Fisheries and Food, London.

Ditylenchus destructor (POTATO TUBER NEMATODE, POTATO-ROT NEMATODE–Fig. 4.2)

Measurements: (from various plant hosts, after Goodey, 1952b)

Females: L = 0·69–1·89 (1·07) mm; a = 18–49 (32); b = 4–12 (7); c = 9–30 (17); V = $^{28-76}73\text{–}90^{4-14}$ ($^{54}80^{8}$).

Males: L = 0·76–1·35 (0·96) mm; a = 24–50 (35); b = 4–11 (7); c = 11–21 (14); T = 40–84 (65).

Note—As indicated by the above measurements, adults vary much in size according to their host and/or age. See review by Hooper (1973b).

Description

D. destructor is similar to *D. dipsaci* of which it was thought to be a biological race until Thorne (1945) described it as a distinct species. The main differences

Table 1

Differentiation of Three Species of Ditylenchus *found in Britain*

	Part of host	Tail tip (Fig. 4.3)	Lateral incisures	Length of post-vulval sac relative to vulva-anus distance
D. dipsaci	leaves, stems and flowers	sharply pointed	4	
D. destructor	underground parts of plants especially potato tubers	rounded	6	2/3
D. myceliophagus	mushroom mycelium	rounded	6	1/2

between these two species are in Table 1. *D. destructor* adults tend to look fatter than those of *D. dipsaci* and the overlap of the oesophagus over the intestine is usually more pronounced in *D. destructor* (cf Figs. 4.1 & 4.2), but this feature is variable.

Biology and effects on host

D. destructor is a migratory endoparasite usually of underground parts of plants. It has been reported mainly from temperate regions, especially Europe including West and South Central U.S.S.R. Some 70 crops and weeds and a similar number of fungus species have been recorded as hosts (see Goodey, Franklin & Hooper, 1965 for most records). Unlike *D. dipsaci, D. destructor* is not resistant to desiccation and does not form 'eelworm wool'. Apparently it survives in soil by feeding on fungi and weeds of which field mint (*Mentha arvensis*) and, less commonly, creeping sowthistle (*Sonchus arvensis*) are important reservoir hosts (Goodey, 1952a).

Important crop hosts

Potatoes. D. destructor enters potato tubers through lenticels and initially causes small white mealy spots just below the surface that are only visible if the skin is removed. Infested areas enlarge and coalesce, and light brown lesions, consisting of dry granular tissue, may be visible beneath the skin. As the infestation progresses, the tissues dry and shrink and the skin becomes dry and papery. (Plate III, upper). Internal tissues gradually darken and there are often secondary invasions of fungi, bacteria, mites etc. When stored in moist conditions (i.e., field clamps or general bulk storage) a general rot may ensue and spread to neighbouring tubers causing severe losses.

Although some potato cultivars used in Continental Europe are resistant to *D. destructor* (see Kostina & Zholudewa, 1974) none of the many commercial cvs. tested in the British Isles have shown marked resistance (see review by Hooper, 1973b).

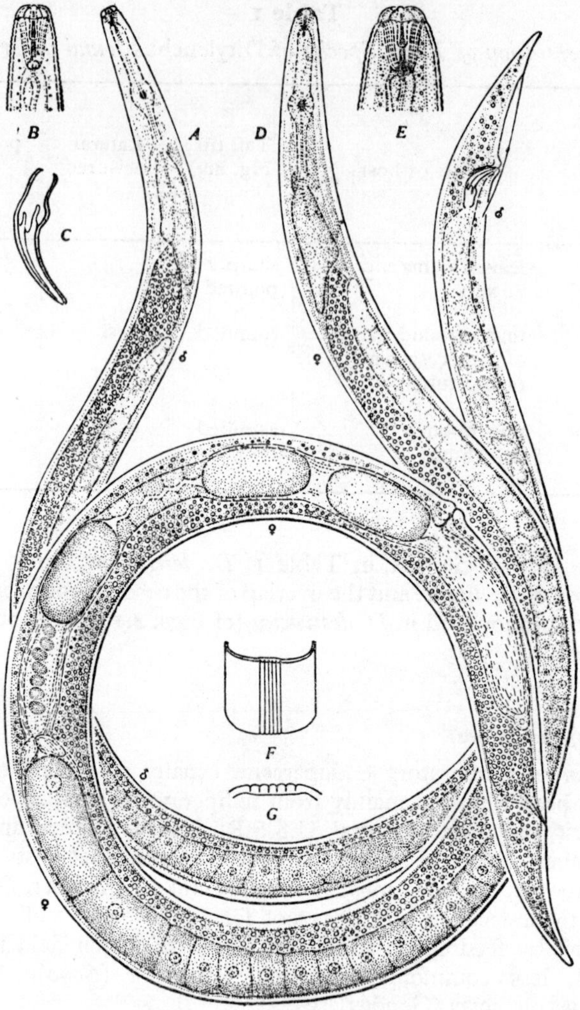

Fig. 4.2. *Ditylenchus destructor*: **A,** male; **B,** anterior end of male; **C,** spiculum; **D,** female; **E,** anterior end of adult female; **F,** section of cuticle at mid-body showing 6 incisures of the lateral field; **G,** cross section of lateral field. (From Thorne, 1945).

Disinfestation of soil is difficult because of alternative weed and fungus hosts and would be expensive for a potato crop (see Chapter 14). Potatoes derived from infested fields should not be planted as seed.

Bulbs—Iris *cvs. Wedgewood, Dutch, English and Spanish*; Tulipa *spp.*

In bulbous iris, infestations cause greyish streaks in the bulb scales usually starting at the base (Plate II, lower) and, as the lesions develop, secondary rotting occurs and the bulbs are eventually destroyed. *D. destructor* is not easily detected in lightly infested bulbs and is therefore readily spread in bulb stocks. Infestations in tulip bulbs, although less common than in iris, cause similar symptoms. Control is by hot-water treatment of dormant bulbs but some kinds of bulbs, especially tulips, may be damaged (see also Chapters 14 & 15).

Other plants sometimes infested include dahlia (infested tubers producing symptoms similar to those on potato), sugar beet and carrots, in which a crown rot may develop. Goodey (1952a) found that lilac (*Syringa vulgaris*), hop (*Humulus lupulus*), *Tigridia pavonia, Gladiolus hybridus* and *Tropaeolum polyphyllum* could all be infested by *D. destructor*. Some of these had been mistakenly considered hosts of *D. dipsaci* although some are attacked by both nematodes. Red, white and alsike clovers (*Trifolium* spp.) are good hosts for *D. destructor* but the nematode apparently causes little damage to the plants. *D. destructor* is also known to destroy the hyphae of cultivated mushroom but the closely related *D. myceliophagus* is more usually responsible for crop losses.

Thorne (1961), Safyanov (1965), Sveshnikova (1967), Webster (1972) and Hooper (1973b) give more information on distribution, host range, etc.

Ditylenchus myceliophagus

Measurements: (ex mushroom, *Agaricus bisporus*; after Goodey, 1958b)

Females: L = 0·6–1·4 mm; a = 23–44; b = 5·4–11·7; c = 10·5–20·5; V = 74–90.

Males: L = 0·6–0·9 mm; a = 21–44; b = 4·5–8·7; c = 9·5–15·3; T = 36–78.

Considerable variation occurs in the size of specimens depending on their age, host, nutrition and environment. See review by Hesling (1974).

Description

D. myceliophagus is similar to *D. destructor*. Some workers have considered them to be the same species but observations by Webster & Hooper (1968), Evans & Fisher (1970c) and Cayrol (1970) clearly show that there are morphological, serological and cytological differences. The main morphological difference is that in *D. myceliophagus*, on average, the post-vulval sac occupies about half or less than half of the distance between the vulva and the anus; whereas in *D. destructor* the post-vulval sac usually occupies about three-quarters of that distance. Also the basal oesophageal glands tend to overlap the intestine less and the posterior third of the tail tends to be relatively fatter in *D. myceliophagus* than in *D. destructor* (see Fig. 4.3). However, these characters vary in both species so much that it is not always possible to identify an individual specimen.

Biology, life cycle and effects on host

D. myceliophagus is a migratory parasite that feeds upon a wide range of fungi and is economically important because it destroys the mycelium of cultivated mushroom (*Agaricus bisporus = A. hortensis*). It has been reported from mushroom growing regions throughout the world, predominantly in temperate areas. Its life cycle is markedly influenced by temperature being completed in about 42 days at 13°C and 18 days at 23°C (Cayrol, 1964). If it is dried slowly, *D. myceliophagus* can survive desiccation but does not form 'eelworm wool' and survives drying less well than *D. dipsaci* (see Perry, 1977). When mushroom

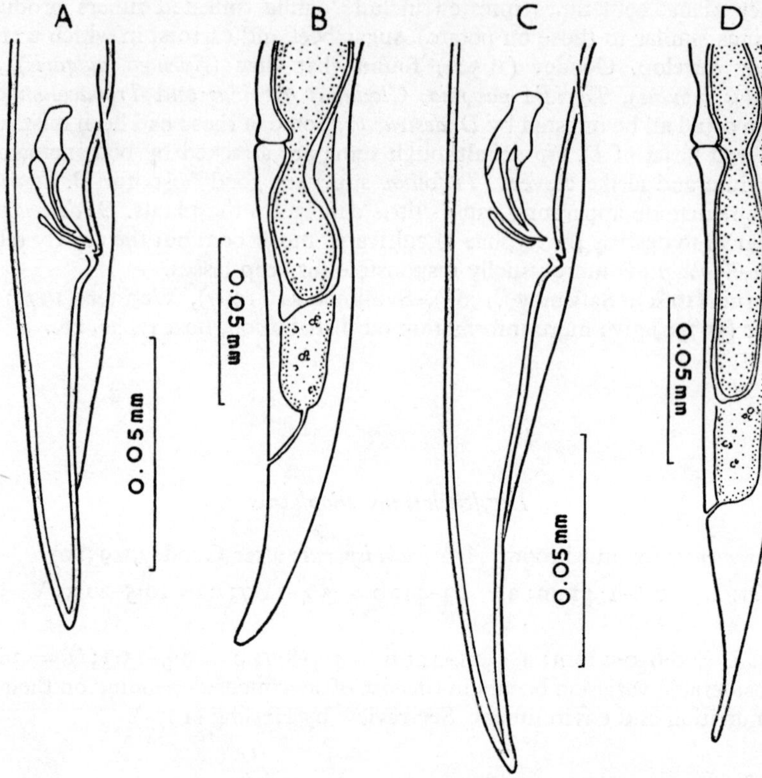

Fig. 4.3. Comparison of posterior parts of *Ditylenchus destructor* and *D. myceliophagus*: **A,** *D. myceliophagus,* male; **B,** *D. myceliophagus,* female; **C,** *D. destructor,* male; **D,** *D. destructor,* female. (From J. B. Goodey, 1965).

mycelium has been destroyed by *D. myceliophagus* the nematodes leave the medium and often form dense sticky clusters on the surface or hang in columns beneath mushroom boxes. In such a state they easily become attached to and dispersed by insects and also get dispersed in drainage water or on implements. When mushroom beds/boxes have become infested no feasible cure is known once the mushrooms are growing. Pretreatment of compost with thionazin gives protection to the end of cropping. Good standards of hygiene are essential for control coupled with 'peak-heating' of the compost before spawning and 'cooking-out' of spent beds after cropping (see Chapter 15). For further information on *D. myceliophagus* see reviews by Hesling (1972 & 1974).

Ditylenchus radicicolus

Measurements: (after Thorne, 1961)

Females: L = 1·2–3·2 mm; a = 24; b = 9–13; c = 19; V = 77.

Males: L = 1·2–2·0 mm; a = 24–30; b = 6·9; c = 15–18.

Description

Females, and to a lesser extent the males, are rather obese, tapering towards each end. Lip region set off by a constriction; spear 12–16 μm long. Basal oesophageal bulb rather broadly pyriform. Developing oocytes mostly in one or two lines, but more than two in the multiplication zone, not arranged about a rachis; ovary may be reflexed once or twice. Post-vulval sac extends about two-thirds the distance from vulva to rectum. Tail conoid with a pointed terminus. The bursa begins just anterior to the spicules and extends three-quarters of the male tail length.

Choi & Loof (1974) placed this species in *Anguina* but its gametocytes are not arranged about a rachis although its obesity suggests *Anguina* spp. Paramonov placed this species in a separate genus, *Subanguina*, see Paramonov (1970), on ecological rather than morphological characters. Until the whole group has been studied in more detail, as indicated on page 26, we prefer to leave it in *Ditylenchus*.

Biology, life cycle and effects on host

D. radicicolus causes characteristic root galls (Plate III, lower) on graminaceous hosts. Goodey (1932) stated that first-stage juveniles invade root tips [this needs verifying because in Tylenchida the second stage is usually infective] and cause the formation of galls within which they moult four times to become adults in about 18–21 days. The galls gradually enlarge due to increase in size of cortical and endodermal cells, the cortex gradually breaks down and the cavities become filled with eggs, juveniles and their parents.

This nematode occurs in various regions of Europe and some parts of North America. It damages barley and rye in Sweden (Kort, 1972) and is often found on the roots of *Poa annua*. Lewis & Webley (1966) found it associated with damage to pasture containing *P. annua* and this population attacked wheat, oats and barley. However s'Jacob (1962) found that a Dutch population heavily infesting *P. annua* caused only a few galls on barley roots whereas a Scandinavian population behaved conversely. Goodey (1932) noted that a population from galls on *Elymus arenarius* differed from some Continental populations in its host range. There seem to be several races of *D. radicicolus* of which Krall & Krall (1970b) recognize six.

Crop rotations excluding known graminaceous hosts seem to keep this nematode under control.

Ditylenchus angustus

This nematode is a serious pest of rice in parts of Africa and Asia. Seshadri & Dasgupta (1975) give a full description and a comprehensive review of its biology etc.

Ditylenchus intermedius

This species is reported sporadically in soil associated with plant roots from temperate regions but there is some doubt about its actual identity. Although

Loof (1961) redescribed a male from de Man's (1880) collection insufficient detail is given to be certain of its status. Loof (1961) considers that a neotype should be designated. *D. intermedius* seems close to *D. myceliophagus* and is possibly a senior synonym.

NOTES ON SOME OTHER *Ditylenchus* SPECIES

D. askenasyi and *D. brenani* attack mosses and induce galls (Steiner, 1936b; Goodey, 1945b). *D. convallariae* was reported from the crown of dormant lily-of-the-valley 'pips' (*Convallaria majalis*)—see Sturhan & Friedman (1965). *D. humuli* damages the buds and roots of hops, light infestations affecting development (Skarbilovich, 1972).

ANGUINA and Related Genera

Nematodes of the genus *Anguina* and their close relatives show more marked adaptation to parasitism than do species of *Ditylenchus*. They cause the formation of well defined galls on above-ground parts of plants, usually on leaves or flower structures, in which the adults develop, mating occurs, the females lay eggs and further development begins. Frequently the form and location of the galls is characteristic of the nematode species. The adult females of *Anguina* species are obese and sluggish, and are usually coiled ventrally when killed by heat. The males are more slender (but stouter than those of *Ditylenchus* species) and, when killed by heat, may be curved with either the dorsal or ventral surface outermost. Typically, mature galls caused by *Anguina* species contain one or several pairs of adults, or their dead remains, plus a mass of eggs or hatched juveniles. Development of the latter stops at the resistant, infective stage which is usually the second, occasionally the fourth stage.

In practice most species of *Anguina* are recognizable by the galls formed on their hosts and the above features of gross morphology. However, separation of the species on morphological features alone is not always easy. Of the 20 or so valid species, two were first described in the 18th century, two in the 19th and nine in the first half of the 20th century; few have been described or revised to present-day standards. Thus the genus needs thorough revision and the taxonomic position of *Anguina* and related genera needs re-examination.

Genus *Anguina* Scopoli, 1777

Anguininae. Females obese, more or less coiled ventrally when killed by heat. Males stout but less so than females; dorsally or ventrally arcuate, or occasionally S-shaped, when killed by heat. Oesophageal glands three, enclosed in swollen basal bulb which may overlap, or partially surround, junction with anterior end of intestine. Ovary or testis of fully mature adults generally reflexed once or twice distally; gametocytes numerous, usually arranged in multiple rows about a more or less well-defined rachis. Obligate plant parasites causing galls on flower structures or leaves of host plants.

Type species: *Anguina tritici* (Steinbuch, 1799) Chitwood, 1935

Type host: *Triticum aestivum* (wheat).

Other species:	Type host
A. agropyronifloris Norton, 1965	*Agropyron smithii*
A. agrostis (Steinbuch, 1799) Filipjev, 1936	*Agrostis tenuis*
A. amsinckiae (Steiner & Scott, 1935) Thorne, 1961	*Amsinckia intermedia*
A. australis Steiner, 1940	*Ehrharta longiflora*
A. balsamophila (Thorne, 1926) Filipjev, 1936	*Balsamorrhiza sagittata*
[1]*A. calamagrostis* Wu, 1967	*Calamagrostis canadensis*
A. chartolepidis Pogosyan, 1966	*Chartolepis biebersteinii*
A. graminis (Hardy, 1850) Filipjev, 1936	*Festuca ovina*
A. graminophila (T. Goodey, 1933) Christie, 1951	*Agrostis tenuis*
A. hyparrheniae Corbett, 1966	*Hyparrhenia collina*
A. klebahni Goffart, 1942	*Primula florindae*
A. kopetdaghica Kir'yanova & Shagalina, 1969	*Cousinia onopordioides*
A. microlaenae (Fawcett, 1938) Steiner, 1940	*Microlaena stipoides*
A. millefolii (Löw, 1874) Filipjev, 1936	*Achillea millefolium*
A. mobilis Chit & Fisher, 1975	*Arctotheca calendula*
A. moxae Yokoo & Choi, 1968 [redescribed by Choi & Loof, 1974]	*Artemisia asiatica*
A. polygoni Pogosyan, 1966	*Polygonum alpestre*
A. poophila Kir'yanova, 1952	*Poa* sp.
[2]*A. pustulicola* (Thorne, 1934) T. Goodey, 1951	Unidentified grass
A. spermophaga Steiner, 1937	*Saccharum spontaneum*
[1]*A. tridomina* Kir'yanova, 1958	*Agrostis canina*
A. tumefaciens (Cobb, 1932) Fil. & Sch. Stek., 1941	*Cynodon transvaalensis*

[1]On the basis of Estonian material, Krall (1970) considers *A. calamagrostis* and *A. tridomina* to be synonyms of *A. graminophila*.

[2]Thorne's (1934) drawing suggests that the genital tract of *A. pustulicola* is of *Ditylenchus* rather than *Anguina* type. But Thorne's specimens were in poor condition, the anatomy of the fore-part of the body was evidently obscure, and the ovary was not described. Otherwise the species looks like a typical *Anguina*.

Anguina tritici (WHEAT GALL NEMATODE—Fig. 4.4 and Plate IV)

This important pest of cereals, formerly much more widespread than at present, was first reported by Needham in 1743. His account of 'worms discovered in smutty corn' is the earliest known record of a plant-parasitic nematode. *A. tritici* has been recorded in most major wheat-growing areas but modern seed-cleaning methods have virtually eliminated it from western Europe, North America, Australasia and U.S.S.R. The last record for England and Wales was in 1956.

Measurements are based on several authors—see Southey (1972).

Females: L = 3–5 mm; a = 21–30; b = 19–25; c = 30–50; V = 88–97; stylet = 9–11 μm.

Males: L = 1·9–2·5 mm; a = 25–30; b = 12–13; c = 14–28; stylet = 9–11 μm.

Second-stage juveniles: L = 0·75–0·95; a = 47–59; b = 4·0–6·3; c = 23–28; stylet = 10 μm.

Eggs: 73–140 × 33–63 μm.

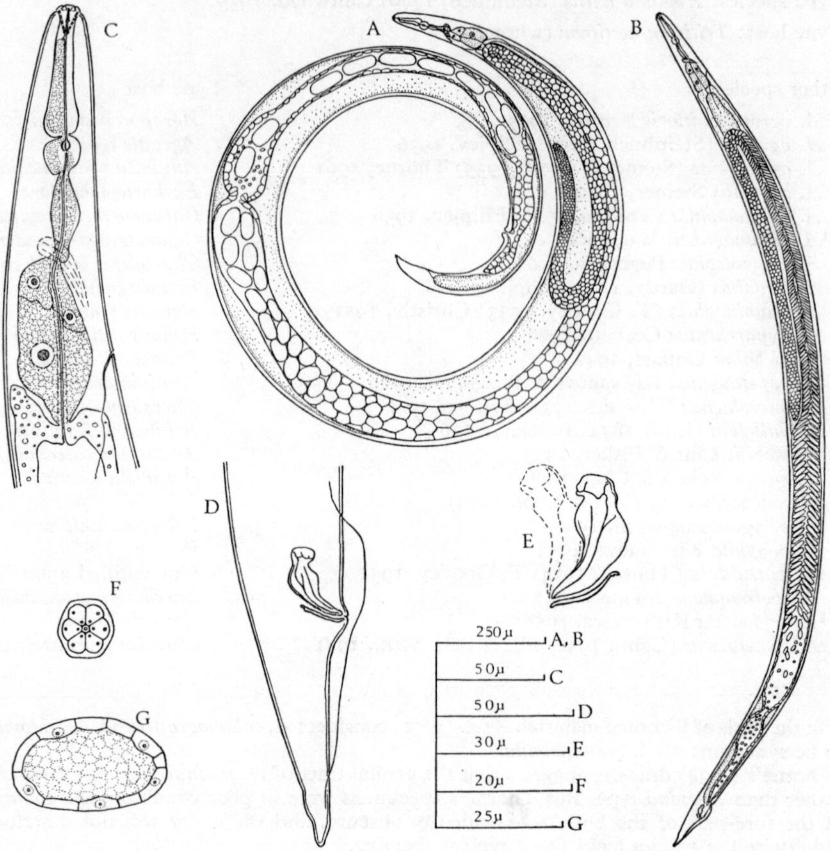

Fig. 4.4. *Anguina tritici*: **A,** female; **B,** male; **C,** female oesophageal region; **D,** male tail; **E,** spicules and gubernaculum; **F,** *en face* view of lip region; **G,** T.S. ovary. (From Southey, 1972, *CIH Descriptions of Plant-parasitic Nematodes*, after Goodey and Thorne).

Description

Cuticular annules very fine, usually visible only in oesophageal region; lateral fields of adults seen only on young specimens, with four or more fine incisures. Lip region low and flattened. Oesophageal glands forming ± pyriform bulb, sometimes irregularly lobed, not overlapping intestine. Tail conoid, tapering to obtuse or rounded tip.

Female obese, spirally coiled ventrally when killed by heat. Isthmus of oesophagus sometimes swollen posteriorly but set off from posterior bulb by a constriction. Anterior branch of genital tract greatly developed; ovary with two or more flexures and very many oocytes in multiple rows around a rachis. Posterior remnant of uterus a simple post-vulval sac. Up to 10 eggs at a time in uterus of gravid females.

Male body sometimes curved dorsally (ventral surface outermost) when killed by heat. Testis with one or two flexures. Spicules paired, stout, arcuate with the capitulum rolled or folded ventrally. Gubernaculum simple, trough-like. Bursa extending from just anterior to spicules to just anterior to tail tip.

Type, and most important, host: *Triticum aestivum* (common wheat).
Other hosts: *Secale cereale* (rye), *Triticum dicoccum* (emmer), *T. monococcum*, *T. spelta* (spelt), *T. ventricosum*.

Biology, life cycle and effects on host

Usually *A. tritici* is first detected by the presence of brown or black, rounded galls ('ear-cockles') among the threshed grain (Plate III). If galls are sown with seed, second-stage juveniles emerge from them in moist conditions, invade host seedlings and feed ectoparasitically on young leaves near the growing point. Infested seedlings are stunted and show abnormal rolling, twisting and crinkling of leaves.

The nematodes normally complete their development only after invading the inflorescence but, rarely, galls containing adults may be formed on leaves. Infested ears are undersized with abnormally spreading glumes. Some or all of the grains are replaced by galls which are shining green at first, later turning dark brown or black in wheat and straw-coloured in rye. The galls are said to be more readily shed than the grains, so many fall out at harvest and infest the soil. The galls contain up to 40 or more adults of each sex, more than most other species of *Anguina*, producing up to 30 000 or more eggs.

A. tritici is diploid ($2n = 38$) and amphimictic. The life cycle follows the normal nematode pattern with four moults. The adults die soon after oviposition and the eggs hatch quickly. The first-stage juveniles do not undergo the first moult within the egg membrane, as in many Tylenchids, but do so soon after hatching so that ripe galls contain a mass of predominantly second-stage juveniles. The latter are extremely resistant to desiccation and have been revived after 32 years storage (Limber, 1973).

A. tritici interacts with *Corynebacterium tritici* in a serious disease of wheat, yellow ear-rot, which occurs mainly in subtropical areas. It is particularly important in India ('tundu' disease—see Gupta & Swarup, 1968) and has been reported also in Western Australia, China, Egypt and Ethiopia. Yellow ear-rot is recognized by the bright yellow bacterial slime on abortive ears and leaf sheaths before the ears emerge. Typical *Anguina* symptoms occur at the seedling stage but few or no galls are formed in infected heads. *A. tritici* also occurs in close association with the bunt fungus, *Tilletia foetida*, on wheat in India (Mathur & Misra, 1961), though interaction is not necessarily implied.

For further details and references, see Southey (1972).

Control

Galls were formerly separated from grain by flotation in brine or plain water but the most effective control is by modern mechanical seed cleaning. In moist conditions, absence of host plants for one year is sufficient to free the soil from *A. tritici* but under conditions of prolonged drought viable galls can persist much longer. Effective control of *A. tritici* also controls yellow ear-rot since the bacteria are infective only in the presence of the nematode. Promising resistance has been found in wheat cultivars in Yugoslavia and Rumania (see p. 324).

Anguina agrostis (BENT-GRASS GALL NEMATODE—Fig. 4.5)

Females: L = 1·5–2·7 mm; a = 17–21; b = 8–11; c = 32–44; V = 87–88; stylet = 8–9 μm.

Males: L = 1·1–1·7 mm; a = 23–38; b = 6–9; c = 20–23; spicules = 35–40 μm.

Second-stage juveniles: L = 0·75–0·8 mm; a = 44–47.

Eggs: 90–150 × 30–50 μm.

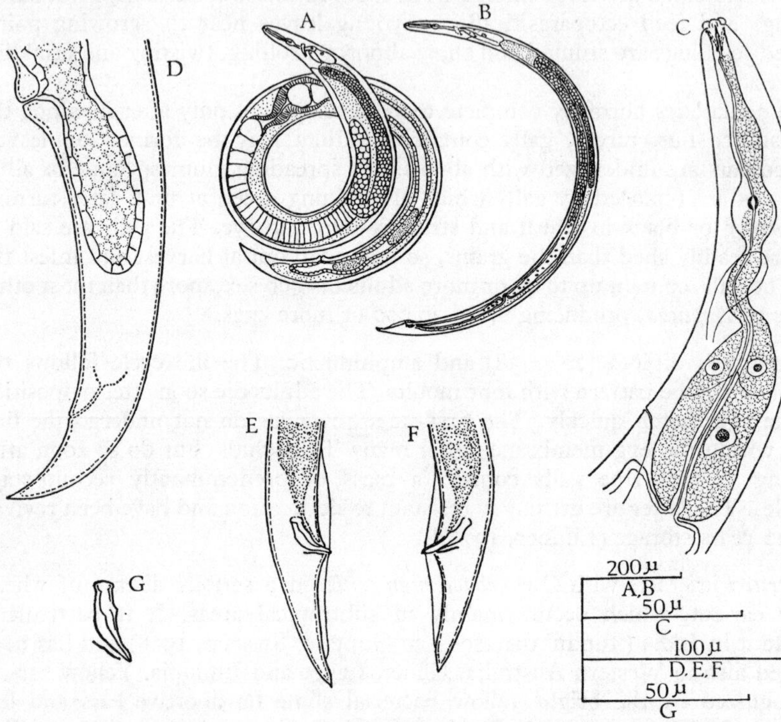

Fig. 4.5. *Anguina agrostis*: **A,** female; **B,** male; **C,** female oesophageal region; **D,** female tail; **E, F,** male tails; **G,** spicule and gubernaculum. (From Southey, 1973, *CIH Descriptions of Plant-parasitic Nematodes,* in part after Goodey).

General morphology similar to that of *A. tritici* but differs in smaller size of all stages, the isthmus of the oesophagus not markedly swollen or set off from the posterior bulb by a constriction, the posterior bulb being pyriform and not lobed, the tail tip sharply pointed, the male curved ventrad when killed by heat and the spicules being more slender in build with little or no ventral folding of the capitulum.

Type host: *Agrostis tenuis*.

A. agrostis forms characteristic cigar-shaped galls, at first green, later dark purple-brown, in florets of various species of *Agrostis* (bent grasses). It has also been reported from other grass genera but different, possibly undescribed, species of *Anguina* may be involved (see Southey, 1973a). Galled florets have the floral parts (glumes, lemma and palea) greatly elongated to three to five or more times normal length, the lemma projecting beyond the glumes as a sharp point (Fig. 4.6). The galls reach 4–5 mm long whereas the normal caryopsis is barely 1 mm.

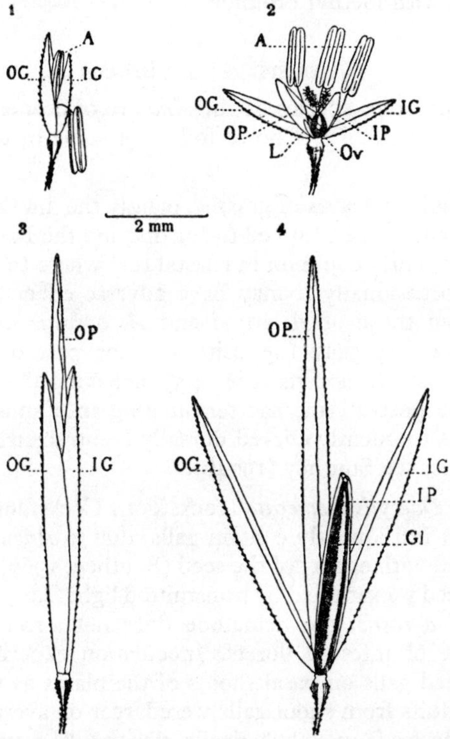

Fig. 4.6. Effects of *Anguina agrostis* on florets of *Agrostis stolonifera*: **1,** normal floret unopened; **2,** normal floret opened to show its parts; **3,** galled floret, unopened; **4,** galled floret opened to show gall and elongated floral structures. **A,** anthers; **Gl,** gall; **IG,** inner glume; **IP,** inner palea; **L,** lodicules; **OG,** outer glume; **OP,** outer palea; **Ov,** ovary. (From T. Goodey, 1930, *Journal of Helminthology*).

A. agrostis is a serious or potentially serious pest in areas where bent grasses are grown for seed, especially N.W. U.S.A. and New Zealand. There have been several reports that *A. agrostis* galls are toxic to livestock (see J. B. Goodey, 1960a). However, these reports related to galls on *Festuca* spp., which may be due to a different species of *Anguina*.

Life cycle generally similar to that of *A. tritici* except that the first stage moults within the egg so it is the second-stage juveniles which hatch, and the latter do not escape from the galls by their own efforts but are expelled in a mass as a result of swelling of a mucoprotein matrix in the presence of water, causing the gall to burst (Collis-George & Blake, 1959).

For further details and references, see Southey (1973a).

Control is by hygienic precautions to prevent spread, and rotation or fallowing in the field to break the life cycle. One year's break from a host is sufficient provided that no bent grasses are allowed to flower. In practice specific herbicides (dalapon or maleic hydrazide) are recommended for this purpose. In lawns and other established turf regular mowing to prevent flowering should provide adequate control if needed. Mechanical seed cleaning is only partially successful in removing infestation from seed; Hague (1963) successfully fumigated infested *Agrostis tenuis* seed with methyl bromide.

OTHER SPECIES OF *Anguina*

Few, if any, of the other species of *Anguina* are of economic importance and most occur on wild plants. Brief notes follow of some of the species found in temperate areas.

A. graminis forms galls on leaves of grasses, mainly the fine-leaved fescue grasses (*Festuca* spp.). It seems to be confined to Europe and the Baltic area of U.S.S.R. In Great Britain it is fairly common in coastal turf where the fine-leaved fescues predominate, and occasionally it may have adverse effect on coastal pastures.

Adults differ from those of *A. tritici* and *A. agrostis* in the more forward position of the excretory pore (opposite anterior part of glandular bulb of oesophagus), and from *A. agrostis* (Fig. 4.5) in having the tail tip rounded or minutely lobed, the post-uterine sac terminating in a small knob of vestigial tissue and the males frequently curved dorsally (ventral surface outermost).

For further details, see Southey (1974b).

Anguina species on *Dactylis glomerata* (cocksfoot). Cocksfoot seed samples from eastern England not infrequently contain galls, due to *Anguina* sp., which have sometimes interfered with export of the seed (Southey, 1969). The galls can only be detected when seed is examined by transmitted light (e.g., on a diaphanascope) because, unlike *A. agrostis,* the nematode does not affect the dimensions or external appearance of infested florets. Inoculation experiments showed that this nematode formed galls on basal shoots of the plants as well as, later, in the inflorescence and adults from shoot galls were larger on average than those from flower galls. The species is morphologically close to *A. graminis.*

A. millefolii causes rounded or irregularly shaped galls on leaves, and occasionally stems or inflorescence, of yarrow (*Achillea millefolium*). (T. Goodey, 1938; Hooper & Doncaster, 1972). It is widely distributed in Europe and U.S.S.R. In England, it occurs at scattered localities but seems by no means common. The species is unusual in producing two generations within the same gall, the adults of the first generation being larger than those of the second; the two generations differ more in size and body proportions than do some different species of *Anguina*. Apart from this feature there are few distinctive morphological characters, one being the very short post-uterine sac.

A. graminophila forms more or less fusiform, purplish galls on the leaf blades of *Agrostis* spp., usually near the leaf bases (T. Goodey, 1933a, b). It differs from most other species of *Anguina* in the more slender form of the adults, the offset lip region, the large rounded stylet knobs, the prominent anterior lip of the vulva with abrupt narrowing of the body just posterior to it, the more anteriorly extended bursa in the male, and the fact that the fourth-stage juveniles, and not the second-stage, are the infective stage. With a few related species (including *A. calamagrostis* and *A. tridomina,* if valid, and *A. tumefaciens*) it forms a rather distinct group within the genus *Anguina* which should possibly be given separate generic rank within the Anguininae.

KEY—Choi & Loof (1974) give a key to the species of *Anguina* recognized by them.

RELATED GENERA

Cynipanguina. Maggenti, Hart & Paxman (1974) described the type and only species *C. danthoniae* causing spherical leaf-galls on a grass, *Danthonia californica*, in California, U.S.A. It differs from all other *Anguina*-like nematodes in having the oesophageal lumen prolonged into a digit-like process which extends from the glandular bulb into the anterior end of the intestine.

Paranguina. First described by Kir'yanova (1955) and redefined by Kir'yanova and Ivanova (1968). According to the latter, this genus differs from *Anguina* in the following features: four oesophageal glands instead of three; both longitudinal and transverse striation of the cuticle; larger amphids (up to $\frac{1}{4}$ the width of the lip region). Type species: *P. agropyri* Kir'yanova, 1955 causing galls on the bases of shoots of *Agropyron repens* in U.S.S.R. Kir'yanova & Ivanova (1968) added five more species from wild plants in U.S.S.R., one as a new combination and three described as new.

Nothanguina. This genus differs from *Anguina* in the absence of a median valvular bulb in the oesophagus and was, therefore, placed in Neotylenchidae, Nothotylenchinae by Whitehead (1959) and Goodey (1963); these categories being raised respectively to superfamily and family rank by Jairajpuri & Siddiqi (1969) (see p. 37). Type species: *N. cecidoplastes* (T. Goodey, 1934) Whitehead, 1959, causing galls on the leaves, stems and flowers of a grass, *Bothriochloa pertusa*, in India.

Chapter 5

Meloidogyne

MARY T. FRANKLIN

Commonwealth Institute of Helminthology, St Albans

Introduction

Root galling due to species of *Meloidogyne* (root-knot nematodes) has been recorded since 1855, causing devastation in many crops, particularly in tropical and subtropical regions. Since the time when crop plants were first grown successively in the same ground these nematodes, which are indigenous in many parts of the world but particularly in the warmer regions, have probably been a potent source of damage and yield loss. The first recognizable published account of their effects is by Berkeley (1855) who described galls, in which he observed 'vibrios' or minute worms, on cucumbers in an English glasshouse. Similar galls containing small worms were described by Licopoli (1875) on *Sempervivum tectorum* in Italy and by Jobert (1878) on coffee in Brazil. The first scientific name applied to root-knot nematodes was *Anguillula marioni* given by Cornu (1879) who examined galls on sainfoin (*Onobrychis viciifolia*) from sandy soil on the banks of the river Loire at Chateauneuf, France (Loiret), but his description is not sufficient to determine the species with certainty. He found similar galls on *Clematis vitalba* and on *Cissus aconitifolia*.

The generic name *Meloidogyne* (from Greek words meaning 'apple' and 'female' in reference to the shape of the females) was given by Goeldi in 1887 to nematodes causing galls on the roots of coffee in the Province of Rio de Janeiro, Brazil. He named the species *M. exigua* and this, being the first adequately described, is the type species of the genus. The name *Meloidogyne* was not generally accepted and for over 60 years root-knot nematodes were usually included in the genus *Heterodera*, first as *H. radicicola* and, from 1932, as *H. marioni*. Three more species were added until, in 1949, Chitwood revised the group, justified Goeldi's name, *Meloidogyne*, and described or redescribed five species and one variety. Since then numerous species of root-knot nematodes have been described, the number now standing at about 33. These are listed in Table 1, from which it will be seen that 11 were first described from crops in temperate climates, so that the genus can no longer be regarded as confined to tropical or sub-tropical areas, as it was in the past.

Table 1

Species of Meloidogyne *with data on Measurements and Host Plants*

	L2 Length		Mean c-value	Characteristic host plants
	mean	range		
kikuyensis de Grisse, 1960	320	(290–360)	11	*Pennisetum clandestinum*
exigua Goeldi, 1887	329	(281–370)	7·5	coffee, tea, *Capsicum, Citrullus*
**artiellia* Franklin, 1961	336	(301–370)	15	*Brassica*, cereals, legumes
**hapla* Chitwood, 1949	337	(312–466)	8	wide range
**ovalis* Riffle, 1963	342	(302–430)	8	*Acer, Betula, Fraxinus, Ulmus*

	L2 Length		Mean c-value	Characteristic host plants
	mean	range		
†*incognita* (Kofoid & White, 1919) Chitwood, 1949	371	(337–403)	8	wide range in warm regions
lordelloi Ponte, 1969	?	(340–380)	?	*Cereus macrogonus*
**deconincki* Elmiligy, 1968	370	(340–400)	8	*Fraxinus, Rosa, Solanum nigrum*
oteifae Elmiligy, 1968	370	(320–400)	8	coffee, *Pueraria*
coffeicola Lordello & Zamith, 1960	?	(337–424)	12	coffee
inornata Lordello, 1956	?	(375–420)	12	*Glycine hispida* (cv. Abura), tobacco, *Anthurium*
microtyla Mulvey, Townshend & Potter, 1975	380	(350–400)	8·2–8·7	grasses
kirjanovae Terentyeva, 1965	392	(359–433)	7	tomato
tadshikistanica Kir'yanova & Ivanova, 1965	399	(350–435)	9	*Pelargonium roseum, Tradescantia*
**litoralis* Elmiligy, 1968	400	(330–450)	11	*Ligustrum*
acronea Coetzee, 1956	401	(354–427)	9	*Eragrostis, Sorghum, Pennisetum, Chloris* (Gramineae)
ethiopica Whitehead, 1968	410	(383–432)	9	tomato, cowpea
indica Whitehead, 1968	414	(381–448)	25	*Citrus*
javanica (Treub, 1885) Chitwood, 1949	417	(387–459)	8·5	wide range in warm regions
**ardenensis* Santos, 1968	417	(372–453)	11	*Vinca, Ligustrum, Fraxinus, Sambucus, Hepatica*
africana Whitehead, 1968	420	(380–470)	11	coffee
mali Itoh, Ohshima & Ichinohe, 1969	418	(390–450)	13	*Malus prunifolia*
**carolinensis* Fox, 1967	429	(367–489)	10	*Vaccinium, Rhododendron* (azalea)
thamesi Chitwood in Chitw., Specht & Havis, 1952	432	(410–476)	8	wide range in warm regions
**naasi* Franklin, 1965	435	(418–465)	6	cereals, grasses, *Beta, Allium*
**graminis* (Sledge & Golden, 1964) Whitehead, 1968	439	(420–510)	6	cereals, grasses
graminicola Golden & Birchfield, 1965	441	(415–484)	6	cereals (esp. rice), grasses
megadora Whitehead, 1968	451	(413–548)	8	*Coffea* spp. and other Rubiaceae (in Angola)
**ottersoni* (Thorne, 1969) Franklin, 1971	?	(430–500)	?	*Phalaris arundinacea, Hordeum vulgare*
lucknowica Singh, 1969	492	(410–575)	12	*Luffa cylindrica*
arenaria (Neal, 1889) Chitwood, 1949	?	(450–490)	?	wide range in relatively warm regions
brevicauda Loos, 1953	522	(459–590)	24	tea
decalineata Whitehead, 1968	543	(471–573)	11	coffee (Tanzania)

L2 = second-stage larva. c = total length/tail length.
* = species that might be expected to survive outdoors in cool temperate climates.
†See page 113.

Systematics

RELATIONSHIPS

Order	Tylenchida (Oerley, 1880) Thorne, 1949
Suborder	Tylenchina (Oerley, 1880) Geraert, 1966
Superfamily	Heteroderoidea Filipjev & Schuurmans Stekhoven, 1941 (Golden, 1971)
Family	Meloidogynidae Skarbilovich, 1959 (Wouts, 1973)
Genera	*Meloidogyne* Goeldi, 1887 *Meloidoderella* Khan, 1972 *Meloinema* Choi & Geraert, 1974

Meloidogyne is by far the most important genus in the family Meloidogynidae, one of the four families at present recognized in the superfamily Heteroderoidea. In the same family, *Meloidoderella* A. M. Khan, 1972 closely resembles *Meloidogyne* but differs in the position of the excretory pore in the female and in the female cuticle which becomes dark brown to black, apparently forming a cyst of which the wall is marked with a clear zig-zag pattern and irregular punctations. *Meloidoderella indica* was found on tomato near Aligarh, India. The only other genus, *Meloinema,* (type *M. kerongense* Choi & Geraert, 1973 from lilac in Korea) was placed in Meloidogynidae but differs in important respects from *Meloidogyne,* e.g., young, vermiform females are found free in the soil, have the vulva 92–95 per cent of the distance from the anterior end and have paired opposed gonads. When they attach themselves to roots the posterior part enlarges and in the two specimens found the vulva is 82 and 95 μm from the terminus and the anus 32 and 40 μm from it. The anterior part of the body was not found but apparently was not swollen. The males also differ markedly from those of *Meloidogyne.* (This genus appears to be very close to *Nacobbodera* of Golden & Jensen (1974), type species *N. chitwoodi* on Douglas fir and other conifers in Oregon, U.S.A.; the authors placed their genus in Nacobbidae). Husain (1976) has erected separate subfamilies, Meloidoderellinae and Meloineminae, within Meloidogynidae for *Meloidoderella* and *Meloinema* respectively.

DIAGNOSIS OF *Meloidogyne*

Female swollen, subspherical, sedentary; cuticle thin, annulated; tail absent, anus and vulva terminal, surrounded by a characteristic pattern of striae on the cuticle; excretory pore opposite or anterior to metacorpus. Basal plate of head skeleton hexaradiate with the two lateral sectors wider than the subventrals and subdorsals. Mouth spear less than 25 μm long, slender, with well-developed basal knobs; procorpus and metacorpus not amalgamated; procorpus cylindrical followed by spherical metacorpus with well-developed musculature and valve plates. Gonads paired, much elongated and convoluted. Oesophageal glands overlap intestine ventrally. Eggs laid in a gelatinous matrix, usually unembryonated, and none retained in the female's body. First moult within the egg; second-stage larvae hatch and are infective. Male vermiform, migratory; spear and cephalic framework well developed, basal plate of the same pattern as in the female. Short, bluntly-rounded tail; no caudal alae; one or two testes;

paired, fairly slender spicules and simple gubernaculum. Undergoes metamorphosis.

Second-stage larvae vermiform, infective, migratory. Head sectors as in female, the lateral ones broader than the others; stylet slender, knobbed, less than 20 μm long. Third and fourth larval stages swollen, sedentary within roots, with blunt terminus and no spear; they develop within the cuticle of the second stage of which they retain the tail spike.

Morphology

MATURE FEMALES (Fig. 5.1)

Globular with a short 'neck' often bent at an angle to the long axis of the body. Vulva and anus terminal, sometimes on a slight protuberance, but the outline of the body terminus is usually smooth. Females appear pearly white but in fact the cuticle is transparent, smooth for the most part and glistening, with variously patterned striae surrounding the terminal region. These striae represent the annules and lateral fields of the larva and form fingerprint-like patterns that are often used for specific identification. The arrangement of the features in the area that includes the vulva and anus, together with the surrounding striae and the phasmids, is known as the perineal pattern or posterior cuticular pattern; the terminology of the different parts is given in Fig. 5.2. Photographs of typical perineal patterns of seven of the earliest described species of *Meloidogyne* are given by Taylor *et al.*, 1955. Annules are present in the cylindrical neck region, which expands posteriorly to join the main body and narrows abruptly at the anterior extremity to the relatively small lip region bearing one, two or three annules. Excretory pore opens on anterior part of neck, often close behind base of mouth spear. Cuticle varies in thickness from 3 or 4 μm to 32 μm in different species and different parts of the body.

Mouth spear slender, 12–15 μm in most species, dorsally curved and with well-developed basal knobs. The dorsal gland duct opens into the oesophageal lumen close behind the spear knobs. The cylindrical procorpus is constricted where it joins the large, well-developed, muscular metacorpus situated towards the base of the neck. There are prominent valve plates in the metacorpus. The glandular region consists of a broad lobe comprising two sub-ventral glands and one dorsal, and lies over the intestine latero-ventrally, close behind the metacorpus. In actively feeding females there is an ampulla in the duct of the dorsal oesophageal gland immediately adjacent to its junction with the oesophageal lumen. During feeding an exudate may be seen emerging from the spear tip; in water it appears to form a precipitate (Bird, 1968). The intestine becomes a syncytium without a lumen which, together with the two coiled, elongated ovaries, fills the swollen body. The rectum loses its connection with the intestine and instead serves as duct for the gelatinous material produced by six huge, unicellular glands which line the posterior part of the body and secrete the matrix that surrounds and encloses the deposited eggs (Maggenti & Allen, 1960). Each gonad is much coiled and reflexed and consists of an ovary with a short germinal zone followed by a region of growth with oocytes in tandem. The oviduct is a short zone four cells long and is followed by the spermatheca, a slightly expanded spherical region, and the uterus. The two uteri join posteriorly to form a common vagina which opens at the slit-like vulva (Triantaphyllou, 1962).

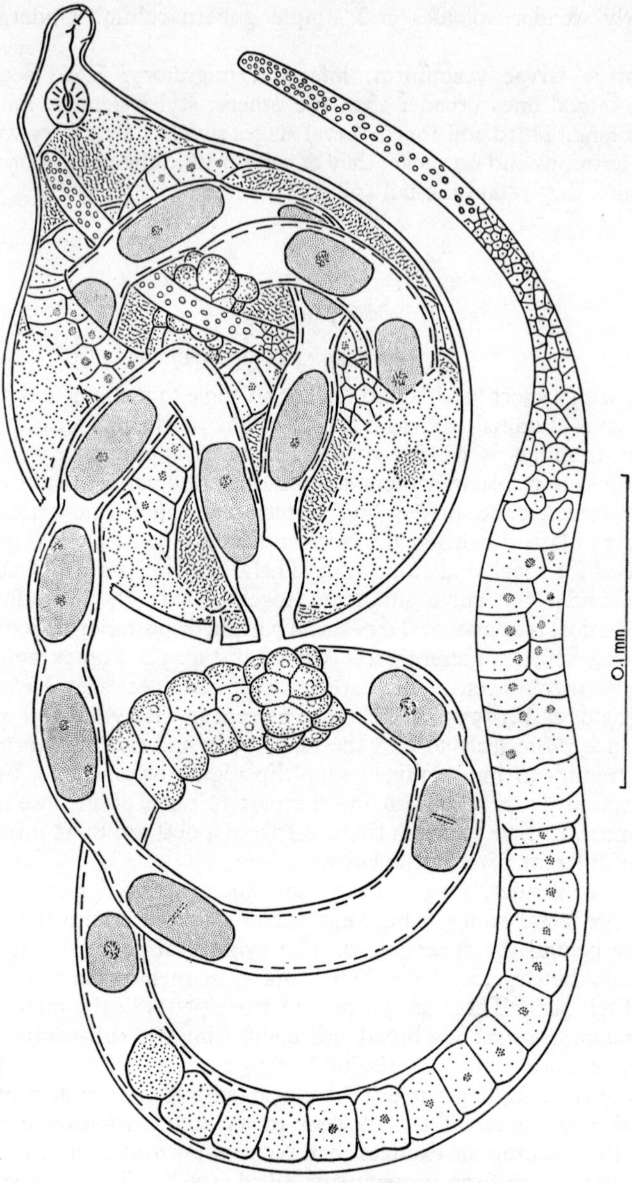

Fig. 5.1. Diagram of whole female of *Meloidogyne* showing ovary dissected.

MALES (Fig. 5.3A–C)

Vermiform, up to 2 mm long, tail short (c = 90–150), bluntly rounded; hind region of body spirally twisted, often through 180°. Cuticle clearly annulated usually with four lines on the lateral field which is sometimes irregularly crossed by the transverse striae; phasmids in the lateral field close to cloaca, posterior to it in some species. Head not sharply offset, truncate cone-shaped with variable numbers of post-labial annules; a circum-oral disc, or head cap, can be clearly

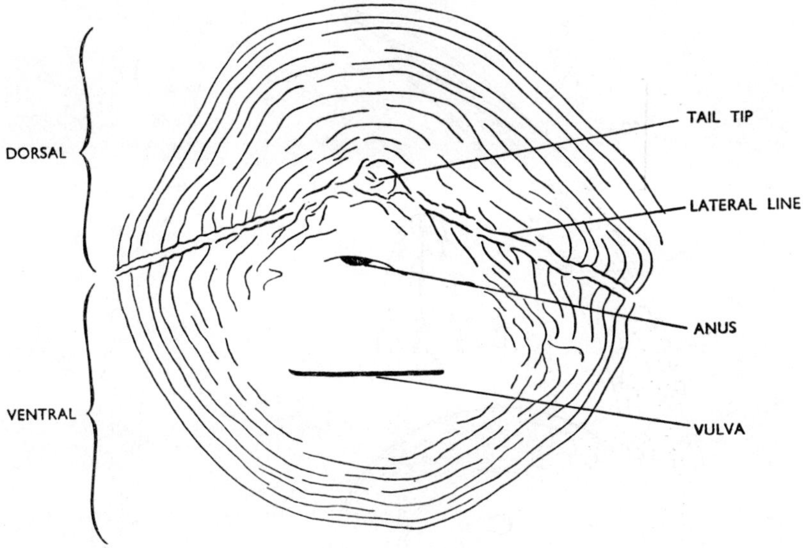

Fig. 5.2. Diagram of perineal pattern of *Meloidogyne* female showing features used in identification.

differentiated from the post-labial annules in dorso-ventral views of heads by the presence of the amphid openings and ducts.

Mouth spear 18–25 μm long with well developed knobs; behind lie the cylindrical procorpus, fusiform metacorpus with central valve-plates, short isthmus and elongated gland lobe overlying the intestine ventrally for about four body-widths. The nerve-ring encircles the isthmus. Spicules slender, ventrally curved, varying in length from about 25 to 33 μm measured along the curve; gubernaculum 7–11 μm long in lateral view. There are one or two testes. Males are often rare; intersex males common in some species, especially in overcrowded roots.

LARVAE

The first stage has a blunt tail and moults in the egg; the second stage hatches and invades suitable host roots. This free-living stage is that most commonly found as it emerges from the egg into the soil. It is slender, varying from 320 to 540 μm long, with more or less pointed tail, and lightly built spear 10–15 μm long (Fig. 5.4C). The head is truncate cone-shaped to rounded, the framework delicate; there is a variable number of post-labial annules, usually two or three (Fig. 5.4C). The cuticle is annulated and usually has four lines in the lateral field. The larval tails vary in length and shape in different species, for example the ratio c varies from about 10 in *M. hapla* to nearly 15 in *M. africana* (Fig. 5.5). The oesophagus resembles that of the male, on a smaller scale, and in some species the rectum is inflated. *Meloidogyne* larvae may be distinguished from those of *Heterodera* by their more slender form, markedly more delicate spear and flatter head (Fig. 5.4). Morphometric information on the larvae of 33 species is given in Table 1. More information on morphology and biology of those

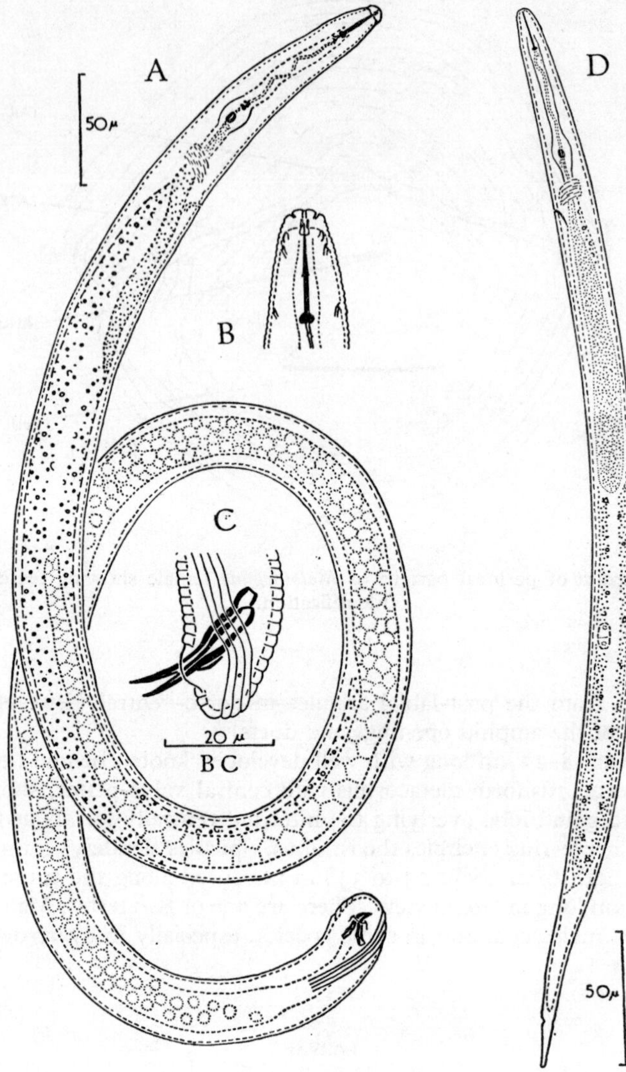

Fig. 5.3. **A–C,** typical male of *Meloidogyne*: **A,** whole worm; **B,** head dorso-ventral; **C,** tail; **D,** second-stage larva. (Drawings are of *M. naasi* from Franklin, 1965).

species known to occur in north temperate regions, whether out-of-doors or in glasshouses, is given later in this chapter.

The morphological characters mainly used in separating species are: in the female—posterior cuticular pattern; in the larvae—length, shape and length of tail (Fig. 5.5), rectum, head annules, stylet knob shape; in the male—head shape and annules, lateral field, stylet knob shape and size, spicule shape. Esser, Perry & Taylor (1976) tabulate morphometric and morphological data for 35 species of *Meloidogyne*.

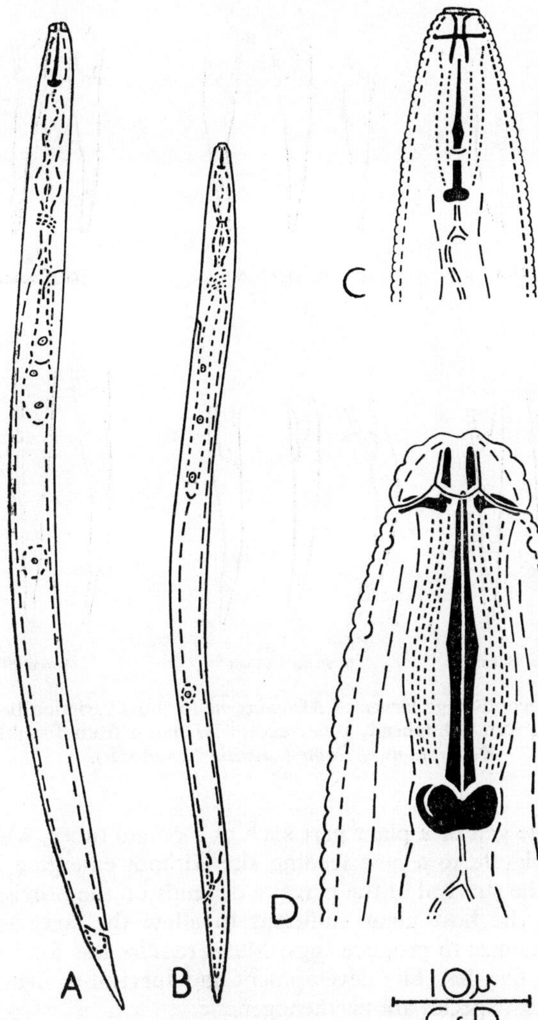

Fig. 5.4. Second-stage larvae of *Meloidogyne* and *Heterodera* for comparison. A, *Heterodera schachtii*; **B,** *Meloidogyne hapla*. (**A** and **B** to same scale.) **C,** anterior region of *Meloidogyne*; **D,** anterior region of *Heterodera*.

Biology

Meloidogyne spp. are all obligate sedentary parasites of higher plants, generally in the roots but occasionally in stems and leaves (e.g., saintpaulia, clover, balsam), and they usually cause galling of the host tissues. The success of a population of *Meloidogyne* in maintaining itself depends on conditions being favourable during both the free-living and parasitic phases of the life-cycle. In the free-living phase the infective second-stage larva must hatch, migrate through the soil, and find and invade a suitable plant root, unless the eggs are

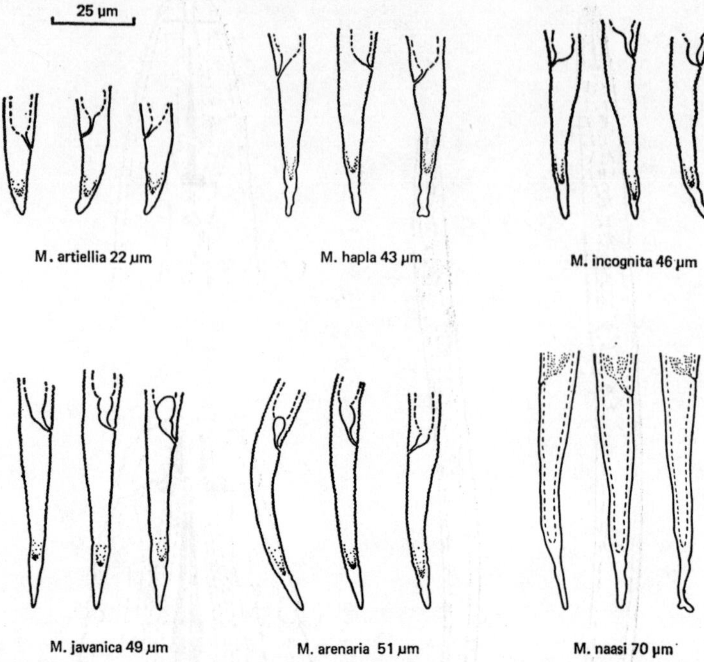

Fig. 5.5. Tails of second-stage larvae of *Meloidogyne* to show variation between species; all to same scale. (From Whitehead, 1968, except *M. naasi* from Franklin, 1973, *CIH Descriptions of Plant-parasitic Nematodes*).

produced in a large gall or a plant part such as a potato tuber, when the larvae may hatch and migrate to a new feeding site without emerging into the soil. Within the host, the survival of the parasite depends on the provision of shelter and nutrition by the host plant sufficient to allow the larva to develop to maturity and the female to produce eggs. Males require less food and space for development than females. The development and survival of males seem to be unimportant as most species are parthenogenetic.

FREE-LIVING PHASE

Meloidogyne eggs in infested soil are present in egg-masses, i.e., they are embedded in the gelatinous matrix consisting of an irregular meshwork of glycoprotein material in which the spaces probably contain water. In dry conditions the water evaporates and the dehydrated matrix shrinks and hardens and, in *M. javanica*, changes from colourless through yellow to reddish-orange. The hardened matrix probably inhibits hatching of the contained eggs by its pressure on them, and may also serve to prevent water loss from the eggs themselves and thus protect them in drought conditions (Bird & Soeffky, 1972). Egg masses may be released into the soil by rotting of the galled tissue or may remain embedded in either live or dead roots; in galls on woody roots such as grapevine or rose they may be well protected from adverse conditions and nematicides. While a minimum amount of moisture must be present for hatching to take place, larvae can hatch over a wide range of soil moistures.

Hatching does not require the stimulus of root exudates, though there is evidence that they can increase the number of larvae emerging. Viglierchio & Lownsbery (1960) found in *in vitro* experiments that tomato root diffusate stimulated hatching of *M. incognita, M. hapla* and *M. javanica*. Hamlen *et al.* (1973) found a greater hatch of *M. incognita* in soil moistened with root exudate fractions from lucerne than with distilled water, but there was a hatch of over 50 per cent in the latter and it is considered that lucerne root exudates would have little effect under field conditions. There is evidence, at least in *M. incognita,* of partial diapause when development of the embryos and hatching come to a halt, even when conditions are favourable for hatching (de Guiran, 1974).

Once hatched, the larvae move through the soil in the moisture films covering the soil particles, and in free water. The optimum conditions for movement are when most of the soil pores have drained of water (= field capacity), resulting in the presence of adequate oxygen, and the temperature is that favoured by the species, usually about 20–30°C. Soil particle size affects movement by its influence on moisture capacity, oxygen and pore size. It has been estimated that nematodes can move fastest in the pores formed by soil particles whose diameters are about one third to one half the nematode's length. (See also Chapter 2).

There is some evidence that larvae are attracted to roots; Lownsbery & Viglierchio (1961) found that larvae of *M. hapla* accumulated around roots of tomato, oats and rye but did not invade the last two. Larvae of *M. incognita* have been found to move to depths of 120 cm in soil while following roots of tomato and cucumber (Bird, 1969).

Newly hatched larvae have food reserves stored in the intestinal cells, lipid globules accounting for about 43 per cent and protein about 23–24 per cent of the material stored (Van Gundy *et al.,* 1957). The reserves are gradually consumed while the larvae are free-living, consumption being rapid at high temperatures (25–30°C), in dryish soils and in oxygenated solutions. The larvae therefore survive longer when activity and motility are reduced at lower temperatures and high soil moisture, so that food reserves are not depleted. Activity is increased in the presence of host roots (Van Gundy *et al.,* 1957).

Entry into plant roots is achieved by thrusts with the mouth spear, usually in the regions of cell elongation and root hairs or at the base of lateral roots. Sometimes the larvae feed on the epidermal root cells before entering the cortex, occasionally causing slight swelling. Penetration is apparently helped by secretions from the sub-ventral oesophageal glands (Bird, 1967).

PARASITIC PHASE

After penetration the larva moves through the cortex inter- and intracellularly and having reached a favourable position, usually close to the pericycle and not already colonized by other larvae, it settles and begins to feed on the cells surrounding the head. The cell walls are punctured by the mouth spear and secretions from the sub-ventral oesophageal gland ducts are probably introduced, the gland ducts, which are at first dilated, becoming much reduced after one or two days. In another two or three days the oesophageal glands increase threefold in size, the median bulb enlarges and becomes spherical and the larva is obviously broader. As might be expected, with increase in girth the nematodes lose their ability to migrate and invade roots (Bird, 1967), though larvae that

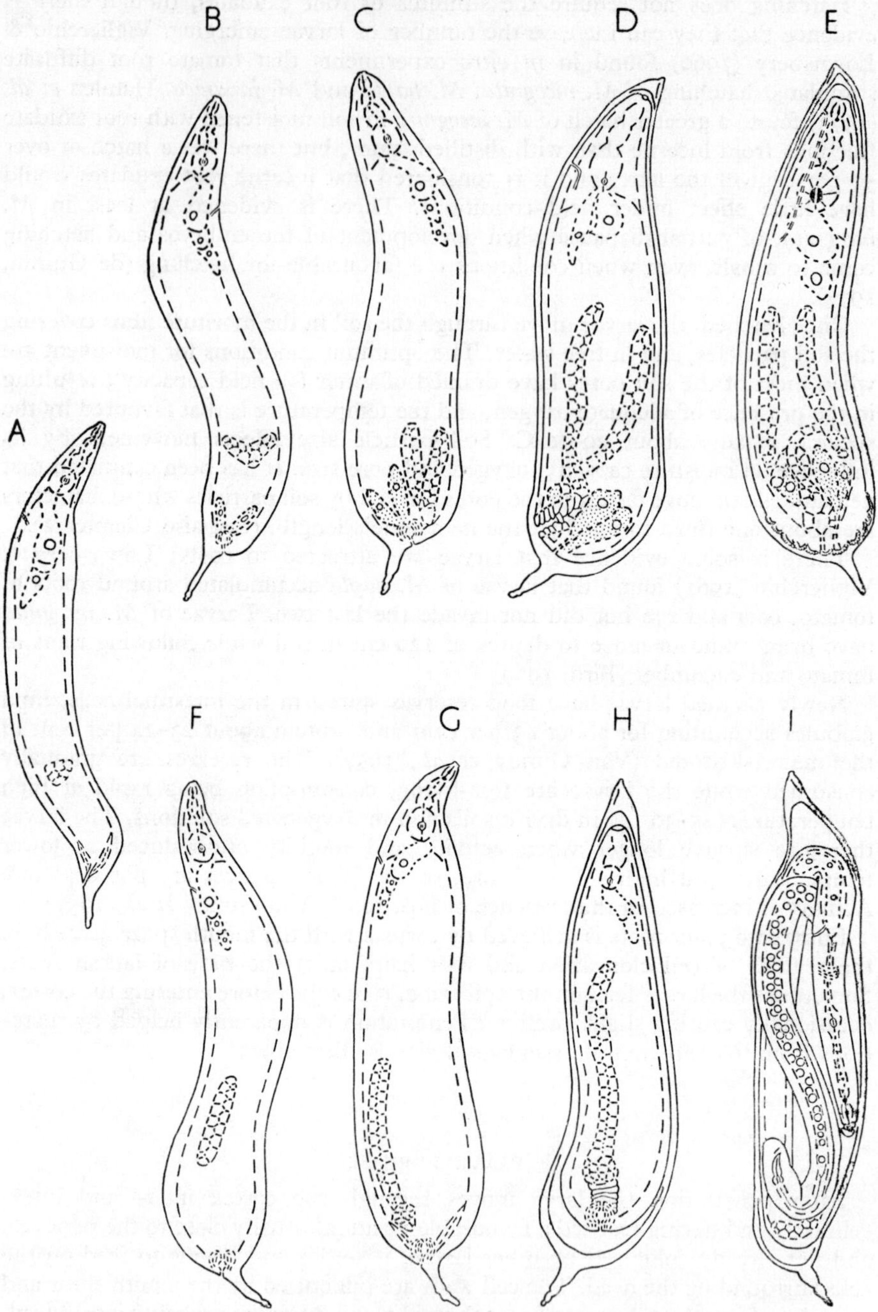

Fig. 5.6. Development of *Meloidogyne incognita*: **A,** sexually undifferentiated second-stage parasitic larva. **B–E,** development of female: **B,** sexually differentiated second stage; **C,** late second stage about to moult; **D,** fourth stage; **E,** adult just after fourth moult. **F–I,** development of male: **F,** sexually differentiated second stage; **G,** late second stage about to moult; **H,** early fourth stage; **I,** adult just after fourth moult. **E** and **I** show second, third and fourth cuticles. (After Triantaphyllou & Hirschmann, 1960).

have entered unfavourable hosts in which they cannot feed or grow may move out after four to six days and invade another root (de Guiran, 1960). The sexes can be differentiated by the end of the second stage, females having a two-branched Y-shaped rudimentary gonad and matrix gland cells surrounding the rectum, while the male gonad is unbranched and spicule primordia can be distinguished at the developing cloaca (Fig. 5.6).

Growth and development in the roots proceeds in two phases, as shown by Bird (1959) for females of *M. javanica* and *M. hapla* and by Triantaphyllou & Hirschmann (1960) for *M. incognita* (Fig. 5.7). After establishment at the feeding site the developing female increases steadily in size for about two weeks. It then undergoes three moults in quick succession, losing the spear and oesophagus which are not reorganized until after the final moult when the reproductive organs are also developed. The second-stage cuticle remains until the final moult, the pointed tail forming a spike on the rounded posterior end, a useful character for distinguishing *Meloidogyne* from *Heterodera* in the developing stages. The nematode cannot feed in the 3rd and 4th larval stages and there is no increase in size (Fig. 5.6A–E), but after the final moult feeding recommences and growth is rapid for another 10 days. The growth curves for *M. javanica* females were shown by Bird (1972a) to vary directly with the growth rate of the host-plant roots, and females continued to grow even after the start of egg-laying. Egg-laying begins about a week after the final moult, the female continuing to feed on the giant cells surrounding her head. Several hundred eggs are laid by each individual, nearly 3000 having been recorded for one female on excised roots in culture.

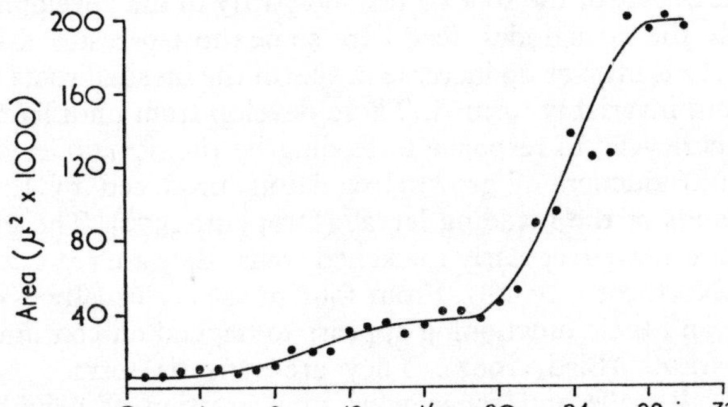

Fig. 5.7. Graph showing rate of growth of females of *Meloidogyne javanica* in tomato roots. (After Bird, 1959).

Males also develop from the second moult to maturity without feeding and Bird (1959) followed their development *in vitro*. Freshly dissected late second-stage larvae kept in distilled water in a perfusion chamber developed to adult males in seven days. On the second day, after the second parasitic moult, one 4th-stage larva doubled its length in five hours, moving vigorously within the cast cuticles. The formation of spear and spicules was observed on the 4th and 5th days and there was evidence of the final moult at this time. After a period of quiescence the male became active again and, using its stylet to penetrate the moulted cuticles, emerged on the 7th day (Fig. 5.6A, F–I).

Although males are found in most species of *Meloidogyne*, reproduction is nearly always parthenogenetic. The ratio of males to females depends on the food available to the parasites: in light infestations on favourable host plants all the larvae develop into females, but if the infestation is heavy and the larvae are crowded, or if the host is unsuitable and fails to provide adequate nutrition, many become males. In many species reproduction is still parthenogenetic in these circumstances and the males appear to be functionless. Intersex males with two testes and an incipient vulva are produced when an initially favourable environment becomes unfavourable, resulting in sex reversal after development of larvae into females has begun. This may happen if the host plant dies or is severely damaged, resulting in a drastic cutting down of the nematodes' food supply (Triantaphyllou, 1960).

The rate of multiplication of a root-knot population depends on the health and suitability of the host plant, the temperature, the density of larvae in the roots and the occurrence of favourable conditions during the free-living stage. A single generation takes a minimum of four to six weeks and in warm conditions there may thus be several generations a year if host roots are continuously available. In a cool climate there is usually only one generation.

Host Reaction

GALLING

The notorious galling of roots infected by *Meloidogyne* is due partly to hyperplasia and hypertrophy of the root cortex and partly to the development of giant cells on which the nematodes feed. In some host-parasite associations of *Meloidogyne* there is little or no increase in size of the invaded roots but syncytia, or giant cells, are invariably formed. These develop from undifferentiated cells, usually of the pericycle, in response to feeding by the nematodes, possibly as a result of the introduction of growth regulators produced by the subventral oesophageal glands of the invading larvae (Dropkin, 1972). The giant cells are multinulceate and have irregularly thickened walls; they are regarded as transfer cells (Jones & Northcote, 1972b). From four to six are usually associated with each nematode and their functioning appears to depend on continuous stimulation by the parasite (Bird, 1962). They are syncytia formed partly by the breakdown of cell walls and consequent amalgamation of neighbouring cells (Bird, 1972b, 1973). This is followed by synchronous mitoses of the nuclei and irregular thickening of the walls as occurs in transfer cells. The response of susceptible plants to root-knot nematodes is reviewed by Bird (1974). (Plate V, top left).

The cross-sectional area of syncytia associated with female *M. javanica* in bean roots studied by Bird (1972a) averaged about 27×10^4 μm^2 four weeks after the nematode started to feed, when the average cross-sectional area of the nematode was about the same. After four weeks growth the syncytia declined in size while the nematode began to lay eggs and continued to expand, reaching an average of 35×10^4 μm^2 at seven weeks. In tomato both syncytia and nematodes were smaller but again the syncytia declined in size after three weeks while the nematodes continued to grow. The area of the syncytial nuclei and their DNA content also declined just before egg laying started, two to three weeks after invasion.

Heavy galling of plants results in stunting of top growth and relatively greater root to shoot ratio. Infested plants show increased susceptibility to water stress and there is increased synthesis of proteins in the galls with disturbance of the normal transport of substances between root and shoot. In the field galled roots are more readily damaged by other microorganisms than are healthy roots. Evidence suggests that exudates from infected roots affect the microflora in the rhizosphere; germination of chlamydospores of *Fusarium oxysporum* was stimulated and numbers of actinomycetes were reduced around tomato roots infected with *M. javanica* (Bergeson *et al.*, 1970).

In plant hosts that respond to invasion by the development of large galls the cortex forms the bulk of the gall tissue by cell proliferation. The swollen female nematodes are completely embedded and their egg sacs may or may not protrude on the root surface. Galls caused by *M. hapla* are usually small and discrete, associated with proliferation of rootlets (Plate V, top right). Similar, but even smaller, galls are induced by *M. microtyla* on tomato (Mulvey *et al.*, 1975). In contrast, those due to *M. naasi* on graminaceous plants such as *Lolium* or barley tend to be elongated, merging gradually with the root; sometimes they affect one side of the cortex more than the other causing the roots to become hooked, especially near the root tips (Plate V, lower left). The root-knot species found in warm conditions, e.g., *M. incognita* and *M. javanica* on tomato or cucumber in glasshouses, cause irregular often massive deformations, the galls sometimes giving the roots a beaded appearance (Plate V, lower right). However, no obvious galling may be present in infected sugar-cane roots or in naturally fleshy stems such as those of saintpaulia or in tubers such as cyclamen. Galls may be found on aerial parts of plants if there is adequate surface moisture, allowing larvae to move onto the plant and invade, as demonstrated by Wong & Willetts (1969) in tomato and bean.

RESISTANCE

Plants show varying degrees of susceptibility to parasitism by root-knot nematodes and many instances are known of wide variations between different cultivars of a single crop plant. A simple classification by Dropkin & Nelson (1960) takes into consideration the growth of both host plant and nematode:

		Host growth Good	*Host growth* Poor
Parasite growth	Good	tolerant	susceptible
Parasite growth	Poor	resistant	intolerant

The lucerne cultivar African is an extreme example of resistance in which the parasite is unable to grow at all. Reynolds *et al.* (1970) found that larvae of *M. incognita* invaded the roots of African and of the susceptible cultivar Lahontan, in similar numbers but within three or four days they left the roots of African, having caused no apparent damage, while galls developed in Lahontan.

A lesser degree of resistance was demonstrated in *Cucumis ficifolius* and *C. metuliferus* by Fassuliotis (1970). Larvae of *M. incognita* entered the roots in similar numbers to those entering the susceptible *C. melo* but delay in development was evident after nine days; by the 26th day 78 per cent of the nematodes in *C. melo* were saccate females as compared with 37 and 22 per cent in the two resistant species. Resistance was associated also with an increase in the proportion of males, and with the relatively small size of the giant cells associated with the retarded females as compared with those surrounding the heads of the successful females.

An intolerant reaction is demonstrated in plants that develop severe necrosis in response to nematode invasion. In the tomato cultivar Y-91 which is resistant to *M. incognita*, a necrotic reaction appeared 12 hours after inoculation. Investigation of the hypersensitive reaction suggested that it could be reversed by applying cytokinins which enabled larvae to develop to maturity in a normally resistant cultivar (Dropkin *et al.*, 1969). Working on the parasitism of *M. javanica* in resistant peach, Kochba & Samish (1971) suggest that the application of NAA or kinetin preserves the juvenile nature of syncytia, usually lost in a resistant cultivar, so that the nematodes can continue to feed and grow.

Resistance may also be overcome by increase in temperature as was shown first by Holtzmann (1965) and subsequently by Dropkin (1969b). Both found that tomato cultivars resistant to *M. incognita* at 25 to 28°C were relatively susceptible above 30°C. Irizarry *et al.* (1971) found that cultivars of *Phaseolus vulgaris* that were resistant to *M. incognita* and *M. hapla* at 25°C lost resistance at 30° although at this temperature they were still resistant to *M. arenaria*.

PATHOTYPES

Sasser (1972) in a study of numerous populations has shown that physiological races, or pathotypes, exist in several species of *Meloidogyne*. He studied the reaction of nine test plants to 11 populations of *M. incognita*, 15 of *M. javanica*, 10 each of *M. hapla* and *M. arenaria*, and two of *M. exigua*. His results showed variations in pathogenicity within the first four species. Several pathotypes of *M. exigua* have been noted in Brazil (Lordello, 1972). Similarly, Southards & Priest (1973) demonstrated six physiological races amongst 17 isolates of *M. incognita* tested on six plant species, Ogbuji & Jensen (1972) distinguished five pathotypes amongst 14 populations of *M. hapla* from northwestern U.S.A. and Kirby *et al.* (1975) found differences in host preference within three species from Florida. Several workers have reported the development of aggressive pathotypes when resistant cultivars of crops are grown frequently in soil infested with the species to which they were supposedly resistant (Riggs & Winstead, 1959; Triantaphyllou & Sasser, 1960; Graham, 1969). The apparent ease with which root-knot nematodes can become adapted to new host cultivars means that extensive testing with a large number of nematode populations is necessary in the search for resistance.

MELOIDOGYNE Species in Temperate Zones

Root-knot nematodes have long been known in tropical regions where they may destroy crops over large areas. This is partly brought about by the rapid multiplication of the nematodes when two or more susceptible crops are grown in succession in conditions favourable to the nematodes. More recently root-knot rematodes have been recognized in the field in temperate regions where at least 11 species have been described; they are indicated in Table 1, but not all of them are known pathogens of cultivated crops. Of the 'tropical' species, three are often found in heated glasshouses in temperate climates where they can cause severe damage if control measures are not adequate.

Brief descriptions follow of species that may be encountered either in the field or under glass in temperate climates.

GLASSHOUSE SPECIES

Meloidogyne incognita

This species is the commonest and most widespread in tropical and sub-tropical regions and also the most variable morphologically. (Figs. 5.8, 5.10 and Plate VI).

Morphology—Female. Perineal pattern with more or less wavy striae, arch high or medium, often trapezoid; lateral fields usually inconspicuous, marked by forking and discontinuities in the striae; phasmids small and fairly close together (21 μm); vulva about as wide as inter-phasmidial distance (22 μm). The head has two, occasionally three, annules behind the labial disc and stylet 13–16 μm long with rounded knobs that may be extended laterally.

Chitwood (1949) recognised a variety within *M. incognita* (later designated subspecies) that he named *acrita,* distinguished by the smoother striae, flatter dorsal arch and more offset stylet knobs of the female. Triantaphyllou & Sasser (1960), using single egg-mass isolates, concluded from study of the perineal pattern that division into subspecies was unwarranted.

However, Esser, Perry & Taylor (1976) claim that the distinction holds good and they raise the subspecies *M. i. acrita* to specific rank as *M. acrita.* Their opinion is based partly on Terent'eva (1967), who found significant differences in the height of the male head between *M. i. incognita* and *M. i. acrita,* and partly on Chitwood (1949) who described the larva of *M. i. acrita* as having de Man ratio a = 22–28 and an unswollen rectum while that of *M. i. incognita* had a = 29–33 and a swollen rectum. Chitwood gave ranges without means and did not mention the number of larvae measured; Terent'eva measured 11 and eight male heads of *M. i. incognita* from two localities and six of *M. i. acrita* from a third locality. It is considered unwise to base specific determination of nematodes solely on morphometrics, at least without taking into account the known variability between populations of the same species from different localities and different host plants. The differences between the perineal patterns of these two taxa have been shown to be unreliable but the shape of the rectum in the larva may be a good character for separating them.

An aberrant form of pattern known as 'type E' (Dropkin, 1953) occurs in small numbers in some populations. It is distinguished by having a number of striae encircling the terminus and passing between the anus and vulva.

Male. Head forming a high truncate cone with a variable number of annules, from one to five on the lateral sectors (Fig. 5.10A–D). Stylet 23–33 μm with knobs somewhat drawn out laterally; spicules 29–40 μm along arc; lateral field with four incisures, outer bands areolated; tail terminus not striated; phasmids opposite cloaca or slightly in front.

Second-stage Larva. Length 337–403 (371) μm, one of the smaller species (see Table 1); tail 38–55 (46) μm long, an elongated cone with minutely rounded terminus; c = 6·9–10·6 (8·1) (Fig. 5.5). Head with two annules on subdorsal and subventral sectors, three on lateral sectors. Stylet 9·6–11·7 (10·5) μm long with prominent, rounded knobs. Lateral fields with four incisures, outer bands partly areolated. Rectum inflated.

Biology. *M. incognita* has been used extensively in investigations of the biology of the genus. The life-cycle takes 30–60 days according to temperature and suitability of the host. Optimum temperatures for hatching, root invasion and development vary according to the origin of populations, but survival at extreme temperatures varies less. Exposure for 15 days in water at 5° or 35°–40°C killed at least 80 per cent of the eggs of populations from both warm and cooler climates but once within the host the nematodes can continue to develop at temperatures unfavourable to free larvae and eggs (Dao D., 1970). Bergeson (1960) found that eggs and larvae could survive in soil at 10°C for more than a year but Sayre (1963) found that this species did not survive the winter in soil or in the roots of asparagus out-of-doors in south-western Ontario. It is unlikely that it would survive out-of-doors in northern Europe.

Distribution and Hosts. *M. incognita* has been recorded in all the warmer regions of the world where it affects most cultivated plants. In Britain it is a serious pest of glasshouse-grown cucumbers and tomatoes and also damages ornamental plants such as cyclamen, saintpaulia, begonia, balsam and occasionally chrysanthemum, but not ferns.

A full account of this species is given by Orton Williams (1973).

Meloidogyne javanica

A widespread species in warm countries and in glasshouses in northern regions, but less common than *M. incognita*. (Figs. 5.8, 5.10 & Plate VI).

Morphology—Female. Perineal pattern characterized by clearly marked lateral fields consisting of two parallel lines with a clear area between, running from the phasmid anteriorly; tail terminus indicated by a small, irregularly circular area; striae usually smooth or slightly wavy, the dorsal arch of medium height. Phasmids 27 μm apart, about equal to vulva width, which averaged 26·2 μm in six populations from Tanzania (Whitehead, 1968); this is somewhat wider than that of *M. incognita*. Head with one annule behind cap, stylet 14–18 (15) μm with rounded knobs.

Male. Male intersexes common, vermiform but with traces of a vulva. Head in lateral view showing a projecting first annule distinctly set off from the second annule which may be incompletely divided into two. In dorsal and ventral views head outline smooth without obvious annules (Fig. 5.10E–G). Stylet 20–23 (21·2) μm with rounded knobs; spicules 20·9–31·7 (26·7) μm along arc; lateral field with four lines, outer bands sometimes partly areolated; phasmids at cloacal level or slightly in front.

Second-stage Larva. 387–459 (417) μm long; tail 36–56 (49) μm long, an elongated cone with finely rounded tip; c = 7·3–11·1 (8·5). Head in lateral view with three annules, but only one seen in dorsal or ventral view. Stylet 9·4–11·4 (10·4) μm long with rounded knobs. Lateral fields with four lines. Rectum inflated (Fig. 5.5).

Biology. Although less frequently reported in British glasshouses than *M. incognita, M. javanica* can be equally damaging. In tropical and sub-tropical conditions it is frequently found mixed with *M. incognita*. The results of extensive surveys by Whitehead (1969) in East Africa, where both species are common, indicated that *M. incognita* is more successful in hotter soils and *M. javanica* in cooler, wetter ones. As with *M. incognita*, the optimum temperature for development varies in different populations. Daulton & Nusbaum (1961) demonstrated thermotypes in *M. javanica*. They studied three populations in a field experiment in North Carolina where, at a depth of 10 cm, the temperature ranged from 31°C in August to 0°C in January. A local population retained appreciable viability after 98 days (from January) whereas only a few individuals of a Rhodesian population survived as long and a population from Georgia (U.S.A.) was only slightly more resistant to low temperatures. In southern California where mean soil temperatures were never below 11°C, *M. javanica* was able to survive over winter (Thomason, 1962).

Distribution and Hosts. M. javanica has been reported from all warm areas of the world and has a very wide host range including tobacco, sugar-cane (the type host), fruit trees, banana and vegetables. Nearly 800 plant species were listed as hosts in Goodey's host list (Goodey *et al.*, 1965) and many more have been recorded since. In Britain it is found only in glasshouses.

For full description, see Orton Williams (1972a).

Meloidogyne arenaria

Originally described on peanut in Florida, *M. arenaria* is less frequently recorded than *M. incognita* or *M. javanica*. (Figs. 5.8, 5.10 & Plate VI).

Morphology—Female. Perineal pattern circular to broadly oval with the longer axis parallel with the vulva. Striae usually smooth, widely spaced, with low to medium dorsal arch and occasionally, a 'wing' on one side; lateral field indicated by slight irregularities of the striae. Vulva 17–32 (23·6) μm wide; phasmids 28–31 μm apart, at least as far apart as vulval width. Head with one post-labial annule; stylet length 14·4–15·8 (15·4) μm; knobs rounded, with backward sloping anterior faces.

Male. Head low, rounded, the labial disc and first annule almost amalgamated and separated by a deep constriction from two (occasionally one) basal annules (Fig. 5.10J, K). Stylet 20–24 μm long with rounded knobs; spicules 31–34 μm; lateral field with four lines, the outer bands fully areolated and the innermost one partly so; phasmids about one tail length in front of cloaca; tail striated nearly to terminus.

Second-stage Larva. Length greater than that of many species, 450–490 μm; tail about 50 μm long, of typical shape; c = 6·0–7·5. Stylet 10 μm with rounded knobs; head with wide, flat cap followed by two annules. Rectum inflated (Fig. 5.5).

The above characters are based on Chitwood's (1949) description of a population from peanut and variations can be expected in other populations. Detailed descriptions of this species are lacking, but Dropkin (1953) gives an analysis of measurements of perineal patterns from a stock population and from a population derived from a single juvenile. He found no differences between them in the parameters that he measured.

Biology. M. arenaria has been studied rather less than the other important species of root-knot nematodes. It reproduces by mitotic parthenogenesis, although

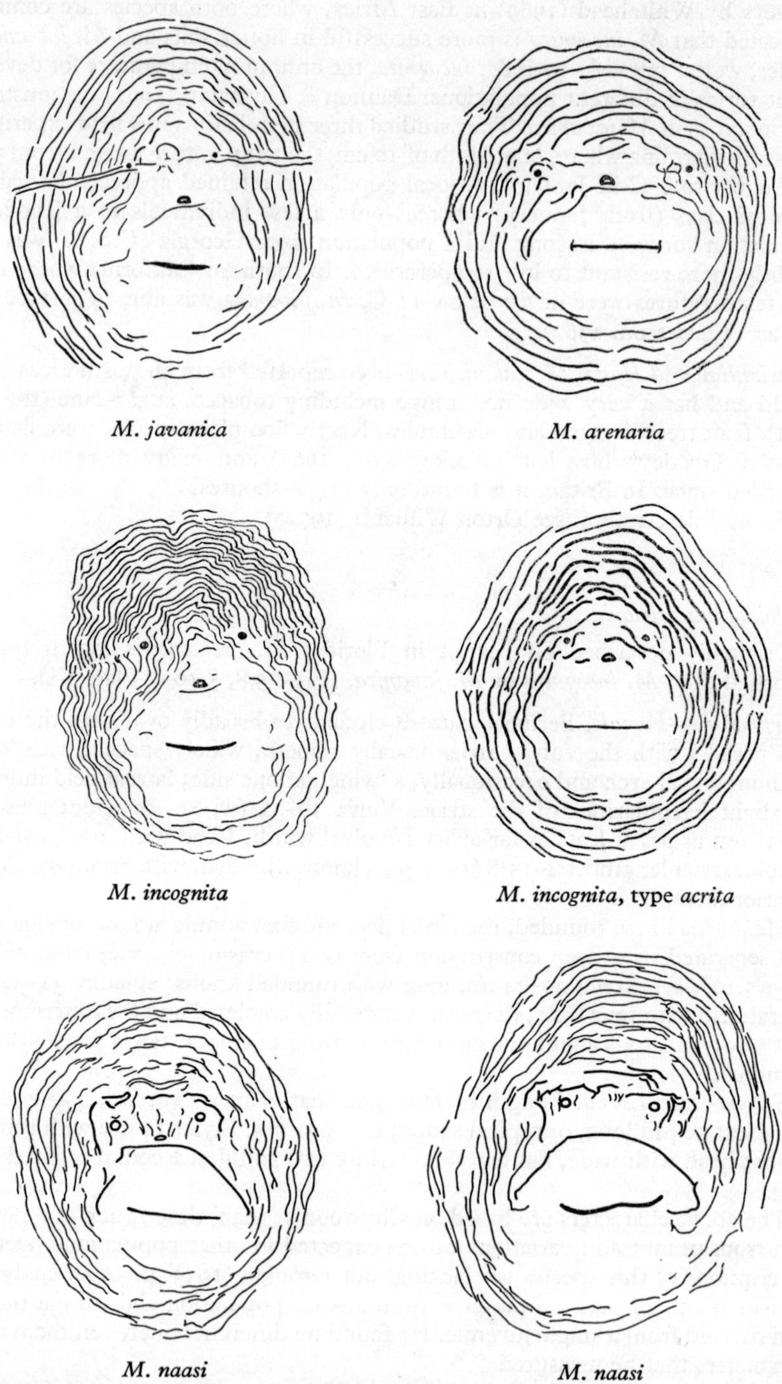

Fig. 5.8. Diagrams of perineal patterns of *Meloidogyne* spp. (*M. naasi* from Franklin, 1965).

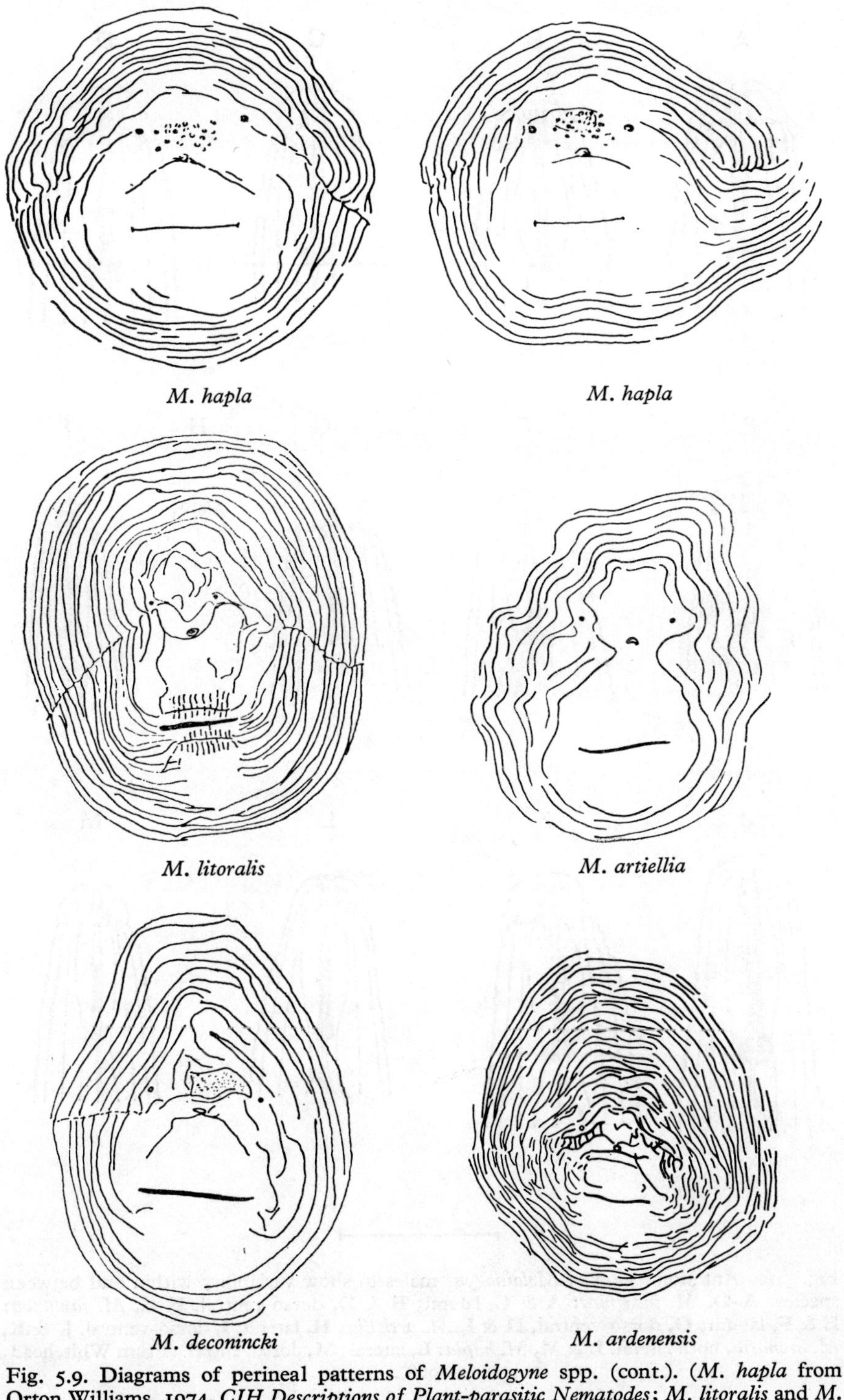

Fig. 5.9. Diagrams of perineal patterns of *Meloidogyne* spp. (cont.). (*M. hapla* from Orton Williams, 1974, *CIH Descriptions of Plant-parasitic Nematodes*; *M. litoralis* and *M. deconincki* from Elmiligy, 1968; *M. ardenensis* from Santos, 1968).

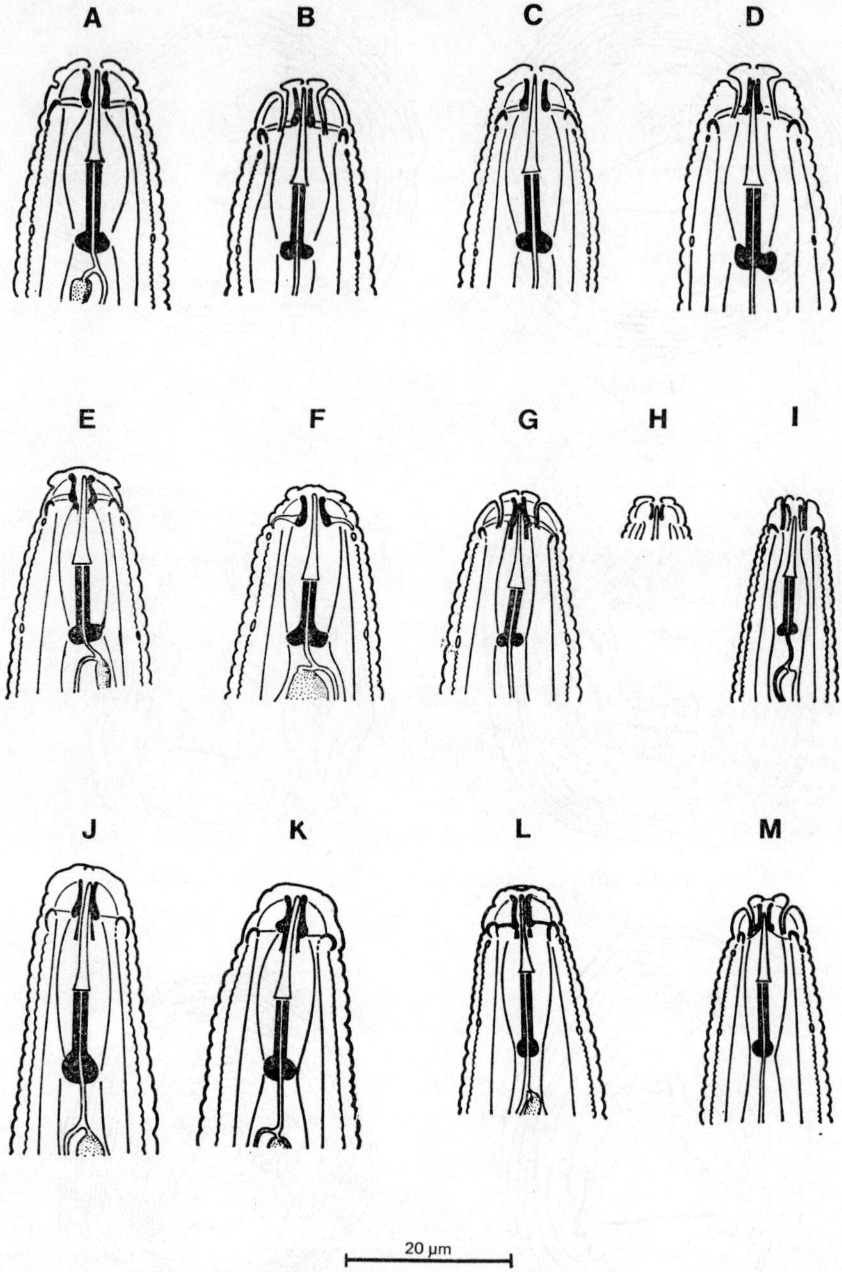

Fig. 5.10. Anterior region of *Meloidogyne* males to show variability within and between species: **A–D,** *M. incognita*: **A** & **C,** lateral; **B** & **D,** dorso-ventral. **E–G,** *M. javanica*: **E** & **F,** lateral; **G,** dorso-ventral. **H** & **I,** *M. artiellia*: **H,** lateral; **I,** dorso-ventral. **J.** & **K,** *M. arenaria*, both lateral. **L & M,** *M. hapla*: **L,** lateral; **M,** dorsoventral. (From Whitehead, 1968).

males are not uncommon. It can be differentiated from *M. incognita* by differences in enzyme patterns in the females (Triantaphyllou & Hussey, 1973).

At an average soil temperature of 20·4°C females on antirrhinum reached the egg-laying stage in 30 days and second-generation juveniles were found in the root tips 36 days later. In tomato, eggs were laid in 39 days. Under the same conditions *M. incognita* and *M. hapla* developed at about the same rate on these hosts (Tarjan, 1952). At temperatures of 20, 25 and 30°C *M. arenaria* reproduced well on tomato but above 32·6°C very few galls and egg-masses were produced, unlike *M. incognita* and *M. javanica* which continued reproduction at 35°C (Thomason & Lear, 1961).

As with the other more widespread species of root-knot nematodes, several pathotypes have been recognized in *M. arenaria.* Sasser (1972) found variations in pathogenicity among 10 populations from different parts of the world which he tested on a series of crop plants.

Distribution and Hosts. M. arenaria often occurs in mixed populations with *M. incognita* (the perineal patterns of the two species are sometimes liable to confusion) and *M. javanica.* It occurs in the field in warm temperate zones, e.g., in a reclaimed sandy area of Egypt and in the south of France. It is a serious pest of vegetable crops in southern France and Italy and appears to be endemic in the northern Mediterranean basin, being more widespread there than *M. incognita.* Damage to flower crops in the Côte d'Azure and the Italian Riviera over the past 50 years is attributed to *M. arenaria,* but it is unusual to find it in the Near East and in North Africa. It appears to be less pathogenic than *M. incognita*: on heavily attacked tomato plants it may cause yield losses of 30 to 40 per cent compared with 75 per cent on plants that appear equally galled with *M. incognita* (Ritter, 1972). Among other hosts recorded are *Aphelandra, Sansevieria, Maranta, Gerbera, Petunia,* carnation, *Primula,* violet, almond, peach, plum, olive, fig, grapevine, forage beet and lettuce.

See also Orton Williams (1975).

OUTDOOR SPECIES

Meloidogyne hapla

This species was originally described by Chitwood (1949) from potatoes on Long Island, New York. He noted that it had been studied there in 1936, and was probably present twenty years earlier. It is known as the northern root-knot nematode, the earlier records being from more northerly latitudes in America.

Morphology—Female. (Fig. 5.9 & Plate VI). In the spherical female the terminal region sometimes protrudes slightly. Perineal pattern with a low, rounded arch; tail area between phasmids and anus stippled; the lateral field marked by a single, somewhat broken line, or by the junction of striae of the dorsal arch with lateral extensions of the ventral striae, forming a 'wing' on one or both sides; dorsal striae generally smooth. Vulva short, about 20 μm. Stylet 10–13 μm long, with rounded, inconspicuous knobs; two annules behind head cap; neck short, 122 μm in 20 specimens, compared with 155 and 198 μm in *M. incognita* and *M. javanica* respectively (Whitehead, 1968).

Male. Head truncate cone-shaped to rounded, not offset, with one annule behind the asymmetrical head cap; stylet 17·3–22·7 (20·0) μm with small rounded knobs; spicules 21·6–28·1 (25·7) μm along arc; lateral fields with four incisures, outer bands areolated, central band occasionally crossed by annulation; phasmids at level of cloaca. (Fig. 5.10L, M).

Second-stage Larva. Length 312–355 (337) μm, one of the five shortest species recorded (Table 1); head not offset, truncate cone-shaped, the cap followed by three annules on sublateral and one on lateral sectors; stylet 7·9–10·9 (9·7) μm with rounded knobs; tail 33–48 (43) μm; c = 7·3–10·2 (7·9) μm; terminus variable, sometimes bifid (Fig. 5.5). Lateral fields with four incisures, irregularly crossed by annulations. Rectum not inflated.

Biology. M. hapla can thrive in a wide range of temperatures. Dao D. (1970) compared a population from potato from the mountains of Venezuela with one from *Berberis* from the Netherlands; he found that activity started at 10°C and both populations reproduced readily on tomato at temperatures from 15 to 35°C, higher temperatures being more favourable. This species is more resistant to cold than *M. incognita*, eggs and larvae surviving well in water at 5°C and some of the Dutch population surviving in soil for 10 days at −10°C; the Venezuelan population remained viable for only three days at this temperature but for 10 days at −5°C. In the field the nematodes usually overwinter as egg masses. Brzeski & Hendricks (1971) found that maximum hatching from egg masses in the laboratory was at 22–27°C and none occurred at 5° or 37°C. They also found stimulation by cucumber root diffusates and inhibition of hatch by leachates of some soils. The effects of soil temperature, moisture and oxygen on hatching, invasion and development of *M. hapla* on lettuce were investigated by Wong & Mai (1973a, 1973b, 1973c). They concluded that in cool, wet soil hatching and invasion are hindered and the host crop may escape early damage. Once invasion has taken place, if temperatures are high, the nematodes develop rapidly and in a regime of 27°C by night and 32°C by day eggs can be produced in 20 days. In onion grown at 26°C egg masses were found from 28 to 35 days after inoculation with larvae (Smith & Mai, 1965).

Olthof & Potter (1972) showed that, in microplots with 18,000 larvae per kg of soil, cabbage yields were reduced by 9 per cent, cauliflower by 24 per cent, lettuce and potato by 46 per cent and onion by 64 per cent. In outdoor plots in Germany infestation spread rapidly in the second season from an infested plot into uninfested soil growing lettuce, and more slowly in plots growing carrots (Stein, 1965). Sugar beet losses of 20 per cent have been recorded in Yugoslavia (Grujičić & Paunović, 1971).

The existence of host-range differences among populations of *M. hapla* has been demonstrated by Sasser (1972) who found that two of eight populations did not infest strawberry and one of 10 did not infest tobacco. Tests of 14 populations from northwestern states of U.S.A. on 12 crop plants enabled Ogbuji & Jensen (1972) to distinguish five pathotypes. This emphasizes the difficulty in devising crop rotations to avoid damage by this nematode.

On white clover in New Zealand phosphorus utilization was reduced by *M. hapla* infestations associated with *Heterodera trifolii* and growth was impaired (Widdowson *et al.,* 1973; Healy *et al.,* 1973). A search for legume varieties resistant to the nematodes showed some tolerance in lines of subterranean clover and lucerne (Yeates *et al.,* 1973).

Galls due to *M. hapla* are typically small and associated with abnormal numbers of lateral roots often giving a matted appearance to the host's root system. When tubers such as potatoes are invaded, however, there may be little outward sign of infestation. The distribution of such infested planting material has probably contributed to the spread of the nematode.

Distribution and Hosts. M. hapla is occasionally found in glasshouses on crops such as lettuce, tomato, chrysanthemum and rose, but more frequently occurs out-of-doors. It occurs in U.K., France, Denmark, Norway, Sweden, Yugoslavia, in U.S.A., Canada and U.S.S.R., at high altitudes in South America, South, Central and East Africa, in Pakistan, Australia and New Zealand. It has appeared on tomato varieties grown for their resistance to *M. incognita* in southern France. In England what was probably this species was recorded on parsnip and carrot by Triffitt in 1931 from a field in Wiltshire, and the nematode recorded by Cornu (1879) on *Onobrychis viciifolia* in the Loire valley in France (see p. 98) was probably this species (Ritter, 1972). The recorded hosts are numerous including root crops such as potato, carrot, parsnip and turnip; vegetables such as peas, beans, brassicas, celery, spinach, parsley, lettuce and onion; forage crops such as clovers, sainfoin and lucerne; fruits, e.g., strawberry and currants, and very many ornamental plants, shrubs and common weeds. Twenty grasses and cereals were found to be resistant in Australia (McLeod, 1971) but cultivars of, for example, oats may be slightly susceptible (Gillard, 1961). Many cucurbits are also resistant.

For a full account of this species, see Orton Williams (1974).

Meloidogyne naasi

This, the second most important root-knot nematode occurring out-of-doors in temperate climates, was first described as a distinct species on barley (Franklin, 1965). (Fig. 5.8, Plate VII).

Morphology—Female. The relatively small spherical females with prominent necks are often found in the swollen roots in groups surrounded by confluent gelatinous egg masses. The characteristic perineal pattern has a fairly low dorsal arch and is roughly circular; it is formed from broken striae that are usually well-marked round the prominent phasmids. A distinct stria runs across the pattern from side to side over the anus. Vulval slit 17–25 (22) μm wide; phasmids 12–25 (18) μm apart; lateral fields indistinguishable. Head with two annules; stylet slender 11–15 (13) μm long with backwardly sloping knobs. Excretory pore on 7th to 11th annule behind the head.

Male. Of variable length (860–1316, mean 1148 μm) with head 9 μm wide by 4 μm high and having three post-labial annules (Fig. 5.3). Lateral field with four incisures with the outer bands irregularly areolated. In heat-relaxed specimens the posterior tenth of the body is spirally twisted through 180°; tail hemispherical, shorter than wide with phasmids at level of cloaca. Stylet 16–19 (18) μm with rounded knobs. Median oesophageal bulb fusiform, with four or five small vesicle-like structures arranged irregularly around the lumen anterior to the valve plates. One testis; spicules typical for the genus, each terminating in a simple point, 25–30 (28) μm long; gubernaculum 4–8 (6) μm.

Second-stage Larva. Length 418–465 (435) μm, distinguished by the long, attenuated tail of 52–78 (70) μm with $c = 6{\cdot}2$, and terminus sometimes irregular (Fig. 5.5). Lateral field with 4 incisures, not areolated; hemizonid immediately in front of excretory pore and four or five annules behind nerve ring. Stylet 13–15 (14) μm long, with backwardly sloping knobs. Median oesophageal bulb fusiform with vesicle-like structures as in the male. Oesophageal gland long, extending to about 40 per cent of body length from anterior end.

Biology. The life-cycle from egg to egg of a population from Illinois, U.S.A., cultured on wheat at 26°C by day and 20°C by night took six to seven weeks

(Siddiqui & Taylor, 1970a) while in the field in England on spring barley it took 12 to 14 weeks (Franklin *et al.*, 1971). In field soil in England hatching reached a maximum in late April to early May when soil temperature exceeded 10°C; in the laboratory eggs were induced to hatch during winter by chilling infested soil for a week at 0 or 10°C before incubating at 20°C (Franklin *et al.*, 1971). On spring barley only one generation a year is possible, but on ryegrass there may be a partial second generation. Reproduction is usually by meiotic parthenogenesis although males are commonly found in the egg sacs.

The galls formed on barley in response to invasion by *M. naasi* are spindle-shaped or club-shaped and sometimes cause the roots to form hooks or spirals. On ryegrass rather inconspicuous cylindrical galls are common.

Distribution and Hosts. M. naasi occurs in northwestern Europe, and in U.S.A. In England and Wales the principal hosts are barley, ryegrass and, more rarely, sugar beet which can be damaged in the seedling stage. Ryegrass appears unaffected but barley can be severely galled in early growth, especially in poorly drained soil. Infestation has also been recorded on onion, lucerne, wheat, various grasses and weeds.

A full description of this species is given by Franklin (1973).

Meloidogyne artiellia

This species was first recorded in Norfolk, England, on oats. It is probably of little economic importance. (Fig. 5.9, Plate VII).

Morphology—Female. Mature females are small, about 0·7 × 0·4 mm, gradually tapering to the neck, and incompletely embedded in the swollen roots but covered by the gelatinous egg sac. Perineal pattern typically an elongated figure-of-eight shape bordered on the inside by somewhat thickened ridges. Lateral lines are not conspicuous; vulva 15–22 (17) μm wide and phasmids usually slightly closer together than this. The distance between vulva and anus is about three times the distance between anus and tail terminus.

Male. Males resemble those of most other species in the genus; there appear to be one small and one broad post-labial head annule (Fig. 5.10H, I). The lateral field has four main incisures in the hind part and a fifth central one in the mid-body region, with the fields partly areolated.

Second-stage Larva. Recognizable by its small size: in the type population the length is 334–370 (354) μm with a = 22–25·5 (23·5), and the short bluntly conical tail is 24·5 μm long (c = 13–16, mean 14·3) (Table 1, Fig. 5·5).

For full morphological descriptions see Franklin (1961) and Whitehead (1968).

Biology. Little is known of the biology of *M. artiellia*; observations in Spain suggest that, in that country at least, it needs a cool period to complete its life cycle and that it increases rapidly towards the end of the growth of the host plant, which is usually an annual. (Tobar Jimenez, 1973).

Distribution and Hosts. M. artiellia has been recorded on wheat in Greece and has been found in France and Spain. In England it has been observed to feed on brassicas, broad bean, clover, lucerne and black medick, apparently doing little harm. It is widespread in some areas in Spain where it reproduces well on wheat, barley, *Cicer arietinum* and *Vicia monanthos*, poorly on oats, and not at all on *Lathyrus cicera* (Alcala *et al.*, 1970; Tobar Jimenez, 1973). Other plants reported to be resistant include carrot, sugar beet, tomato, onion, strawberry, lettuce, potato, sainfoin, dwarf bean, rye and grasses.

Meloidogyne ardenensis

Described by Santos in 1968 from *Vinca minor* in an uncultivated area in Leicestershire, England; has been found also on *Ligustrum vulgare, Sambucus* sp., *Fraxinus excelsior, Hepatica angulosa, Astilbe* sp., *Saxifraga cortusifolia* and, in Scotland, on *Lonicera nitida* in a garden. The infestation on ash (*Fraxinus excelsior*) was found on seedlings in the Fen district of Cambridgeshire. It is not known to cause damage.

The perineal pattern is characterized by the well-marked, smooth, relatively unbroken striae forming a high, flattened arch; the striae at the tail terminus are distinctly whorled and the lateral fields visible for a short distance near the terminus (Fig. 5.9, Plate VII). The second-stage larva averages 417 μm long and has a relatively short tail, 32–45 (39) μm long with c = 8·6–12·3 (10·6); the hemizonid is unusual in lying behind the excretory pore. A detailed morphological description is given by Santos (1968).

Meloidogyne deconincki

This species was found on *Fraxinus excelsior, Rosa* sp. and *Solanum nigrum* growing in the State University garden at Gent, Belgium; in experiments mature females developed in carrot, peanut, pepper and tomato but not in barley, beet, maize, ryegrasses, oats, wheat or cotton.

M. deconincki differs from *M. ardenensis* in having, amongst other features, distinct punctations in the tail area of the perineal pattern which has a high, flattish arch, small phasmids 25–31 (27·5) μm apart and vulva 20–28 (25) μm wide. Lateral fields sometimes marked by breaks in the striae (Fig. 5.9, Plate VII). In the second-stage larva the hemizonid is in front of the excretory pore and the tail is longer and more attenuated than in *M. ardenensis*; body length 340–400 (370) μm, with tail about 47 μm and c = 6·8–9·8 (7·8). For full description see Elmiligy (1968); other records are lacking.

Meloidogyne litoralis

Found on *Ligustrum* sp. on sand dunes near Ambleteuse on the north coast of France. The perineal pattern is composed of smooth striae forming a high, rounded arch with disordered striae at the terminus, lateral field marked by an irregular or zig-zag line, phasmids 13–20 (17) μm apart and vulva 18–29 (22) μm wide. Prominent lines run inwards towards the ends of the vulva (Fig. 5.9, Plate VII).

Second-stage larva 330–450 (400) μm long with tail about 38 μm and c = 9·7–11·8 (10·6). No other information available.

OTHER SPECIES

Other species of *Meloidogyne* from cool temperate climates but not recorded in Europe are:

M. ovalis on *Acer saccharum* and other *Acer* spp., *Betula* spp., *Fraxinus* spp., *Ulmus* spp. and *Celtis occidentalis* in Wisconsin, U.S.A. (See Riffle, 1963; Whitehead, 1968).

M. ottersoni on *Phalaris arundinacea* in Wisconsin, U.S.A., described by Thorne (1969). The genus in which it was originally placed, *Hypsoperine,* was later synonymized with *Meloidogyne*. This species has also been recorded on barley.

M. mali in northern Japan on *Malus prunifolia*, other *Malus* spp., *Prunus* × *yedoensis*, *Rosa* hybrid, *Vitis vinifera*, *Morus bombycis*, *Castanea crenata*, *Acer palmatum* and *Trifolium repens* (Itoh *et al.*, 1969).

M. microtyla on *Festuca rubra*, *Dactylis glomerata*, oats, barley, wheat, rye, timothy and *Trifolium repens* in southwestern Ontario, Canada (Mulvey *et al.*, 1975).

A species of *Meloidogyne* has been found on marram grass (*Ammophila arenaria*) at several places round the coast of Britain but has not yet been formally described and named. It was briefly described by Triffitt (1929) who thought that it was *Heterodera schachtii*, but her drawings and description strongly suggest that her specimens were *Meloidogyne*.

CHAPTER 6

Cyst Nematodes: Morphology and Identification of *Heterodera, Globodera* and *Punctodera*

J. J. HESLING

Glasshouse Crops Research Institute, Littlehampton

The nematodes described in this and the next Chapter are characterized by the transformation of the mature female into a tough cyst, usually filled with embryonated eggs. The resistant cyst with its contents is often the life-stage most readily available for species diagnosis, and so it is inevitable that much cyst-nematode classification and identification have been based on cyst characters. Until recently all cyst nematodes (except *Sarisodera,* a genus excluded from this Chapter) were in a single genus, *Heterodera,* and the increasing number of species led to their informal classification into groups, each named after a 'typical species', e.g., *schachtii* group, *rostochiensis* group, etc. This grouping caused some confusion but this was partly resolved when *Heterodera* was formally reorganized by Mulvey & Stone (1976) into three genera, viz. *Heterodera, Globodera* and *Punctodera. Heterodera,* however, appears still to contain more or less distinct species-groups.

The recognition of the new genera as well as of many new species of Heteroderids has made possible the first speculations, based on substantial evidence, about the possible phylogeny of the Heteroderidae (Wouts & Sher, 1971; Stone, 1975a). Fig. 6.1, based on Wouts, 1973a, 1973b, 1974; Luc, Germani & Netscher, 1973; and Stone, 1975a, shows a suggested phylogeny which indicates the probable relationship between cyst-forming nematodes and their near relatives. A statement (based on Mulvey & Stone, 1976) of the systematic position of cyst nematodes is given below, followed by a detailed diagnosis of *Heterodera* and brief, comparative, diagnoses of *Globodera* and *Punctodera.*

Systematic Position

Order	Tylenchida (Oerley, 1880) Thorne, 1949
Suborder	Tylenchina (Oerley, 1880) Geraert, 1966
Superfamily	Heteroderoidea Filipjev & Schuurmans Stekhoven, 1941 (Golden, 1971)
Family	Heteroderidae Filipjev & Schuurmans Stekhoven, 1941 (Skarbilovich, 1947) amend. Wouts, 1973
Subfamily	Heteroderinae Filipjev & Schuurmans Stekhoven, 1941 amend. Skarbilovich, 1947 & Wouts, 1973
Genera	*Heterodera* Schmidt, 1871 amend. Mulvey & Stone, 1976
	Globodera Skarbilovich, 1959 (Mulvey & Stone, 1976)
	Punctodera Mulvey & Stone, 1976
	Sarisodera Wouts & Sher, 1971 amend. Luc, Germani & Netscher, 1973, Mulvey & Stone, 1976

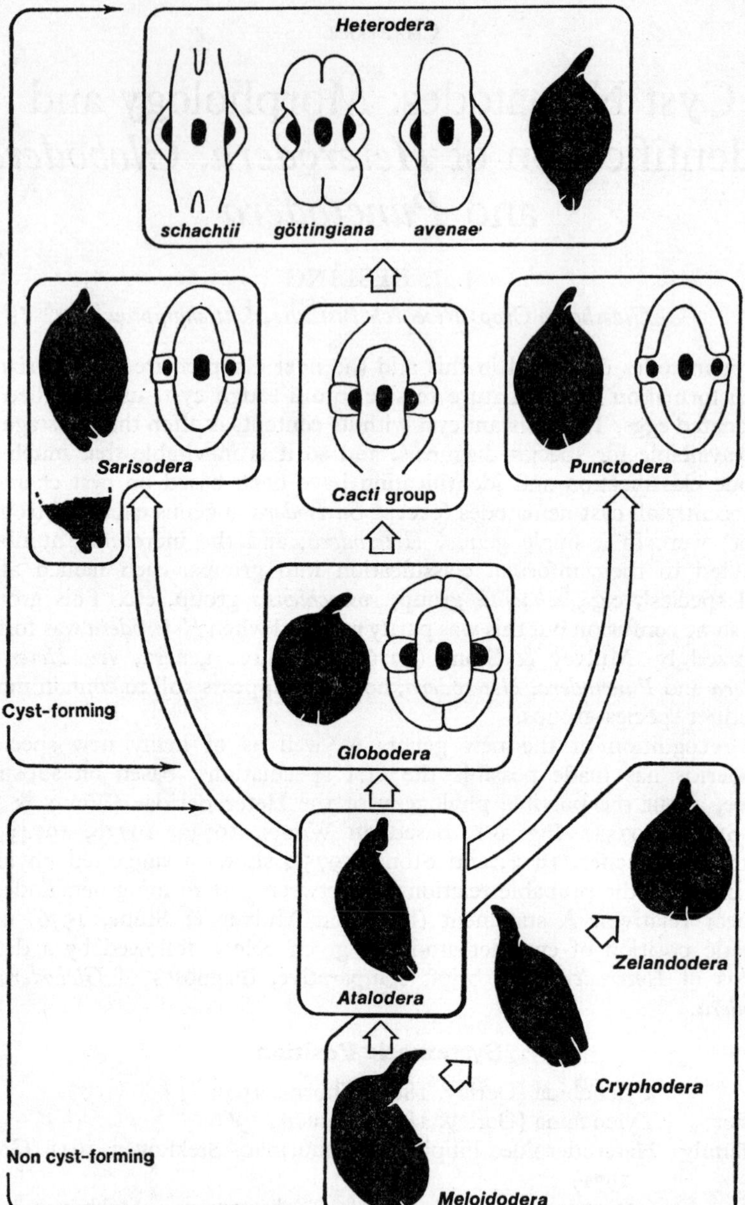

Fig. 6.1. Diagram of a suggested phylogeny within the family Heteroderidae. The phylogeny is based on the form of the female and the position of anal and vulval apertures (black silhouettes), and on the form of the head of the second-stage larva viewed *en face* by the scanning electron microscope.

Diagnoses of the Genera

Adapted from Mulvey & Stone (1976)

TYPE GENUS *Heterodera*

Mature female and cyst—body ovoid to globose with protruding neck and more or less protruding terminal (vulval) cone, i.e., lemon-shaped. Paired ovaries. Amphimictic or parthenogenetic. *Cyst*—brown to black; cuticle with surface ridges forming lace-like or zig-zag pattern; underbridge, vaginal remnant, and bullae, present or absent. Cyst ambifenestrate, bifenestrate, or circumfenestrate; no anal fenestra. Vulval slit terminal on cone, length 10–60 μm; vulval papillae absent. Anus in 'dorsal' surface of vulval cone. *Second-stage larva*—vermiform with oesophageal glands filling body width; stylet < 30 μm long, fairly stout with well formed knobs. Tail conical, taper abrupt (short cone) or extended (fine cone). Oral disc and lips sub-circular to elongate, boundaries unclear. *Male*—body vermiform with axial twist. Testis single. Labial disc of low profile. Stylet fairly stout with well-formed knobs. Tail length less than half body width, bursa absent. Spicule tips single or bifid.

GENUS *Globodera*

Diagnosis as for *Heterodera* but with the following differentiating features: *Mature female and cyst*—body spheroidal with protruding neck, terminal protuberance absent. Amphimictic. *Cyst*—brown; cuticle with surface ridges forming lace-like pattern which may be characteristically different on perineum. Underbridge, vaginal remnant, and bullae rarely present. Cyst circumfenestrate. Vulval slit within terminal fenestra, length < 15 μm, usually lost on cysts from soil. Vulval papillae present; anus sub-terminal. *Second-stage larva*—as *Heterodera*, but oral disc sub-circular with distinct lip sectors. *Male*—as *Heterodera*, but spicule tip single.

GENUS *Punctodera*

Diagnosis as for *Heterodera* but with the following differentiating features: *Mature female and cyst*—body globose, or ovoid tapering to protruding neck, terminal protuberance absent, i.e., globe or pear-shaped. Believed amphimictic. *Cyst*—light brown; cuticle with surface ridges forming zig-zag or lace-like pattern. Underbridge and vaginal remnant absent; bullae present or absent. Cyst circumfenestrate at anus and vulva; fenestrae equal in size; anus offset toward ventral margin of anal fenestra. Vulval papillae absent. *Second-stage larva*—as *Globodera* but oral disc elongated dorso-ventrally. *Male*—as *Globodera*.

Gross Morphology of Females and Cysts

In all three genera considered in this Chapter the female becomes a swollen sac of eggs, and her cuticle hardens to form a tough protective cyst around them. As it matures, the cyst changes in colour from white to various shades of brown; in some species the cyst is covered with a crusty white substance of variable thickness—the subcrystalline layer (p. 132). The colour-changes in the maturing cyst are due to the action of polyphenol oxidase on polyphenols in the cyst wall which, although tough and often opaque, remains permeable to chemical substances and to dissolved oxygen (Ellenby, 1946a, 1946b, 1955). While attached to the root, the newly-formed cyst is normally full of eggs which are tightly packed in a mucoid mass. If the entire content of a new cyst is released into water it swells immediately, suggesting that there is considerable turgor pressure within (Plate VIII, top); this may in part explain its saccate form and the extrusion of the eggs of some species into a mucoid 'egg-sac'.

When the cyst drops off the root into the soil it is dead (apart from its egg content); within a few weeks the mucoid material surrounding the eggs disappears, probably as a result of microbial activity, and the subcrystalline layer is lost. Such a cyst may lie in soil for many years; in course of time it may be damaged by predators, implements and physical changes in the soil, and it may be eroded by action of soil microbes and chemicals. The eggs in it may hatch, the emigrating larvae probably damaging some of the more delicate structures in the cyst as they leave (Thorne, 1961). Thus, cysts extracted from soil vary considerably in appearance and content, and the identification of an old, battered and empty cyst may be difficult or impossible because it is sometimes necessary to examine in detail several characters of the cyst, eggs and larvae. These characters are listed and discussed below.

CYST SHAPE

There are two basic shapes of cyst: round or ovoid, and lemon-shaped (Fig. 6.2).

Fig. 6.2. Typical shapes of cysts: left, *Globodera*; centre, *Punctodera punctata*; right, *Heterodera*.

Round or ovoid cysts. (Genera *Globodera* and *Punctodera*): These differ in their development from lemon-shaped (= *Heterodera*) cysts. Wilson (1971) compared the growth patterns of *G. rostochiensis* (round) and *H. schachtii* (lemon-shaped), and showed that the difference in their ultimate shapes was due more to relative lack of growth in length of the round cyst than to increase in breadth. She also showed that round cysts whose growth had been arrested in some way tended to depart from the spherical shape, while large well-developed round cysts were more truly spherical. This finding emphasizes the need for caution when describing the typical shape of a cyst.

The largest *Globodera* cysts are almost spherical except for an anterior protuberance which is the remains of the neck of the adult nematode. Smaller cysts, whose development has perhaps been curtailed, may be ovoid so that they resemble the cysts of *Punctodera punctata* (see below). About 10 species of *Globodera* and three of *Punctodera* have been described (page 145).

The cysts of *P. punctata* are typically elongate-ovoid, about $1\frac{1}{2}$ times as long (excluding neck) as broad, but sometimes almost spherical. The genus *Punctodera* contains two other described species, *P. chalcoensis* and *P. matadorensis,* both of which have near-spherical cysts, and there may be others.

Lemon-shaped cysts. (Genus *Heterodera*): Species in this genus form cysts with an anterior protuberance—the neck, and a posterior protuberance—the vulval cone.

While retaining their basic lemon shape the cysts show wide variation of form and symmetry. Descriptions of the form of lemon-shaped cysts, particularly of the degree of prominence of the vulval cone (see below), are usually subjective, and detailed studies by several workers have confirmed that no lemon-shaped cyst can be positively identified by its shape alone (Jones, 1952; Taylor, 1957; Goffart, 1960a & b; Oostenbrink, 1960b). However, it is recognized that certain species tend to form a cyst of 'typical' shape, e.g., *H. galeopsidis* tends to form a symmetrical, rounded-lemon-shaped cyst with an obtuse vulval cone. 'Typical' shapes, and differences of shape, are more difficult to describe than to observe; their recognition comes from long experience of the species. Attempts to demonstrate shape differences by cyst measurements have met with limited success; e.g., Jones (1950) showed that there are differences between the regression lines of breadth on length in *H. galeopsidis* and *H. schachtii,* and Sandner (1962) showed that there are differences in the degree of taper of the posterior end of cysts of *H. galeopsidis* and *H. trifolii.* Statistical analysis of several measurements from very many cysts of each of the lemon-shaped species might reveal other specific differences between certain cyst features, but such laborious methods are of no use in routine advisory work; there are considerable practical difficulties in measuring cysts (Jones, 1952) and in most species at least the smaller cysts show aberrant forms (Thorne, 1961) (Fig. 6.3). Difference in cyst shape between some species is very slight and, in such cases, measurements may fail to show any difference (Hirschmann, 1956).

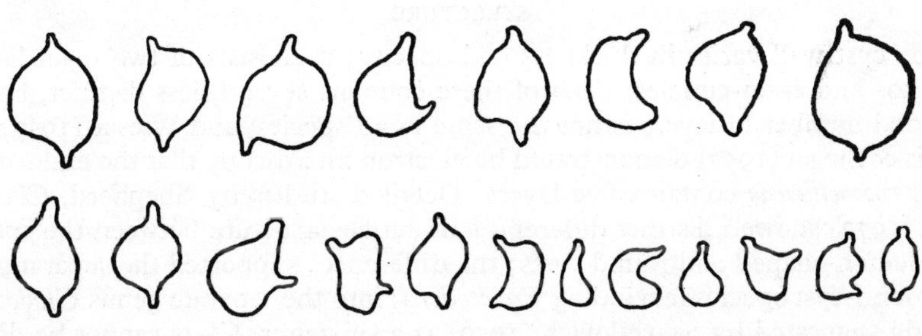

Fig. 6.3. Outlines of cysts of *Heterodera schachtii* showing a wide range of size and shape, including aberrant forms. (From *Principles of Nematology* by Gerald Thorne. Copyright 1961 McGraw-Hill Book Co. and used with their permission).

SIZE

Most of the foregoing criticisms of the use of shape in the identification of cysts apply also to the use of cyst size. Cysts of some species tend to be larger than those of others, and this difference can be used to eliminate the smaller species if the cyst in question is quite large. Thus, cysts of *H. carotae, H.*

cruciferae and *H. goettingiana* rarely reach the size of the largest cysts of *H. avenae* or *H. schachtii*. It must be remembered that those species which have the largest cysts always produce many smaller cysts which cannot, of course, be identified by their size.

COLOUR

There are two kinds of colour difference between species of cyst nematodes. The first is manifest in the colour sequence that occurs as the female 'matures', a sequence known as chromogenesis and used as a differentiating character (Guile, 1966, 1967, 1970). Thus, some species change from white through yellow to brown, while others have no distinct yellow phase. The second kind is the difference in colour of cysts extracted from soil.

The yellow phase of females in some species is due to a colour change in the body contents and not in the cuticle; the yellow phase of *G. rostochiensis* is an example of this (Franklin, 1951). Later colour changes, i.e., from yellow to brown, take place in the cuticular layers which form the cyst wall, and the deepening of the brown colour probably continues for some time after death of the female. Within a species the larger cysts are generally darker, but the eye is easily deceived when observing colour of cysts because it fails to ignore the cysts' translucence, reflectivity and surface texture.

Normal 'new' cysts of some species have a 'typical colour' (although this is probably always linked subconsciously with the optical properties of the cyst wall), but there is great variability and it is only after much experience that cyst colour can be used in identification (Hijner, Oostenbrink & den Ouden, 1953).

Features of the Cyst Wall

STRUCTURE

The cyst wall varies in thickness and opacity; it consists of two main layers, the exo- and endo-cuticles. Each of these contains several, less distinct, layers; the total number of layers is not the same in all species, and Wieser (1953) and Ferris & Siegel (1957) demonstrated by electron microscopy that the endocuticle of *G. rostochiensis* contains five layers. Detailed studies by Shepherd, Clark & Dart (1972) showed distinct differences in cuticle structure between the species with lemon-shaped and round cysts; the differences supported the separation of the round-cyst species (excluding *Punctodera*) into the separate genus *Globodera*, already suggested by Skarbilovich (1959) as a subgenus. Cysts cannot be differentiated routinely by the number of cuticular layers because few species have been studied and the techniques needed are too elaborate. However, the thinness and translucence of a cyst wall may be of some confirmatory use, e.g., in cysts of *P. punctata* the wall is often so thin that the eggs are visible within the cyst (Franklin, 1951).

TEXTURE

'Texture' refers to the appearance of the cyst surface when viewed under low magnification (×20 – ×40) in incident light; it varies from dull-matt (often seen in *H. humuli*, *H. carotae* and on small cysts of other species) to the shiny smoother surface of new cysts of some species, e.g., *H. goettingiana*. Old specimens invariably lose much of their 'newness' and shine; cyst texture is variable within a species, and so is of limited use in cyst identification.

PATTERN

Jones (1952) noted that cyst-wall patterns were of two types—a reticulate, rugose or zig-zag pattern (on the cyst surface), and a punctate pattern (immediately sub-surface); the surface pattern will be considered first.

Surface pattern. The derivation of the surface pattern is not clear. Franklin (1939) and Wieser (1953) suggested that it is derived from cuticular annulations of the larva. Gunther and Kämpfe (1967) showed that a ripple pattern is already visible on the mid-body of fourth-stage larvae of *H. schachtii*, and that there are fine, closely-spaced longitudinal striations on the cuticle of the third-stage larva which become modified into a zig-zag pattern at the rounded larval tail. Young cysts (adult females) of *H. cruciferae* (Mackintosh, 1960) and *H. schachtii* (Gunther & Kämpfe, 1967) show a rugose exocuticle when sectioned, but in some other species such sections apparently show a smooth surface (Wieser, 1953). At low magnifications the external cuticular pattern contributes much to cyst texture (see above); at high magnifications the pattern appears as ridges = hills, and furrows = valleys, grooves, sulci. In some species this impression is emphasized by the optical properties of the cyst wall, which are readily demonstrable by phase-contrast microscopy. Careful focussing does not always reveal the limits of surface contours, for the corrugations are sometimes more shallow than the axial resolution of the best light microscope. The external cuticular pattern is an ideal subject for scanning electron microscopy (S.E.M.), which, while rarely failing to show ridges and furrows, confirms their frequent shallowness.

The external cuticular pattern is of use in identification, the perineal* pattern being particularly useful for the identification of some *Globodera* species. Comparisons of whole-cyst patterns are difficult to use because of the great variability within and between cysts of each species. In one of the earliest attempts at cyst grouping, Taylor (1957) used the type of external cuticular pattern to place into groups the 16 species that were then known; today there are more than 60 species, including *species inquirendae*, and the variation now encountered has made the form of the external cuticular pattern relatively less helpful for classification.

On *Globodera* cysts the perineum is a limited area which thereby permits more intensive studies and comparisons of its cuticular pattern than are possible on the whole cyst. The number, size and form of the ridges of the perineal pattern have proved useful for species identification (Miller & Gray, 1968, 1972; Green, 1971; Stone, 1973; Hesling & Ellis, 1974) but there is considerable within-species variation, the full range of which is not known. Some published descriptions of the external cuticular pattern are based on observations of living adult females rather than cysts, which are dead and may be superficially eroded. The difference in the appearance of the pattern on these two life stages is slight and, at present, usually unimportant.

---

*perineum = region between anus and vulva.

In S.E.M. studies of the external cuticular pattern and of other features of *Globodera* cysts, Green (1971) focussed attention on the cyst's 'terminal area', which lies between the vulva and the tail. Its size, and the number and form of the ridges on it were used in a key to separate *Globodera* species. More recent work shows that the key is imperfect, but the potential of these characters of the terminal area for species differentiation remains.

Punctations. In the subsurface layers of cysts of many species there are punctations arranged irregularly or in near-parallel rows. Punctations are very variable, even on one cyst; they may be extremely fine and faint, or coarse and heavily pigmented. They have been interpreted in many ways (Hesling, 1965) for, according to magnification and focus, and under different systems of light microscopy, they can appear as dark or light spots, as pits, or as raised tubercles. However, it seems that they are minute radial rods of optically dense and refractive material that extend only partially through the thickness of the cyst cuticle; they are usually viewed end-on when a piece of cyst-wall is examined *en face*. Because of their great variability both the punctate and the external cuticular pattern should be used with great caution in cyst identification.

THORNLETS

Thornlets (Russian ШИПИКИ), usually less than 1 μm long, were described by Kir'yanova (1959) on the external surface of cysts then identified as *Heterodera* (now *Globodera*) *tabacum* and later (Kir'yanova, 1963) referred to as *H. pseudorostochiensis*. They seem not to have been observed on any other species, nor on *G. tabacum* from elsewhere. Ellis & Hesling (1974) observed minute thorn-like projections on a few cysts of *G. rostochiensis*; on closer observation the projections were found to be remnants of fungal micronemes passing through the cyst-wall from hyphae within the cyst.

THE SUBCRYSTALLINE LAYER (S.C.L.) (Plate VIII, lower)

This is found only on new cysts, and varies from a short-lived white bloom (*H. cacti*, see Southey, 1957) to a thick chalky-looking crazed deposit (*H. avenae*). The S.C.L. is generally considered to be protective, and there have been several suggestions to explain its nature and formation; some of these, plus a very detailed account of its composition, were given by Brown, Callow, Green, Jones, Rayner, Shepherd & Williams (1971). The S.C.L. usually consists of two layers; the outer layer fractures into pyramidal blocks presenting an irregular polygonal appearance. One component of the S.C.L. is a fatty acid which seems to be produced by an unidentified, possibly symbiotic, fungus which utilizes waste products exuded through the cuticle of the adult nematode.

Kir'yanova (1969, 1970) uses the form of the S.C.L. as 'a supplementary feature for species-definition', but the great variability in the S.C.L., as shown by Brown *et al.* (1971), and its brief existence, must limit its usefulness. However, the clear presence or seeming absence of the S.C.L. may be helpful in eliminating from consideration species known to differ (in their S.C.L.) from a new cyst whose identity is in question.

Features of the Cyst Posterior

(i) *Heterodera* SPECIES—VULVAL CONE AND ASSOCIATED STRUCTURES

Shape and form

The prominence at the posterior end of lemon-shaped cysts is the vulval cone (Figs. 6.4 & 6.5). Its shape and degree of prominence differ; thus, the cone may be obtuse (*H. galeopsidis*) or narrow (*H. schachtii*), and it may be so insignificant that the cyst at first glance seems spheroidal or ovoid (*H. cruciferae*).

Cone-top and fenestration

Vulval bridge

At the apex of the cone is the vulval slit; this is bordered by the vulval labia which form parapets of a 'bridge' (the vulval bridge) which crosses a thin-walled area (fenestra) on the cone-top. The vulval bridge thus divides the fenestra into two semifenestrae. With one exception (on young cysts of *H.*

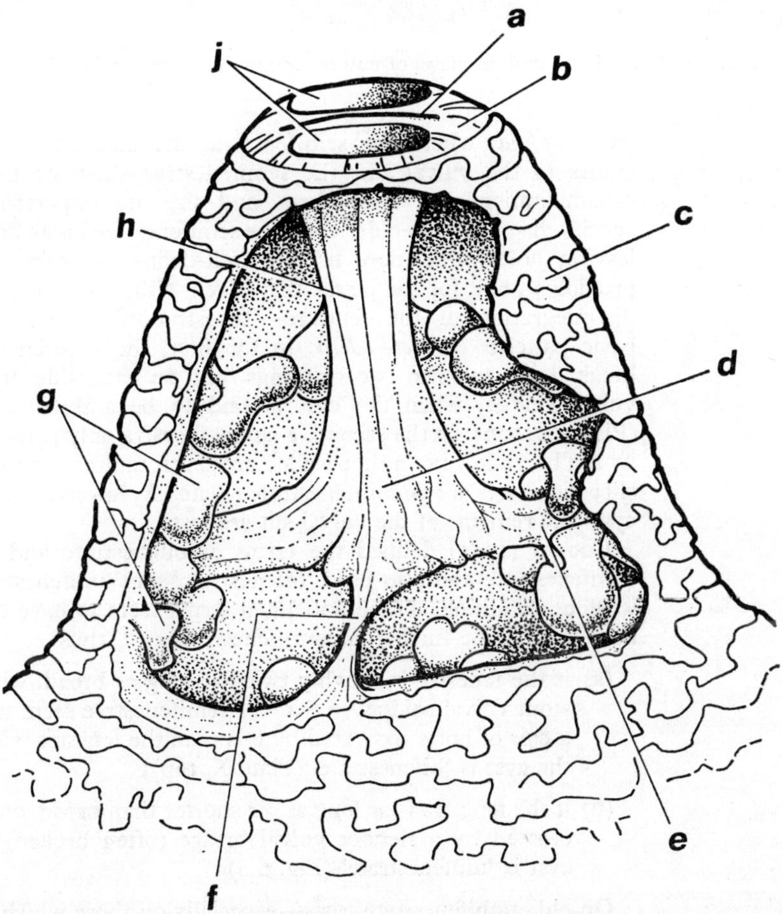

Fig. 6.4. Diagrammatic view of cyst cone-top in, e.g., *Heterodera schachtii*: a = vulval slit on vulval bridge, b = basin, c = external cuticular pattern, d = cervix, e = under-bridge, f = Mulvey's bridge, g = bullae, h = vaginal remnant, j = semifenestrae.

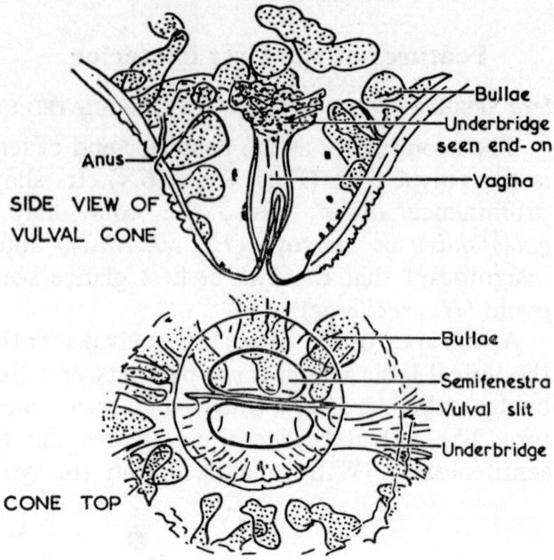

Fig. 6.5. Lateral and end-on views of vulval cone of *Heterodera galeopsidis.*

graminophila, where the semifenestrae are striated at right angles to the vulval slit) the semifenestrae show no useful ornamentation, but their shape and size are important in species diagnosis. On older cysts the semifenestrae break down, leaving holes with more or less well-defined margins, and provide an exit for the larvae that hatch within the cyst. The distintegration of the fenestra—'fenestration'—follows two basic patterns in *Heterodera*: in the first, the semifenestrae break down to form two openings, one on each side of the vulval bridge, when the cyst is said to be ambifenestrate (Plate IX, top); in the second, a single fenestra only is formed by the breakdown of a thin-walled area surrounding the vulval slit; such cysts are circumfenestrate (Plate IX, lower); in these cysts the remains of the vulval slit are lost.

Cooper (1955) defined the terms ambifenestrate and circumfenestrate and also used 'bifenestrate' and 'unifenestrate' without definition. These undefined terms seem to have been used to describe further an ambifenestrate cyst, thus:

(a) if the fenestra is roughly twice as long as broad, with a stout vulval bridge, so that the semifenestrae seem to be a pair of holes separated by less than the fenestra width, the cyst is 'bifenestrate' (Plate X, top);

(b) if the fenestra is as long as, or shorter than broad, and is crossed by a slender vulval bridge (often broken) the cyst is 'unifenestrate' (Fig. 6.6).

On old, ambifenestrate cysts—especially on those which are unifenestrate, the vulval bridge may be lost and the fenestra then appears to be a roughly circular hole in the cone-top; such cysts seem to be circumfenestrate, but it is often possible

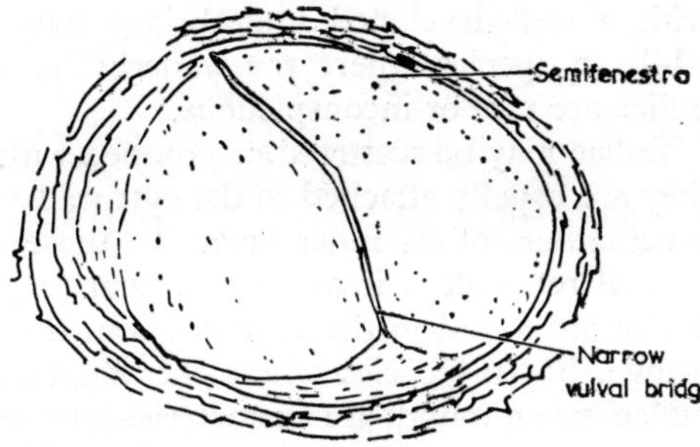

Fig. 6.6. End-on view of fenestral region of *Heterodera carotae* cyst. Note slender vulval bridge which may disintegrate to leave an almost circular fenestra.

to detect the remains of the vulval bridge to confirm their original ambifenestrate nature.

Type of fenestration is important in species diagnosis.

Basin

On cysts of several species the fenestra is encircled by a more or less distinct oval or near-circular band of cuticle whose outer margin abuts the external cuticular pattern of the cyst. This circumfenestral area is known as the basin (Cooper, 1955), though in fact it rarely is a concavity with the fenestra as the centre of the 'bowl'. The basin is particularly clear on some species, e.g., *H. humuli, H. glycines, H. galeopsidis,* and sometimes has wrinkle-like markings called 'rivulets' (Cooper, 1955); on *H. goettingiana* the markings sometimes form a brickwork-like pattern. The distinctness of the basin, its width (relative to the fenestra), and its shape (circular or oval), are useful for cyst identification.

Vaginal remnant or sheaf-like organ

In certain species the remnant of the vagina persists in the vulval cone; the remnant is flattened dorso-ventrally, and ridges or folds in its walls may give it a striated appearance when it is viewed dorso-ventrally (Plate X, lower) or terminally (as an optical cross-section (Mulvey, 1959a)). The combined effect of the striations and the 'waisted' shape of the remnant give it the appearance of a sheaf of grain, hence the name 'sheaf-like organ' (A. L. Taylor, personal communication). In lateral view the vaginal remant is narrow and opaque, and striations may not be evident; if it is twisted the sheaf-like appearance can be lost (Plate XI, top). Proximally, the vagina appears to thicken to form the 'basal bulb' (Cooper, 1955) or 'cervix' (Thorne, 1961) which is often hollow in newly formed cysts but later becomes solid and is linked to the cyst wall by two arms extending across the vulval cone. These arms lie directly beneath and parallel to the vulval bridge and form the

Underbridge

'underbridge' (Plates X, lower & XI). The ends of the underbridge, and in some species (e.g., *H. cajani*) the outer margin of the cervix, are much forked or divided, and the forks or divisions often join conspicuous single or branched knob-like

Bullae

structures termed 'bullae' (Cooper, 1955). Generally, species with a well-developed underbridge have prominent bullae, while in species where the underbridge is absent or weak, bullae are rare or inconspicuous.

Bullae may be scattered or grouped within the vulval cone; they are usually attached to the cyst wall and seem to be solid evaginations of its inner layer. In those cysts in which the vaginal remnant is invariably present (e.g., *H. schachtii*), the bullae are found in the cone at about the level of the underbridge, but in species where the vaginal remnant is absent, bullae, when found, are characteristically clustered in the cone immediately beneath the fenestra (e.g., *H. avenae*) (Plate XII, top).

'Mulvey's bridge' and other features

Mulvey (1959a) recorded another bridge-like structure crossing the vulval cone in newly-formed brown cysts of *H. trifolii*. This bridge, possibly of hardened muscle tissue, ran at right angles to the underbridge and was found just below the level of the fenestra (Plate XII, lower). It does not persist in old cysts. Thorne (1961) figured and described the vulval cone of *H. avenae* and showed a complex muscle system supporting the inner cone structures which included a peculiar spheroidal vaginal chamber that forced the bullae against the body wall and crowded them back near the fenestra (Fig. 6.7). The muscular structures and vaginal chamber are not found in old cysts.

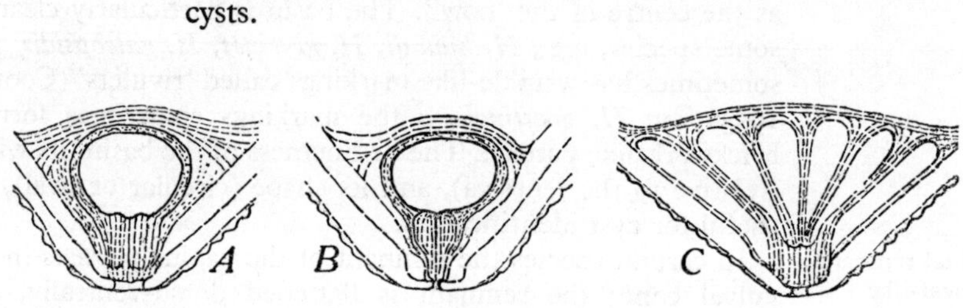

Fig. 6.7. Anatomy of the vulval cone of females of *Heterodera avenae*. **A, B,** dorsal and lateral views of cone showing spheroidal vaginal chamber; **C,** muscle system supporting inner cone structures. (From *Principles of Nematology* by Gerald Thorne. Copyright 1961 McGraw-Hill Book Co. and used with their permission).

(ii) *Globodera* SPECIES

On *Globodera* cysts the taxonomist is concerned in particular with that terminal region of the cyst extending from the posterior 'pole' (i.e., diametrically opposite the cyst neck) to a latitude just anterior to the anal pore. Within this terminal region lie the vulva, vulval fenestra and fenestralia, basin (where present), anal pore, perineum, and the 'terminal area' defined by Green (1971).

Terminal area (Green, 1971)

The terminal area is a distinct surface of transversely ridged cuticle containing the anus and lying between the vulva and the tail (see Fig. 6.8); it is laterally delimited by two terminal bands formed where the ridges that encircle the vulva and those encircling the tail cross. The terminal area may be long or

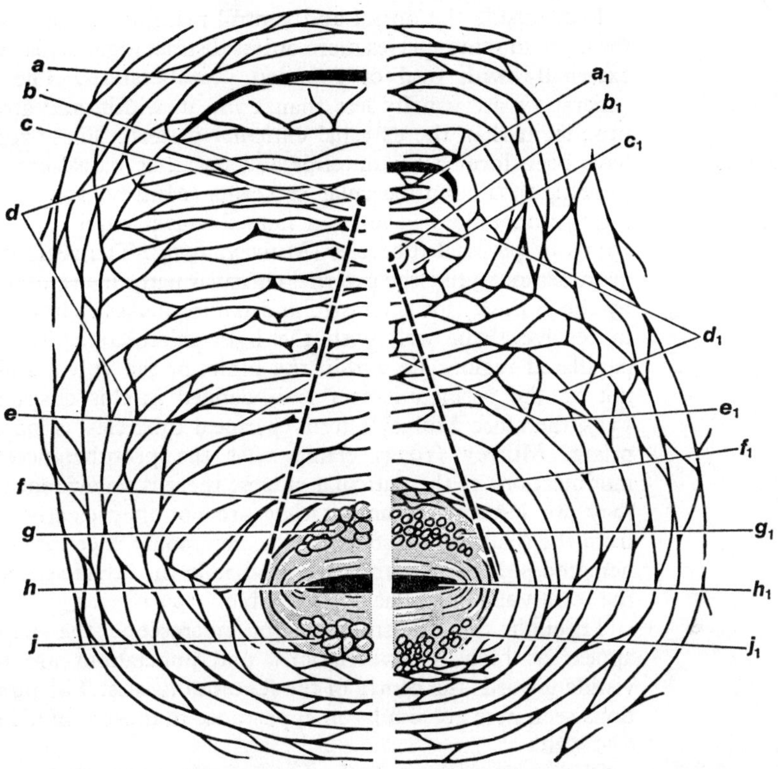

Fig. 6.8. Morphology of terminal area of *Globodera* females: left side of diagram depicts terminal area of *G. pallida*, right side *G. virginiae* and *G. solanacearum* a = tail, b = anus, c = circumanal grooves, d = terminal band, e = perineal pattern, f = perineal ridge adjacent to vulva, g = posterior vulval crescent and papillae, h = vaginal aperture (vulval slit), j = anterior vulval crescent and papillae. The bold broken line depicts the side of the anal-vulval triangle. (Adapted from Green, 1971).

short, broad or narrow, depending on the configuration of the circles. The size of the terminal area is characteristic of the species, but difficult to define.

Vulva

The vulva is the external female genitalia (Kenneth, 1963) and, on round cysts, is that region around the vulval slit extending to the external cuticular pattern.

Fenestra

Fenestral tissue

The central, very translucent, part of the vulva is the fenestra, and before fenestration this is a thin sheet of tissue, 'fenestral tissue', perhaps of two or more layers. Before fenestration the vaginal opening (= vulval slit) is a slit or oval-shaped aperture in the centre of the fenestra; minute differences in the shape and ornamentation of the slit (on young cysts only) are used in species differentiation (Miller & Gray, 1972). The junction of the cyst wall and the translucent area is the fenestral margin and is clearly seen by transmitted light. Fenestrae may be circular or oval and differ in size; shape and size are diagnostic features.

Just outside the dorsal and ventral margins of the fenestra there are in some species more or less distinct regions decorated externally with oval or rounded protuberances. The protuberances are usually less than 1 μm in width, and are less prominent than the external cuticular pattern. These regions have been termed the anterior and posterior vulval crescents (Green, 1971) or tubercular crescents (Mulvey, 1973), while the protuberances have been named vulval papillae (Green, 1971), or perineal tubercles (Mulvey, 1973). The regions and protuberances, like the external cuticular pattern, are best seen by the S.E.M., for their size and nature makes it impossible to resolve them well with the light microscope. The two papillated regions lie within the limits of the vulva, and are not on the perineum (as this is defined in this chapter, p. 131), therefore Mulvey's term 'perineal tubercles' is inappropriate. Mulvey (1973) claims that the protuberances are modifications of the cuticular ridges; there is no evidence that they are innervated, and if they are not they clearly differ from the (usually) sensory papillae commonly encountered on nematodes. However, they are papillae in the broad sense, and the term vulval papillae is acceptable.

Vulval crescents

Vulval papillae

Through the transmitted-light microscope the papillae appear as blobs or punctations, and punctations are often visible towards the centre of the fenestral tissues. The punctations seen here are similar in appearance to those visible in the cyst wall.

Observations through the light microscope, which, unlike the S.E.M., takes account of colour and translucence, suggest, in accord with Mulvey (1973), that the vulval papillae are mainly modifications of the ridge and furrow pattern which seems to taper-off over the vulval margin, while the more central punctate fenestral tissue may be a thin extension of the subsurface punctation layer in the cyst wall. The punctate fenestral tissue rarely extends to the vulval slit, and on some species is delimited short of the slit by a transfenestral bar or bridge which lies beneath the fenestral tissue. The transfenestral bar and the punctate layer are particularly clear on *G. pallida,* and stand out as 'D'-shaped shelves of tissue—'fenestral shelves' (Hesling & Ellis, 1974). These shelves are often visible at magnifications of 100 × or less. Fenestral shelves and vulval crescents are not exactly the same.

Transfenestral bar

Fenestral shelves

On undamaged, unfenestrated specimens of some species the centremost tissue may be bridged by two very slender bars resembling the hyaline parapets (Cooper, 1955) of the vulval bridge of some lemon-shaped cysts. The central fenestral tissue is very fragile and is frequently absent on cysts from soil. At early fenestration the broken or curled edge of the centremost fenestral tissue may seem like hyaline parapets guarding an oval or near-circular vaginal opening, but in these cases the true opening—a slit or oval—has been lost.

On cysts of some species there is a more or less distinct

circumfenestral zone or basin; the outer limit of the basin usually lies directly beneath that of the vulva. The basin (which is only seen by transmitted light) is the region where the cyst wall tapers from its normal thickness at the circumference of the basin to that of the thin fenestra; it is because of this and, frequently, because it is lesser pigmented than the body of the cyst that it appears as a circumfenestral band with a translucence greater than that of the rest of the cyst wall.

Perineum, and Granek's ratio

The perineum is that region between the vulva and the anal opening; it is easily observed on preparations of the posterior parts of round cysts (on lemon-shaped cysts the perineum is on the side of the vulval cone and it is therefore difficult to present it for examination with the transmitted light microscope). The length of the perineum (the slight curvature is ignored when measuring) and the number of ridges of the external cuticular pattern on the perineum, are important in cyst identification.

The length of the perineum, plus the width of the dorsal end of the basin (if present) is the distance 'B' of Granek's ratio (Granek, 1955, 1968). Granek's ratio is defined (Hesling, 1973) as:

$$\frac{\text{Distance from anus to nearest edge of fenestra}}{\text{Length of fenestra}}$$

Granek's ratio is useful in species diagnosis, but there is considerable overlap between species in its values (Mulvey, 1960a; Jones, 1962; Ellis, 1970; Green, 1972; Ellis & Hesling, 1974).

Perineal pattern

The ridges and furrows of the perineum form a pattern (perineal pattern) that can be used in cyst identification (Green, 1971; Hesling & Ellis, 1974). There are two basic forms: (i) a maze-like or reticulate pattern (usually found when the perineal length exceeds 60 μm), and (ii) a pattern of parallel ridges (perineal length usually less than 50 μm). In some species the number of perineal ridges and the distance 'B' are correlated (Hesling & Ellis, 1974): the number of ridges can be used in species diagnosis.

Differences between the maze-like perineal patterns are claimed to be helpful in cyst identification, but the variability of these patterns is great and their appreciation somewhat subjective. Miller & Gray (1972) used the variability and relative breadths of perineal grooves and ridges in species diagnosis. Grooves and ridges are often too shallow to be resolved accurately with the light microscope and so cannot be measured; however, an *estimate* of their breadths can be made by dividing perineal length by the number of ridges or grooves, but comparisons of relative breadths of ridges and grooves must be subjective. On *G. millefolii* there are concentric perianal rings of grooves and ridges (Mulvey, 1972).

Vulval bodies

Wilson (1968) first described and illustrated the 'endocuticular' vulval bodies apparently attached to the fenestra of white 'cysts' (adult females) of *G. rostochiensis*. Vulval bodies were also seen in females of *G. tabacum*, *H. schachtii* and *H. cacti*; they were less developed in the last two species. In further studies, Mulvey (1973) showed that the arrangement and number of the vulval bodies differed with species and could be used to separate *G. rostochiensis* from *G. solanacearum*, *G. tabacum* and *G. virginiae*. The function of vulval bodies is unknown.

(iii) *Punctodera* SPECIES

The terminal region of cysts in this genus is distinctive because of the presence of two separate and similar-sized fenestrae around vulva and anus, and the obscure and very short vulval slit (see Diagnosis p. 127). The terminal area has not been studied in such detail as that of *Heterodera* and *Globodera* but *P. matadorensis* is distinguished from the type species *P. punctata* by the presence of well developed bullae between vulval and anal fenestrae.

Gelatinous Matrix

The 'egg sac' or gelatinous matrix is a mass of jelly which is found on the vulval end of developing cysts; it should not be confused with the brown 'cement' globules which often appear on mature females about to turn brown (Jones, 1950). The relative size of the matrix and cyst varies—some species commonly have a matrix which almost covers the posterior 'hemisphere' of the cyst, while in other species the matrix is small or apparently absent. It is probable that cysts of most or all species of *Heterodera* and *Globodera* possess a matrix—for many years it was not recorded in potato cyst nematodes but, in cultures of *G. rostochiensis* on potato slices, Williams (1957) demonstrated that a small matrix was, in fact, produced. In many species a variable number of eggs is laid into the matrix but in some, e.g., *H. avenae*, eggs are never found within it, and it has been suggested that in this species the vulval slit is too small to permit their passage.

The origin of the matrix in cyst nematodes is not completely known. In comparison, the gelatinous matrix of the closely related *Meloidogyne* (Chapter 5) is secreted by large cells around the rectum (Maggenti & Allen, 1960) and in *Tylenchulus* from the excretory pore (Maggenti, 1962). The egg sac is probably protective; it is noteworthy that it is a frequent feature of other nematodes in which the female is saccate and sedentary—*Trophonema*, *Trophotylenchulus*, *Rotylenchulus* and some *Nacobbus*. In bisexual species of cyst nematodes it is common to find one or more male nematodes within the egg sac. Insemination and fertilization might be hindered when the vulval cone is coated and the uterine tract is filled with the viscous egg sac material, but mating may precede egg sac production, which occurs in some species shortly after the female has ruptured the root cortex (Hirschmann, 1956); in *H. cyperi*, cysts with eggs may develop completely beneath the root epidermis (Golden, Rau & Cobb, 1962). Mating is not a necessary precursor of egg sac formation because the egg sac is found in parthenogenetic species, e.g., *H. trifolii*. The egg sac does not persist on cysts which have fallen from the roots into the soil.

Cyst Contents: Eggs and Larvae

EGG

Eggs are found in cysts on roots, in the egg sac and in cysts free in the soil; they differ in size, shape and in their length: breadth ratios. These differences may be not only between but within a species. Egg length and cyst size are not correlated (Müller, 1958). Egg shape may be dependent on the way the larva is folded within the egg (it is commonly folded four or five times but in *G. leptonepia* has seven folds). Egg shape is unreliable in species identification because it may be altered by treatment of the eggs (e.g., osmotic effects or drying) and by the development of the larva within. Egg size and length: breadth ratio may be useful confirmatory characters in differentiation but many species have eggs of similar size. Giant eggs may be found (Mulvey, 1959b); they are clearly anomalous and represent a small percentage of eggs in a cyst. The interval between egg formation and egg hatch varies, but only in exceptional circumstances could this difference be useful in diagnosis, for egg hatch is influenced by many factors (see p. 160 *seq.*).

LARVA

Second-stage, infective larvae which hatch from the egg may be useful in species diagnosis. Larvae may be prematurely released from eggs by gentle pressure—in this way they are often artificially 'hatched' when a cyst is dissected for examination; when so released they may be the first stage, which differs from the second stage. Larvae exhibit a wide range of difference in several features, but there is considerable overlap which limits the use of larval characters in identification. Features which have been studied and sometimes found useful are listed below:

Length
Width

$$\text{'a'} = \frac{\text{Length}}{\text{greatest width}}$$

Head cap—height, width and number of annulations
Cephalic skeleton—form and stoutness
Stylet length and stoutness; size and form of knobs
Position of opening of dorsal oesophageal gland
Position of excretory pore
Length of true tail (anus to tail tip)
Length of clear part of tail
Form of tail
Ratio of clear tail length to stylet length
Lateral field—width and number of incisures

Larvae exhibit a wide range of size between *and within* species. Fenwick & Franklin (1951) showed how larval measurements may be affected by the previous treatment of larvae, and they emphasized the need to use a careful and uniform method of measurement. Kämpfe's (1962a) studies confirm these findings. There is no doubt that fixatives and mountants affect nematode dimensions, and may make some structures (e.g., stylets) almost invisible; care is therefore needed when comparisons are made with old specimens in slide collections. Stone (1971) suggested that heat and cold formalin, respectively, should become standard methods of killing and fixing larvae that were to be measured.

For a reliable estimate of larval length it is necessary to measure several larvae from several cysts; Hirschmann (1956), comparing *H. trifolii* and *H. glycines*, found that it was most economical to measure three larvae from each of 36 cysts. Fenwick & Franklin (1951) found (with reservations) that a homogeneous population of cysts could be sorted into one of four groups by measurement of four larvae from each of four cysts.

Larval features have been studied in relatively few species; in some, differences in the form of certain organs, and in fine detail, are only visible with the aid of the finest microscopes. Recently, the scanning electron microscope has revealed hitherto unresolvable detail of the head cap (Stone, 1973; 1975a) (Plate XIII), while intensive studies using light microscopy have shown characters of the anterior region that are useful in species diagnosis (Webley, 1970; Evans & Webley, 1970; Webley, pers. comm.). Plate XIII illustrates some of these characters, and especially the form of the stylet and stylet knobs.

Among larvae there is usually a small percentage of abnormal forms. Both giant and dwarf larvae have been recorded; other abnormalities may occur and their range has been illustrated by Mulvey (1960c), and by Kämpfe (1962a) who claimed that the larval tail is morphologically its most labile part.

It is often difficult to determine the anterior limit of the clear part of the larval tail; tail deformity is common, while in some species there is a diagnostic tail-kink or constriction (Webley, pers. comm.). Counting the incisures and bands in the lateral field is exacting work for their clarity depends on specimen turgor and fixation; lateral fields are most reliably appraised by the scanning electron microscope.

Cytology

The number and nature of the chromosomes have been studied in a few species but the results have been of little use in taxonomy because the chromosomes are extremely small, skilful techniques are needed to demonstrate them, and they can only be examined under the finest microscopes. The usefulness of chromosome studies in taxonomy seems to be in disclosing whether a specimen has the usual somatic number of 18 chromosomes (found in amphimictic species), has 27 (found in triploid species that reproduce by mitotic parthenogenesis), or is an aneuploid, reproducing by meiotic parthenogenesis (*H. betulae*). Cytological studies have led to speculation about phylogeny within the genus *Heterodera* (*sensu stricto*), and to the thesis that some parthenogenetic species are aneuploid or triploid variants of existing diploid amphimictic species, e.g., *H. sacchari* and *H. leucilyma* seem to be polyploids derived from *H. oryzae* (Triantaphyllou, 1970). Studies of the relative content of DNA in nuclei of some of the Heteroderidae shows that those species with chromosome numbers equivalent to 3n are truly triploid, and that increased chromosome numbers have not resulted through chromosome fragmentation or other chromosomal rearrangements (Lapp & Triantaphyllou, 1969).

Host Ranges

Species of *Heterodera* have different plant hosts; some cyst-nematode species attack a limited group of plants, others parasitize many plants in different families, with overlap of host ranges. Species with non-overlapping host range could perhaps be identified by culturing them on their known hosts. One cyst-nematode species may successfully parasitize several unrelated plant species but not always closely related species or different cultivars of the same species; host tests should, therefore, not be limited to one cultivar.

Species Identification by New Taxonomic Methods

SEROLOGICAL METHODS

The use of a gel-diffusion technique by Webster & Hooper (1968) is referred to in Chapter 7, p. 167; they divided seven cyst-nematode species into two serologically distinct groups. From preliminary investigations Mabbott (1968) suggested that serological techniques could be useful to separate the (then) two 'strains' of '*H. rostochiensis*' (*G. rostochiensis* pathotype A and *G. pallida* pathotype B). Scott & Riggs (1971) demonstrated the serological similarity of two races of *H. glycines*, and the serological unrelatedness of this species and *H. betulae*. Immunoelectrophoresis of nematode extracts revealed a greater number of precipitin bands than did gel-diffusion.

ELECTROPHORETIC METHODS

Several electrophoretic methods using different gels, coupled with micro-densitometry, have been used in the quantitative and qualitative analysis of nematode proteins. These methods have distinguished genera, species, and sub-species or races, but the taxa distinguished had already been brought to notice by more conventional methods; it seems that new species have not yet been discovered by electrophoresis. Thus Trudgill & Carpenter (1971), Greet (1972), Trudgill & Parrott (1972) and Stone & Williams (1974) used polyacrylamide-gel electrophoresis to distinguish *H. avenae*, *G. pallida* and some American species of *Globodera*. It has shown that there is much protein-sharing between species and between genera of nematodes, and that the method can detect differences between different isolates of a single parthenogenetic species (Evans, 1971). Dickson, Sasser & Huisingh (1970) compared the soluble protein analyses of four nematode genera including *H. glycines*; they found that the species of host plant had no significant effect on protein profiles. However, Ishibashi (1970) claimed that the host species and its health, as well as the age of the nematodes, influenced the protein pattern. Such apparent contradictions are difficult to explain but it is well known that protein profiles may be influenced by treatment during and prior to the 'run'. Dickson, Huisingh & Sasser (1971) concluded that disc electrophoresis of enzymes may be a more sensitive chemo-taxonomic tool than general protein analysis. It seems that, with further development and more standardization, electrophoresis could be an important, though hardly routine, method in taxonomy.

MORPHOMETRICS WITH STATISTICAL ANALYSIS

Taxonomic data can now be handled conveniently by electronic computers, and complex statistical analyses have been used to elucidate known or suspect intertaxal relationships or differences, and to suggest new ones (e.g., Dixon & Ryland, 1964; Green, 1971; Øydvin, 1974). A possible advantage of introducing the precision of mathematics into taxonomy is that defined, albeit arbitrary, boundaries can be erected to delimit taxa (see Bird, 1971b), but the problem of the variable quality of data with its contained variation due to human error will always remain. By linking genetic and morphometric studies, Netscher & Pernes (1971) demonstrated the establishment of significant morphological differences in two inbred lines of the bisexual *H. oryzae*, and they warn that it is important to determine variability in natural populations before adopting a taxonomy based on quantitative characters.

Lists of Species

Genus *HETERODERA* Schmidt, 1871 (lemon-shaped cysts)
(arranged by morphological grouping)

- vulval slit < 25 µm
  - circum-fenestrate
    - *H. betulae* Hirschmann & Riggs, 1969
    - *H. cacti* Filipjev & Schuurmans Stekhoven, 1941
    - *H. estonica* Kir'yanova & Krall, 1963
    - *H. thornei* Golden & Raski, 1977
    - *H. weissi* Steiner, 1949
  - bifenestrate
    - *H. amaranthi* Stoyanov, 1972
    - *H. avenae* Wollenweber, 1924
    - *H. hordecalis* Andersson, 1975
    - *H. iri* Mathews, 1971
    - *H. latipons* Franklin, 1969
    - *H. longicaudata* Seidel, 1972
    - *H. mani* Mathews, 1971
- vulval slit > 30 µm
  - bullate
    - *H. cajani* Koshy, 1967
    - *H. elachista* Ohshima, 1974
    - *H. fici* Kir'yanova, 1954
    - *H. galeopsidis* Goffart, 1936
    - *H. glycines* Ichinohe, 1952
    - *H. lespedezae* Golden & Cobb, 1963
    - *H. leucilyma* Di Edwardo & Perry, 1964
    - *H. oryzae* Luc & Brizuela, 1961
    - *H. rosii* Duggan & Brennan, 1966
    - *H. rumicis* Pogosyan, 1961
    - *H. sacchari* Luc & Merny, 1963
    - *H. schachtii* Schmidt, 1871 [GENOTYPE]
    - *H. sonchophila* Kir'yanova, Krall & Krall, 1976
    - *H. trifolii* Goffart, 1932
    - *H. zeae* Koshy, Swarup & Sethi, 1970
  - abullate
    - *H. carotae* Jones, 1950
    - *H. cruciferae* Franklin, 1945
    - *H. cyperi* Golden, Rau & Cobb, 1962
    - *H. goettingiana* Liebscher, 1892
    - *H. graminis* Stynes, 1971
    - *H. graminophila* Golden & Birchfield, 1972
    - *H. humuli* Filipjev, 1934
    - *H. longicolla* Golden & Dickerson, 1973
    - *H. mothi* Khan & Husain, 1965
    - *H. urticae* Cooper, 1955 [Redescribed by Mathews, 1971]

Species inquirendae

H. aquatica Kir'yanova, 1971
H. arenaria Cooper, 1955 in Kir'yanova & Krall, 1971
H. bifenestra Cooper, 1955
H. cardiolata Kir'yanova & Ivanova, 1969
H. chaubattia Gupta & Edward, 1973
H. graduni Kir'yanova in Kir'yanova & Krall, 1971
H. limonii Cooper, 1955
H. major[= *avenae*] var. *arenaria* Cooper, 1955
H. medicaginis Kir'yanova in Kir'yanova & Krall, 1971
H. methwoldensis Cooper, 1955
H. oxiana Kir'yanova, 1962
H. paratrifolii Kir'yanova, 1963
H. polygoni Cooper, 1955
H. salixophila Kir'yanova, 1969
H. tadshikistanica Kir'yanova & Ivanova, 1966
H. turcomanica Kir'yanova & Shagalina, 1965
H. ustinovi Kir'yanova, 1969
H. vigni Edward & Misra, 1968

Nomina nuda

H. indocyperi Husain & Khan in Khan, Siddiqi, Khan, Husain & Saxena, 1964 [= *H. mothi*]
H. scleranthii Kaktina, 1957
H. sonchophila Kir'yanova & Borozdina in Kir'yanova, 1969 [= *H. sonchophila* Kir'yanova, Krall & Krall, 1976]

Genus *GLOBODERA* Skarbilovich, 1959 (round cysts)

G. achilleae (Golden & Klindic, 1973) Mulvey & Stone, 1976
G. artemisiae (Eroshenko & Kazachenko, 1972) Mulvey & Stone, 1976
G. leptonepia (Cobb & Taylor, 1953) Mulvey & Stone, 1976
G. mali (Kir'yanova & Borisenko, 1975) Mulvey & Stone, 1976
G. millefolii (Kir'yanova & Krall, 1965) Mulvey & Stone, 1976
G. pallida (Stone, 1973) Mulvey & Stone, 1976
G. rostochiensis (Wollenweber, 1923) Mulvey & Stone, 1976 [GENOTYPE]
G. solanacearum (Miller & Gray, 1972) Mulvey & Stone, 1976
G. tabacum (Lownsbery & Lownsbery, 1954) Mulvey & Stone, 1976
G. virginiae (Miller & Gray, 1968) Mulvey & Stone, 1976

Species inquirendae

G. mirabile (Kir'yanova, 1971) Mulvey & Stone, 1976
G. pseudorostochiensis (Kir'yanova, 1963) Mulvey & Stone, 1976

Genus *PUNCTODERA* Mulvey & Stone, 1976 (round or ovoid cysts)

P. chalcoensis Stone, Moss & Mulvey, 1976
P. matadorensis Mulvey & Stone, 1976
P. punctata (Thorne, 1928) Mulvey & Stone, 1976 [GENOTYPE]

Other, Possibly New, Species Mentioned in Literature

Pogosyan (1962) mentioned a 'cabbage-form' of *H. schachtii* found on cabbage, on *Thlaspi arvense* and on *Polygonum aviculare*. It differs from *H. schachtii* in size of cyst, larva and egg, presence of dark brown bullae, a reddish-brown colour of the cyst, structure of the cyst cuticle, and details of fenestralia; measurements are given. In the same paper Pogosyan stated that cysts on *Scandix iberica* were suspected to be a new species close to *H. cruciferae*. Edward & Misra (1968) reported *Heterodera* larvae in soil in Uttar Pradesh, India; two species were noted, one similar to *H. glycines*, the other to *H. trifolii*. A possibly new species was recorded by Riffle (1969), who reported *Heterodera* (*sens. lat.*) larvae associated with roots of conifer seedlings. In the same year Stoyanov reported a new cereal cyst-nematode on wheat in Bulgaria (the species shows some similarity to *H. latipons*), while a *Heterodera* sp. recorded on *Sesamum orientale* by Yadav & Verma (1971) was later identified as *H. cajani* (Stone, pers. comm.).

In 1972 two *Heterodera* spp. close to *H. trifolii* were found on carnations in Southern France (Cuany & Dalmasso, 1972), and another on sisal in Kenya (Anon., 1972). Jones (1972) referred to the 'rye cyst-nematode'.

Pathotypes and Species Complexes

The term 'pathotype' is used for variants of cyst-nematode species which differ from one another in their ability to multiply on clones or hybrids of crop plants that have been selected or bred for their resistance to some populations of the nematode (see also Chapter 16). The preferred use of the name 'pathotype' is explained by Cole & Howard (1966) and Howard (1972). It is usually difficult or impossible to distinguish pathotypes by conventional morphological methods, and prolonged and detailed investigation may show that what at first seems to be a pathotype complex within a species is in fact a species complex (e.g., *G. rostochiensis* and *G. pallida*). Several species/pathotype complexes are known in cyst nematodes, thus the round cyst nematodes are particularly confused, containing complexes of pathotypes within species whose true identity remains arguable. Somewhat similar situations are found with *H. avenae*, *H. trifolii* and *H. glycines*. Infraspecific variation in the latter species is well documented in the U.S.A. (Ross, 1962; Smart, 1964; Miller, 1966, 1969, 1971; Riggs, Slack & Hamblen, 1968) and the terminology and identity of the infraspecific forms in this case were explained by Golden, Epps, Riggs, Duclos, Fox & Bernard (1970).

Cereal cyst nematode (*H. avenae sens. lat.*) comprises a complex of pathotypes and, perhaps, species. At least six pathotypes of *H. avenae* are known (Cook & Williams, 1972); three of these (nos. 1, 2 and 3, British notation) have been studied more than the rest (Cook, 1975) and the results indicate that pathotype 3 is probably a separate species close to *H. mani*.

Acknowledgements

I am grateful to J. F. Southey, Dr. A. R. Stone and D. Webley who contributed much to this Chapter by their comments and by material help.

Key to HETERODERA Species, Based Mainly on Features of Cysts and Second-stage Larvae

	Cyst bullate	1
	Cyst abullate	10
1.	Cyst bifenestrate	2
	Cyst ambifenestrate	3
2.	Cyst vulval slit $<$ 20 μm long	8
	Cyst vulval slit $>$ 20 μm long	*H. fici*
3.	Fenestral length $<$ 39 μm	*H. schachtii*
	Fenestral length $>$ 39 μm	4
4.	Second-stage larva (L2) clear-tail length $1\frac{1}{3}$ × stylet length	5
	L2 clear-tail length equal to or slightly longer than stylet length	6
5.	Fenestra length 41–46 μm, males absent	*H. galeopsidis*
	Fenestra length 58–77 μm, males present	*H. rosii*

6. L2 spear knobs as wide as high, forward-pointing	*H. trifolii*
L2 spear knobs wider than high	7
7. L2 spear length 28·7–30 μm	*H. rumicis*
L2 spear length 22–24 μm	*H. glycines*
8. Clear tail length/stylet length of L2 2·5 (1·9–3·1); length of true and clear tail of L2 94 μm (83–103 μm) and 62 μm (51–75 μm) respectively; cyst pale brown	*H. iri*
Clear tail length/stylet length of L2 < 1·9	9
9. Anterior faces of stylet knobs of L2 deeply concave; cyst with distinct underbridge	*H. mani*
Anterior faces of stylet knobs of L2 shallow concave; cyst without underbridge	*H. avenae*
10. Cyst circumfenestrate	*H. cacti*
Cyst bifenestrate	11
Cyst ambifenestrate	14
11. Cyst with prominent underbridge	12
Cyst without underbridge	13
12. Cyst vulval slit < 9 μm long	*H. latipons*
Cyst vulval slit 17–25 μm long	*H. hordecalis*
13. Cyst vulval slit > 30 μm long	*H. humuli*
Cyst vulval slit 18 μm long	*H. longicaudata*
14. Underbridge usually present	15
Underbridge absent	16
15. Fenestra near-circular, brick-like pattern on basin	*H. goettingiana*
Fenestra often broader than long; semifenestrae kidney-shaped	*H. cruciferae*
16. Large egg sac with eggs; length of L2 454 μm (430–475 μm)	*H. carotae*
Small egg sac without eggs; length of L2 541 μm (467–615 μm)	*H. urticae*

*Key to GLOBODERA Species Based on Features of Cysts and Second-stage Larvae

Anal area encircled by cuticular rings	*G. millefolii*
Anal area not encircled by cuticular rings	1
1. Second-stage larva (L2): anterior faces of spear knobs pointing cephalad	2
L2: spear knobs evenly rounded, anterior faces neither clearly pointing cephalad nor sloping caudad	3
L2: anterior faces of spear knobs sloping caudad	4
2. Cyst perineal pattern maze-like	*G. virginiae*
Cyst perineal pattern of more or less parallel lines	*G. pallida*
3. Granek's ratio 2·2 (1·3–3·8)	*G. solanacearum*
Granek's ratio 1·5 (0·9–2·8)	*G. tabacum*
4. Granek's ratio 4·6 (2·7–8·9)	*G. rostochiensis*
Granek's ratio < 2	5
5. Fenestral length 16 μm (12–18 μm)	*G. achilleae*
Fenestral length 25 μm (20–32 μm)	*G. artemisiae*

*Insufficient information was available for *Globodera mali* (distinguished within the genus by its host plant, apple (*Malus domestica*)) to be included in the key.

Ta

A Compendium of some Characters of Larvae

As far as possible data are from original descriptio

Species	Second-stage larva: Stylet length µm	Body length µm	True tail length µm	Clear tail length µm	Anterior face spear knobs (↑ forward slope / ←→ rounded / ↓ rearward slope)	Distance of d.o.g.o. behind spear base µm
ACHILLEAE	25 (24–25)	492 (472–515)	55 (47–63)	27 (20–32)	←→	5·7 (5·6–6·1
ARTEMISIAE	23 (18–29)	413 (357–490)	40 (33–61)	23 (18–28)	↓	?
MILLEFOLII	?	?	?	?	?	?
PALLIDA	24 (21–26)	484 (440–525)	52 (46–62)	24	↑	2·7, 2·9
ROSTOCHIENSIS	22 (21–23) 23 (19–26)	468 (425–505)	44 (42–50)	24	↓	2·6
SOLANACEARUM	24 (22–26)	497 (390–577)	50 (33–60)	24 (16–30)	←→	5·5 (5–7)
TABACUM	24 (22–26) 22	477 (410–527)	52 (46–59)	28 (22–31)	←→	5·5 (4–7)
VIRGINIAE	24	504 (419–592)	50 (34–61)	27 (19–35)	↖↗	6·1 (3–9)

*Except *G. leptonepia,* which is scantily described.

*ts of Globodera Species**

asurements greater than 10 μm given to nearest μm

Granek's ratio	No. of lines on dorso-ventral axis of perineal pattern	Distance 'B' of Granek's ratio μm	Fenestral length μm	Fenestral width μm	Fenestral shape	Vulval slit length μm
	Cyst fenestralia and perineum					
6 (1·3–1·9)	< 14	27 (22–34)	16 (12–18)	?	?	5 (2·8–5·6)
o (0·8–1·7)	?	26 (19–47)	25 (20–32)	23 (14–28)	oval	14 (12–15)
1 (1·8–2·2)	?	65	20–28 31	15–28	?	35**, curved 14
2 (2–3)	< 14	44 (22–67)	21	19	oval	11 (9–15)†
6 (2·7–8·9) 5 (2–7)	> 14	60 (37–77) 68 (29–116)	15 (8–20) 18	18	circular	9 (6–11)†
2 (1·3–3·8) 7 (1·3–2·8)	< 14	48 (31–84)	24 (15–32)	21 (12–28)		9·4 ± 1·5† (5–13) 10–11
5 (0·9–2·8) 8 (0·7–5·2) 7 (1·0–2·2) 3 (1·0–1·5)	< 14	39	28–30; 33	28	oval	10–11
8 (1·5–4·2) 7 (2·4–8·9)	10 on para-type, 10–12 on 5 other specimens selected at random.	52 (28–85)	20 (13–28)	18 (13–25)	oval	9·2 (8–11)† 9–10

This value seems anomalous. †Adult ♀.

Ta

Characters of Males of Species of Heterod

When possible, measurements are give
parentheses. Where two or more figu
by a semicolon, these repres

Species	Length (mm)	Max. width (μm)	'a' value	Spear length (μm)	Spear-knobs; slope of ant. face	Spicule (μm)
AVENAE	1·36; 1·38		39	26–29	knobs rounded	36
CACTI	1·01			26–29 (28)		33·5
CAROTAE	1·0; 1·1	31·4		29	rearward	
CRUCIFERAE	1·18			25		
FICI	0·87–0·90	25–32	29–36	30		30–35
GLYCINES	1·1–1·4 (1·3)	27–31 (29)		26–28 (27)	knobs massive, rounded	34–37
GOETTINGIANA	1·21–1·29 (several authors)	29–40	51	28		33; 38
GRAMINIS	0·57–0·92 (0·78)	17–29 (24)	27–40 (33)	21–24 (23)	rounded, concave anteriorly	31–35 (3
HORDECALIS	0·81–1·39 (1·11)	26–32 (28)	30–48 (39)	24–28 (27)	not rear-ward; slightly concave anteriorly	34–42 (3
HUMULI	1·03; 0·85			31; 27		32; 35

nd in Temperate Areas

served ranges followed by means in
sets of figures are given, separated
a from different authors.

Spicule hape; tip	Guberna-culum (μm)	Head shape; no. of annules	Lateral field	Hemi-zonid; no. of annules ant. to exc. pore	Cephalids, no. of annules behind head	Excretory pore	Tail; length (μm)
unt or uncate; p rolled		5 ann.					
dentate; uncate		not offset; small, dome-shaped					
identate		12 × 8 μm deep; 7–8 ann.					
identate		4 ann.; 6 ann.			anterior 3rd–4th; posterior 8th–10th		
nt. 3rd ylindrical; est tapered	8 (single rod)	7–8 μm deep				post. to nerve ring	6 μm; 5 terminal papillae
identate	9·9–12·5 (11·4)	hemi-spherical 6–7 ann.; 4–5 ann.	4 lines ext-ending round tail	3–8	near 6th and 8th; near 2nd	144 ± 11 μm from ant. end	3·7–8·4 (5·3); 1·7–5·5 (3·5)
uncate	10·9	5 ann.					
rcuate; identate	9–12 (11)	offset; 9 × 4 μm deep; 4–5 ann.	4 lines; outer bands wider; irreg. areolate	7	ant. 2nd; post. 7th–8th	83–145 (115) μm from anterior	cloaca near-terminal
visted identate	9–13 (10·5)	offset; 4–5 ann.	4 lines; outer bands wider; areolate	6–8	ant. 2nd–3rd; post. 6th–7th	140 μm from anterior	tip smooth
		11 × 6 μm deep; 4–5 ann.					

Ta

Species	Length (mm)	Max. width (µm)	'a' value	Spear length (µm)	Spear-knobs; slope of ant. face	Spicule (µm)
IRI	1·06–1·45 (1·29)	33–40 (38)	35–40 (37)	28–33 (30)	knobs rounded; rearward slope	36–43 (3
LATIPONS	0·96–1·41 (1·17)	25–33 (29)	41	22–29 (27)	concave anteriorly	32–36 (3
MANI	1·10–1·75 (1·39)	26–33 (29)	41–58 (49)	22–30 (26)	sl. concave anteriorly; rearward slope	35–43 (3
ROSII	0·84–1·64 (1·22)	25·42 (32)	28–41 (39)	28–31 (30)	rounded	36
SCHACHTII	1·5; 0·8–0·9	20–29; 28–42	32–48 (several authors)	23; 30	rounded	33
TRIFOLII	0·63–0·98; 1·32					
URTICAE	0·93–1·43 (1·20)	28–40 (35)	34	24–30 (27)	spear strong; usually rearward slope	30–42 (3

:ontinued)

Spicule shape; tip	Gubernaculum (μm)	Head shape; no. of annules	Lateral field	Hemizonid; no. of annules ant. to exc. pore	Cephalids, no. of annules behind head	Excretory pore	Tail; length (μm)
isted; dentate	11–14 (12) simple	offset; 12 × 7 μm deep; 5 ann.	4 lines; outer bands crenate and areolate	9	ant. 2nd; post. 7th	172–202 (191) μm from anterior	rounded, short
arcuate, pered; isted steriorly	8, trough-shaped	offset; 12 × 6 μm deep	4 lines; outer bands irreg. areolate	3 ann. wide, 2 body-widths behind median bulb	ant. 2nd–3rd; post. at mid-stylet level		
isted; dentate	9–14 (12) simple	sl. offset; 11 × 6 μm deep; 5 ann.	4 lines; outer bands crenate and areolate	7	ant. 2nd–3rd; post. 7th–8th	139–195 (165) μm from anterior	truncate, short
rved ntrally & t.; tip otched	10·8, simple plate	offset; 3–4 ann.	4 lines			61–63μm from anterior	short, rounded; 4 lateral indentations
dentate	c. $\frac{1}{3}$ length of spicules	hemi-spherical 3–5 ann.	4 lines ext. round tail		ant. 2nd; post. 6th–8th	approx. midway between oes. bulb & post. margin of subv. oes. gl.	v. short
dentate	$\frac{1}{3}$ length of spicules	4–5 ann.					
isted; tip athulate; dentate	shallow boat-shaped; beaded margin	offset; 13 × 7 μm deep; 4–6 ann.	4 lines, end near tail tip; outer band crenate and sl. areolate	11 μm ant. to exc. pore	ant. 2nd–3rd; post. at mid-stylet level	162 μm from anterior	rounded

Ta

Characters of Males of Species of Globodera and Punctoc

(Measurements set

Species	Length (mm)	Max. width (µm)	'a' value	Spear length (µm)	Spear-knobs; slope of ant. face	Spicule (µm)
G. ACHILLEAE	1·05–1·22 (1·15)	34–39 (37)	28–34 (31)	27–29 (28)	rearward	30–37 (3
G. PALLIDA	1·20; 1·05	28·4; 29·3 (at exc. pore)		28; 27	rearward	36
G. ROSTOCHIENSIS	0·89–1·27 (1·08)	31–46 (39) (28 at exc. pore)	22–36 (27)	25–27 (26)	spear strong; knobs rounded, rearward slope	32–39 (3
G. SOLANACEARUM	0·75–1·45 (1·11)	22–40 (26)	43	24–27 (26)	ant. face rearward; base rounded	29–35 (3
G. TABACUM	0·7–1·4			24–27	spear strong; knobs rounded	
G. VIRGINIAE	0·85–1·33 (1·16)	23–33 (27)	27–52 (43)	24–29 (27)	knobs rounded; rearward slope	30–35 (3
P. PUNCTATA	0·9–1·3					34–35

ıd in Temperate Areas

n Table 2)

Spicule hape; tip	Guberna-culum (µm)	Head shape; no. of annules	Lateral field	Hemi-zonid; no. of annules ant. to exc. pore	Cephalids, no. of annules behind head	Excretory pore	Tail; length (µm)
cuate; tip ınded	11·2–12·9 (11·8)	offset; 10 × 5 µm deep; 5 ann.	4 lines; not areolated	2			v. short, cloaca extends beyond terminus
›ut, cuate; tip tire	11, rod-shaped; simple	rounded, offset; 12 × 6–7 µm deep; 6–7 ann.	4 lines; not ext. round tail; areolate in places	2–3	ant. 2nd–4th; post. 6th–9th	15% L from anterior	truncate
arcuate; › entire	10–14 (12)	sl. offset, hemi-spherical, 12 × 7 µm deep; 6 ann.	4 lines	2	ant. 2nd; post. 5th		1·7–6·7 (4·4) variable shape
arcuate; › entire	12·4	scarcely offset; 11 × 6 µm deep; 7 ann.	4 lines ext. round tail; outer band irreg. areolate	1–5	post. 8th–10th	14·9% L from anterior	2–8 (4·3); short, slight protuber-ance
pered; untly ›inted		6 ann. + labial disc		adjacent to exc. pore			0·2–3·6 µm
cuate; › entire	12·5	12 × 6µm deep; 7 ann.	4 lines ext. round tail; outer band irreg. areolate	1–4	ant. 2nd; post. 7th	15·1% L from anterior	0·5–5·0 (2·7)
ender, tip tire; lges rolled		4 ann. + cap	4 lines ext. round tail			2 body-widths post. to nerve ring	

CHAPTER 7

Cyst Nematodes: Biology of *Heterodera* and *Globodera*

T. D. WILLIAMS

Rothamsted Experimental Station, Harpenden

The cyst-forming nematodes of the genera *Heterodera* and *Globodera* are primarily root parasites though they may also occur on underground stems, e.g., stolons and tubers of potato (Oostenbrink, 1950) and, exceptionally, on leaves of white clover (Ross, 1960). The adult males of both genera are vermiform and free-living whereas the females are sedentary in, or attached to, the root. The female body wall tans to form a resistant brown cyst in which many second-stage larvae, each enclosed in an egg membrane, can remain dormant for several years. In the two genera there are at present a total of some 45 valid species (see Chapter 6), many of which attack important crops. A number of others await full taxonomic study and description, being at present distinguished mainly by association with a particular host plant or location.

Distribution and Spread

Although *Heterodera* and *Globodera* species are generally regarded as parasites of crops in cool temperate areas they are increasingly encountered in lower latitudes, albeit at higher altitudes, e.g., *G. rostochiensis* in India (Jones, 1961), Algeria and Tunisia; *H. schachtii* in Golden Valley, California may achieve an exceptional five generations per year. *H. glycines,* the soya-bean cyst nematode, occurs widely in the southern U.S.A. and in Egypt and Asia, while *H. sacchari* and *H. oryzae* are tropical pests of sugar cane and rice respectively. *G. rostochiensis* occurs widely in the main potato growing areas of the British Isles, throughout Continental Europe and Western U.S.S.R., around the Mediterranean basin, Iceland, India (higher altitudes), South Africa, the Andean regions of South America and Central America. It is also known from East and West Canada as well as isolated occurrences in the U.S.A. More recently it has been found in Japan and New Zealand. Another potato cyst nematode, *G. pallida,* is distributed widely in Western Europe and Peru, and also occurs in Iceland, Ecuador, Colombia, Northern Bolivia, India and New Zealand (Evans, Franco & de Scurrah, 1975; Stone, 1975b). Unlike *G. rostochiensis,* it has not yet been reported from the U.S.S.R., North America or Japan. Its South American range appears to be more southerly than that of *G. rostochiensis. H. avenae,* the cereal cyst nematode, is known from Western Europe, Norway, Greece, Western Siberia, U.S.A., Canada, India, Israel, Japan, Peru, South and West Australia, Victoria, and New South Wales. The distribution of all cyst-nematode species and their rate of development and reproduction are affected by temperature, rainfall and soil type as well as host crops (Jones, 1975a, c). Spread in the field through their own activity is slow, perhaps about one metre/annum, but cultivation and other forms of farm transport, including movements of transplants, tubers, soil washings, and packaging material, greatly accelerate the spread of foci of infection. The movements of man, birds and other animals, wind, flood and irrigation water can also aid in dispersion.

Life History

Raski (1950) described the development of the type species, *H. schachtii*; a schematic representation of the life cycle of a *Heterodera* species is shown in Fig. 7.1. As the embryo increases in length following gastrulation it begins to move within the egg-shell (a) and two or three flexures develop. When the first moult occurs (in the egg) the stylet begins to form at the anterior end. The second (invasive) stage larva (b) emerges from the egg, using its stylet to cut a slit in the shell, and leaves the cyst to enter the soil. *H. iri* larvae always emerge tail first; the stylet plays no part in hatching (Laughlin, Lugthart & Vargas, 1974). Most *H. schachtii* eggs are retained in the cyst, only a few passing into the egg sac (h & i). The size and content of the egg sac varies with the species; *H. carotae* and *H. cruciferae* have a large egg sac containing half the eggs, in *H. glycines* about one-third of the eggs are extruded. Fewer eggs are extruded in *H. trifolii*, *H. galeopsidis* and *H. goettingiana* while none escape in *H. avenae*, *H. humuli* and *Punctodera punctata*. *G. rostochiensis* produces only a small gelatinous extrusion (Williams, 1957). The second-stage larva usually enters a host root just behind the growing tip or where a lateral root emerges and migrates until its head reaches the pericycle. Depending on temperature the second stage is completed in about seven days from entry and the 2nd moult occurs. The third-stage larva (c) has a well developed rectum, and genital primordia appear. Female larvae (g) have paired ovaries, the male (d) a single testis. The third stages are complete in about 10 or 11 days after entry.

As male development proceeds, elongation of the male larva occurs within the cuticle of the third and fourth larval stage (e). The fourth moult is complete by about the 15th day when the adult male is still coiled within the third larval cuticle. There are three flexures so the adult male (f) is about four times the length of the third-stage larval cuticle. The inset (f) shows the male spicules in enlargement; the tips are differently notched according to species and have been used in diagnosis. The males emerge about 20 days after larval entry and become free living. They move towards those females whose vulval ends have protruded through the root. Many males may surround each female and multiple matings can occur (Green, Greet & Jones, 1970). Males are short lived, remaining active for nine or ten days (Evans, 1970).

The third-stage female (g) is shorter than the second stage and the twin ovaries can be distinguished. These increase rapidly in length at the time of the third moult. The fourth-stage larva is flask-shaped and is about 0·4 mm long. The body cavity becomes almost filled with the developing ovaries. During the fourth moult the reproductive system opens to the exterior via the formation of the vulva. By the thirtieth day the ovaries have obliterated other organs within the body cavity and are full of eggs (h). Shortly after the females rupture the root cortex, the egg sac begins to form in lemon-shaped species. If removed it reforms and appears to be a constant feature. It is easily lost when females are removed from roots because it adheres to soil particles.

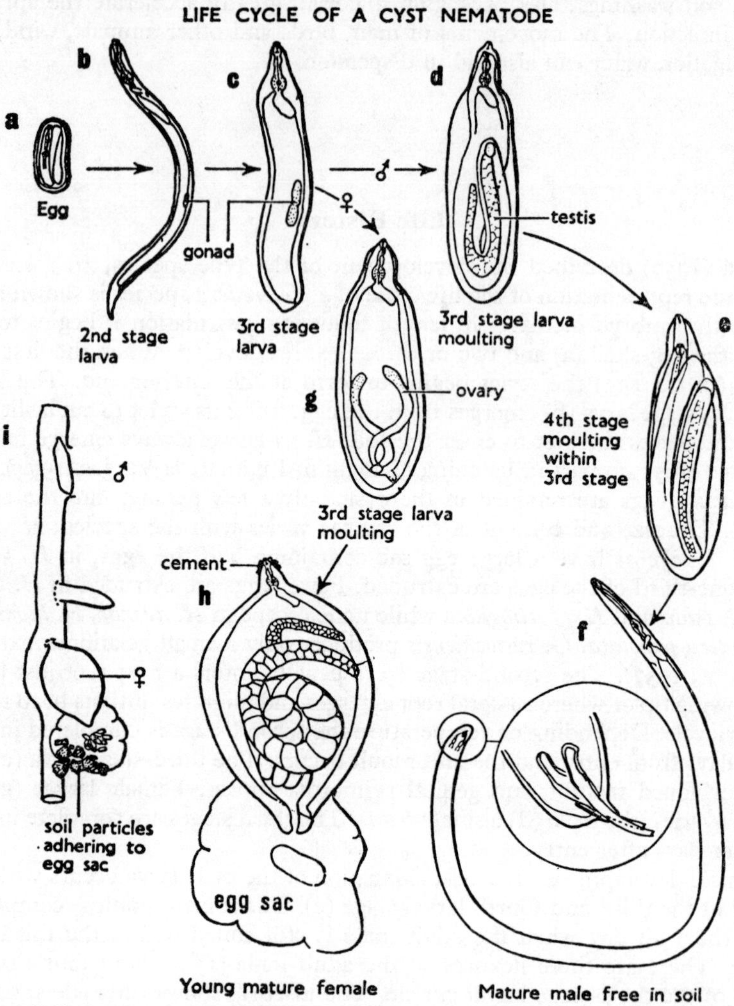

Fig. 7.1. Life cycle of a cyst nematode, *Heterodera* sp. (From Jones & Jones, 1974).

Cytogenetics and Reproduction

Riley & Chapman (1957) found that *G. rostochiensis* was diploid (2n = 18). Cotten (1960) found metaphase figures with nine, ten or eleven pairs of chromosomes in *G. rostochiensis* females, those with nine predominating; only a few had eleven pairs. Mulvey (1960) reported that *H. schachtii*, *H. cruciferae*, *H. goettingiana* and *H. avenae* were diploid (2n = 18). *H. trifolii* has 27 unpaired chromosomes at metaphase I indicating that the species is triploid. Giant eggs of *H. trifolii* were found by Mulvey (1959b) following impregnation of *H. trifolii* by males of *H. schachtii* but union of sperm and egg nuclei was not observed; polyploidy was inferred, as in giant *G. rostochiensis* eggs which produced giant larvae (Onions, 1953). Mulvey (1960b) found that oogenesis in *H. schachtii* and *H. avenae* was similar. Oogonia multiply by mitosis and increase in size. Chromosomes are not usually discrete at this stage and their numbers are determined later, at metaphase (Triantaphyllou, 1971). Oogenesis and the chromosome numbers of 12 bisexual *Heterodera* and *Globodera* species are compared by Triantaphyllou (1975). The oocytes move to the oviduct and thence to the spermatheca where the small amoeboid sperm are stored. Only spermatids and spermatozoa are found in adult males, the sperm output is therefore limited. Chromatin in *Globodera* sperm passes through a finely stranded phase during maturation; this phase does not occur in *Heterodera* (Shepherd, Clark & Kempton, 1974). One sperm enters each oocyte and the chromatin mass of the sperm is resolved into nine chromosomes. During meiotic division two polar bodies are given off. The egg and sperm pronuclei form a fusion nucleus with 18 chromosomes. Cleavage then follows. Mitotic parthenogenesis occurs in *H. trifolii, H. galeopsidis, H. sacchari* and *H. lespedezae*—one polar nucleus and egg pronucleus are formed, both with the somatic chromosome number. No sperm is needed for activation of the oocytes.

In *H. betulae* where $n = 12$ or 13 instead of the basic *Heterodera* $n = 9$, meiotic parthenogenesis occurs. Oogenesis and spermatogenesis are similar to those of amphimictic species. Two divisions take place in the maturation of the oocytes and two polar bodies are produced. Mature eggs contain the haploid number of chromosomes. Activation of oocytes is probably by substances produced in the glandular part of the uterus. In *H. betulae* an endomitotic division occurs during prophase which results in the doubling of the chromosome number to $2n = 24$ before the first cleavage division. Parthenogenesis has not been observed in *G. rostochiensis* or *G. pallida*. Den Ouden (1960b), Ellenby (1957) & Williams (1957) found no eggs in females reared in isolation.

Evidence suggests that the sex ratio (to be distinguished from sex determination) in *Globodera* spp. may be environmentally controlled, more males being produced when the population of larvae inside roots is dense (Ellenby, 1954a; den Ouden, 1960b). Kerstan (1969) found that equal numbers of males and females of *H. schachtii* develop in single-larva inoculations and that an increase in male: female ratio under stress is caused by a reduction of females *not* an increase in males. Johnson & Viglierchio (1969) reached the same conclusion. Increasing inoculation densities of *H. glycines* increased the male: female sex ratio (Koliopanos & Triantaphyllou, 1972). Sex expression in *Heterodera* species is almost always genetically controlled, not modified by environmental influences as is the sex ratio. An exception may be *G. rostochiensis* in which single-larva inoculations produce few males and many females (Trudgill, 1967) whereas crowded conditions and other stressful situations result in a high male: female ratio. Antimetabolites can also affect the sex ratio; d-l methionine is most effective in decreasing numbers of females of *G. rostochiensis* (Evans & Trudgill, 1971) while d-l alanine is most active in the case of *H. avenae* (Prasad & Webster, 1967). Thornley and Hesling (1973a, b) compared mathematical models with biological findings on the assumption either that (a) the sex of *G. rostochiensis* is irreversibly determined by nutrient status early in development, or that (b) the sex of the larva is genetically determined and the sex ratio is varied by differential entry or mortality of the males and females in the roots. They concluded that a decision between these alternative hypotheses could not be made on available data. Environmental determination of sex expression is much more common in *Meloidogyne* species because of their predominantly parthenogenetic nature (Davide & Triantaphyllou, 1967a, b). Triantaphyllou (1960) found evidence that sex in *Meloidogyne* may be reversed when conditions change for the worse during development of second- or third-stage larvae. Males with two testes develop, presumably from the twin primordia of the female gonads (Triantaphyllou, 1973).

Though hybridization between species and/or pathotypes of *Globodera* spp. and *H. avenae* sensu lato can occur, Parrott (1972) found incompatibility between the then 'pathotypes' of *Heterodera rostochiensis* where single matings were more successful within than between 'pathotypes'. This would be the first step towards genetic isolation of sibling species. Stone (1973) erected the new species *H.* (= *G.*) *pallida* from populations previously regarded as pathotypes of *H. rostochiensis*. Andersen (1965) found that hybridization between two pathotypes of *H. avenae* was possible. Crosses between various pathotypes of *H. glycines* also occur. Green & Miller (1969) found that experimental crosses between pathotypes of *G. rostochiensis* and *G. tabacum, G. virginiae, G. mexicana* and *G. solanacearum* (then Osborne's cyst nematode) could be made, though in most cases there was a measure of incompatibility. *H. glycines* and *H. schachtii* have also been hybridized (Potter & Fox, 1965) but the progeny was only capable of limited development. Fox (1967) failed to cross *H. glycines, H. carotae* and the then Osborne's cyst nematode because the female of each species failed to attract males of the others. Yeates (1970) also found that *H. carotae, H. cruciferae* and *H. goettingiana* failed to interbreed *in vitro*. Green & Plumb (1970) reviewed the relationships of some *Globodera* and *Heterodera* species as indicated by the male attractants emitted by their females, and Green & Greet (1972) studied the location of the secretions that attract males of *G. rostochiensis* and *H. schachtii* to their females.

Factors Affecting Hatching

Species of *Globodera* and *Heterodera* exhibit varying degrees of dependence on root exudates to stimulate hatching of larvae—*G. rostochiensis, H. humuli, H. carotae* and *H. cruciferae* are highly dependent on root exudates for substantial hatch to occur in the laboratory (Fenwick, 1949; Winslow, 1955a), though at least a 30 per cent decrease in numbers may occur in the field in the absence of a host crop (Oostenbrink, 1950). *H. goettingiana* has not so far been stimulated to hatch *in vitro* although in the field it responds to the presence of host roots (Shepherd, 1963). *H. avenae,* in earlier studies, was not found to respond to cereal root exudates (Winslow, 1955a; Hesling, 1957) but these tests were conducted at temperatures of 20–25°C the range usually used in *G. rostochiensis* work. Williams & Beane (1972) found that at lower temperatures, 10–15°C, more appropriate to conditions found in the field at the time of autumn or spring invasion, a marked response to root exudates occurred. Species such as *H. schachtii, H. trifolii* and *H. galeopsidis,* which have wide host ranges are less dependent on stimulation by root exudates and hatch more readily in water. While, in the main, hatching is stimulated by exudates from host plants there are exceptions. Thus *H. schachtii* is stimulated by the non-hosts *Hesperis matronalis* and *Coronopus squamatus* while *H. cruciferae* is not stimulated by *Coronopus squamatus* which is a host (Jones, 1950; Winslow, 1955a). *G. rostochiensis* is also stimulated to hatch by some non-host clones of *Solanum tuberosum* ssp. *andigena* (Williams, 1958). There is no correlation between resistance to female nematode development and the production of active root exudates.

The constitution of the natural hatching factors for *G. rostochiensis* is unknown. The main difficulties in identification are their instability and the problem of achieving fractions of sufficient purity though substantial progress has been made. Complete analysis must await pure isolates which can be examined by chromatography, electrophoresis and other structure determining techniques, and the production of a volatile derivative for gas chromatography and mass spectrometry (Clarke & Hennessy, 1975). The hatching factor(s) are soon inactivated in soil (Fenwick, 1956); this lessens their possible value in control.

Artificial hatching agents exist—*G. rostochiensis* is stimulated by anhydrotetronic acid (Calam, Todd & Waring, 1949) and *H. schachtii* by rivanol (Winner, 1957) and by many dyes (Shepherd, 1962a). Zinc salts stimulate many species to hatch though they inhibit the hatch of *H. avenae*. Many metal ions stimulate the hatch of *H. schachtii* and vanadate ions are particularly effective in hatching *G. rostochiensis* (Clarke & Shepherd, 1966). Picrolonic and flavionic acids hatch *H. cruciferae* and *G. rostochiensis* respectively. Clarke & Shepherd (1964) found that 31 out of 283 compounds hatched *H. schachtii* as well as or better than beet root exudate. Many hatching agents appeared to act as direct or indirect electron acceptors. Clarke & Shepherd (1968) summarized the results of testing 444 compounds for their ability to stimulate hatch of *G. rostochiensis*. Greet (1975) showed that five species of *Globodera* differed in their response to five artificial hatching agents and suggested that the differences might have some diagnostic value. Although some of these substances hatch *Globodera* and *Heterodera* species in numbers equal to the natural agents they do not work at such extreme dilutions—5 parts in 10^8—as did an impure natural product isolated by Clarke & Widdowson (1966). The activity of root exudates is calculated as a logarithmic activity (L.A.) value based on hatch in a series of geometric dilutions (Fig. 7.2) (Shepherd, 1962b).

Root exudates are not the only factors which influence the rate of hatching. Soil moisture and soil aeration (closely linked because of the relationship between soil moisture and pore space) are both important, as is temperature. These aspects and their interrelationships are reviewed by Jones (1975a, c). Once inside the host root the developing cyst nematode larva is no longer subject to fluctuating soil moistures, instead it is enclosed by the more stable environment of the host tissue. Only when the adult males emerge do soil conditions again affect movement.

Wallace (1954, 1955c) found that maximum emergence of larvae of *H. schachtii* in fallow soils or sand occurs at the hydrostatic pressure deficiency at which most of the pore spaces are emptied of water. This is near the point of inflection of the moisture characteristic for the soil or sand concerned and depends on pore size (Fig. 2.1, page 48). Wallace (1956) also found that adding quite small amounts of clay (with very small particle size) decreased the hatch. The speed and extent of movement of *Heterodera* species in soil depends on particle size, pore size, thickness of water films, diameter of the nematodes, their length, and the amplitude and frequency of wave forms along the body (Wallace, 1958a, b; 1959, 1960, 1968; Jones, 1975a). Encysted eggs of *Globodera* and *Heterodera* species differ considerably in their ability to survive desiccation and this appears to bear some relation to the proportion of larvae that hatch in water. Eggs of *G. rostochiensis* which hatch only slowly in water can be stored within the cysts for months or years in an air-dry condition. By contrast, many larvae of *H. schachtii* hatch in water and emergence from cysts is considerably lessened by air drying. Winslow (1955a) and Hesling (1956) found that even short periods

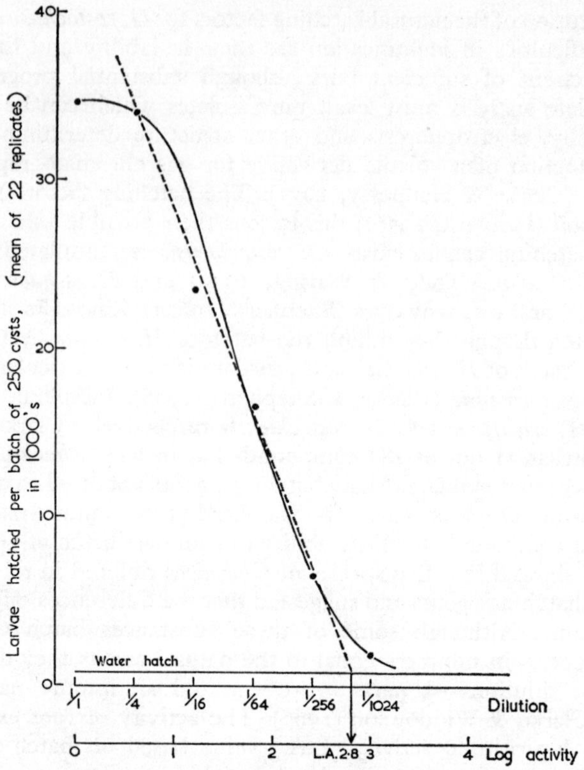

Fig. 7.2. Typical hatch/dilution curve for *Globodera rostochiensis* with potato root exudate, showing the derivation of the L.A. value. (From Shepherd, 1962b).

of drying inhibited subsequent *H. avenae* hatch whereas Duggan (1960) found that *H. avenae* survived for three months in air-dried soil but not six months. In contrast, an Australian population showed little loss of viability of eggs in air-dried soil after 13 months (Meagher, 1970). *H. avenae* from Rajasthan, India, can withstand two years dry storage out of soil and yet hatch freely when placed in water. The greater resistance of Indian and Australian populations to desiccation may be adaptations to the local very dry conditions which prevail for part of the year, though Meagher (1975) considers that differences in technique and age of cysts may account for the conflicting observations so far recorded. Meagher also found that eggs of *H. avenae* could be stored at low relative humidity for five years in the laboratory and survival was enhanced by storage at low temperatures (5°C). Populations of *H. glycines* in fallow soil remained infective for 90 months at near field capacity, for 38 months in dried soil and for 19 months in flooded soil (Slack, Riggs & Hamblen, 1972). More precise information about the water content of nematodes can be gained using the interference microscopy methods of Ellenby (1968) or the more convenient immersion refractometry technique (Ellenby, 1975).

For many *Globodera* and *Heterodera* species the optimum temperature for *in vitro* hatch is about 25°C (Shepherd, 1962b). Chitwood & Buhrer (1946) found that potatoes in the field on Long Island, U.S.A. were heavily invaded at 16°C while Evans (1969) found many *G. rostochiensis* at all stages of development between soil temperatures of 10–15°C and he concluded, with Jones & Parrott (1969), that British populations were more cold-adapted. Bishop (1955) and Wallace (1955b) found that for *G. rostochiensis* and *H. avenae* alternating temperatures of 25°C and 15°C resulted in greater emergence than continuous exposure to 25°C. Although alternating temperatures may be stimulatory, the activity rate of temperature regulated processes predicted from mean temperatures underestimates the true rate (Currie, 1973). Dunn (1954, 1962) showed that the temperature of storage of *G. pallida* prior to exposure to root exudate strongly affected subsequent hatch at 15°C. Cysts kept at 29°C for five weeks did not start to hatch for 20 days, but then the hatch was similar to that from cysts kept at 23°C whereas the hatch at the latter temperature was 20–30 times greater than from those stored at 15°C. Dunn stresses the importance of the storage temperature prior to exposure to diffusate in contrast to the work of Bishop (1955) who varied the temperatures *during* immersion in diffusate. The response of *H. avenae* to temperature has been much studied in Britain, Canada and Australia (Cotten, 1962; Fushtey & Johnson, 1966; Banyer & Fisher, 1971a, b; Meagher, 1975). Some of the relationships are complex and differ according to locality. Fushtey & Johnson found that in Canada a minimum period of eight weeks at 0–7°C was a prerequisite for hatch and that a following temperature of 15°C gave the best hatch. Essentially this was similar to the earlier findings of Cotten (1962) in Britain. Banyer & Fisher (1971a) distinguish between temperature regimes governing pre-hatch maturation and hatch itself while, in contrast to the sequence described by Fushtey & Johnson, Banyer & Fisher (1971b) found a type of dormancy in some encysted eggs when exposed to 20°C following eight weeks at 7°C; a similar dormancy occurred in the field in the Australian spring.

Recent literature (since 1971) on the physical and chemical factors affecting hatching of cyst nematodes is comprehensively reviewed by Clarke & Perry (1977).

Seasonal Effects

Even when conditions appear favourable some species become dormant—few *G. rostochiensis* larvae hatch *in vitro* when cysts are left in soil until August whereas if taken in before the end of July and stored dry, hatching in the winter is normal (Fenwick & Reid, 1953). Earlier, Calam, Raistrick & Todd (1949) had reported dormancy irrespective of whether cysts were kept at constant (23°C) or fluctuating temperatures from August. Cunningham (1960) found dormancy was the outcome of conditions prevailing in moist soil; dry laboratory conditions prevent its onset. Winslow (1956) found that, in many *Heterodera* species, *in vitro* hatch was slightly lower in winter and the dormancy which occurred was induced by earlier conditions. The dormancy could be shortened by storing infested soil at high temperature but the cessation of dormancy in the spring was not necessarily associated with the rise in soil temperatures. The dormancy noted by Winslow and Calam *et al.* had already set in before they extracted their cysts; Fenwick & Reid (1953) found none because they extracted their cysts before the onset of this phase. Shepherd & Cox (1967) preferred the term 'facultative diapause' to describe this situation where only removal from the field before August prevents a later reduction of hatch *in vitro*.

Host-parasite Relationships

At the onset of hatching, *G. rostochiensis* larvae cut slits in the egg-shell by the very precise placement of stylet thrusts when stimulated by natural hatching factors. Co-ordination of stylet movement was less good when hatching was induced by sodium metavanadate (Doncaster & Shepherd, 1967; Doncaster, 1974).

Widdowson, Doncaster & Fenwick (1958) first showed that *G. rostochiensis* larvae were attracted to roots under sterile conditions. The most attractive areas were the root tip, zone of elongation and root-hair zone. Aqueous root washings attracted *G. rostochiensis* and *H. schachtii* (Wallace, 1958c; Weischer, 1959). Most of the materials known to be released by roots fail to attract larvae but CO_2 in water does attract *H. schachtii* (Johnson & Viglierchio, 1961). Bird (1960) found, however, that root tips were more attractive than 'CO_2 only' sources. Soils are strongly buffered against change and most organic materials in aqueous solutions are probably soon degraded by micro-organisms. Wallace (1960) found that moisture gradients in sand did not affect the attraction of *G. rostochiensis* to host roots. Larvae of *H. schachtii* in an electric field move to the anode (Jones, 1960). Bird (1960) found that reducing agents attract *H. schachtii* but he also considered that the preferred parts of the root might be selected by a tactile sense.

Mankau & Linford (1960) observed that *H. trifolii* penetrated clover roots within 15 minutes of contact. Doncaster & Shepherd (1967) found that *Globodera* and *Heterodera* larvae perforate the most rigid parts of the cortical cell walls in the same way that they leave the egg. In *H. cruciferae* stylet thrusts are more frequent when cell contents escape. Before egg hatch, granular secretions occur in all three oesophageal gland ducts. After a week in the host the dorsal oesophageal gland enlarges and the subventrals (previously the most active) become smaller. Dorsal gland secretions possibly promote syncytial (transfer cell) development; they have this role in giant cell formation induced by *Meloidogyne* (Bird, 1968). The final permanent feeding site is usually at the endodermis. Ingestion of food is achieved by the pharynx behaving as a 'leaky valve', in this way the 'pump' can operate with only one metacorpal outlet valve and the net flow during pulsations is rearward into the intestine (Doncaster, 1971). Passive ingestion of food has not so far been demonstrated in any Heteroderid species but it does occur in species of *Hexatylus* and *Ditylenchus*.

The formation of syncytia is an integral part of the host-parasite relationship of *Globodera* and *Heterodera* species (Plate XV). The syncytia are sites of high metabolic activity, they were first noted by Treub (1887). Cell wall dissolution to form syncytia is well documented in *Globodera* and *Heterodera* infections (Mankau & Linford, 1960; Endo, 1964, 1965). There is no evidence of nuclear division or fusion. Enlarged nucleoli and irregular nuclei are common. Sembdner (1963) observed abundant mitochondria, Golgi bodies, protoplastids and dense endoplasmic reticula in syncytia in tomato roots galled by *G. rostochiensis*. The role of syncytia in acting as transfer cells was first determined by Jones & Northcote (1972a). Wall protuberances form in cells adjacent to the conducting elements and as expansion proceeds sieve elements are crushed. The transfer cells form as a response to solvent flow. Jones & Dropkin (1975) have further demonstrated, using a KOH digestion and scanning electron microscope technique, the extensive scale of cell-wall breakdown in syncytia formed by *H.*

glycines and *G. tabacum*. They consider that the enzymes responsible for cell-wall dissolution are of plant, not nematode origin. Endo (1964) found that the persistence and nature of syncytia in soya-bean roots infected by *H. glycines* depended on the feeding habits of the male and female. Syncytia induced by males were of shorter duration than those induced by females which were more extensive and had denser cytoplasm related to the longer feeding and development time. Trudgill (1967) established similar correlations for *G. rostochiensis* in stressed and unstressed potato plants.

The Nature of Resistance

Plants exhibit varying degrees of resistance towards *Heterodera* and *Globodera* species. The following scheme, based on Jones (1954) and Shepherd (1959) lists the categories in which plants can be placed.

1. Immune—larvae do not invade.
2. Resistant—invaded.
   (*a*) Larvae fail to develop past the invasive stage.
   (*b*) Some larvae develop to third stage.
   (*c*) Some larvae develop into mature males but females absent or very few.
   (*d*) Abundant males; females few, not prolific.
   (*e*) Females few but highly prolific.
3. Susceptible—many highly prolific females produced.

These categories are not rigid and considerable variation can occur. Most of our knowledge of these relationships within the genera is from *G. rostochiensis*, *G. pallida*, *H. avenae*, *H. schachtii* and *H. glycines*.

Larvae may readily enter resistant plants (Williams, 1956) though Dropkin & Webb (1967) found that fewer *Meloidogyne hapla* larvae entered resistant tomato plants. Host necrosis near the nematode is often considered as an expression of resistance; it is not a feature of *Globodera* and *Heterodera* on efficient hosts since the parasites are so well adapted. Ross (1958) considered the response of a resistant soyabean cultivar to be based on hypersensitivity. Wilski & Giebel (1966) found that larvae of resistance breaking pathotypes of *G. rostochiensis* produced only small amounts of B-glucosidase with consequently smaller production of a toxic aglycone from a non-toxic phenolic glucoside in the host. Trudgill (1974) suggested that d-l tyrosine influenced the sex of *G. rostochiensis* (see also p. 159) via phenolic pathways and the role of phenols in resistance. Applied three days after invasion it increased the proportion of larvae becoming male without affecting total numbers or the differential death rate. Giebel & Stobiecka (1975) consider a decreased ratio of proline to hydroxyproline in potato root suppresses cell hypertrophy and enhances resistance to *G. rostochiensis*. The roles of auxins, enzymes, phenolics and proteins in nematode resistance are reviewed by Giebel (1974). The most fundamental expression of resistance is the extent to which the root undergoes syncytial development in response to larval invasion. Little is known of any biochemical (enzymatic?) processes which may influence syncytial development.

Host susceptibility may be dependent on successful molecular mimicry by the parasite. Common antigenic properties appear to be a feature of compatible host-pathogen interactions between *Meloidogyne hapla* and cotton and soya-bean

(McLure, Misaghi & Nigh, 1973). Jones (1975b) suggested that a gene-for-gene relationship might exist between potato cyst nematodes and resistant potatoes with an antibody/antigen relationship between host cells and nematode saliva or between other gene products. Jones & Parrott (1965) advanced the hypothesis that dominant genes for resistance to *G. rostochiensis* in potato hybrids ex. *Solanum andigena* and *S. demissum* are matched by recessive genes in nematodes able to multiply on them. Males of all genetic constitutions (AA, Aa, aa) can mature in the roots of resistant plants but only recessive (aa) females can mature. In *H. avenae*, Andersen (1965) found that races 1 and 2 would interbreed and that race 2 was dominant to race 1. Andersen & Andersen (1973) demonstrated a linkage between a gene in barley conferring resistance to *H. avenae* pathotypes 1 and 2 and a lack of purple colour at nodes, ligules and awns.

Biochemistry and Physiology

Clarke *et al.* (1967) and Clarke (1968) analysed the dried egg shells of *G. rostochiensis* and the cyst wall. Smith & Ellenby (1967) studied changes in the free amino-acids in *G. rostochiensis* cysts. Aspartic and glutamic acids and serine increased with increasing maturity while proline, glycine, valine, alanine, methionine and isoleucine decreased. They also considered the yellow pigment in females to be a quinoid tyrosine derivative. Ellenby (1946a) had earlier found that polyphenol oxidase was involved in the tanning of cysts. The sub-crystalline layer of *H. mani, H. avenae* and *H. goettingiana* consists of cosanoic acids and their calcium salts, and probably represents an inert deposit derived from excess sugary wastes excreted through the cuticle by feeding females (Brown *et al.*, 1971). Most work on enzymes has been done on genera other than *Heterodera*. Conventional enzyme analysis is difficult due to the minute quantities available and the impermeability of the intact cuticle to many reagents. Homogenate analysis may not represent the situation in the whole organism though histochemical localization has been successful in some cases. Goffart & Heiling (1962) detected invertase, pectinase and amylase in *G. rostochiensis*. Ellenby & Smith (1968) found transaminases in the cyst walls of *G. rostochiensis*.

Aist & Riggs (1969) found alanine, aspartic acid, glutamic acid, glycine and serine in washings from entire *H. glycines* larvae; asparagine, glutamine, leucine, isoleucine, lysine, methionine sulphoxide, threonine, tyrosine, valine and methionine were additionally found in extracts from crushed larvae. α-amino-butyric acid, histidine, phenylalanine and proline were emitted by whole live adults, and arginine, cystathionine, hydroxyproline and ornithine were found only in extracts of crushed adults. The work of Giebel, Wilski, & Stobiecka has been referred to in connection with the role of enzymes in resistance mechanisms.

Clarke & Hennessy (1976) found glycogen and oligosaccharides in the larvae and cyst wall of *Globodera rostochiensis* and trehalose in the egg perivitelline fluid. Whole eggs contained all three substances but the larvae contained very little trehalose. Prior to hatching, following stimulation by hatching agents, the energy requirements of the larvae increase sharply. Trehalose is found in many helminths and glycogen commonly occurs as a reserve polysaccharide.

To supplement traditional taxonomic methods, based on morphology, biochemical and physiological attributes have been used to group cyst nematodes (see Chapter 6). Examples are the use of serology (p. 143), attraction of males by female secretions (see p. 160), electrophoretic analysis of proteins (p. 143), specificity of host range and effects of root exudates (Winslow, 1954, 1955), and the phylogeny of host angiosperm families (Krall & Krall, 1970). Using immunodiffusion serology techniques Webster & Hooper (1968) placed six species into two groups. One contained *H. schachtii, H. trifolii* and *G. rostochiensis,* the other *H. cruciferae, H. goettingiana* and *H. carotae.* On this basis, *G. rostochiensis* appears to have more affinity with bullate lemon-shaped species than with abullate kinds. The sources of the antigens are uncertain—excretory products may be involved, also the collagen (Clarke, Cox & Shepherd, 1967) in the cuticle of *G. rostochiensis* larvae and egg shells.

Host Ranges

While most species of *Globodera* and *Heterodera* are limited in their hosts to one or few related plant families, some, especially *H. schachtii, H. trifolii* and *H. galeopsidis,* have hosts in many families. *G. rostochiensis* and allied round-cyst nematode species occur almost entirely in only one family, the Solanaceae; *H. avenae* is similarly confined to one family, the Gramineae. Sometimes the host ranges of two species partially overlap, thus both *H. schachtii* and *H. cruciferae* can thrive on Brassicas and many other cruciferous genera whereas only *H. schachtii* reproduces on Chenopodiaceae and only *H. cruciferae* on the crucifer *Coronopus squamatus.* An unusual relationship is that of American populations of *H. schachtii* on tomato (Golden & Shafer, 1959) where a degree of ectoparasitism occurs (Steele, 1971a). Steele (1971b) also found numerous larvae of *H. schachtii* in roots of sunflower, parsley, egg-plant and sweet-pea though few if any mature females survived. Host lists for *Heterodera* species have been prepared by Jones (1950), Winslow (1954), and Oostenbrink (1955). First recorded hosts are also listed by Goodey, Franklin & Hooper (1965). Further details of hosts, bionomics, etc. of *Globodera* and *Heterodera* species are given in the Commonwealth Institute of Helminthology Descriptions of Plant-parasitic Nematodes.

The Biology and Economic Status of Some Agriculturally Important GLOBODERA and HETERODERA spp. of Temperate Areas

Some species are polyphagous and weeds may play a part in supporting populations in the absence of a host crop but others, e.g., *G. rostochiensis* and *G. pallida,* have restricted crop and weed host ranges in Europe, though not in Latin America. The number of generations per year depends on factors such as duration of the host crop, soil temperature and moisture (Jones & Jones, 1974; Jones, 1975c). In light, open soils moisture can be particularly limiting for larval movement and invasion and, later, drought can intensify the symptoms and damage caused by a spring invasion. This has been studied experimentally for *H. avenae* by Dixon (1963) who found that adequate spring moisture followed by summer drought conditions was followed by much worse crop damage than a dry spring and moist summer regime.

THE POTATO CYST NEMATODES (*Globodera rostochiensis* and *G. pallida*)

These are closely related species which only sparingly hybridize in the laboratory (Parrott, 1972; Kort & Jaspers, 1974); there is no evidence of fertile offspring being produced in mixed field populations. They are most readily distinguished by the colour of the maturing females, golden yellow for *G. rostochiensis*, white or cream for *G. pallida* (see also Chapter 6). Both are confined to the Solanaceae though there are many hosts within this economically important plant family. Careful evaluation of earlier literature is essential since the separation of *G. pallida* (Stone, 1973) may alter the interpretation of earlier research findings. *G. rostochiensis* was probably introduced from the Andes into Europe some time after 1850 and has spread into most potato growing areas. Soil populations were so increased by continuous potato growing during and after two world wars that losses became severe (Spears, 1968; Jones, 1965; Jones, 1970; Jones, 1972a). This species can be stored dry for many years and can persist in the absence of a host crop for 10 or more years. There is usually only one generation per year but there can be a limited second in Europe (Oostenbrink, 1950; Evans, 1969). Brown (1969) found that potato losses increased linearly with pre-crop nematode numbers. On average, each 10 eggs/g soil decreased yields by 1 tonne/ha over the range 0–150 eggs/g. Jones (1966) considers that the logarithm of the pre-planting population gives the best relationship with yield loss. Increase of the nematode is density dependent, decrease under fallow or non-host crops is density independent. The populations of potato cyst nematodes behave fairly consistently from year to year because the soil temperature is similar when larvae establish themselves, the environment in the root is independent of soil moisture and many other conditions of cultivation and choice of cultivar change little (Jones, 1974). In the absence of hosts the decline rate is of the order 30–50 per cent per annum (Huijsman, 1961). Den Ouden (1960a) showed that this decline was not a steadily continuing process; most of the annual decrease followed a short period of spontaneous hatching in the spring. The economic threshold for crop loss from *G. rostochiensis* and *G. pallida* is between 20–30 eggs/g soil. When *G. rostochiensis* males and females rupture the root cortex the severity of *Verticillium* wilt is increased (Corbett & Hide, 1971; Hide & Corbett, 1973).

In addition to the stunting effects of potato cyst nematodes, Trudgill *et al.* (1975a, b) found that heavily infested plants contained less K and P but more Ca and Na. Nitrogen status was little affected and did not appear to be a limiting factor. Evans, Parkinson & Trudgill (1975) found that stomatal resistance and negative water potentials were increased by nematode attack. Irrigation lessened this stress. Maris Piper resisted water stress better than Pentland Dell, particularly when infested by *G. rostochiensis*; the improvement over Pentland Dell was less marked when *G. pallida* was the pathogen.

For most cyst nematodes a four- to six-course rotation is usually sufficient to ensure that populations fall to safe levels. For *G. rostochiensis* only potatoes need to be considered but for beet, cereal and pea cyst nematodes more than one host crop is involved.

Resistant potato cultivars, e.g., Maris Piper, Ulster Glade, bred from wild or cultivated clones of Andean potatoes are now grown extensively in Britain, and corresponding cultivars exist in Holland and Germany. (See Chapter 16). These allow increased frequency of potato cropping. Without such cultivars, 4 or 6 course rotations with or without nematicides are necessary (Jones, 1966,

1970; Jones & Parrott, 1968; Howard, 1972; Jones & Jones, 1974). Nollen & Mulder (1969) have developed a programme in Holland whereby potatoes may be grown twice in four years on infested land providing one of the crops is resistant and the soil is fumigated with D-D once in four years.

For further details, see Chapter 14, also Stone (1973), Jones & Jones (1974).

BEET CYST NEMATODE (*Heterodera schachtii*)

This nematode has a very wide host range, including many major crops and weeds, which makes control by rotation more difficult. Plants attacked are sugar and fodder beet, mangolds, red and spinach beet, brassicas, mustard, cress, rhubarb and numerous members of the Caryophyllaceae, Amaranthaceae, Labiatae, Portulacaceae, and Scrophulariaceae (Jones, 1950; Winslow, 1954; Steele, 1965). The decline rate in the absence of hosts may be more than 60 per cent per annum since *H. schachtii* hatches freely in water (Winslow, 1954; Jones, 1956). *H. schachtii* was first recognized as the cause of beet-sickness in Germany in 1859 following intensive beet cultivation near the sugar factories, a sequel repeated in Britain from 1925 onwards. Rotation of host crops, including brassicas, was encouraged from 1935 in Britain and enforced through the Sugar Beet Eelworm Order of 1943 (Petherbridge & Jones, 1944) and its subsequent revisions. Yield losses have been effectively minimised but the nematode has continued to spread through cultivations and other movements of soil. The accepted economic threshold for *H. schachtii* on sugar beet is 10 eggs/g air-dried soil. Although some wild species of *Beta* have marked resistance to *H. schachtii* (Swink, 1954; Shepherd, 1958; Golden, 1958) breeding for resistance has proved difficult and has not yet produced cultivars suitable for commercial use in Europe. Because of the eight-month duration of the sugar-beet crop, *H. schachtii* can achieve two generations per year in Western Europe and, exceptionally, five in California. See also Jones & Dunning (1972); Franklin (1972).

CEREAL CYST NEMATODE (*Heterodera avenae*)

This nematode has probably been long established in North West Europe. Grasses are far less efficient hosts than cereals but can enable populations of *H. avenae* to maintain themselves in the absence of cereals (Winslow, 1954; Hesling, 1958b; Stone, 1960). It is widespread, occurring in the Middle East, India, Southern Australia and North America as well as Europe. It hatches fairly freely in water but cereal root exudates can further stimulate *in vitro* hatch at lower temperatures (Williams & Beane, 1972). Typical field symptoms are the 'knotting' and proliferation of spring-cereal seminal roots and consequent reduction of nodal root growth (Williams & Salt, 1970). Dixon (1969) established regressions for the effects of *H. avenae* on spring wheat, barley and oats. He found that for every increase of 10 eggs/g soil before sowing there was an approximate loss per hectare of 376 kg oats and 188 kg wheat, but only 75 kg/ha of barley.

A recent survey by entomologists of the South-east Region of the Agricultural Development and Advisory Service showed that *H. avenae* was present in 94 per cent of the 357 fields sampled; 26 per cent of the fields had more than 10 eggs/g soil, 13 per cent had over 20 eggs/g, and 4 per cent more than 40 eggs/g (Anon., 1975a). Gair (1965) listed the population levels at which different degrees of damage to winter and spring cereals occur. In general, winter crops are less damaged though they may be capable of increasing nematode populations more than spring crops. Often however, *H. avenae* populations stabilize and even decline under continuous cereal cropping (Stone, 1968; Gair *et al.*, 1969). In his survey of barley fields in England and Wales, Stone found that pre-crop *H. avenae* populations of more than 20 eggs/g soil tended to fall at all levels of cereal yield. Williams (1969) showed that formalin removed a restraint on *H. avenae* in a Bedfordshire population that tended to stabilize at about 10 eggs/g. The role of fungal parasites of *H. avenae* females and eggs is being further investigated (Kerry, 1974, 1975). Under certain conditions, not yet fully understood, large populations (> 200 eggs/g) of *H. avenae* occur exceptionally on the thin downland chalk soils of South and South-west England. Substantial losses (2·5 t/ha) of the more tolerant barley crop occur at these sites and, as chemical control is not economic, these outbreaks are best combatted using resistant barley cultivars like Ansgar, Sabarlis and Tyra which suppress the development of females of pathotypes 1 and 2. (See Chapter 16). Usually only one generation per year occurs even though winter crops may be in the ground for about nine months. Several pathotypes of *H. avenae* exist throughout the world (Kort, Dantuma & van Essen, 1964; Cook & Williams, 1972; Jones, 1972b; Brown, 1974). Some of these may prove to be new species, as did *G. pallida,* and it is evident that *H. avenae* is but one of a complex of allied species including *H. iri, H. mani, H. hordecalis* and other as yet undescribed forms—see Chapter 6. See also Empson (1965); Williams & Siddiqi (1972).

PEA CYST NEMATODE (*Heterodera goettingiana*)

The pea cyst nematode is widely distributed in North West Europe and is well established in Eastern and Central England. There are many infestations in gardens and allotments. It attacks field and garden peas, broad and field beans, vetches, soya-beans and lentils (Jones, 1950; Winslow, 1954). Peas are especially sensitive, root growth being reduced but without the proliferation of lateral roots that is so often a feature of *Heterodera* attack. Severe nitrogen deficiency often follows the complete suppression of nitrogen-fixing nodules on pea roots, so that rapid yellowing and death of the haulm ensues (Oostenbrink, 1955). Fungal infections frequently aggravate the nematode damage and total pea crop failure can occur (Winfield, 1965). Field beans are more tolerant and less efficient hosts (Jones & Moriarty, 1956; Moriarty, 1963). *H. goettingiana* can persist at damaging levels for 10 years or more in the absence of a host crop (Brown, 1958). Although hatching of larvae in field soils is stimulated by host roots, exudates *in vitro* have failed to stimulate emergence (Shepherd, 1963). See also Stone & Course (1974); Anon. (1971).

OTHER EUROPEAN SPECIES

There are several other species of *Heterodera* attacking field crops in Britain and other Western European countries; some, restricted in host range, can cause damage in limited areas, others with wider host ranges nevertheless cause little crop damage.

The cabbage cyst nematode (*H. cruciferae*) attacks all brassicas and many other crucifers. The carrot cyst nematode (*H. carotae*) causes damage to crops in the black fens and light soils of East Anglia; carrots (both cultivated and wild) are the only known hosts despite the prevalence of other umbelliferous weeds. The clover cyst nematode (*H. trifolii*), does not often cause detectable crop damage; it has hosts in the Leguminosae, Caryophyllaceae, Polygonaceae, Labiatae and Scrophulariaceae. The three foregoing species are all capable of completing two generations in a year. *Heterodera humuli* is found on hops, nettles and hemp but so far no damage has been attributed to it.

SOYA-BEAN CYST NEMATODE (*Heterodera glycines*)

Although not known in Europe and apparently adapted to a higher temperature range than European *Heterodera* species, *H. glycines* is a pest of major economic importance in Asia, U.S.A. and Egypt. The 'yellow-dwarf' disease of soya-bean was first associated with a cyst-forming nematode in Japan in 1915 but not described as a new species until 1952 (Ichinoe, 1952). Ichinohe (1961) further described the morphology and bionomics of this nematode. It was found in N. Carolina, U.S.A. in 1954 and has since been found in six other Southern and South-eastern States. Several generations can be completed in a single growing season and losses may be so great that the harvest is less than the seed sown. *H. glycines* causes poor nodulation and early yellowing of soya-bean plants—cf *H. goettingiana* on peas. Several highly resistant cultivars of *Glycine max* are known (Ross & Brim, 1957; Chambers & Epps, 1967). New resistant cultivars became available in the U.S.A. from 1967 and have helped curtail the yield losses estimated at that time to be as much as 30 million bushels. As in the case of resistant oats invaded by *H. avenae*, the yield of some resistant soya-beans is lessened by *H. glycines* invasion but succeeding crops benefit from the decreased number of nematodes. Other hosts are lespedeza, snap bean, vetches and lupin, as well as some Labiatae, Carophyllaceae and Scrophulariaceae. *H. glycines* is closely related to *H. trifolii* but does not reproduce on clover and can be distinguished morphologically at the second-larval stage (Hirschmann, 1956). It also is closely allied to *H. schachtii* (Triantaphyllou, 1970; Stone, 1975a).

CHAPTER 8

Aphelenchoides and Related Genera

MARY T. FRANKLIN
Commonwealth Institute of Helminthology, St. Albans

The superfamily Aphelenchoidea (Fuchs, 1937) Thorne, 1949 was erected for nematodes with the tylenchid type of oesophagus comprising mouth spear, cylindrical procorpus, swollen muscular metacorpus with valve plates, and terminal glandular region variously lobed and usually overlapping the intestine. Members of the Aphelenchoidea differ from the Tylenchoidea in having the dorsal gland duct opening into the anterior part of the metacorpus and not into the lumen of the procorpus, as in the Tylenchoidea. Females of Aphelenchoidea always remain vermiform and have a single prodelphic ovary and the vulva in the posterior third of the body.

Four families are recognized in the Aphelenchoidea. The main differentiating characters of the three families which contain species likely to be found in association with cultivated crops are set out in Table 1. The fourth family, Entaphelenchidae Nickle, 1970, contains obligate parasites of insects and is not dealt with here. Each of the families Aphelenchidae and Paraphelenchidae has one genus, *Aphelenchus* and *Paraphelenchus* respectively; these nematodes feed on fungi and may at times cause damage in mushroom cultures. The family Aphelenchoididae contains numerous genera of which *Aphelenchoides* and *Rhadinaphelenchus* include plant-parasitic species.

APHELENCHIDAE (Fuchs, 1937) Steiner, 1949

Aphelenchus avenae Bastian, 1865

Originally described by Bastian in 1865, a fuller description with designation of a neotype was published by Goodey & Hooper (1965) and the life-history and larval development were studied by Hechler (1962). The following description is taken from these authors and from Hooper (1974a). (See Fig. 8.1).

Morphology—17 Females (topotypes): $L = 551–858$ (mean 670) μm; $a = 25–39$ (33); $b' = 7{\cdot}2–9{\cdot}2$ ($8{\cdot}3$); $b = 4{\cdot}9–6{\cdot}9$ ($5{\cdot}9$); $c = 27–35$ (31); $V = {}^{29–46}74–78^{9–14}$ (${}^{38}76{\cdot}5^{11}$).

Body cylindrical, slightly curved ventrally when killed by gentle heat, tapering at both ends. Head low, about twice as wide as deep, with three or four faintly visible annules. Body annules 1 to 2 μm wide; lateral field usually with 12 longitudinal lines in the mid-body region: the number decreases anteriorly and posteriorly as shown in Fig. 8.1I, J. Excretory pore, hemizonid, hemizonion and cephalids as shown (Fig. 8.1H). Vulva slightly protruding; anterior lip of anus distinctly prominent; tail bluntly rounded, 1·5 to 2·5 anal body-widths long, with sub-terminal phasmids (Fig. 8.1C). Mouth stylet 14–15 μm long with slight basal thickenings. Procorpus cylindrical, slightly constricted at its junction with

Table 1

Differentiating Characters of the Families of Aphelenchoidea (Fuchs, 1937) Thorne, 1949 associated with Fungi and Higher Plants

APHELENCHIDAE (Fuchs, 1937) Steiner, 1949	PARAPHELENCHIDAE (Goodey, 1951) Goodey, 1960	APHELENCHOIDIDAE (Skarbilovich, 1947) Paramonov, 1953
Oesophageal glands overlap intestine mainly dorsally	Oesophageal glands in a spatulate bulb abutting intestine	Oesophageal glands overlap intestine dorsally or dorsolaterally
Stylet with or without small basal thickenings	Stylet as in Aphelenchidae	Stylet base without thickenings, with slight thickenings or with distinct knobs
Spicules slender; V-shaped gubernaculum	Spicules slender with large apex and posteriorly directed rostrum; V-shaped gubernaculum	Spicules rose-thorn shaped, variable. Usually no gubernaculum but sometimes a small similar structure is present
Well-developed caudal alae with ribs	With or without caudal alae	No caudal alae, or with a small ala at terminus
Mycophagous in soil or in decaying plants	Mycophagous in soil	Plant-parasitic, mycophagous, predatory; in soil, aerial parts of plants, sea and fresh water, lichens, or associated with insects
Aphelenchus	*Paraphelenchus*	Numerous genera including *Aphelenchoides*, *Rhadinaphelenchus*, *Seinura*, *Laimaphelenchus*, *Bursaphelenchus*

the prominent, broadly oval, muscular metacorpus. The bulb is 20 × 13 μm in the neotype. The glandular region of the oesophagus envelops the anterior end of the intestine irregularly (compare Fig. 8.2), with the long dorsal lobe generally on the left side of the body and sometimes up to 100 μm long. Gonad outstretched anteriorly, lying on the right side of the intestine with oocytes partly in a single, partly in a double row, the vagina thick-walled and a post-vulval uterine sac reaching about half-way to the anus. The structure of the gonad is described in detail by Geraert (1972).

9 Males: L = 632–807 (739) μm; a = 27–33 (29); b′ = 8·1–9·6 (8·5); b = 5·5–5·9 (5·7); c = 24–29 (27); T = 48–67 (60).

Males are uncommon in most natural populations but occur in cultures under certain conditions (see p. 176, under Biology). In general structure they resemble the females but have 10–12 incisures on the lateral field and the stylet is 12–14 μm (Hechler, 1962). The peloderan bursa, supported by four pairs of papillae, reaches from the level of the spicule heads to the tail terminus. The spicules are 24–29 μm long, slender, cephalated, ventrally curved and narrowing to a slender

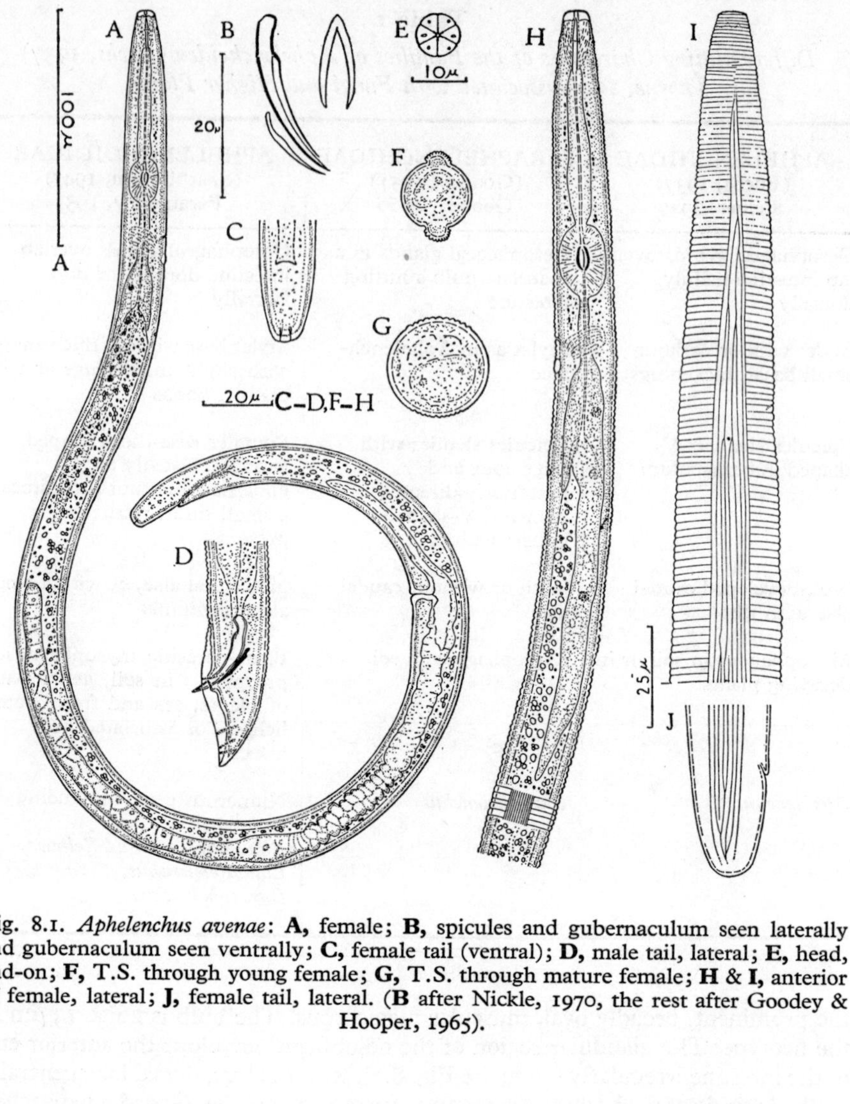

Fig. 8.1. *Aphelenchus avenae*: **A,** female; **B,** spicules and gubernaculum seen laterally and gubernaculum seen ventrally; **C,** female tail (ventral); **D,** male tail, lateral; **E,** head, end-on; **F,** T.S. through young female; **G,** T.S. through mature female; **H** & **I,** anterior of female, lateral; **J,** female tail, lateral. (**B** after Nickle, 1970, the rest after Goodey & Hooper, 1965).

point. The gubernaculum is 11–12 μm (Hechler, 1962) to 14 μm long (Goodey & Hooper, 1965) and V-shaped when viewed ventrally (see Fig. 8.1B).

The eggs are 70–80 × 25–35 μm, and first-stage larvae moult within them. A fully developed stylet and metacorpus are first seen in the second-stage larvae which hatch and begin feeding. Subsequent larval stages increase in size, apart from which they differ only in the number of incisures on the lateral field which is given by Goodey & Hooper (1965) as four in the second stage, six in the third stage and eight in the fourth stage, all counted in the mid-body region, the numbers becoming gradually fewer at either end. Measurements of larval stages from a population cultured on *Pyrenochaeta terrestris* are given by Hechler (1962) and from one on mushroom mycelium by Goodey & Hooper (1965). As might be expected, there are size differences between populations from different food sources (Evans & Fisher, 1970a). Other morphometric differences are

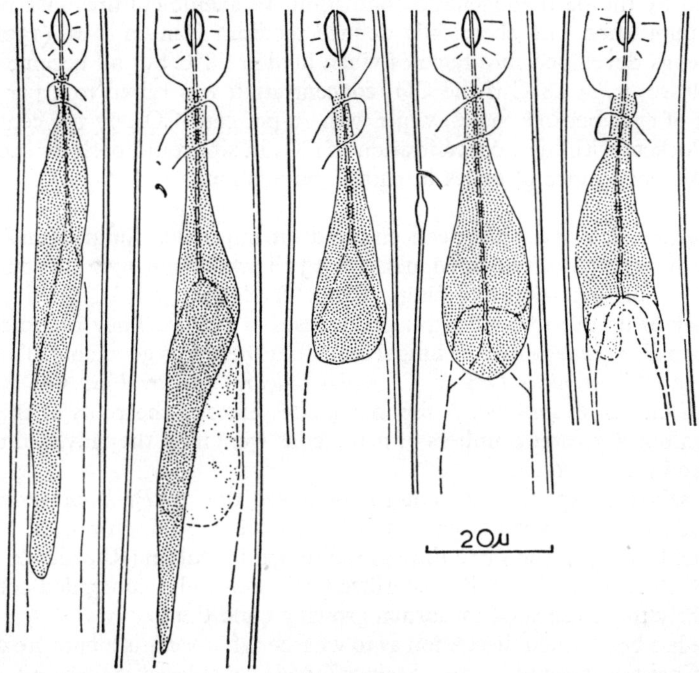

Fig. 8.2. *Aphelenchus avenae*: Oesophageal region showing variations in overlap of glandular region. (After Goodey & Hooper, 1965).

associated with length of feeding period (Fisher, 1970), and Monoson (1971) records the effects of the species of food fungus and the temperature on a number of morphometric values.

Biology. *A. avenae* is common in soil and rotting plant materials where it feeds on fungi. Males are rarely found in most wild populations, which are parthenogenetic, but an amphimictic population is known in Western Australia (Evans & Fisher, 1970b). The nematode is easy to culture *in vitro* on a number of fungal species and for this reason it has frequently been used as a laboratory animal, e.g., as food for predaceous nematodes (Hechler, 1962) and for studying feeding behaviour (Fisher & Evans, 1967; Doncaster, 1971). Hechler (1962) found that the generation time at 28°C on *Pyrenochaeta terrestris* was six days, while Goodey & Hooper (1965) found it to be 11–12 days at 20°C on *Botrytis cinerea*. On cultures of *Rhizoctonia solani* maximum numbers were greater and were reached sooner at 25 and 30°C than at lower temperatures (Evans & Fisher, 1970a).

Fisher (1969, 1972) studied copulation, fecundity and rate of reproduction in both amphimictic and parthenogenetic populations. Females of a parthenogenetic population required a continuous supply of food to produce eggs and produced the same average number on three different fungus species. When males were placed with parthenogenetic females copulation took place but far fewer eggs were laid than in the absence of males. As the proportion of males to females increased from one to eight per four females, the number of eggs laid decreased, whereas with amphimictic females it increased. Very few males were produced by the parthenogenetic population. In axenic cultures of a parthenogenetic population in a chemically defined medium Hansen *et al.* (1972) found that all eggs developed into females when held at 28°C but all became males at 30°C. However, at 28°C if the CO_2 concentration was raised to 5 per cent, 60 per cent of eggs became males, while with 10 per cent CO_2, 77 per cent became males. Pockets with high concentration of CO_2 in soil could perhaps account for spasmodic occurrence of males in natural populations.

Food sources. A. avenae has been cultured on numerous fungi, many of them plant pathogens (Mankau & Mankau, 1963; Townshend, 1972). Because of its feeding on phytopathogenic fungi, Klink & Barker (1968) investigated the possibility of its use to control fungal diseases of plants. They found that with 4000–6000 *A. avenae* per ml of fungal inoculum there was good control of disease caused by *Rhizoctonia solani* on *Phaseolus vulgaris* and by *Fusarium oxysporum* f. sp. *pisi* on *Pisum sativum.* A few nematodes entered the roots at this rate of inoculum but if greater numbers of nematodes were used they invaded the roots and reproduced in them.

In glasshouse experiments root rot of maize due to *Pythium arrhenomanes* was greatly reduced when large numbers (125 000 per 6-inch pot) of *A. avenae* were added to soil at the same time as the fungal inoculum (Rhoades & Linford, 1959). Such results, obtained in sterilized soil and under controlled conditions, are unlikely to be repeated in normal growing conditions.

There has been much discussion as to whether *A. avenae* is a parasite of higher plants (Christie & Arndt, 1936). Steiner (1936) considered it to be a facultative parasite of phlox. Decker (1962) found it in the roots of 28 plants and noted that it was more numerous in roots in the presence of *Pratylenchus.* However, when Mankau & Mankau (1963) used inocula of 100 000 *A. avenae* on seedlings of bean and sweet orange growing in sterile soil in 6-inch pots no lesions were found in the roots, and in two and six months respectively the nematodes had disappeared. These workers conclude that the nematode is an obligate fungivore and when found in lesions is probably feeding on fungi.

Barker & Darling (1965) were able to culture *A. avenae* on callus tissue of several plants, including tobacco and tomato. Reproduction was best on undifferentiated tissues and the callus became resistant if it was grown for two or three weeks before inoculation. Presumably *A. avenae* cannot easily penetrate intact tissue but is probably able to feed on healthy cells adjacent to fungal lesions. Feeding and reproduction on tomato roots was observed in experiments by Chin & Estey (1966) who also found that *Verticillium* wilt was more severe when the nematodes were present. Feeding, but not reproduction, was also recorded by these authors on roots of maize, oats, beet, turnip, cabbage and radish.

In culture *A. avenae* has been shown to feed on fungi and it suppressed the formation of some mycorrhizal associations with *Pinus resinosa* but it did not enter the pine roots or mycorrhizae (Sutherland & Fortin, 1968).

The only important crop with which *A. avenae* is closely associated is mushroom (*Agaricus bisporus*). In cultures it has been shown to increase rapidly on mushroom mycelium and to destroy it (Hooper, 1962) but, although frequently present in commercial mushroom beds it seldom builds up to damaging numbers. This may be because the temperature of the beds does not reach the optimum for the nematode (Hesling, 1972), which is 28–30°C for most populations studied, or possibly because of competition from other compost-inhabiting nematodes that are better adapted to the conditions.

PARAPHELENCHIDAE (Goodey, 1951) Goodey, 1960

Paraphelenchus myceliophthorus Goodey, 1958 (Fig. 8.3)

Morphology—50 Females: L = 580–820 (699) μm; a = 22–34; b = 4·1–6·6; c = 13–24; V = $^{26-55}71-78^{9-15}$.

The body of both sexes is finely annulated and has six lines on the lateral field; head not offset, flat and about one third as high as broad; tail conical with rounded terminus. Head skeleton consisting only of a stellate stylet guide; stylet without basal thickenings, about 16 μm long, the anterior conical part about 5 μm. Procoprus slender, cylindrical; metacorpus an ovoid muscular bulb

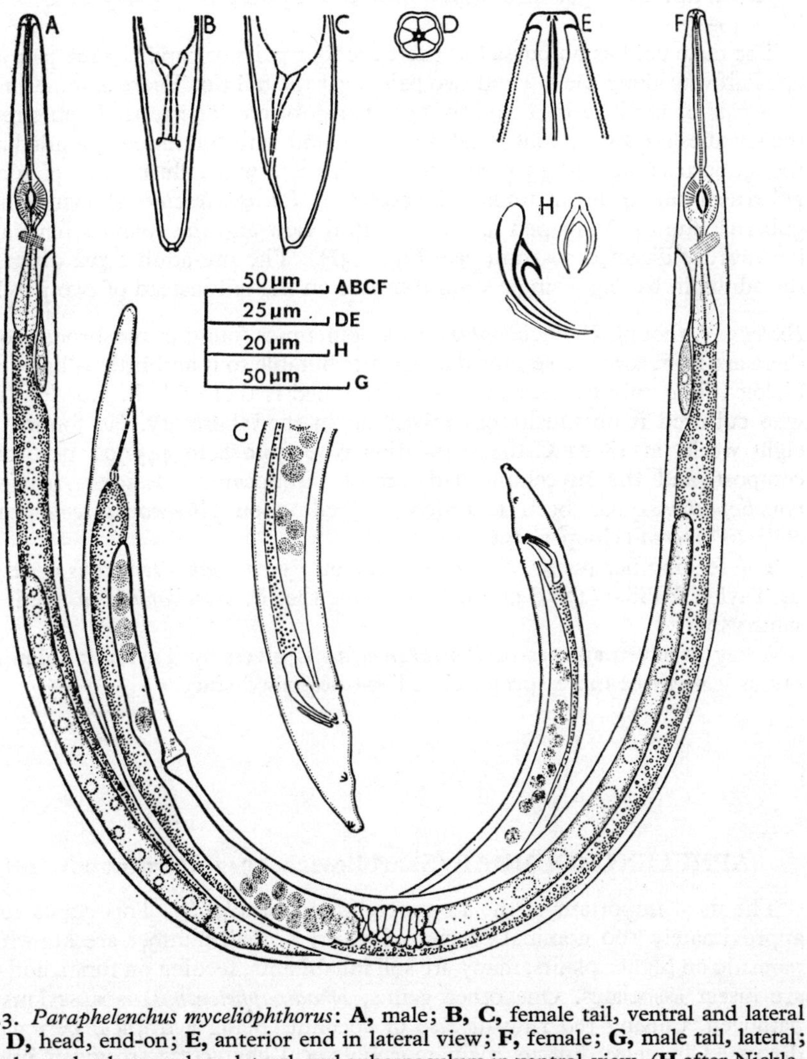

Fig. 8.3. *Paraphelenchus myceliophthorus*: **A,** male; **B, C,** female tail, ventral and lateral views; **D,** head, end-on; **E,** anterior end in lateral view; **F,** female; **G,** male tail, lateral; **H,** spicules and gubernaculum, lateral, and gubernaculum in ventral view. (**H** after Nickle, 1970, the rest after Goodey, 1958 and 1963).

with relatively large valve plates, the dorsal oesophageal gland opening in the anterior part and the ventral glands opening behind the valve. The spatulate glandular region of the oesophagus abuts on to the intestine, the junction being slightly displaced dorsally by the prominent excretory cell which is connected to the excretory pore by a distinct duct (Goodey, 1959). The hemizonid is four or five annules behind the excretory pore and extends over three to four annules; prominent deirids can be seen in the lateral field at the level of the hemizonid. The nerve ring encircles the isthmus close behind the metacorpus. In the female the tail ends in a pair of small papillae and there is a rectum about twice the anal body-width long (Fig. 8.3B, C). The gonad is prodelphic and a post-vulval sac extends about two thirds of the distance to the anus. Sperms may be present in the uterus and the post-vulval sac.

50 Males: L = 560–820 (691) μm; a = 23–32; b = 3·9–5·9; c = 14–30; T = 46–47

The male tail has no bursa but there are four pairs of papillae, one pair adanal, one half way along the tail and two pairs near the tail tip. There is sometimes one or a pair of papillae in a slightly pre-anal position. The testis is outstretched; the spicules are 28 μm long, slightly curved and with prominent rounded apices; the gubernaculum is 13·5 μm long. Nickle (1970) illustrates spicules and gubernaculum from an unnamed species of *Paraphelenchus* showing that the gubernaculum is V-shaped in dorso-ventral view and the spicules have a sharp backwardly directed rostrum (see Fig. 8.3H). The pre-adult larva differs from the adults in having a single ventral mucro on the tail instead of two papillae.

Biology. Although *P. myceliophthorus* is sometimes found in mushroom compost there are no records of serious damage attributable to it and little is known of its biology. The only experimental work recorded is that of J. B. Goodey (1960b) who cultured it on mushroom mycelium in the laboratory. He found that in eight weeks at 18–21°C the population rose from 8 to 443 700 per 100 g of compost and the mycelium had almost disappeared. Besides *Agaricus*, *P. myceliophthorus* has been recorded as feeding on *Alternaria tenuis* and on *Botrytis cinerea* (Hooper, 1962).

The feeding behaviour of *P. acontioides* on *Pyrenochaeta terrestris* is described by Taylor & Pillai (1967) and it is probable that *P. myceliophthorus* feeds in the same way.

A key to seven species of *Paraphelenchus* is given by Taylor & Pillai (1967) but at least three more species have been described since.

APHELENCHOIDIDAE (Skarbilovich, 1947) Paramonov, 1953

The most important of the 17 genera is *Aphelenchoides*. This genus contains approximately 100 nominal species of which a small number are known to be parasitic on higher plants, many are soil inhabitants, feeding on fungi, and others are insect associates. One other genus, *Rhadinaphelenchus*, is a serious plant pathogen, causing red ring disease of coconut. Being a tropical pest it is not discussed in detail but its main morphological characters are given briefly at the end of this chapter.

APHELENCHOIDES Fischer, 1894

These are slender nematodes with finely annulated cuticle and two, three or four lines in the lateral field; the head is usually offset. The tail in both sexes is more or less tapering, conical, with the terminus bluntly rounded or digitate with a simple or variable mucro. Male tails are characteristically curved or hooked when killed by heat, without a bursa but with three pairs of papillae, one adanal and two ventro-submedian between anus and terminus. The hexaradiate head framework is lightly sclerotized; the stylet has basal knobs or thickenings. The oesophagus has a characteristically prominent ovoid median bulb, with well-developed valve plates, and the glandular region is in the form of an elongated lobe lying dorsally over the intestine. The female gonad is short and outstretched, or long and reflexed one or more times, with the oocytes in one or more rows. The vulva is in the posterior third of the body. There is usually a post-vulval sac which varies in size and often serves as a spermatheca. The male gonad is usually straight, the spicules paired, rose-thorn shaped with more or less prominent apex and rostrum (Fig. 8.4). There is no gubernaculum.

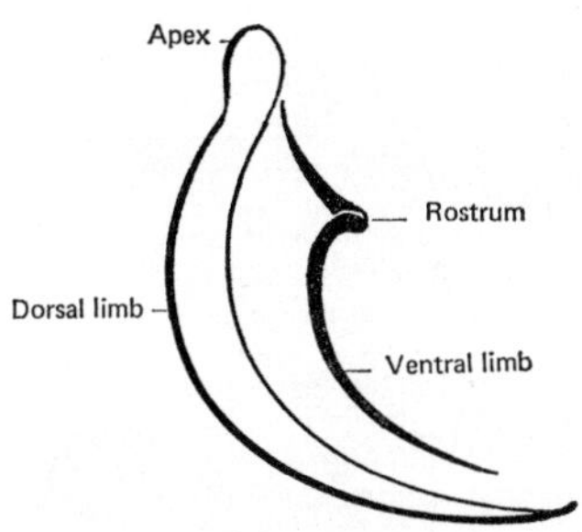

Fig. 8.4. Diagram of aphelenchoid spicule seen laterally.

The following species are plant parasites, all attacking the above-ground parts of their hosts.

- *A. ritzemabosi*, a leaf nematode attacking chrysanthemums and numerous other cultivated and weed hosts
- *A. fragariae*, a leaf nematode, also with numerous hosts
- *A. blastophthorus*, scabious bud nematode
- *A. besseyi*, rice white-tip nematode and cause of strawberry summer-dwarf disease, or crimp, in U.S.A.
- *A. subtenuis*, mainly on narcissus but also on other bulbous plants; not common
- *A. lilium*, on lily in Japan

On cultivated mushroom *A. composticola* often causes serious damage.

A detailed description is given of *A. ritzemabosi* and differentiating morphological characters for the others.

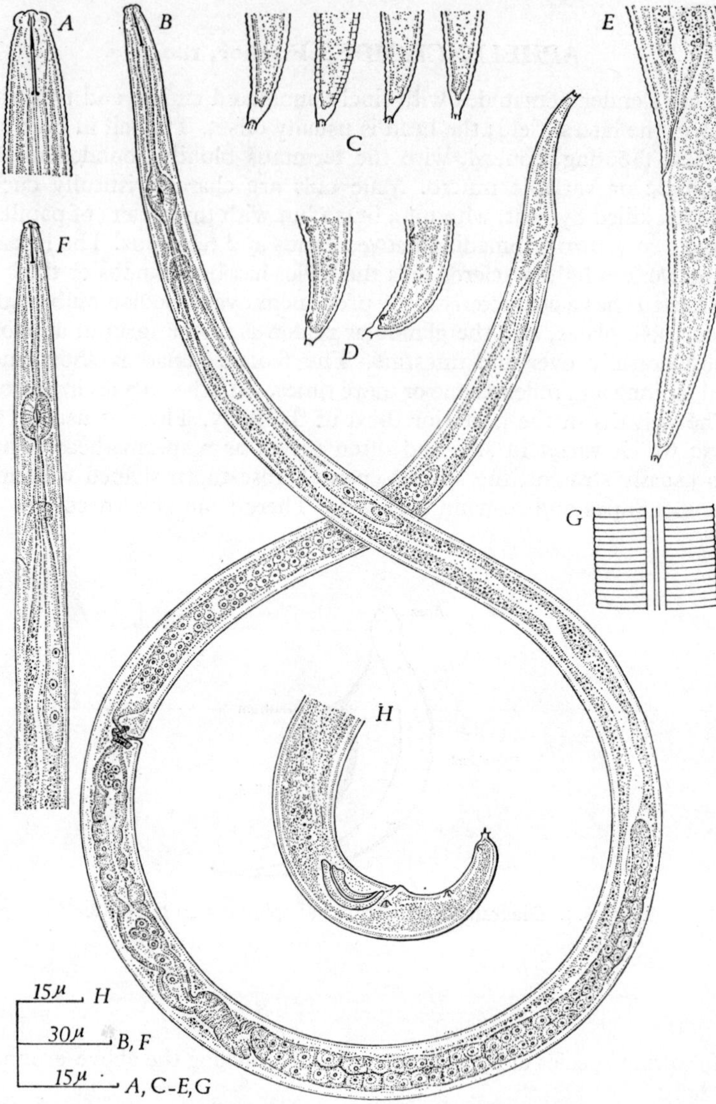

Fig. 8.5. *Aphelenchoides ritzemabosi*: **A,** female, head end; **B,** female; **C,** female tails; **D,** male tails; **E,** female, posterior end; **F,** female oesophageal region; **G,** lateral field; **H,** male posterior end. (From Siddiqi, 1974, *CIH Descriptions of Plant-parasitic Nematodes*).

Aphelenchoides ritzemabosi (SCHWARTZ, 1911) STEINER & BUHRER, 1932 (Fig. 8.5)

This description is largely taken from Siddiqi (1974b).

Morphology—Females: L = 770–1200 μm; a = 40–45; b = 10–13; c = 18–24; V = $^{48-33}66\text{–}75^{14-17}$. (Measurements from Allen, 1952).

Body slender with distinct annules; lateral field with four lines; lips offset, convex in outline; tail conical, tapering gradually to the terminus which has a mucro with two to four fine points of irregular length and not noticeably splayed out (i.e., not stellate). Hexaradiate, weakly developed head skeleton; mouth spear about 12 μm long with distinct basal knobs. Procorpus slender; metacorpus well-developed, muscular, with distinct valve plates centrally placed; glandular lobe of the oesophagus overlying the intestine dorsally, extending well back to a point five or six times the greatest body-width behind the metacorpus. Nerve ring one to two body-widths behind the metacorpus and excretory pore half to two body-widths behind it. Vulva slightly prominent; gonad with oocytes in more than one row; post-vulval sac prominent, extending more than half way to anus, often packed with sperm.

Males: L = 700–930 μm; a = 31–50; b = 10–14; c = 16–30; T = 35–64. Males are usually common. When killed by gentle heat they have characteristically hooked tails, shaped like a walking-stick handle. General morphology as in female. The spicules are smoothly curved, lacking a prominent apex or rostrum; the dorsal limb is 20–22 μm long and the ventral slightly shorter and more slender. The arrangement of the caudal papillae is typical of the genus. The mucro is simpler than that of the female, often with a single point nearer the ventral than the dorsal side.

Biology and Hosts. A. ritzemabosi is known principally as the chrysanthemum nematode but it has been recorded in the above-ground parts of nearly 200 plants in at least 40 families, usually living as an endoparasite but sometimes ectoparasitically in unopened buds. Unlike other economically important species of *Aphelenchoides, A. ritzemabosi* has not so far been cultured on fungi.

Serious damage to chrysanthemum is caused by the nematodes feeding on the cells of the mesophyll, after entering the leaves through the stomata. The cells are killed and brown necrotic areas develop which are at first angular, being delimited by the veins (Plate XVI), but later spread over the whole leaf. The nematodes multiply in the leaf and, when conditions become adverse, they emerge through the stomata and move over the surface of the plant to enter a fresh leaf. For this they need a film of moisture over the plant surface. The disease usually starts at the base of the plant and progresses upwards, nematode movement being favoured by wet and misty conditions. Eventually flower buds may be invaded and are deformed or killed. If chrysanthemum cuttings are infested the nematodes may feed in the terminal bud, on the surface of the young leaves and growing point, causing subsequent stunting and distortion and formation of rough brown feeding areas on the surface of leaves and stems (Southey, 1952); sometimes the growing point of infested cuttings is killed. At the end of thc growing season or when the air is dry the nematodes remain in the diseased leaves which may become dry and, if detached, can be spread by wind. French & Barraclough (1962) found that, in leaves kept dry for three years at 4° or 7°C, 33 per cent and 8 per cent respectively of nematodes of all stages could be revived. Under normal outdoor conditions nematodes in leaves or soil in heavily infested plots did not survive the winter and had all disappeared by early April. However, they can survive in buds of dormant chrysanthemum stools and invade the young shoots when growth is resumed (Hesling & Wallace, 1961a). Many common weeds such as *Stellaria media* and *Senecio vulgaris* serve as reservoir hosts in the absence of the chrysanthemum crop. Differences in susceptibility between chrysanthemum varieties were investigated by Hesling & Wallace (1961b) and Wallace (1961) but have not been put to practical use due to the wide and constantly changing range of commercially available cultivars.

Among other host plants are dahlia, delphinium, gaillardia, phlox, primula (Plate XVII), carnation and zinnia and the annual aster, *Callistephus chinensis,* the seeds of which can carry dry but viable nematodes (Brown, 1956). Besides those mentioned above, the recorded weed hosts include *Galium aparine*, *Bellis perennis* and *Cerastium vulgatum* (Goodey, Franklin & Hooper, 1965).

In addition to ornamental plants, *A. ritzemabosi* is a parasite of two important fruit crops, strawberry and black currant. In both of these it is mainly ecto-parasitic. It enters the buds in summer and remains inactive throughout the winter. In early spring it feeds on the unexpanded leaves so that, subsequently, when growth starts, the leaves and flower buds are seen to be deformed (Plate XVI) or destroyed. Heavily infested black currant buds may be killed, leaving bare branches. The later development of secondary buds often gives the bushes a characteristic, excessively branched appearance. The scale leaves and aborted leaves of infested buds are invaded by *A. ritzemabosi* and on falling to the ground they carry the nematodes with them; these may then enter weeds and re-infest the black currants later in the season (Brown & Franklin, 1953).

The *Aphelenchoides* disease in strawberry is complicated by the frequent association of *Corynebacterium fascians* with the nematodes. The feeding of *A. ritzemabosi* alone in strawberry buds results in symptoms of dwarfing, production of alaminate leaves and roughened brown feeding areas on the leaf surface at the base of the midrib (Plate XVI). The centre of the plant, including the flower buds, may be killed, resulting in 'open-centre' plants. When *C. fascians* and *A. ritzemabosi* are present together, symptoms described as 'cauliflower disease' are produced in some cases, probably depending on the virulence of the strain of bacterium. In this disease there is proliferation of axillary buds with many of the leaves being reduced to swollen petioles. The condition was investigated by Pitcher & Crosse (1958) who concluded that the nematodes alone may cause disease symptoms consisting of feeding areas and alaminate leaves that die and result in open-centre plants. *C. fascians* may also be present but plays no part in the formation of the alaminate leaves. The mainly bacterial disease which is manifested in cauliflower-like symptoms, is initiated by the bacteria that are carried into the buds by the nematodes and there stimulate growth of dormant meristems, but the symptoms may be modified by the nematodes. In the field *C. fascians* is almost universally present but there are many strains, of which only a few can provoke cauliflower disease symptoms in strawberry (Pitcher, 1963). Runners from diseased plants are almost certain to carry any pathogens present in the mother plants.

Aphelenchoides fragariae (RITZEMA BOS, 1891) CHRISTIE, 1932 (Fig. 8.6)

Morphology. Female: L = 450–890 μm; a = 36–63; V = 64–71
Male: L = 480–860 μm; a = 40–63; stylet = 10 μm

A. fragariae may be distinguished from *A. ritzemabosi* by the following characters (see Figs. 8.7 & 8.8):

(i) the body is more slender
(ii) the head is relatively straight-sided
(iii) the male tail is less sharply hooked
(iv) the spicules have a more prominent apex and rostrum
(v) there are two lines in the lateral field compared with four in *A. ritzemabosi*
(vi) the excretory pore is usually closer to the metacorpus
(vii) the tail in the female is more sharply conical and the mucro is usually simple

The value of the mucro as a reliable differentiating character is doubtful as some populations indistinguishable from *A. fragariae* in other features have a more or less 'tufted' mucro.

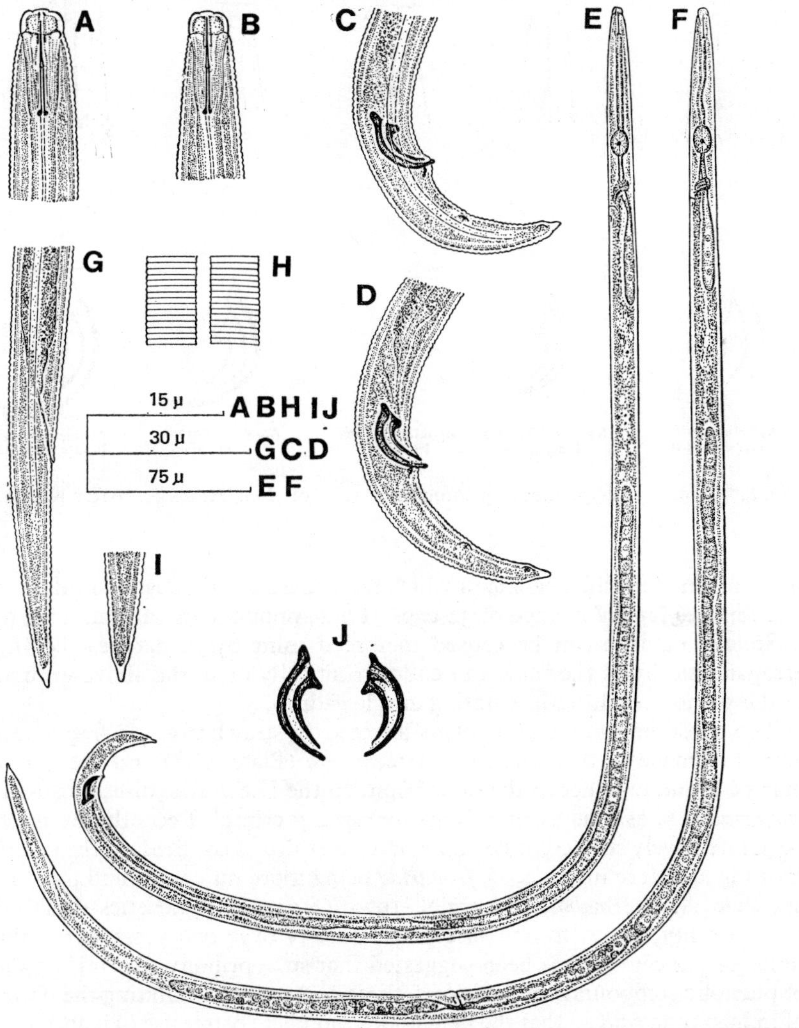

Fig. 8.6. *Aphelenchoides fragariae*: **A,** female, head end; **B,** male, head end; **C, D,** male tails, lateral; **E,** male; **F,** female; **G,** female hind end; **H,** lateral field; **I,** female terminus; **J,** spicules, lateral. (From Siddiqi, 1975, *CIH Descriptions of Plant-parasitic Nematodes*).

Biology. A. fragariae is a common and widespread species recorded mainly in moist conditions on ornamental plants and on strawberry. More than 200 flowering plants and nearly 100 ferns were already known to be hosts in 1965 (Goodey, Franklin & Hooper, 1965) since when many others have been recorded. The nematode feeds chiefly within leaves after passing through the stomata, causing brown blotches delimited by leaf veins (Plate XVII), as does *A. ritzemabosi,* and can cause serious losses in glasshouse-grown ferns and in forced *Lilium* spp. Damage to begonia, in which both the leaves and the fleshy stems may be heavily invaded, to pot-grown *Ficus, Saintpaulia,* gloxinia and *Impatiens* may also be serious. Strumpel (1967) gives an account of the biology and life-history of *A.*

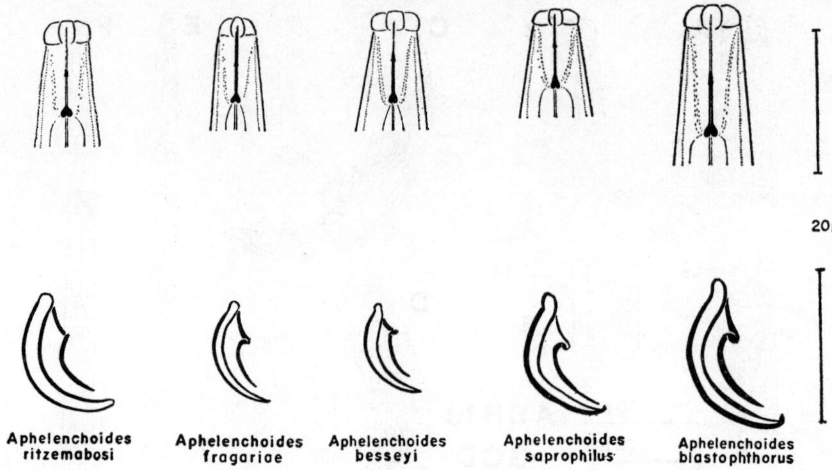

Fig. 8.7. Heads and spicules of 5 common species of *Aphelenchoides*. (After Nickle, 1970).

fragariae in Lorraine begonia in which the generation time is 10–11 days at 18°C and females lay an average of 32 eggs. Thus, populations can build up rapidly.

Serious damage can be caused to forced mint by nematodes feeding both ectoparasitically in the buds and endoparasitically in all the above-ground parts of the plants, resulting in stunting and leaf-drop.

The most important crop plant affected is strawberry. *A. fragariae* causes similar damage to that due to *A. ritzemabosi* (Plate XVI) and the two species may be found together in the same plant. In the U.S.A. the disease caused by *A. fragariae* is known as 'spring dwarf' or 'spring crimp'. Recently the disease has been intensively studied in Poland where one third of the fields in the strawberry-growing areas are infested, *A. fragariae* being more numerous and more important than *A. ritzemabosi* (Szczygiel, 1970). Strawberry varieties differ in their susceptibility; losses to one susceptible variety over two years were calculated to be 22 per cent. It has been suggested that susceptibility depends on the level of phenolic compounds in the leaves, these compounds inhibiting the destruction of indoleacetic acid so that the nematodes can make better use of it and reproduce to a greater degree than in resistant varieties in which IAA is destroyed more rapidly (Szczygiel & Giebel, 1970). *A. fragariae* does not survive long in soil in the absence of plants. Szczygiel & Hasior (1971) found that the longest survival time was three months in dry or fresh plants buried 15 cm in soil, and if nematodes were added to soil in water suspension they persisted for only four weeks. When *A. fragariae* survives in a field in the absence of strawberries it is probably on weeds such as *Stellaria media*.

A. fragariae, unlike *A. ritzemabosi,* is capable of feeding and reproducing on fungus and has been cultured on *Alternaria citri*.

Investigations into the population changes of *A. fragariae* in strawberry plants stored at different temperatures and a technique for assessing the degree of infestation have been published by Hirling (1971, 1972). When plants were stored in an unheated glasshouse or in a room at 14–15°C the nematode numbers remained fairly constant, but at 0–2°C they increased by 131–188 per cent and at −2 to −1°C by 56 per cent. The most reliable way of estimating nematode

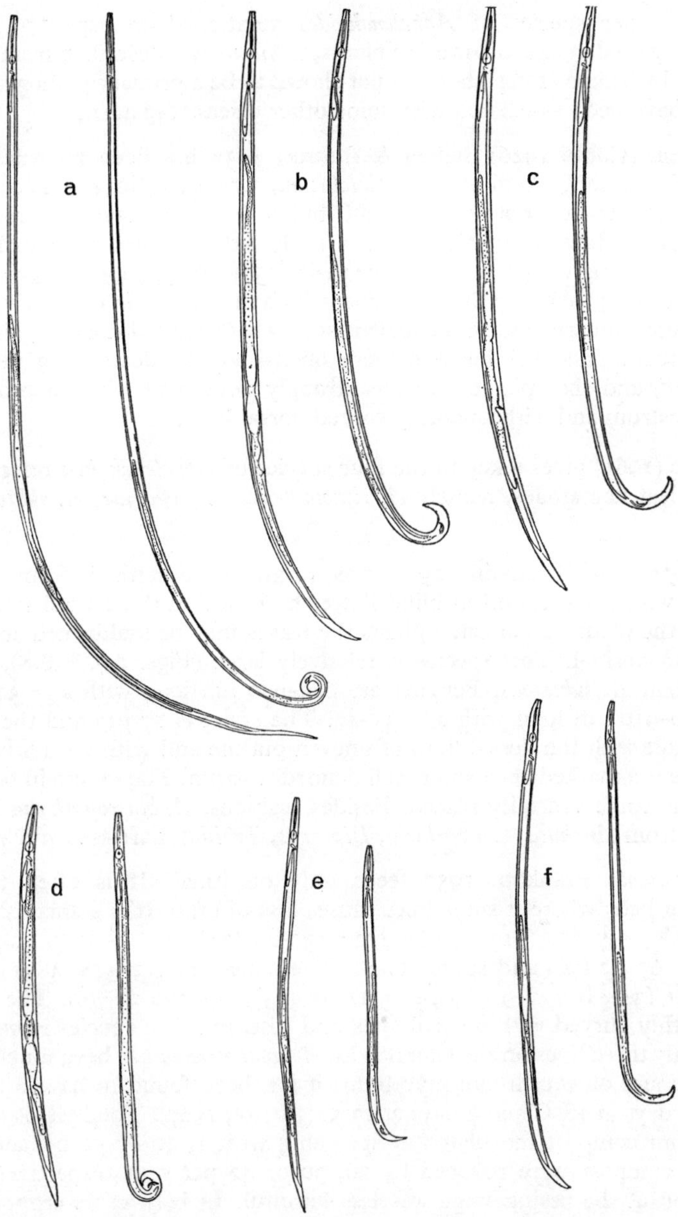

Fig. 8.8. Whole males and females, drawn to same scale, of: (a) *Rhadinaphelenchus cocophilus,* (b) *Aphelenchoides ritzemabosi,* (c) *A. blastophthorus,* (d) *A. saprophilus,* (e) *A. fragariae,* (f) *A. subtenuis.* (After Nickle, 1970).

numbers in plants was found to be by extracting the nematodes from the crowns of 30 plants (excluding old leaves and inflorescences, leaving about 20 g of tissue) taken in spring and extracted in dilute H_2O_2 in a Baermann funnel.

A more detailed account of the morphology and biology of *A. fragariae,* with notes on its control, are given by Siddiqi (1975).

Of the other species of *Aphelenchoides* mentioned on page 179 as being associated with damage to flowering plants, *A. lilium* was described from diseased lily bulbs by Yokoo (1964), but was not shown to be a primary pathogen; it may possibly have been associated with some other disease organism.

A. subtenuis (Cobb, 1926) Steiner & Buhrer, 1932 has been recorded chiefly from bulbs, among them *Allium, Colchicum, Crocus,* tulip and narcissus. In narcissus it invades both leaves and bulbs and can cause necrotic areas (Aitkenhead & Janson, 1970). This is a fairly large species, the females being 820–1300 μm long with a = 44–66, the males 750–1010 μm long and a = 54–66. It can be distinguished by the tail shape, which in the female is bluntly rounded with a ventral mucro, and in the male is somewhat rounded dorsally, also with a ventral mucro. The tail forms a more open hook on death than that of *A. ritzemabosi,* and the spicules are less strongly developed with no pronounced apex or rostrum and with smoothly curved dorsal limb.

Thorne (1961) gives a key to the four species of *Aphelenchoides* referred to as bud and leaf nematodes, namely *A. ritzemabosi, A. fragariae, A. subtenuis* and *A. besseyi.*

A. blastophthorus Franklin, 1952 was originally described from *Scabiosa caucasica* where it is found in blind flower buds and in the central buds in the crown of the plants. In affected plants the leaves may be malformed and darker green than normal. This species is relatively large (Figs. 8.7 & 8.8), but less slender than *A. subtenuis.* Females are 680–900 μm long with a = 32–47 and males 670–910 μm long with a = 35–47. The stylet is 17 μm and the spicules stoutly built with the dorsal limb of uneven outline and with a slightly hooked tip. There is a marked apex and a well defined rostrum. The mucro in both sexes is a simple spike ventrally placed. Besides scabious, *A. blastophthorus* has been recorded from *Anchusa, Convallaria, Dipsacus, Trollius,* narcissus and violet.

A. composticola Franklin, 1957 feeds only on fungi. It is often found in mushroom beds where it sometimes causes loss of crop. It is a smallish species with females 447–614 (517) μm long, a = 30–42 (34); b = 8–10 (9), c = 11–17 (14), V = 67–72 (70) and spear 11 μm. The males are 412–518 (489) μm long, a = 28–41 (34), b = 7–9 (8·5), c = 11–20 (15), spicules 21 μm. The spicules are smoothly curved with a small apex and rostrum. The species is unusual in having only three lines on the lateral field. *A. composticola* has been much studied in culture and on mushroom mycelium; it has been found to have a life-cycle taking 10 days at 18°C and 8 days at 23°C (Cayrol, 1967). Total yields of mushrooms from compost inoculated at spawning with 1, 10 or 50 nematodes per 100 g of compost were reduced by 26, 30 or 42 per cent respectively, while inoculation at the casing stage was less harmful. In both cases cropping continued for 12–14 weeks (Arrold & Blake, 1968). In earlier pot experiments (Goodey, 1960b) mushroom mycelium was completely destroyed in 53 days by nematodes added to the compost at spawning at the rate of 30 per 100 g of compost and cultured at 18·5°C.

A. saprophilus Franklin, 1957 is another fungivorous species often found in rotting plant material but there is no evidence of association with any disease of higher plants. It is of stouter build than *A. composticola* and has four lines on the lateral field. The spicules are quite heavily built with an unevenly curved dorsal limb slightly knobbed at the tip like that of *A. blastophthorus* but smaller (about 22 μm measured along the dorsal limb against 27 μm in *A. blastophthorus*) (Fig. 8.7). It is easily cultured on various fungi.

A. besseyi Christie, 1942 is primarily a parasite of rice and therefore not encountered in cooler regions. It is the cause of the disease known as 'white tip'. Its other main crop host is strawberry in which it gives rise to the 'summer dwarf' or 'summer crimp' disease which occurs along the Atlantic coast of U.S.A. from Florida to Maryland. It has also been reported on strawberry in Queensland, Australia, and on several flowering plants in Hawaii, the Philippines and Japan. The lateral field has four lines and tails of both sexes have a stellate mucro. Spicules are smoothly curved, without a prolonged apex and with small rostrum (Fig. 8.7). For more details see Franklin & Siddiqi (1972). Fortuner (1970) gives a key to eleven species of *Aphelenchoides* with stellate mucrones.

Rhadinaphelenchus cocophilus (subfamily Rhadinaphelenchinae Husain & Khan, 1967) is related to *Aphelenchoides*. It differs in having a heavily sclerotized head framework, an elongated metacorpus twice as long as wide, a vulval flap, and, in the male, a small terminal bursa and a gubernaculum-like structure. Both sexes are extremely slender with 'a' value approaching 100 and length about 1 mm. The male tail is strongly hooked on death (Fig. 8.8). This nematode is a parasite of coconut and other palms and the cause of the devastating 'red ring' disease on which much work has been done. See Brathwaite & Siddiqi (1975) for a full account of this nematode covering morphology, biology and control.

CHAPTER 9

Migratory Soil Tylenchida

D. C. M. CORBETT

*Rothamsted Experimental Station, Harpenden**

The migratory soil Tylenchida include more known plant-parasitic nematode genera than any other order. In this chapter, several of these genera are described, but it would not be possible within the space available to describe all of them, especially since three of the genera as formerly recognized, *Tylenchus, Tylenchorhynchus* and *Criconemoides*, have recently been split, respectively, into six, four and eight genera. The Tylenchida are distinguished by a stylet, usually with basal knobs, and an oesophagus with a dorsal oesophageal gland opening close behind the stylet base; two subventral glands open into the lumen of the oesophagus about half way along its length within the median bulb, if this structure is present. The three glands are contained within a sac-like posterior oesophagus that abuts or overlaps the intestine in ways that characterize different genera.

Where the life cycle of nematodes described in this chapter has been investigated it has been found to be simple. Eggs are laid singly, in roots or in soil, and the first moult is in the egg with three subsequent moults to produce adults (except in *Helicotylenchus multicinctus* where the first-stage juvenile is free and the first moult takes place shortly after hatching). The length of the life cycle varies from five to six days at 25°C for *Tylenchus* (*Cephalenchus*) *emarginatus* to 100 days at 23°C for *Rotylenchus robustus*, but most take about a month at the most favourable temperature; at lower temperatures in the field the complete life cycle will take much longer.

Short notes are given on the biology of genera prevalent in temperate areas. No attempt has been made to be exhaustive; the notes are merely intended to represent present ideas on the biology of species in the genera in question. To simplify this, wherever possible references are given to recent review papers in which more information is available.

TYLENCHUS Bastian, 1865 and Related Genera (Fig. 9.1)

SYSTEMATIC POSITION: Tylenchida: Tylenchoidea: Tylenchidae: Tylenchinae.

Until recently *Tylenchus* included *Tylenchus, Cephalenchus, Aglenchus, Filenchus, Lelenchus,* and *Malenchus* as subgenera but these were all raised to generic rank by Geraert (1970). A key to these and other closely related genera is given below.

KEY TO *Tylenchus* AND CLOSELY RELATED GENERA
(Adapted from Bello, 1971)

1.	Body narrowing abruptly immediately behind vulva . .	*Malenchus*
	Body narrowing gradually behind vulva	2

*Now at Agricultural Research Council, London.

2. With a ventral vulval membrane that covers the vulva to the dorsal side . . . *Pleurotylenchus*
   Without a ventral vulval membrane 3
3. Males without bursa . . . . *Miculenchus*
   Males with bursa 4
4. Spermatheca not separated from ovary . . . . *Cephalenchus*
   Spermatheca separated from ovary. 5
5. With lateral vulval membrane . . . . *Aglenchus*
   Without lateral vulval membrane 6
6. Tail relatively short and ventrally curved . . . . *Tylenchus*
   Tail relatively long and generally straight. 7
7. Strongly annulated but small in size (0·3–0·6 μm) . . . . . *Filenchus*
   Finely annulated . . . . *Lelenchus*

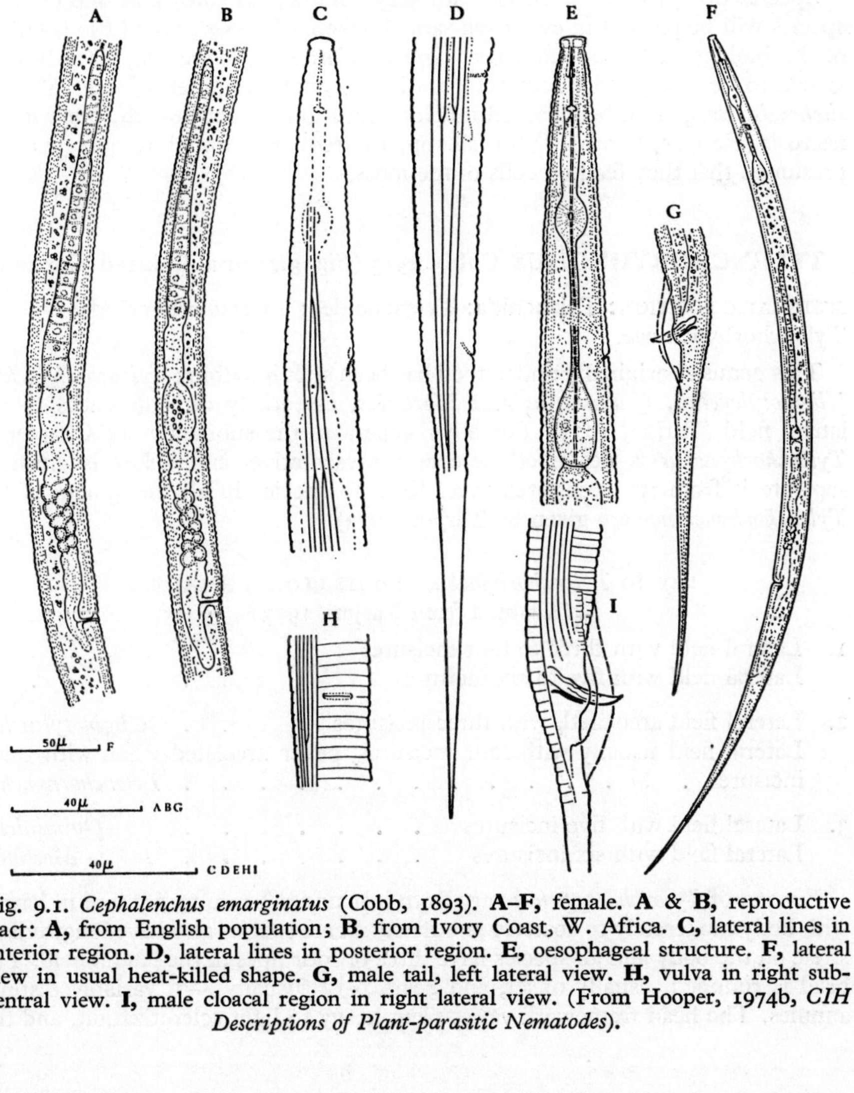

Fig. 9.1. *Cephalenchus emarginatus* (Cobb, 1893). **A–F,** female. **A & B,** reproductive tract: **A,** from English population; **B,** from Ivory Coast, W. Africa. **C,** lateral lines in anterior region. **D,** lateral lines in posterior region. **E,** oesophageal structure. **F,** lateral view in usual heat-killed shape. **G,** male tail, left lateral view. **H,** vulva in right subventral view. **I,** male cloacal region in right lateral view. (From Hooper, 1974b, *CIH Descriptions of Plant-parasitic Nematodes*).

Females within these genera are about 0·3–1·2 mm long. The body of most species is variably annulated and ends in a long thin tail. A lateral field is present with two to six lines. The head is usually at least slightly offset and very faintly annulated, and has no head skeleton. The spear is about 6 to 18 μm long, and usually knobbed. The oesophagus consists of a short procorpus running into a well defined median bulb, followed by a long narrow isthmus ending in a pear-shaped basal bulb which abuts the intestine. The excretory pore is usually about opposite the basal bulb. The vulva varies in position from about 45–78 per cent of body length, but is mostly between 55–65 per cent. There is a single anterior ovary and a small to large post-vulval sac. Reproduction may be sexual or asexual.

Males are about the same size as females and resemble them in organization of the head and oesophagus. Males, except in *Miculenchus*, have a small adanal bursa, which may have unequal alae, and spicules that are about 9 to 23 μm long. Type species: *Tylenchus davainei* (Bastian, 1865) Filipjev, 1934

Species of *Tylenchus sensu lato* are very common in soils, and one or more species will be present in every soil sample taken. However, very little is known of the biology of these nematodes. *Cephalenchus emarginatus* has recently been shown to be a migratory root ectoparasite, feeding on epidermal cells of *Picea sitchensis*. Large numbers caused much diminution of root growth, but without necrosis (Hooper, 1974b). *Tylenchus* spp. are frequently found in moss, and it is presumed that they feed on cells of the moss.

TYLENCHORYHNCHUS Cobb, 1913 (Fig. 9.2) and Related Genera

SYSTEMATIC POSITION: Tylenchida: Tylenchoidea: Tylenchorhynchidae: Tylenchorhynchinae.

This genus as originally understood has been split into four: *Tylenchorhynchus, Uliginotylenchus, Quinisulcius* and *Merlinius,* essentially on differences in the lateral field (Tarjan, 1973). For other genera in the subfamily see Chapter 1. *Tylenchorhynchus* is described here as representative, and a key is given to separate it from its closest relatives. Keys to species in all the genera of the *Tylenchorhynchinae* are given by Tarjan (1973).

KEY TO *Tylenchorhynchus* AND ITS CLOSEST RELATIVES
(Adapted from Tarjan, 1973)

1. Lateral field with three to four incisures 2
   Lateral field with five to six incisures 3
2. Lateral field areolated, with three incisures . . . . *Uliginotylenchus*
   Lateral field usually with four incisures, never areolated when with three incisures . . . . *Tylenchorhynchus*
3. Lateral field with five incisures . . . . *Quinisulcius*
   Lateral field with six incisures . . . . *Merlinius*

Females of *Tylenchorhynchus* vary from just under ½ mm to 1·7 mm in length with body moderately to coarsely annulated. Some species also have longitudinal striae on the body and all species have three or four lines in the lateral field. The head is rounded, usually offset, and bears 1–13 (usually 4–7) variably distinct annules. The head framework varies from heavy to light sclerotization, and the stylet is 13–39 μm (usually 16–27 μm) long, with well formed basal knobs. The oesophagus consists of a long narrow procorpus, followed by a well formed median bulb running into a long narrow isthmus that terminates in a pear-shaped basal bulb abutting the intestine. The excretory pore is usually opposite the anterior end of the basal bulb of the oesophagus. The vulva is between 42 and 61 per cent of the body length, and there are two opposed, outstretched gonads, each usually with a spermatheca.

Males resemble females in the fore part of the body, but they have spicules that are 12–32 μm long, and a gubernaculum 5–18 μm long. The bursa originates anterior to the cloaca, and extends to the tail tip. Phasmids are conspicuous on the bursa.

Type species: *Tylenchorhynchus cylindricus* Cobb, 1913

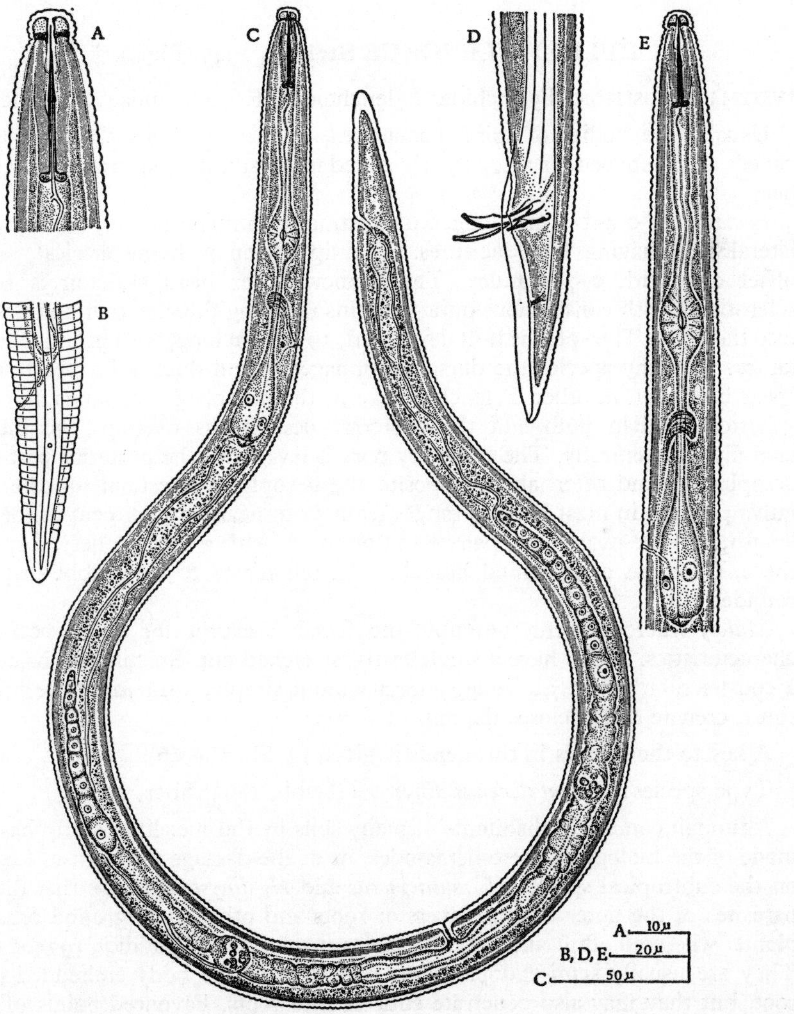

Fig. 9.2. *Tylenchorhynchus cylindricus* Cobb. **A–C,** neotype female; **D, E,** male from cotton soil in California. **A,** head end. **B, D,** tail ends. **E,** oesophageal region. All lateral views. (From Siddiqi, 1972a, *CIH Descriptions of Plant-parasitic Nematodes*).

Tylenchorhynchus spp. and related species are common in soil samples. They are called 'Stunt nematodes' because of the stunting of attacked plants. All stages feed actively on epidermal cells of roots and on root hairs. Cell function is affected and roots of some crops (e.g., *Lolium perenne, Phaseolus radiatus, Brassica campestris* and *Pisum* sp.) can be discoloured light to dark brown after feeding by *T. dubius*: root and shoot weight of several host crops are decreased by feeding of these nematodes (Sharma, 1971; Bridge, 1974; Siddiqi, 1976; Loof, 1974a).

T. dubius, common in European soils, is found in association with fungi in root diseases of several crops: where both fungi and nematodes are present disease is worse than where either is alone (Bridge, 1974).

HELICOTYLENCHUS Steiner, 1945 (Fig. 9.3)

SYSTEMATIC POSITION: Tylenchida: Tylenchoidea: Hoplolaimidae: Rotylenchinae.

Usually the bodies of spiral nematodes, as these and *Rotylenchus* spp. are called, are C-shaped, or more tightly coiled into a spiral, when relaxed by gentle heat.

Females are 0·4–1·1 mm long, with distinctly annulate bodies, usually with a lateral field having four incisures. The lip region is hemispherical, slightly offset and with 0–7 annules. The framework or head skeleton is heavily sclerotized, with conspicuous outer margins reaching three or four annules back into the body. The spear is well developed, 19–38 μm long, with prominent basal knobs. In many species the dorsal oesophageal gland duct is further from the spear base than in other Tylenchid genera; the oesophagus contains an oval to spherical median bulb and the posterior oesophagus overlaps the intestine laterally and ventrally. The excretory pore is level with the posterior part of the oesophagus, and often about opposite the oesophago-intestinal junction. The vulva position in most females ranges from 54 to nearly 70 per cent of the body length, and the ovaries are paired and opposed, with a small spherical spermatheca, offset, in each genital branch. The tail tapers to a variably shaped or rounded tip.

Males, where present, resemble the females, except for their specific sex characteristics. Males have a single testis, stretched out. Spicules cephalated by a constriction, 16–31 μm long, gubernaculum simple, 5–11 μm. The bursa is short, crenate and encloses the tail.

A key to the species in this genus is given by Sher (1966).

Type species: *Helicotylenchus dihystera* (Cobb, 1893) Sher, 1961.

Although common inhabitants of many soils in Europe, little study has been made of the biology of these nematodes, or of the damage they cause. Research on the subtropical species *H. multicinctus* and *H. dihystera* shows that they are parasites of the outer cortical layers of roots and other underground organs of plants, where all adult and juvenile stages may be found (Siddiqi, 1972b; 1973). They are usually semi-endoparasites, with part of the body embedded in the root, but they may also penetrate entirely into roots. Favoured points of entry appear to be behind root tips and at the junction of lateral and main roots. All juvenile and adult stages feed, producing small discoloured lesions in the cortex of the root or other underground organ attacked. The presence of *H. nannus*

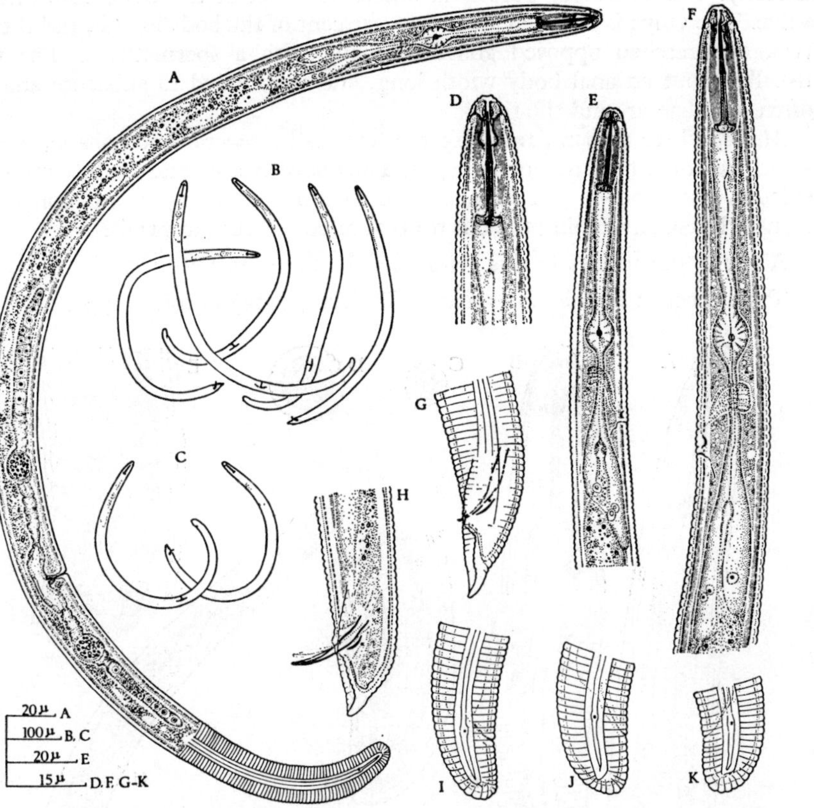

Fig. 9.3. *Helicotylenchus multicinctus* (Cobb). **C–F, H** & **I,** topotypes; remainder specimens from banana from Samoa Island. **A, B, C,** adults in relaxed body posture. **D,** head end, female. **E,** oesophageal region, female. **F,** oesophageal region, male. **G, H,** tail ends, male. **I–K,** tail ends, female. (From Siddiqi, 1973, *CIH Descriptions of Plant-parasitic Nematodes*).

(= *H. dihystera*) worsens wilt by bacteria in carnations and tomatoes (Stewart & Schindler, 1956; Libman *et al.*, 1964).

ROTYLENCHUS Filipjev, 1936 (Fig. 9.4)

SYSTEMATIC POSITION: Tylenchida: Tylenchoidea: Hoplolaimidae: Rotylenchinae.

Females are about 1–2 mm long, distinctly annulated, with four incisures in the lateral field, and when killed by gentle heat usually form a spiral. The head is hemispherical, with four to seven distinct annules, and a strongly sclerotized head skeleton. Characteristically in this genus the basal head annule has longitudinal striations giving it a tiled appearance. The spear is well developed, 30–50 μm long, and has large rounded basal knobs. Unlike *Helicotylenchus*, the dorsal gland outlet is close behind the spear base. The oesophagus has a narrow procorpus running into a well developed oval median bulb, followed by a narrow isthmus and a saccate posterior oesophagus that overlaps the intestine mainly dorsally. The excretory pore is at about the level of the oesophago-intestinal valve. The vulva is between 50 and 60 per cent of the body length, and there are two, outstretched opposed gonads, usually with a spermatheca. The tail is usually about an anal body width long, and is rounded to subacute and most often annulate around the tip.

Males, where present, resemble the females in general body shape, size, and structure of the head and oesophagus. They have arcuate and slightly cephalated spicules about 30–40 μm long, a gubernaculum about 15–25 μm long, and a crenate bursa enveloping the tail tip that appears acute in profile.

A key to the species in this genus is to be found in Sher (1965).

Type species: *Rotylenchus robustus* (de Man, 1876) Filipjev, 1936.

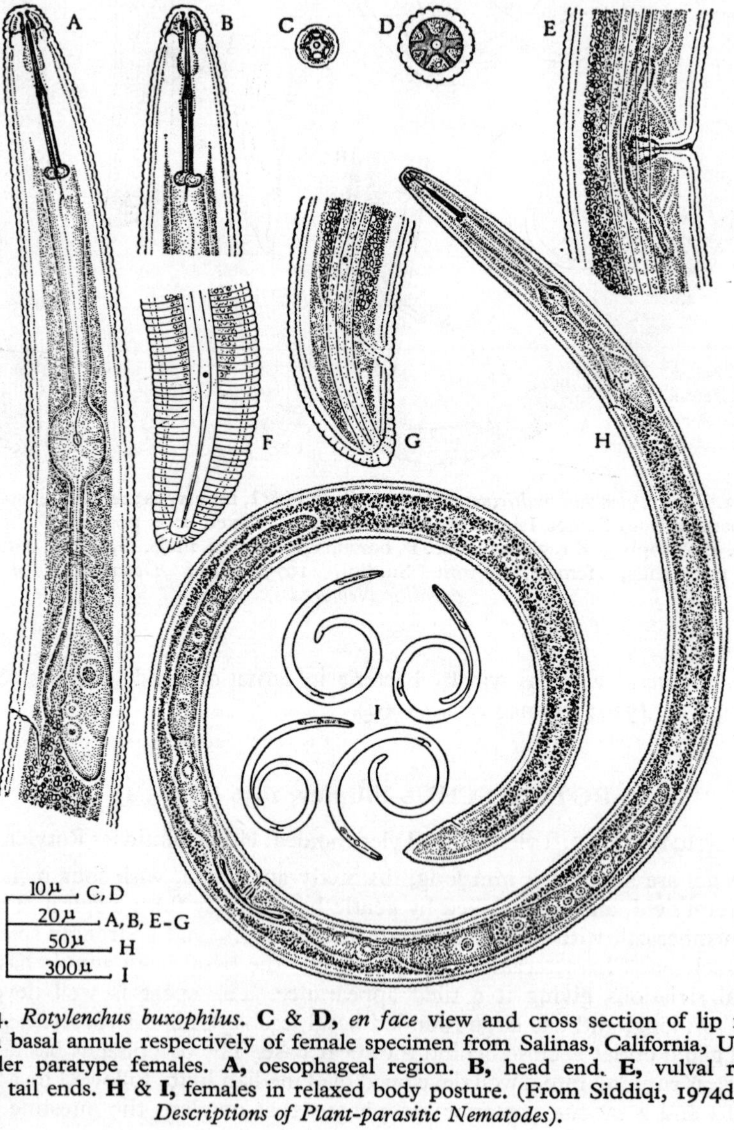

Fig. 9.4. *Rotylenchus buxophilus*. **C** & **D,** *en face* view and cross section of lip region through basal annule respectively of female specimen from Salinas, California, U.S.A.; remainder paratype females. **A,** oesophageal region. **B,** head end. **E,** vulval region. **F** & **G,** tail ends. **H** & **I,** females in relaxed body posture. (From Siddiqi, 1974d, *CIH Descriptions of Plant-parasitic Nematodes*).

Rotylenchus spp. generally prefer lighter, sandy soils, and live as migratory ectoparasites, becoming partially embedded in roots when feeding. They attack roots along the sides and at the tips. Parasitized roots may show small, brown necrotic lesions, that develop into a general decay of the root cortex, with root systems much decreased in size. Above-ground plant parts are stunted and yellowed, and yields of parasitized crops like carrots and lettuce are lessened. When *R. robustus* and *Fusarium oxysporum* f. sp. *pisi* occur together on peas, roots are more extensively and quickly decayed than in the presence of each organism alone: the disease caused by joint attack leads to severe losses and is called 'early yellowing' (Siddiqi, 1972c; 1974d).

PRATYLENCHUS Filipjev, 1936 (Fig. 9.5)

SYSTEMATIC POSITION: Tylenchida: Tylenchoidea: Pratylenchidae: Pratylenchinae.

The genus *Pratylenchus* comprises more than 40 species, many of which characteristically cause lesions on roots of hosts, leading to the common name 'root-lesion nematode'.

Females are slender, and specimens killed by gentle heat are nearly straight, 0·34–0·8 mm long. The body is finely annulated and the lateral field usually has four lines, but up to six or eight may be found in some species. The lip region is slightly set off, flat in front, with rounded outer margins, and may have two, three or four annules, depending on species. The head has a characteristically conspicuous skeleton, and a short but strong spear, 14–20 μm long, with rounded basal knobs. The oesophagus overlaps the intestine ventrally, with an overlap that may be one to two and a half body widths long depending on species. The excretory pore is about opposite the oesophago-intestinal junction. The tail tip may be smooth or crenate, and varies in shape from truncate to pointed. The vulva is about 75–90 per cent of the body length, depending on species, and the gonad is a single, anterior one with a short undifferentiated post-vulval sac. In sexually reproducing species, the spermatheca is conspicuous, spherical to oval in shape, and often full of spermatozoa. In asexually reproducing species, the spermatheca is inconspicuous, and often indistinguishable.

Males may or may not be present, depending on the mode of reproduction of the species. They generally resemble the females, but are smaller, about 300–600 μm. The lateral field, like that of the female, may have four to six or eight lines. Spicules are slender, 12–20 μm long, simple in organization, and the gubernaculum is simple, 3·5–4·5 μm long. The tail is usually about twice the male body diameter, and enveloped by the bursa, which often has a crenate edge.

Type species: *Pratylenchus pratensis* (de Man, 1880) Filipjev, 1936.

A key is given to ease identification of the commonest species, including the three important tropical species, *P. coffeae*, *P. brachyurus* and *P. zeae*.

KEY TO *Pratylenchus* SPP.

1. Head with two annules 2
   Head with three annules 5
   Head with four annules 11

2. Males present, spermatheca oval, usually with sperm, spear more than 15 μm . . . . *P. coffeae* (Zimmerman, 1898) Goodey, 1951
   Males absent or very rare, spermatheca inconspicuous, without sperm 3

3. Spear 19 μm or more, vulva 83–89 per cent *P. brachyurus* (Godfrey, 1929) Goodey, 1951
   Spear less than 19 μm, vulva usually less than 84 per cent . . . 4

4. Vulva 80–84 per cent *P. neglectus* (Rensch, 1924) Chitwood & Oteifa, 1952
   Vulva less than 80 per cent . . . . *P. scribneri* Steiner, 1943

5. Tail tip crenate 6
   Tail tip smooth 8

6. Males unknown, spermatheca inconspicuous, without sperm, post-vulval sac more than one body width . . . . . *P. crenatus* Loof, 1960
   Males present, spermatheca usually with sperm. 7

7. Spermatheca oval to elongate . *P. pratensis* (de Man, 1880) Filipjev, 1936
   Spermatheca spherical . . . . *P. fallax* Seinhorst, 1968

8. Males absent or very rare, spermatheca obscure, without sperm . . 9
   Males present, spermatheca usually with sperm. 10

9. Tail tip broadly rounded or truncate, head continuous with body contour *P. thornei* Sher & Allen, 1953
   Tail tip narrowly rounded to subacute, vulva about 72 per cent *P. zeae* Graham, 1951

10. Spermatheca spherical *P. penetrans* (Cobb, 1917) Chitwood & Oteifa, 1952
    Spermatheca oval . . . . *P. vulnus* Allen & Jensen, 1951

11. Tail concave above subacute tip, spear more than 16 μm *P. goodeyi* Sher & Allen, 1953
    Tail not concave above truncate or narrowly rounded tip, spear less than 16 μm . . . . *P. vulnus*

About 14 species occur in Europe, but not all of them have been associated with damage in the field or in glasshouses. *P. penetrans* is the best known and is found damaging a wide variety of crops (Corbett, 1973). *P. vulnus* (see Corbett, 1974), *P. crenatus* (see Kemper, 1959; Caubel, 1973), *P. convallariae* (see Cayrol & Ritter, 1962) and *P. thornei* (see Oostenbrink *et al.*, 1956) have been shown to damage a number of field crops; *P. coffeae, P. goodeyi* and *P. vulnus* damage hothouse plants (D'Herde *et al.*, 1969; Goffart, 1953; Winfield, 1974), and *P. fallax* and *P. pinguicaudatus* have been shown to damage root tissues *in vitro* but not in the field (Corbett, 1969; 1972). *P. neglectus* is widely distributed, but has not been found unequivocally associated with damage, although large numbers have been recorded on some hosts.

Damage occurs in annual and perennial crops. Spread in perennial crops is in infested material. *Pratylenchus* spp. are migratory root endoparasites, invading roots especially at the tips and at the base of lateral roots branching from main roots (Corbett, 1972). All three free juvenile stages and adults can invade roots

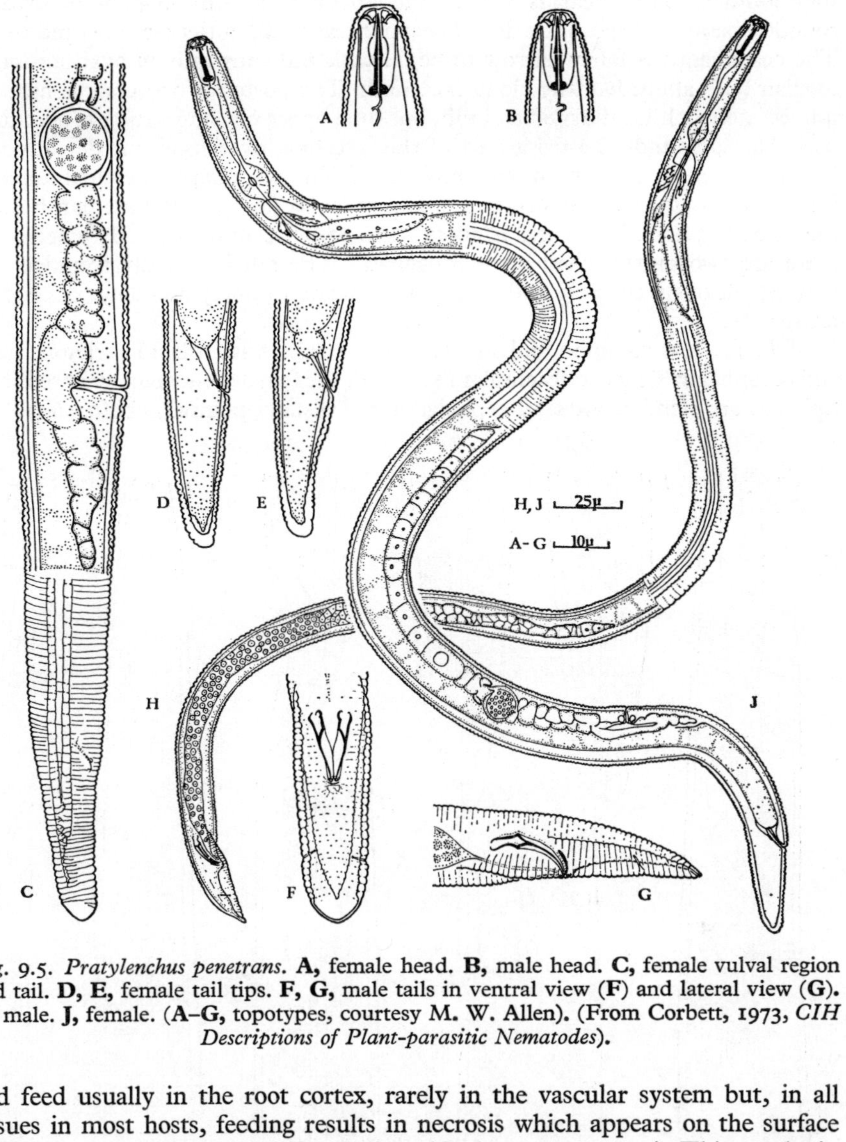

Fig. 9.5. *Pratylenchus penetrans*. **A,** female head. **B,** male head. **C,** female vulval region and tail. **D, E,** female tail tips. **F, G,** male tails in ventral view (**F**) and lateral view (**G**). **H,** male. **J,** female. (**A–G,** topotypes, courtesy M. W. Allen). (From Corbett, 1973, *CIH Descriptions of Plant-parasitic Nematodes*).

and feed usually in the root cortex, rarely in the vascular system but, in all tissues in most hosts, feeding results in necrosis which appears on the surface of roots as brown to black diffuse lesions (Corbett, 1973; 1974). This results in a brown sparse root system, and stunted and chlorotic above-ground parts of plants that yield poorly. Several species are associated with root rotting or wilt fungi in damaging interactions (see Chapter 3).

PRATYLENCHOIDES Winslow, 1958 (Fig. 9.6)

SYSTEMATIC POSITION: Tylenchida: Tylenchoidea: Pratylenchidae: Radopholinae.

Females are about $\frac{1}{2}$ to nearly 1 mm long, distinctly annulated, with four to six lines in the lateral field and the body straight to slightly arcuate in death. The head is strongly sclerotized, hemispherical or flattened in front with three or four annules. The spear is robust and about 16–25 μm long, with broadly rounded basal knobs and the dorsal oesophageal gland outlet close behind them. The oesophagus is fairly narrow to begin with and runs into an oval, muscular median bulb about halfway along its length. The posterior oesophagus narrows immediately behind the median bulb, but soon opens out into a saccate glandular part that surrounds the front part of the intestine, overlapping it dorsally and laterally by never more than one body width. The oesophago-intestinal junction forms a conspicuous oval structure about twice as far from the median bulb as the excretory pore. The vulva is about 54 to 62 per cent of the body length, and there are two outstretched opposed gonads. The tail is broadly rounded and may be variously coarsely and irregularly crenate to finely annulate, depending on species.

Males are less common and smaller than females with a head skeleton, spear and oesophagus structurally similar to those of the female, but somewhat reduced. Spicules are slender, arcuate, cephalated and distally pointed, about 16–28 μm

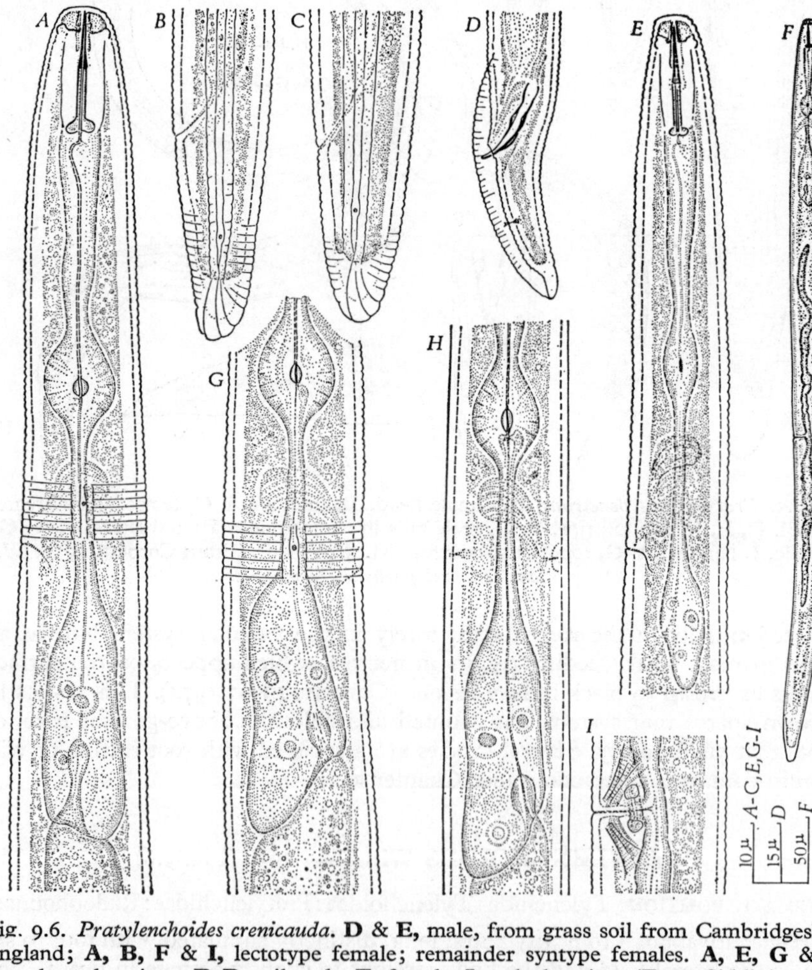

Fig. 9.6. *Pratylenchoides crenicauda*. **D & E,** male, from grass soil from Cambridgeshire England; **A, B, F & I,** lectotype female; remainder syntype females. **A, E, G & H,** oesophageal regions. **B–D,** tail ends. **F,** female. **I,** vulval region. (From Siddiqi, 1974c, *CIH Descriptions of Plant-parasitic Nematodes*).

long, with a simple gubernaculum about 4–10 μm long. The bursa originates near the level of the retracted spicule head, and completely envelops the tail, which is characteristically arched dorsally in mounted specimens.

A key to species in the genus is given by Sher (1970).

Type species: *Pratylenchoides crenicauda* Winslow, 1958

Little is known of the biology of these species, but they are found in soil samples and crop roots, where they can be mistaken for *Pratylenchus* spp. which they closely resemble. They are migratory endoparasites, often found in the cortex of parasitized roots. *P. crenicauda* is a parasite of grass roots (Goodey, 1940); and a species resembling *P. laticauda* is found in cereal roots. *P. laticauda* is a pest of several types of plant in the Netherlands, evoking root lesions and a root rot similar to those produced by *Pratylenchus* with growth diminution of the parasitized plants (Braun & Loof, 1966; Hijink & van Rossen, 1968).

RADOPHOLUS Thorne, 1949 (Fig. 9.7)

SYSTEMATIC POSITION: Tylenchida: Tylenchoidea: Pratylenchidae: Radopholinae.

The genus comprises 23 species, but only one species, *R. similis,* is important as a plant parasite. It is a tropical plant pest, but has recently become an important pest of ornamental glasshouse plants in Europe.

Females are 0·52–0·88 mm long, and when fixed are straight to slightly arcuate ventrally, with a distinctly annulated cuticle. The lateral field has four incisures, stretching from near the median bulb to the tail tip. The lip region is hemispherical, usually with three to four annules, and has a strongly sclerotized framework. The spear is 17–20 μm long with well developed round basal knobs. The median oesophageal bulb is well developed, round to oval, and the posterior oesophagus extends over the intestine in a long dorsal and dorsolateral overlap. The excretory pore is about the level of the oesophago-intestinal junction. The vulva is conspicuous, lying just behind mid-body, and there are paired opposed gonads each with a spherical spermatheca usually filled with small rod-shaped spermatozoa. The tail tapers to a narrowly rounded or indented terminus.

Males differ considerably from females in appearance, having an indistinct oesophagus and greatly reduced and indistinct spear. The lip region is high, rounded and dome-shaped with three to five annules and no visible sclerotized framework. The excretory pore is usually two to three body widths behind the faint outline of the median bulb. There is a single testis, outstretched anteriorly, and a coarsely crenate bursa enveloping about two-thirds of the tail. The spicules are strongly cephalated, about 18–22 μm long with pointed distal ends. The gubernaculum is rod-like and protrudes through the cloaca in fixed specimens.

Colbran (1971) provides a key to species in this genus.

Type species: *Radopholus similis* (Cobb, 1893) Thorne, 1949.

R. similis is an important pest of several tropical crops causing much studied diseases in bananas and citrus (Orton Williams & Siddiqi, 1973). In Europe it has recently been found in the open in the South of France, probably transported there from the U.S.A. on ornamentals (Scotto la Massese, 1967) but it has become more important as a pest of glasshouse crops in Northern Europe (Jakobsen, 1973; Coolen, Hendrickx & d'Herde, 1971).

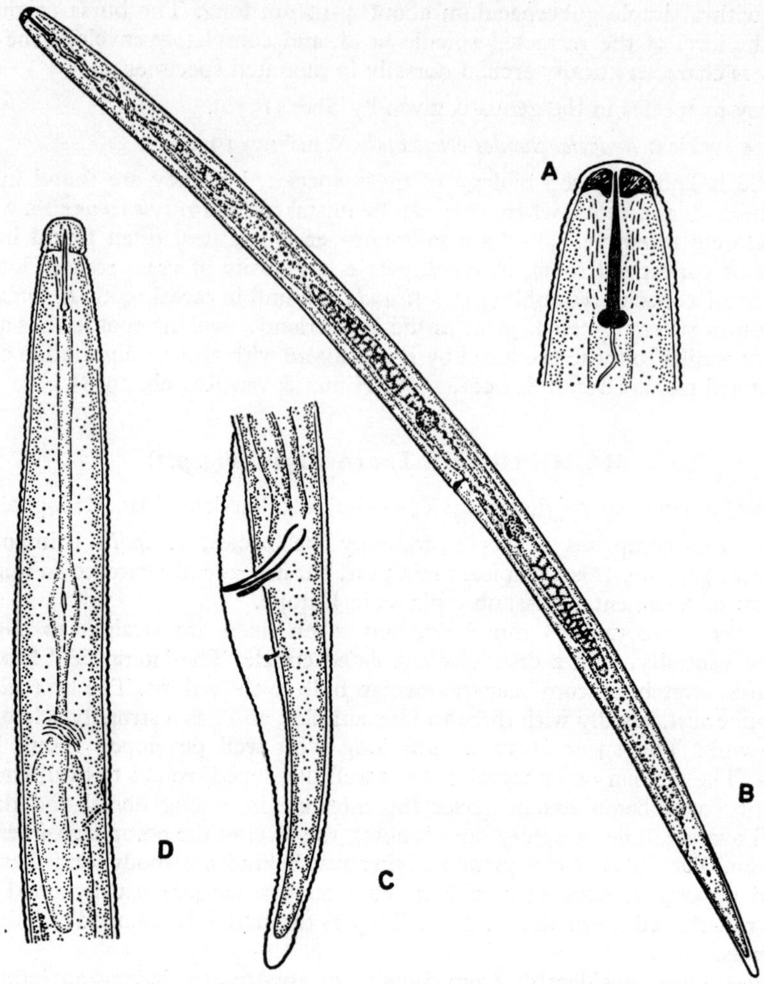

Fig. 9.7. *Radopholus similis* (Cobb, 1893). **A,** head of female. **B,** female. **C,** male tail. **D,** anterior part of male. (From Heyns, 1971).

Reproduction is normally sexual, but can take place in the absence of males. All three free juvenile stages and adults can be found in the soil, but only the juveniles and females feed; males cannot, as their oesophagus and stylet are degenerate. Several races of *R. similis* have been distinguished (Wehunt & Edwards, 1968).

Damage occurs particularly in perennial crops, and spread is in infested host material. It is a migratory root endoparasite that invades roots just behind the tip, and penetrates them within 48 hours. Feeding by the females and juveniles leads to the formation of cavities in the cortex of banana roots, and in the cortex and phloem/cambium ring of the stele in citrus. As feeding progresses cells in the cavities become necrotic with the accumulation of dark coloured substances leading to the appearance at the surface of the root of a dark lesion. Attack leads to stunting of the root system, and stunting and chlorosis of the above-ground parts of the plant.

PARATYLENCHUS Micoletzky, 1922 (Fig. 9.8)

SYSTEMATIC POSITION: Tylenchida: Tylenchina: Criconematoidea: Paratylenchidae.

These are among the smallest soil-dwelling plant-parasitic nematodes.

Females are from 0·18 to 0·6 mm long, rarely longer than ½ mm. The body is finely annulated, and in many species older females assume a strongly ventrally curved position. The head is truncate, conical or rounded in shape, bears three to four barely visible annules and internally has a lightly cuticularized skeleton. The stylet varies in different species from 12 μm to about 120 μm, and appears unrelated to length of the body. Stylets generally consist of a longer front portion and a shorter posterior portion bearing basal knobs. The oesophagus is criconematid, i.e., has a short procorpus merging into a large median bulb behind which is a longer isthmus that expands into a small posterior bulb abutting the intestine. The vulva is posterior, lying between 68–86 per cent, and runs into a single anterior genital branch which always contains a spermatheca. There is no post-vulval sac.

Males exhibit some sexual dimorphism—they are generally thinner than females, and have a stylet and oesophagus that is degenerate, with a smaller spear, or no spear at all, and reduced oesophageal structure. Spicules are nearly straight to slightly curved and cephalated by a constriction. There may or may not be a small adanal bursa.

Raski (1975a, b) provides a key to many species of *Paratylenchus*.*

Type species: *Paratylenchus bukowinensis* Micoletzky, 1922

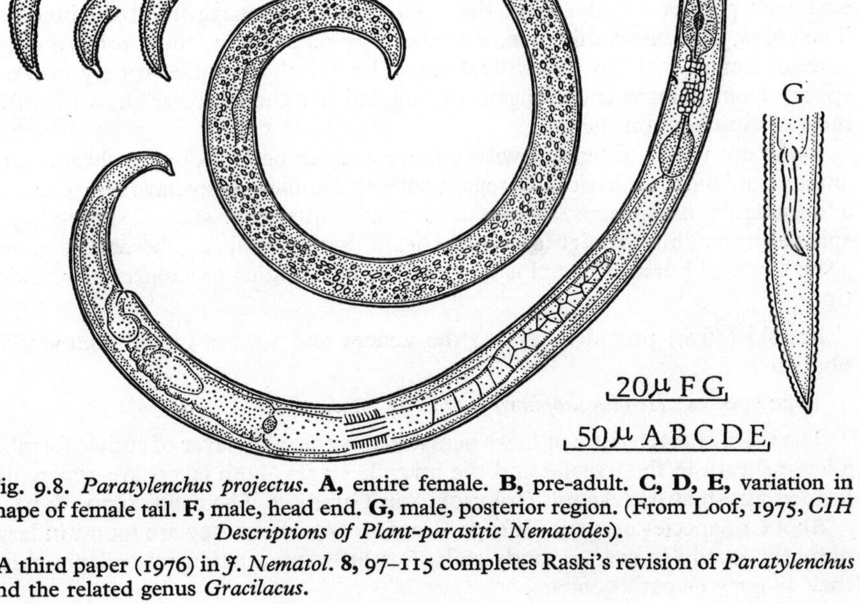

Fig. 9.8. *Paratylenchus projectus*. **A,** entire female. **B,** pre-adult. **C, D, E,** variation in shape of female tail. **F,** male, head end. **G,** male, posterior region. (From Loof, 1975, *CIH Descriptions of Plant-parasitic Nematodes*).

*A third paper (1976) in *J. Nematol.* **8,** 97–115 completes Raski's revision of *Paratylenchus* and the related genus *Gracilacus*.

In unfavourable conditions, moulting ceases at the fourth-stage juvenile, which becomes a survival stage that can endure long periods in the soil. It differs from all other stages in having a very small spear or none, and a degenerate oesophagus, and is not a feeding stage (Fig. 9.8B). It is difficult to recognize as a *Paratylenchus* in this form. Once a host grows near these juveniles, root diffusates stimulate the final moult (Fisher, 1967). Females are attracted to young mature zones of the host's roots, where they feed by inserting the stylet into epidermal cells of the young mature root, or into the base of root hairs. They remain feeding at one position for several days. It is uncertain if feeding by all *Paratylenchus* spp. causes damage to crops, but they can reach very large numbers—one sample I extracted had 250 000/l soil. Reports suggest that root weight may be decreased by *P. projectus* without producing symptoms in the above-ground parts (Loof, 1975b), but *P. neoamblycephalus* causes a decrease in feeder-root number and a darkening of attacked roots, leading to death of lateral roots and a decrease in height and top weight of *Myrobolan* plum seedlings (Braun & Lownsbery, 1975). According to Ritter (1960) very high populations of *Paratylenchus* spp. damage carnations in the south of France.

HEMICYCLIOPHORA de Man, 1921 (Fig. 9.9)

SYSTEMATIC POSITION: Tylenchida: Tylenchina: Criconematoidea: Criconematidae: Hemicycliophorinae.

Females are about 0·75–1·75 mm long, coarsely annulated, with a cuticular sheath attached loosely to the body at the head and vulva. The lip region is bluntly rounded, annulated though sometimes indistinctly so. The spear is usually long, with the anterior, tapered portion most often several times longer than the posterior portion, which terminates in well developed basal knobs, sloping posteriorly. The oesophagus is typically criconematid, with the procorpus and median bulb amalgamated and having a large crescentic valve; the isthmus and posterior bulb form a spatulate region abutting on the intestine. The excretory pore is usually about the level of the oesophago-intestinal junction. The vulva, with protruding lips, is at about 75–90 per cent, and there is a single anterior gonad with no post-vulval sac. The tail shape varies widely between species from pointed and elongate to rounded but the cuticular sheath is often most conspicuous on the tail.

Males are wholly different, with a more slender body lacking a sheath, finer annules and a hemispherical lip region without annules. They have no spear, and a degenerate oesophagus. There is a single anterior testis, a sickle-shaped spicule protruding through a cloacal sheath, with a simple gubernaculum, and a small adanal bursa. The tail is long and slender leading to a narrowly rounded tip.

Brzeski (1974) provides keys to the genera and species in the Hemicycliophorinae.

Type species: *Hemicycliophora typica* de Man, 1921

The main characteristic of these nematodes is the extra layer of cuticle forming a loose sheath in the females and the juvenile stages: both layers are apparently formed afresh at each moult (Johnson, van Gundy & Thomson, 1970a & b).

About 14 species occur in Europe (Loof, 1968) where they are found in large numbers locally in coarse, sandy soils, but little information is available about their biology or pathogenicity.

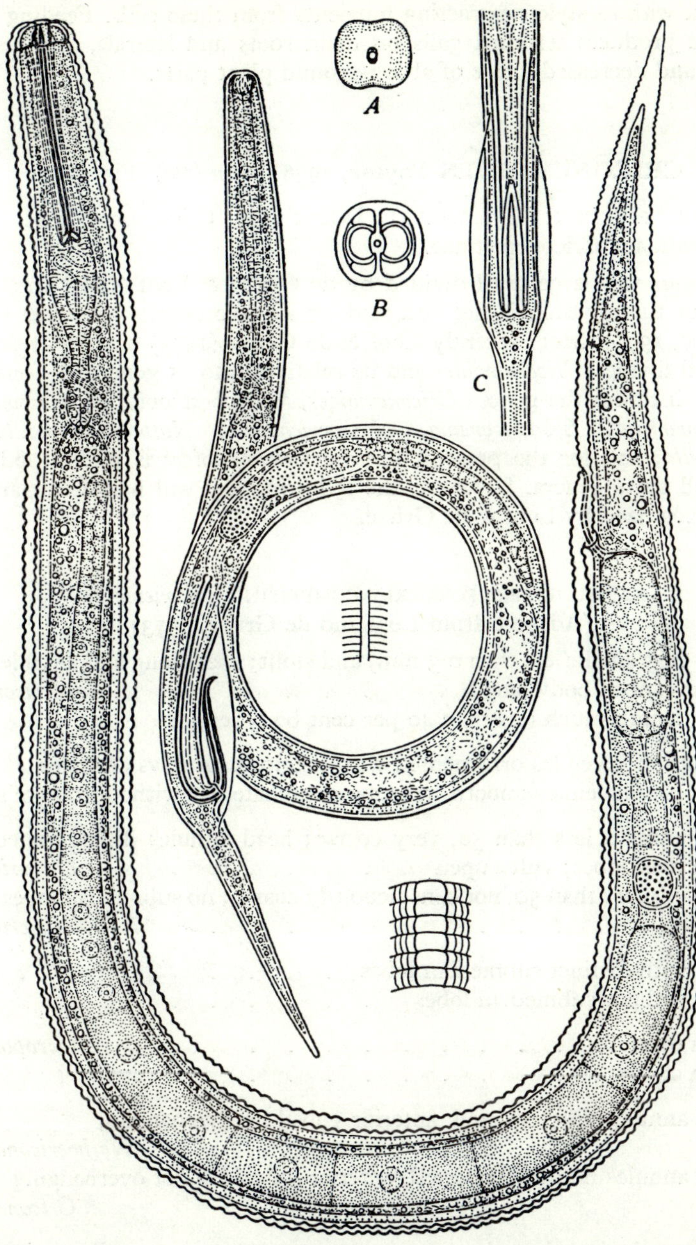

Fig. 9.9. *Hemicycliophora penetrans*. **A,** labial disc of female. **B,** basal pattern of cephalic framework. **C,** male tail, ventral. (From Thorne, 1955).

The most studied species is *H. arenaria* (see Franklin & Stone, 1974) native to sandy soils in California, U.S.A., where it may be a pest of several crops. Males occur, but are not essential for reproduction.

H. arenaria is attracted to, and feeds on root tips. It lives entirely outside its host's roots, penetrating through the root epidermis to two to three cells into the cortex with its stylet, extracting nutrients from these cells. Feeding by this nematode produces terminal galls on main roots and laterals, and results in stunting and decreased yields of above-ground plant parts.

CRICONEMOIDES Taylor, 1936 (*sensu lato*) (Fig. 9.10)

SYSTEMATIC POSITION: Tylenchida: Tylenchina: Criconematoidea: Criconematidae: Criconematinae.

The genus was large until divided by de Grisse & Loof (1965) into several genera but this division is not accepted by all taxonomists (Raski & Golden, 1966; Luc, 1970). More recently Loof & de Grisse (1973) have defended their views, and divided *Criconemoides* and its relatives into 15 genera: *Criconemoides sensu lato* includes the genera *Criconemoides, Macroposthonia, Discocriconemella, Xenocriconemella, Lobocriconema, Nothocriconema, Nothocriconemoides* and *Criconemella,* and for the present purpose *Criconemoides* is understood as including all these genera. However, the following key will help to separate the genera recognized by Loof & de Grisse.

KEY TO GENERA FORMERLY COMPRISING *Criconemoides*

(Adapted from Loof and de Grisse, 1973)

1. Body very small (less than 0·3 mm) and stout; spear long and flexible, about 40 per cent of body length . . . . *Xenocriconemella*
   Spear length much less than 40 per cent body length 2

2. Annules of juveniles ornamented with longitudinal rows of scales . . 3
   Annules of juveniles smooth, irregular or crenate, but without rows of scales 4

3. Body annules less than 50, very coarse; head annules differentiated, with submedian lobes; vulva open . . . . *Lobocriconema*
   Annules more than 50, not conspicuously coarse; no submedian lobes; vulva closed . . . . *Nothocriconema*

4. Head with distinct submedian lobes 5
   Head without submedian lobes 7

5. Vulva open . . . . *Macroposthonia*
   Vulva closed 6

6. Head annules differentiated, anterior vulval lip overhanging *Nothocriconemoides*
   Head annules not differentiated, anterior vulval lip not overhanging *Criconemoides*

7. First annule disc-shaped . . . . *Discocriconemella*
   First annules not disc-shaped . . . . *Criconemella*

Females in *Criconemoides sensu lato* are usually short and stout (0·20–0·75 mm, rarely up to 1 mm) with 33–200 well marked but ornamented body annules. The lip region may or may not be well defined, and anteriorly may bear sub-lateral lobes and labial plates or discs called 'pseudolips'. The stylet is generally long, with a short basal part having knobs, bearing forward pointing processes. It varies in length from about one third to one tenth body length. The oesophagus is characteristically criconematid, with the wide procorpus merging into a swollen median bulb and terminating in a small posterior bulb abutting the intestine. The vulva is transverse situated between 90–92 per cent of body length. The single genital branch is anterior, and most often contains a spermatheca that may or may not have spermatozoa. There is no post-vulval sac. The anus is often obscure and difficult to see. The tail varies in shape, and may be short and conical to broadly rounded.

Males are about the same length as females but differ markedly from them in appearance. They are thinner and more vermiform in shape with a weakly annulated cuticle and have a vestigial spear, or none, with a degenerate oesophagus. They have one anterior testis, a straight to slightly curved spicule cephalated by a constriction, a simple gubernaculum, and an anal sheath through which the spicule is everted. There may or may not be a small adanal bursa.

Type species: *Criconemoides morgensis* (Hofmänner in Hofmänner & Menzel, 1914) Taylor, 1936.

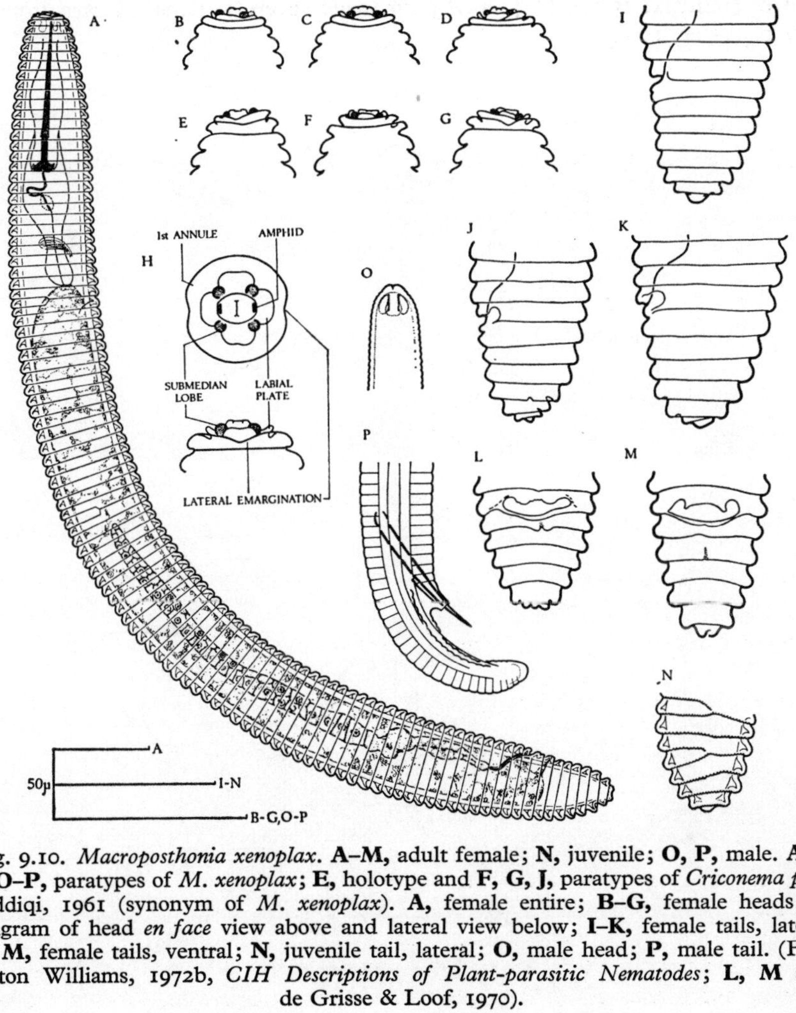

Fig. 9.10. *Macroposthonia xenoplax*. **A–M,** adult female; **N,** juvenile; **O, P,** male. **A–D, I, O–P,** paratypes of *M. xenoplax*; **E,** holotype and **F, G, J,** paratypes of *Criconema pruni* Siddiqi, 1961 (synonym of *M. xenoplax*). **A,** female entire; **B–G,** female heads; **H,** diagram of head *en face* view above and lateral view below; **I–K,** female tails, lateral; **L, M,** female tails, ventral; **N,** juvenile tail, lateral; **O,** male head; **P,** male tail. (From Orton Williams, 1972b, *CIH Descriptions of Plant-parasitic Nematodes*; **L, M** after de Grisse & Loof, 1970).

Little is known about the biology and host-parasite relations of most species of *Criconemoides*. They usually occur discontinuously, mostly in coarse sandy soils and rarely in large numbers. Together with their near relatives, they have a unique mode of movement with muscular contractions passing the length of the body producing forward movement from the grip of the prominent annules on soil particles, wholly unlike the sinuous movements of vermiform nematodes and more like those of earthworms. They are sluggish, and are therefore rarely extracted from soil using the Baermann funnel types of extraction procedure.

All species studied parasitize roots of plants, feeding in the cortex by inserting the stylet into roots near the growing tips, or in other parts of the root (Orton Williams, 1972b; Loof, 1974b). *C.* [= *Macroposthonia*] *xenoplax* added to soil in which young plum plants were growing destroyed root cortical tissue, causing darkening of the roots, increased water stress, lowered nutrient levels in leaves, and decreased plant weight (Mojtahedi & Lownsbery, 1975). *C.* [? = *M.*] *quadricorne* on its own caused no symptoms in pecan trees, but when inoculated with *Fusarium solani* and *Pythium irregulare* it caused a decrease in root weight (Hsu & Hendrix, 1973). *C.* [= *M.*] *curvatum* severely damaged carnations in France (Ritter, 1960).

CHAPTER 10

Plant-parasitic Dorylaimida: Morphology and Identification

D. J. HOOPER
Rothamsted Experimental Station, Harpenden
J. F. SOUTHEY
Plant Pathology Laboratory, Harpenden

Plant-parasitic Dorylaimida are confined to two divergent groups: one, containing the genera *Longidorus*, *Xiphinema* and *Paralongidorus*, is in the family Longidoridae suborder DORYLAIMINA, and the other, containing the genera *Trichodorus* and *Paratrichodorus*, is in the family Trichodoridae superfamily Trichodoroidea in the suborder DIPHTHEROPHORINA. The only known nematode vectors of plant viruses also occur in these two groups, some Longidorids transmitting nepoviruses whereas Trichodorids transmit tobraviruses.

Suborder DORYLAIMINA (de Man, 1876) Pearse, 1942

Superfamily Dorylaimoidea Thorne, 1934
Family Longidoridae (Thorne, 1935) Meyl, 1961.

Longidorid nematodes have a basic Dorylaimoid structure in that they have a hollow axial spear (odontostyle) with an extension (odontophore), a slender anterior oesophagus and a prominent posterior, cylindrical, muscular, oesophageal bulb (Fig. 10.1A). The oesophageal bulb has a dorsal oesophageal gland duct and associated gland near its anterior end, a pair of subventral glands about half-way along the bulb and sometimes a suggestion of a second subventral gland duct which opens towards the base of the bulb. An oesophago-intestinal valve (cardia) is present. Longidorids are distinguished from most other Dorylaims by their very long odontostyle, the nature of the guide ring around the odontostyle and the long, usually relatively narrow, body.

The cuticle generally appears smooth when viewed with a light microscope although a cross-hatched pattern is sometimes visible on the neck and tail regions. The inner layer of the cuticle usually widens at the neck and tail end and has very faint transverse striations that have a radial appearance on the tail.

Hypodermal lateral chords are present, usually occupying about a third to a quarter the mid-body width, but narrowing to a thin line in the oesophageal region and somewhat expanding in the anal region. The chords have a glandular appearance and give rise to pores throughout the length of the body; ventral and dorsal body pores are also sometimes present, especially in the oesophageal region.

Amphid apertures are lateral, sublabial; in *Xiphinema* they are consistently wide with goblet-shaped pouches, whereas in *Longidorus* the openings are very

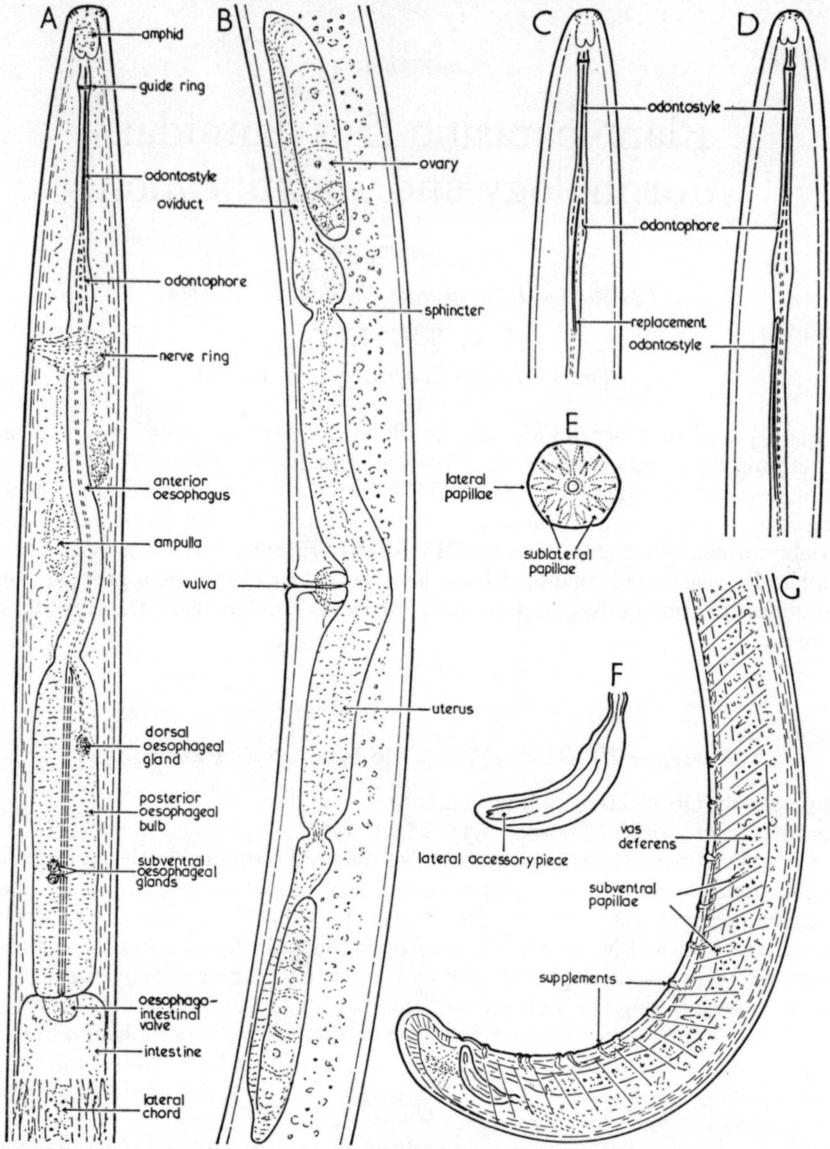

Fig. 10.1. Basic morphology of *Longidorus*: **A,** anterior region of adult; **B,** genital tracts of female; **C,** anterior end of first-stage juvenile; **D,** anterior end of second- to fourth-stage juvenile; **E,** *en face* view; **F,** spicule in lateral view; **G,** male tail, lateral. (From Hooper & Southey, 1973).

obscure pores and the pouches tend to be larger and very variable in shape between species; *Paralongidorus* species have slit-like openings of varying width and pouches of variable shape.

Females basically have a median, transverse, slit-like vulva with relatively short, paired, opposed genital branches, the ovaries being reflexed (see Fig. 10.1B). In some species of *Xiphinema* the anterior branch is degenerate and the vulva may be more anteriorly placed; also the genital-tract structure is generally more complicated than in Longidorinae (see pages 220–221).

Male tails are usually of similar shape to those of females; they have massive, ventrally curved spicules with lateral accessory pieces (Fig. 10.1F) and an adanal pair plus a series of prominent ventro-median supplementary papillae (supplements). The testes are paired and opposed, the anterior testis almost reaching to the middle of the body, and they connect via a common vas deferens to the cloaca. Prominent, oblique, copulatory muscles are present in the supplement region.

There are four juvenile (larval) stages, the first stage emerging from the egg. Juveniles are readily recognized by having a functional and a replacement odontostyle, the latter being situated in the wall of the anterior oesophagus; first-stage juveniles are distinguished from the others in that the anterior part of the replacement odontostyle is embedded within the wall of the odontophore of the functional stylet (cf. Fig. 10.1C & D). First-stage juveniles of some species have a characteristically shaped tail (Figs. 10.2 & 10.3). At moulting, the cuticle, amphid pouch linings, lining of the anterior oesophagus, rectal lining and the functional odontostyle are shed; the previous replacement odontostyle moves forward and becomes attached to a reformed odontophore and, before moulting is completed, a new replacement odontostyle is formed in the anterior oesophageal wall. Thus in juveniles, in the actual process of moulting from second or third stage, three odontostyles are sometimes visible, the one being shed, the new functional stylet and the new replacement stylet. Moulting from the fourth stage to adults is similar except that a replacement odontostyle is not formed, although a vestigial tip (mucro) is sometimes visible in the anterior oesophagus of adults.

Identifications to species are most easily made on adults and/or first-stage juveniles; the characters mainly used are:

Body length and width.
Lip region width and shape, and neck contour.
Shape and size of amphid pouch.
Position of guide ring.
Odontostyle (ost) and odontophore (oph) length.
Dorsal and ventral body pores especially in oesophageal region.
Tail shape (1st stage juvenile often characteristically different from adult).
c' = tail length/anal body width.
V = distance from anterior end to vulva expressed as percentage of total length.
Number and arrangement of preanal supplements on male.
Shape of spicules and lateral accessory pieces.

Subfamily LONGIDORINAE Thorne, 1935

Longidoridae. Guide ring single, prominent, around anterior part of odontostyle. Odontostyle-odontophore junction simple, without prongs. Odontophore plain, without basal flanges. Vulva about median; genital branches paired, opposed with reflexed ovaries. Tails bluntly rounded to roundly conoid. Dorsal oesophageal gland nucleus well posterior to its orifice.

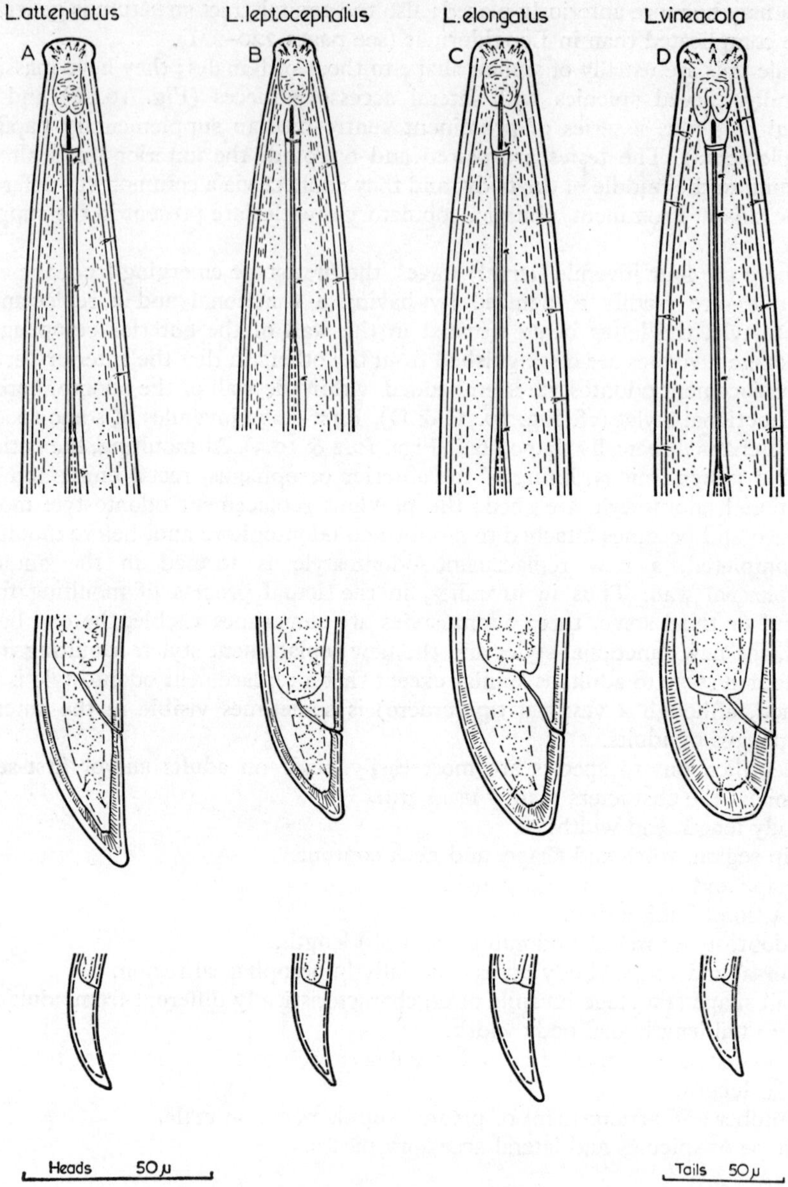

Fig. 10.2. Right lateral views of anterior region, female tail and first-stage juvenile tail of *Longidorus* species. (From Hooper, 1973).

TYPE GENUS: *LONGIDORUS* (MICOLETZKY, 1922) FILIPJEV, 1934)

Type species: *Longidorus elongatus* (de Man, 1876) Thorne & Swanger, 1936.

Syn. *Longidorus monohystera* Altherr, 1953
Longidorus menthasolanus Konicek & Jensen, 1961

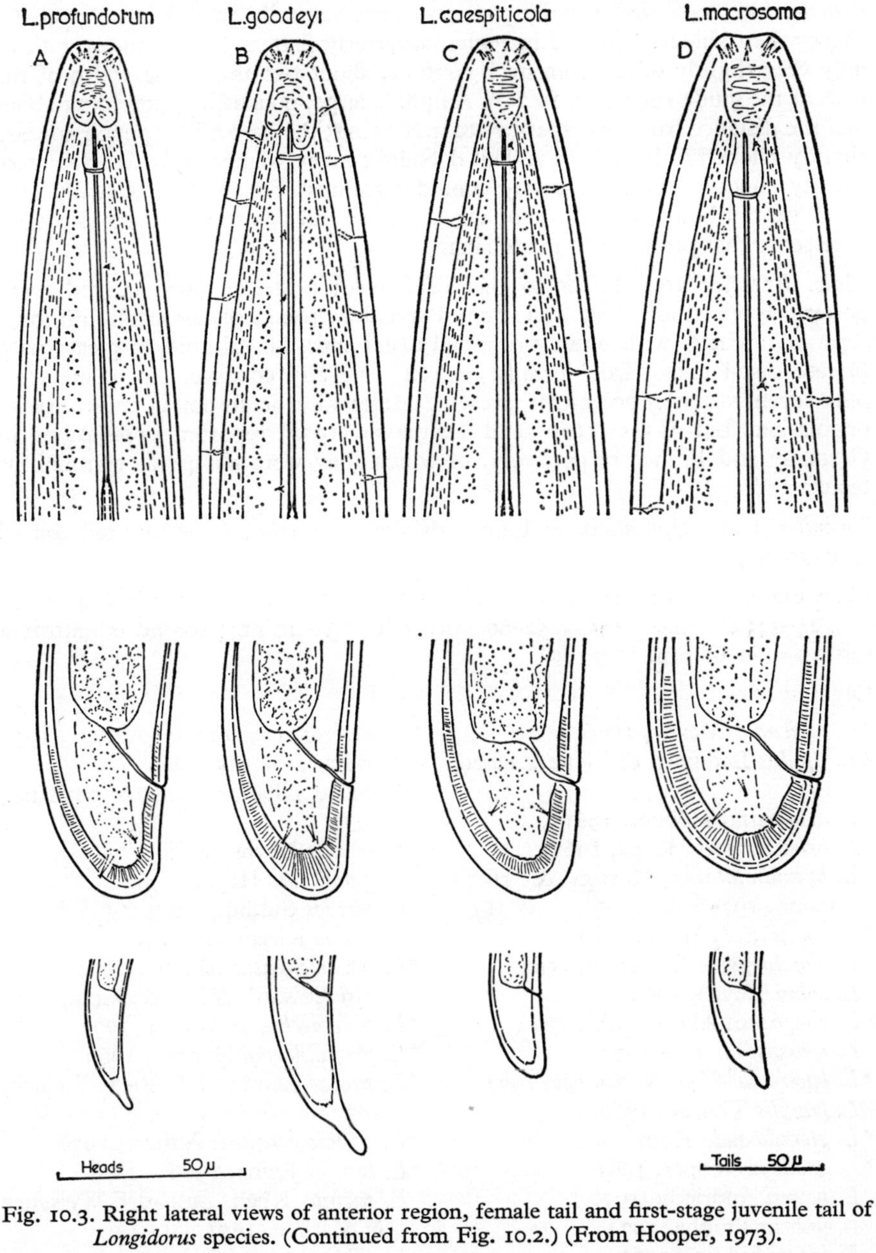

Fig. 10.3. Right lateral views of anterior region, female tail and first-stage juvenile tail of *Longidorus* species. (Continued from Fig. 10.2.) (From Hooper, 1973).

♀♀: L = 4·5–6·4 mm; a = 62–81; b = 9·7–17·5; c = 73–141; c′ = 1·0–1·4; V = 45–53; odontostyle (ost) = 91–102 μm; odontophore (oph) = 34–71 μm; oral aperture to guide ring 30–36 μm.

♂♂: L = 5·0–5·3 mm; a = 93–113; b = 11·8–13·0; c = 89–135; ost = 72–94 μm; oph = 45–58 μm; spicule length = 58 μm.

Female. Moderate size, usually 5–6 mm long, ventrally curved in an open C-shape when killed by heat. Lip region anteriorly flattened, continuous with or only very slightly offset from neck contour, about two thirds the width of the neck at the guide ring (Fig. 10.2C). Amphids large, pouch-like, extending about half the distance from oral aperture to guide ring; the base of the pouch appears slightly bilobed in lateral view. Tail dorsally convex, ventrally flattened or very slightly concave, with a roundly conoid terminus (Fig. 10.2C). Usually four to six ventral body pores in oesophageal region and occasionally a single dorsal pore somewhat posterior to guide ring.

Male. Anterior end similar to that of female; posterior third curved more posteriorly when killed by heat. Paired spicules prominent, about 58 μm long, separate, arcuate, with bluntly rounded distal ends, lateral guiding pieces with furcate distal ends. Tail dorsally convex, ventrally concave. An adanal pair plus seven to ten supplements present. Males rare or absent in most British populations but Wyss (1969) and Dalmasso (1970) reported populations in Germany and France respectively, in which they were almost as common as females.

Juveniles. Lip-region shape and amphids similar to those of adults; tail conoid (Fig. 10.2C).

Measurements of first stage (J.1); L = 1·2–1·4 mm; a = 41–54; b = 4·1–6·2; c = 25–34; c′ = 4–5; ost = 52–60 μm; oph = 31–40 μm; second odontostyle (ost 2) = 62–68 μm.

Other species:

- *L. africanus* Merny, 1966
- **L. apulus* Lamberti & Bleve Zacheo, 1977
- **L. attenuatus* Hooper, 1961
- *L. belondiroides* Heyns, 1967
- *L. breviannulatus* Norton & Hoffmann, 1975
- **L. caespiticola* Hooper, 1961
- **L. closelongatus* Stoyanov, 1964
- *L. cohni* Heyns, 1969
- *L. congoensis* Aboul-Eid, 1970
- *L. crassus* Thorne, 1974
- **L. euonymus* Mali & Hooper, 1974
- *L. fragilis* Thorne, 1974
- **L. globulicauda* Dalmasso, 1969
- **L. goodeyi* Hooper, 1961
- *L. heynsi* Andrássy, 1970
- *L. indicus* Prabha, 1973
- *L. jonesi* Siddiqi, 1962
- **L. juvenilis* Dalmasso, 1969
- *L. laevicapitatus* Williams, 1959
- **L. leptocephalus* Hooper, 1961
- *L. longicaudatus* Siddiqi, 1962
- *L. macromucronatus* Siddiqi, 1962
- **L. macrosoma* Hooper, 1961
- **L. macroteromucronatus* Altherr, 1974
- *L. martini* Merny, 1966
- *L. mirus* Khan, Chawla & Seshadri, 1972
- *L. monile* Heyns, 1966
- *L. moniloides* Heyns, 1966
- *L. nirulai* Siddiqi, 1965
- *L. nudus* Kirjanova, 1951
- **L. paraelongatus* Altherr, 1974
- *L. pisi* Edward, Misra & Singh, 1964
- **L. poessneckensis* Altherr, 1974
- **L. profundorum* Hooper, 1966
- **L. protae* Lamberti & Bleve Zacheo, 1977
- **L. pseudoelongatus* Altherr, 1976
- *L. reneyii* Raina, 1966
- *L. saginus* Khan, Seshadri, Weischer & Mathen, 1971
- *L. siddiqii* Aboul-Eid, 1970
- *L. sylphus* Thorne, 1939
- *L. taniwha* Clark, 1963
- *L. tarjani* Siddiqi, 1962
- **L. vineacola* Sturhan & Weischer, 1964

*Species reported from Europe.

The following are omitted for the reasons given:
L. brevicaudatus (Schuurmans Stekhoven, 1951) Thorne, 1961 and *L. meyli* Sturhan, 1963—*species inquirendae* (see Aboul-Eid, 1970); *L. striola* and *L. tardicauda*, both of Meerzheevskaya, 1951—*nomina nuda* because of inadequate descriptions and absence of differential diagnosis.

COMMON EUROPEAN SPECIES

Longidorus attenuatus Hooper, 1961. (Fig. 10.2A).

Female. L = 6·1 (5·2–7·3) mm; a = 158 (120–210); c′ = 1·7 (1·4–2·0); V = 49 (44–55); ost = 78 (73–84) μm. Body rather narrower than *L. elongatus*. Lip region flattened anteriorly, distinctly expanded and set off from the neck contour; about three-quarters of the width of the neck at the guide ring. Five to eight ventral body pores in oesophageal region; no dorsal pores seen. Tail dorsally convex, conoid.

Male. Usually very rare.

*J.*1—Tail conical; c′ about 4.

L. leptocephalus Hooper, 1961. (Fig. 10.2B).

Female. Lip region characteristically narrow, about half the neck width at the guide ring, slightly more rounded than in *L. elongatus* and very slightly offset from neck contour. Amphids reach about half way to guide ring, somewhat bilobed at the base. Three to five ventral body pores in the anterior oesophageal region and a single dorsal pore about 25 μm posterior to guide ring. Tail dorsally convex, ventrally flattened to slightly concave with a narrowly conoid, almost digitate, terminus.

Male. Undescribed but occasional specimen has been found in some populations.

*J.*1. Tail conical; c′ about 3.

Note. As indicated by Flegg (1967b) there seem to be two basic size forms of this species. In the two populations originally described (Hooper, 1961) length of females is 4·2 (3·5–4·9) mm; a = 96 (78–127); c′ = 1·2 (1·0–1·4); ost = 64 (60–70) μm; V = 54 (52–57) and other populations have subsequently been found with similar measurements. However, populations of larger forms with body length of females about 5·3 (4·7–6·0) mm; a = 99 (83–134); c′ = 1·1 (0·9–1·4); V = 51 (47–55); ost = 71 (60–77) μm seem to be more common. The size ranges of the two forms overlap so it is not always possible to assign individual specimens to a particular form.

L. caespiticola Hooper, 1961. (Fig. 10.3C).

Female. L = 6·7 (5·2–8·2) mm; a = 72 (64–80); c′ = 0·8 (0·6–0·8); V = 53 (48–56); ost = 110 (109–114) μm. Body robust, usually forms open 'C' when killed by heat. Lip region smoothly rounded, almost cone shaped, continuous with neck contour which is convex in the amphid region; lip region width usually just less than half the body width at the guide ring. Amphid pouches large, flask shaped, tapering posteriorly, not bilobed, reaching about two thirds the distance to the guide ring; coiled sensillae usually prominent. Five to six ventral body pores in the anterior oesophageal region; a single dorsal pore about 30 μm posterior to guide ring. Tail dorsally convex, bluntly conoid to rounded.

Male. Anterior end similar to female; posterior third more ventrally curved when killed by heat. Spicules large, 90 μm long; lateral guiding pieces with bifid distal ends. An adanal pair of supplements and a ventral series of 15–20 arranged in a staggered double row. Tail about one anal body width (a.b.w.) long (c′ about 1), dorsally convex, ventrally concave with a bluntly rounded terminus. Males generally common.

*J.*1. Tail bluntly rounded; c′ about 2.

L. goodeyi Hooper, 1961. (Fig. 10.3B).

Female. L = 6·7 (5·6–7·7) mm; a = 85 (67–117); c′ = 0·8 (0·6–1·0); V = 52 (50–55); ost = 101 (96–109) μm. Body robust, often assumes closed 'C' to spiral shape when killed by heat. Lip region bluntly rounded, wide, about half neck width at guide ring, continuous with neck contour. Amphid pouches large, asymmetrically bilobed, the ventral lobe almost reaching to the guide ring. About 10 ventral body pores throughout the oesophageal region; three, sometimes four dorsal body pores opposite anterior half of odontostyle. Tail dorsally convex, bluntly conoid.

Male. Not recorded.

*J.*1. Tail elongate, digitate; c′ about 3.

L. profundorum Hooper, 1966. (Fig. 10.3A).

Female. L = 7·0 (6·0–8·3) mm; a = 105 (81–119); c′ = 0·9 (0·7–1·0); V = 53 (50–58); ost = 97 (91–104) μm. Moderately robust in appearance, usually forms open 'C' shape when killed by heat. Lip region flattened anteriorly, somewhat truncated, continuous with neck contour which is slightly convex in amphid region. Lip region width about one third body width at guide ring. Amphid pouches evenly bilobed extending just over half way to guide ring. Ventral body pores absent; dorsal body pores usually absent in anterior oesophageal region. Tail dorsally convex, sometimes slightly concave ventrally, bluntly conoid.

Male. Tail dorsally convex, ventrally concave, with roundly conoid terminus about three-quarters the a.b.w. long. Spicules about 70 μm long; lateral guiding pieces with bifid distal ends. An adanal pair of supplements plus a ventral series of 10–15 that are mainly in a single row rather than in the usual staggered double row.

*J.*1. Tail ± digitate; c′ = 3–4.

L. macrosoma Hooper, 1961. (Fig. 10.3D).

Female. L = 9·1 (7·9–11·9); a = 95 (77–142); c′ = 0·6 (0·5–0·7); V = 51 (47–56); ost = 130 (119–148) μm. Body long and robust, posterior half ventrally curved when killed by heat. Lip region wide, about half body width at guide ring; continuous with neck contour, flattened to slightly concave anteriorly appearing somewhat truncated. Amphid pouches large, tapering posteriorly (flask shaped), not lobed, almost reaching to guide ring. Two to four dorsal pores and five to eight ventral body pores in the anterior oesophageal region. Tail hemispherical to bluntly rounded.

Male. Posterior third of body loosely coiled when killed by heat. Spicules massive, 105 μm long; lateral guiding pieces narrowed distally. An adanal pair of supplements plus a staggered ventral series of 15 to 19. Tail dorsally convex, ventrally concave, bluntly conoid to rounded. Males usually common.

*J.*1. Tail digitate; c′ about 2.

L. vineacola Sturhan & Weischer, 1964. (Fig. 10.2D).

Female. L = 8·2 (6·9–9·2) mm; a = 137 (120–149); c′ = 0·8–0·9; V = 52 (50–54); ost = 97 (90–101) μm. Body rather like a large form of *L. elongatus*. Lip region expanded from neck contour, flattened anteriorly, about three-quarters neck width at guide ring. Amphid pouches evenly bilobed reaching about two thirds the distance to the guide ring. Eighteen to 27 lateral and 13–17 ventral body pores in oesophageal region; three to six dorsal pores in spear region. Tail dorsally convex, bluntly conoid to rounded.

Male. Spicules 63 μm long; lateral guiding pieces with bifid distal ends. Ventral supplements 14–17 in number. Tail dorsally convex, ventrally concave; c′ = 0·8–1. Males common.

*J.*1. Tail terminus ± conoid; c′ = 2–2·5.

GENUS *PARALONGIDORUS* SIDDIQI, HOOPER & KHAN, 1963, EMEND. ABOUL-EID, 1970

Longidorinae. Lip region may be offset by a constriction at the level of the amphid opening. Amphid openings transverse slits, sublabial, often extending completely across the head. Amphid pouches large, variable in shape; sometimes like those of *Longidorus* spp., others like those of *Xiphinema*. Guide ring, odontostyle and odontophore as in *Longidorus*.

Type species: *Paralongidorus sali* Siddiqi, Hooper & Khan, 1963.

P. maximus is the only species recorded to any extent in Europe and it is therefore described.

Paralongidorus maximus (Bütschli, 1874) Siddiqi, 1964

Syn. *Dorylaimus maximus* Bütschli, 1874
Longidorus maximus (Bütschli, 1874) Thorne & Swanger, 1936

♀♀: L = 7·6–12·4 mm; a = 77–133; b = 12·8–23·7; c = 178–320; V = 36–40; ost = 152–187 μm; oph = 42–70 μm; oral aperture to guide ring 37–47 μm.
♂♂: L = 8·6–10·7 mm; a = 97–110; b = 12·9–14·1; c = 190–237; ost = 171–173 μm; oph = 53–56 μm; spicule length = 100–106 μm.

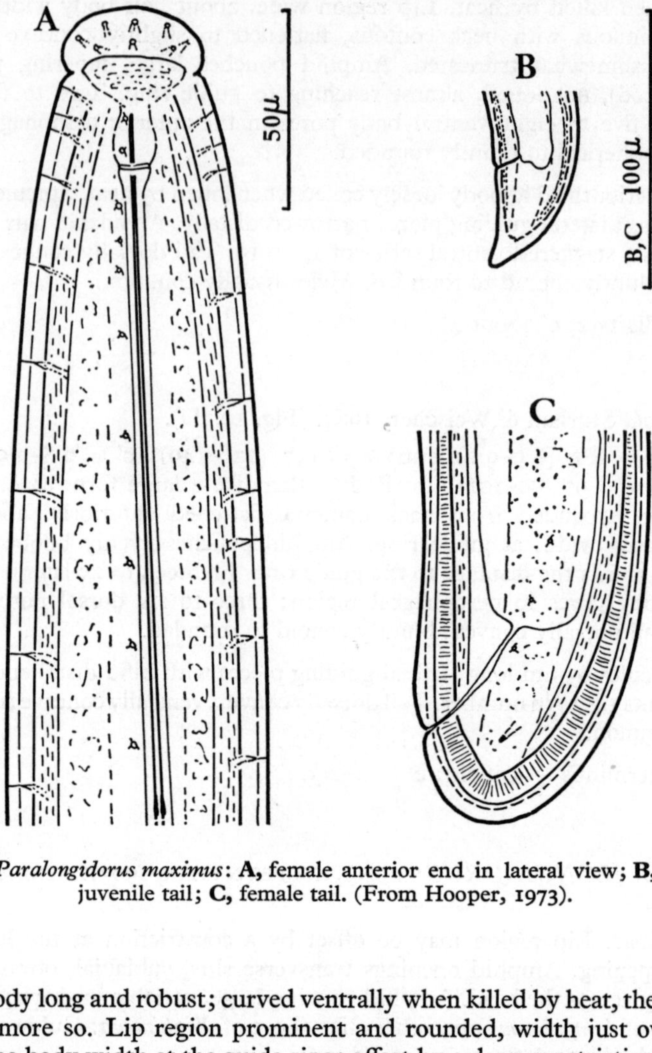

Fig. 10.4. *Paralongidorus maximus*: **A,** female anterior end in lateral view; **B,** first-stage juvenile tail; **C,** female tail. (From Hooper, 1973).

Female. Body long and robust; curved ventrally when killed by heat, the posterior third the more so. Lip region prominent and rounded, width just over three-quarters the body width at the guide ring; offset by a deep constriction. Amphid apertures wide, almost extending across the head when viewed laterally; amphid pouches inverted-stirrup (goblet) shape, reaching almost half way to guide ring (Fig. 10.4A). Tail short, bluntly rounded, about two thirds a.b.w. long (Fig. 10.4C).

Male. Anterior end similar to female, tail end more curved ventrally when killed by heat. An adanal pair and 14 to 15 ventro-median supplements present. Spicules massive, 100–106 μm long, with lateral guiding pieces with pointed distal ends. Males usually extremely rare.

Juveniles. Lip region and amphids similar to those of adults. Tail of first stage rather short, 1·4 a.b.w. long, with a somewhat digitate terminus (Fig. 10.4B). Measurements of first stage:

L = 1·3–2·3 mm; a = 39–49; b = 3·8–8·8; c = 35–52; ost = 73–89 μm; oph = 24–31 μm; ost 2 = 84–106 μm; oral aperture to guide ring 15–25 μm.

Other species:

P. afzali (Khan, 1964) Siddiqi & Husain, 1965
P. beryllus Siddiqi & Husain, 1965
P. boshi Khan, Saha & Seshadri, 1972
P. capensis Heyns, 1967
P. citri (Siddiqi, 1959) Siddiqi, Hooper & Khan, 1963
P. droseri Sukul, 1972
P. epimikis Dalmasso, 1969
P. erriae Heyns, 1965
P. eucalypti Fisher, 1964
P. fici Edward, Misra & Singh, 1964
P. fischeri Heyns, 1972
P. flexus Khan, Seshadri, Weisher & Mathen, 1971
**P. georgiensis* (Tulaganov, 1937) Siddiqi, 1965
P. hooperi Heyns, 1966
P. lutosus (Heyns, 1965) Aboul-Eid, 1970
P. microlaimus Siddiqi, 1964
P. paramaximus Heyns, 1965
P. remyi (Altherr, 1963) Siddiqi, 1965
P. sacchari Siddiqi, Hooper & Khan, 1963
P. similis Khan, Chawla & Prasad, 1972
P. spasskii Heyns, 1972
P. spiralis Khan, Saha & Seshadri, 1972
P. strelitziae (Heyns, 1966) Aboul-Eid, 1970
P. utriculoides (Corbett, 1964) Siddiqi & Husain, 1965
P. xiphinemoides Heyns, 1965

*Species reported from Europe; most of the others are from Africa, India and Australia.

Lamberti (1975) gives keys to *Longidorus* and *Paralongidorus* species.

Subfamily XIPHINEMINAE Dalmasso, 1969

Longidoridae. Amphid openings wide, slit-like, almost extending across base of lip region in lateral view; amphid pouches usually inverted-stirrup (goblet) shape. Base of odontostyle forked (2–3 prongs) at its junction with odonotophore. Odontophore with prominent basal flanges. Guiding sheath around base of odontostyle showing as two guide rings; the posterior is the more prominent being a definite thickening at the base of the sheath whereas the fainter anterior one results merely from folding of the sheath. Vulval position variable; genital tract sometimes having paired, opposed, reflexed branches (didelphic). Some species have a more anterior vulva with a posterior genital branch only (monodelphic) and some also a vestigial anterior branch lacking an ovary (pseudomonodelphic). Tails vary greatly between species—bluntly rounded, rounded with a terminal peg, elongate conoid to filiform. Dorsal oesophageal gland just posterior to its orifice.

TYPE AND ONLY GENUS: *XIPHINEMA* COBB, 1913 (Fig. 10.5)

For practical purposes members of this genus are most easily recognized by the long stylet whose odontophore (spear extension) has three characteristic, radially-arranged flanges that serve as attachments for the stylet protractor muscles. Although odonotostyle and odontophore have separate origins, functionally they form a single unit. Other important diagnostic features of *Xiphinema* are the relatively posterior position of the spear guiding sheath and the wide amphid openings.

Type species: *X. americanum* Cobb, 1913. (See pp. 225–226).

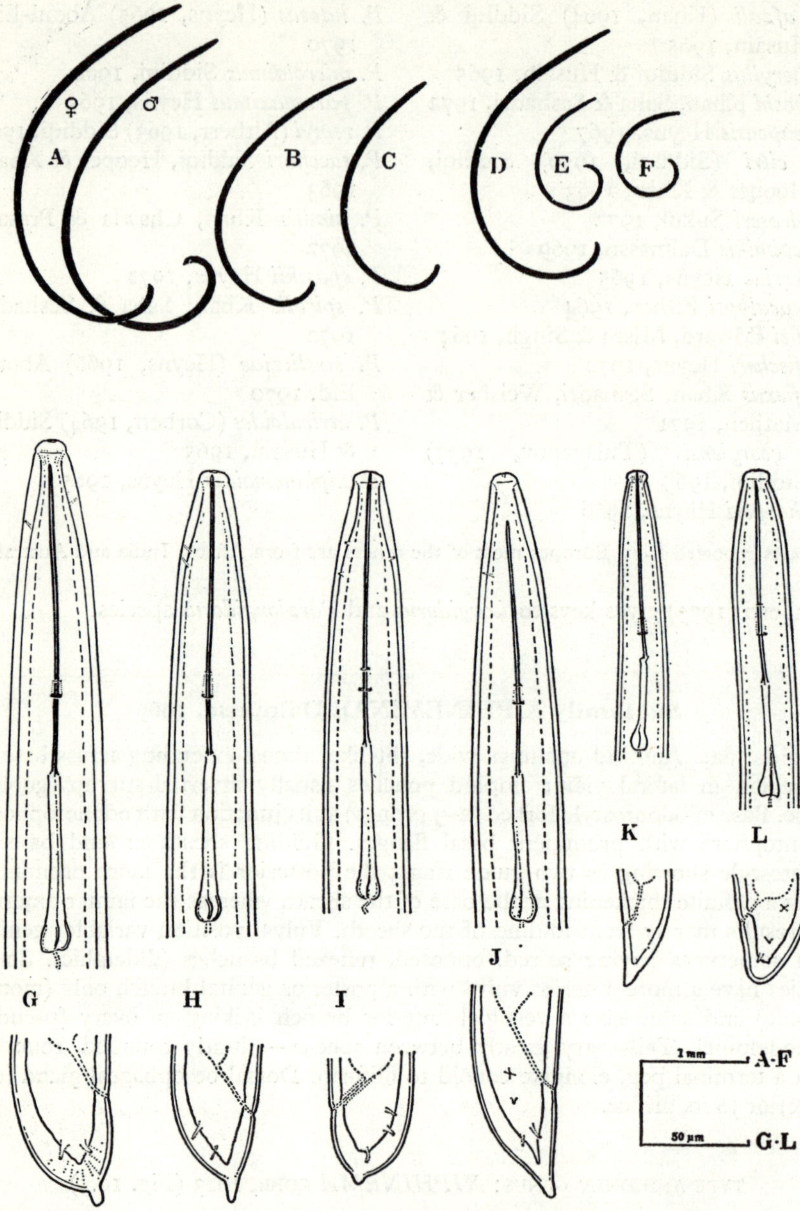

Fig. 10.5. Females of some common, mostly temperate, *Xiphinema* species: **A–F,** body shape when relaxed by heat; **G–L,** anterior portions and tails; **A, G,** *X. diversicaudatum*; **B, H,** *X. index* (Californian population); **C, I,** *X. vuittenezi*; **D, J,** *X. coxi*; **E, K,** *X. mediterraneum*; **F, L,** *X. americanum* (Japanese population). (After Southey, 1973b).

Note added in proof: According to Lamberti (pers. comm.) **F, L** of Fig. 10.5 represent an undescribed species, not *americanum*.

Symptoms of attack by *Ditylenchus dipsaci*: upper photograph, on onion seedlings showing typical 'bloat' effect. Lower photo: on oats (resistant plants on right). (Photo: Rothamsted Experimental Station).

PLATE I

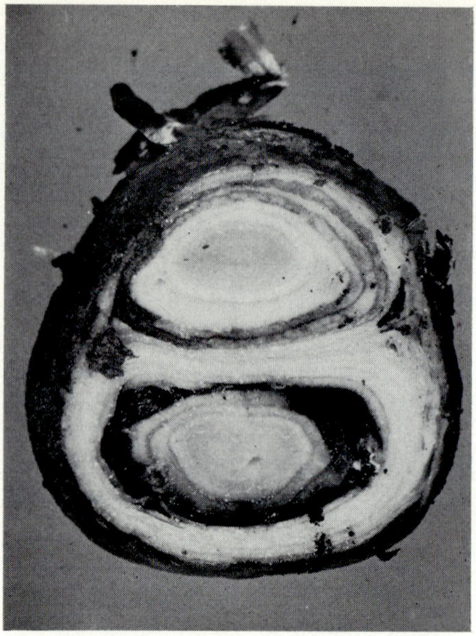

Upper photos: effects of *Ditylenchus dipsaci* on narcissus bulbs; left, cut vertically and right, cut transversely. Lower photo: effects of *D. destructor* on iris bulbs. (Photo: Rothamsted Experimental Station).

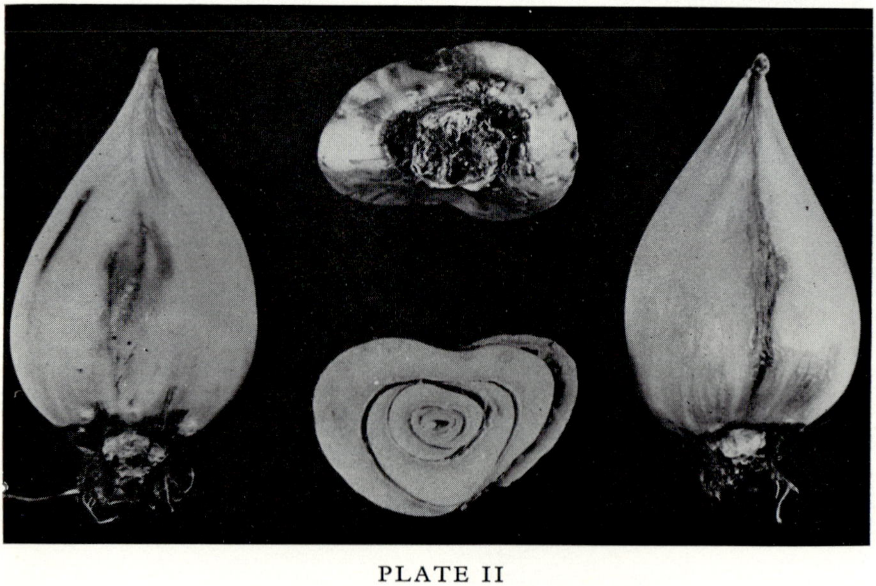

PLATE II

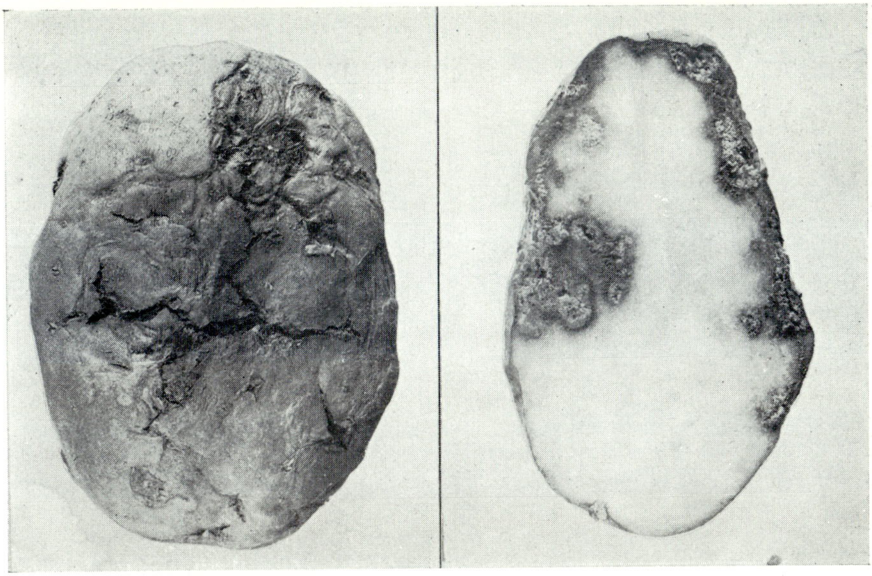

Upper photos: effects of *Ditylenchus destructor* on potato tubers showing, left, external appearance and, right, internal damage. Lower photo: galls on grass roots (*Poa annua*) caused by *D. radicicolus*.

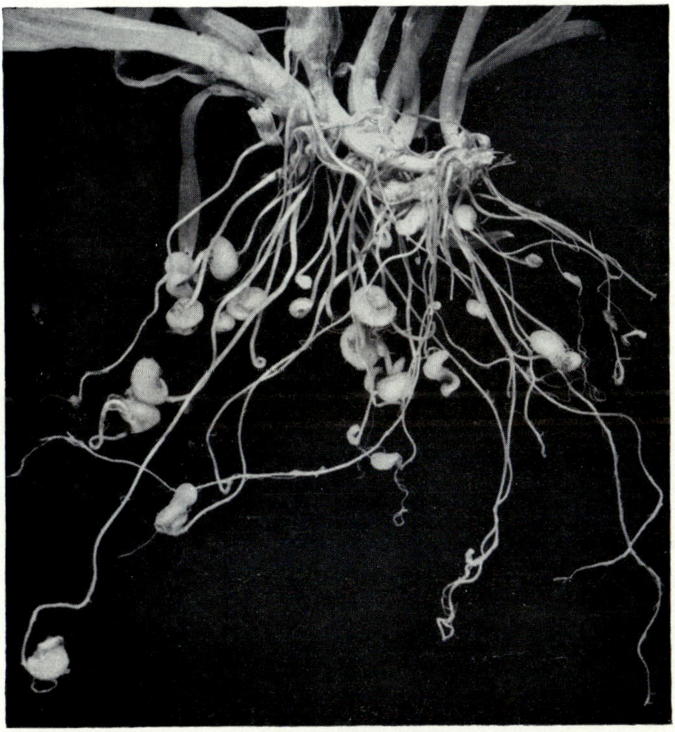

PLATE III

Effects of *Anguina tritici* on wheat at harvest showing, above, healthy and infested ears and, below, healthy grains (left) and galls ('ear-cockles') from infested ears. (Photo: Rothamsted Experimental Station).

PLATE IV

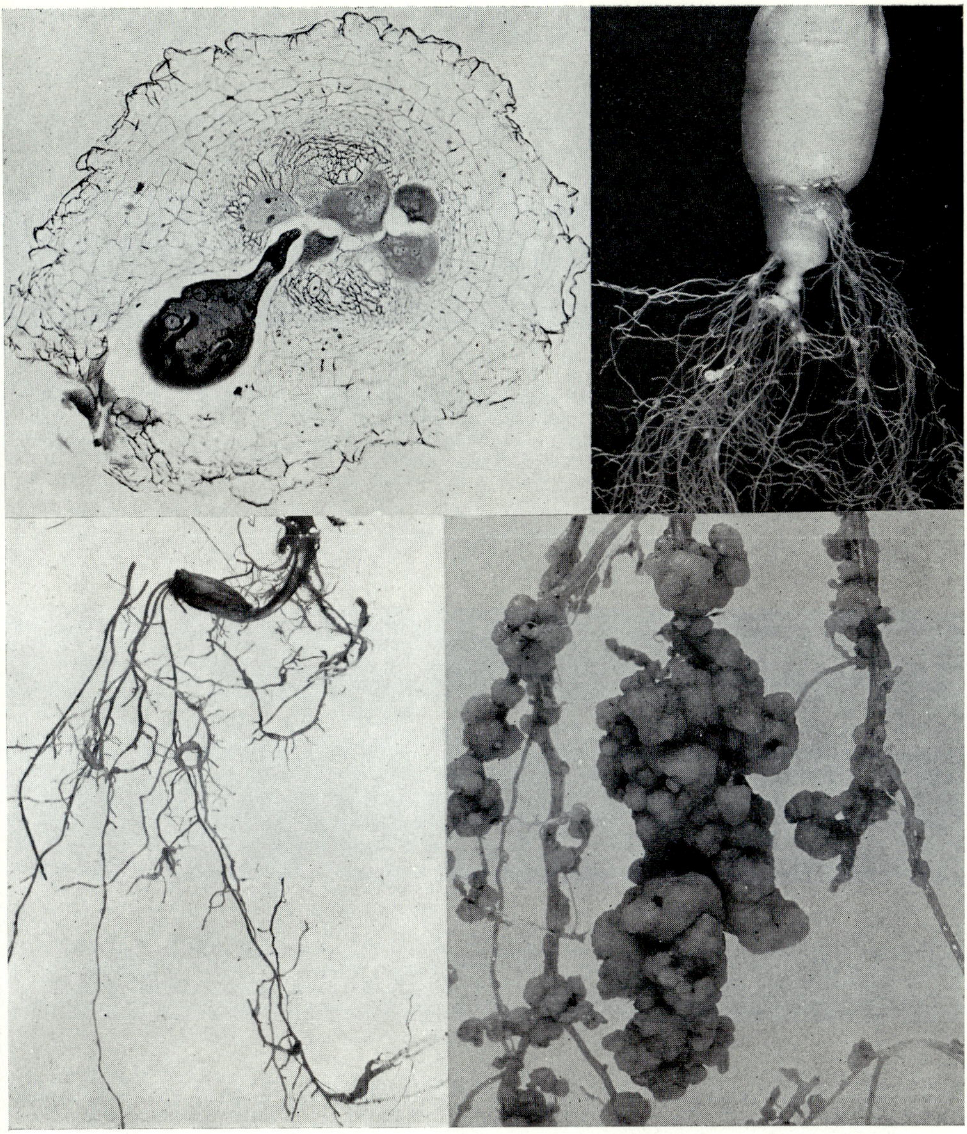

Effects of *Meloidogyne* spp. on roots of host plants: top left, T. S. tomato root gall showing *M. incognita* female and associated giant cells; top right, carrot showing malformation and galls due to *M. hapla*; lower left, root system of barley with galls caused by *M. naasi*; lower right, galls on cucumber roots due to *M. incognita*. (Photos: Rothamsted Experimental Station).

PLATE V

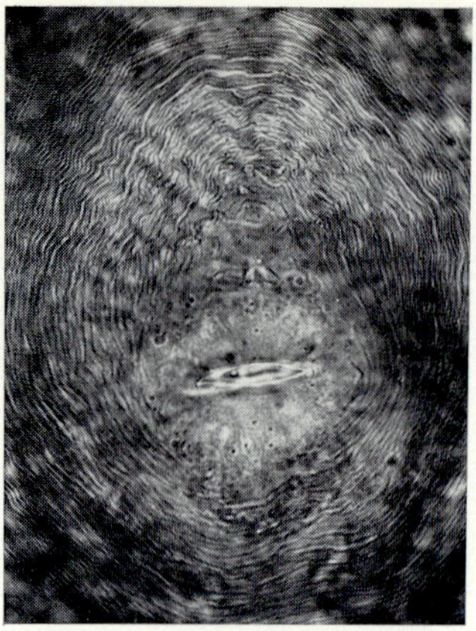

M. incognita

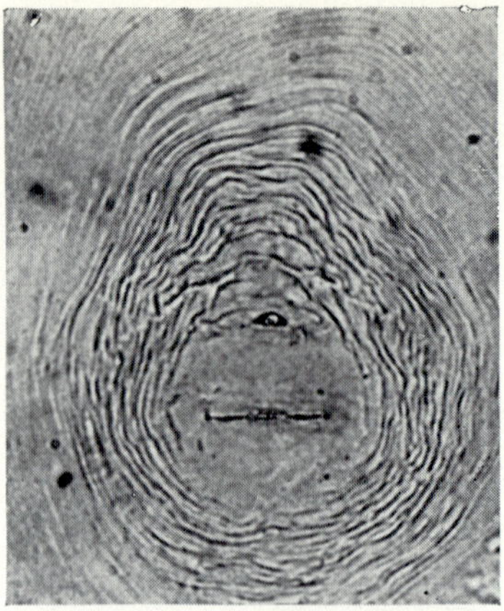

M. incognita ('*acrita*' *type*)

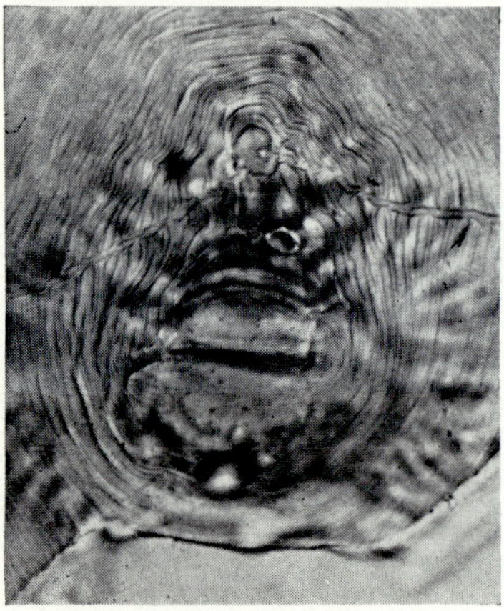

M. javanica

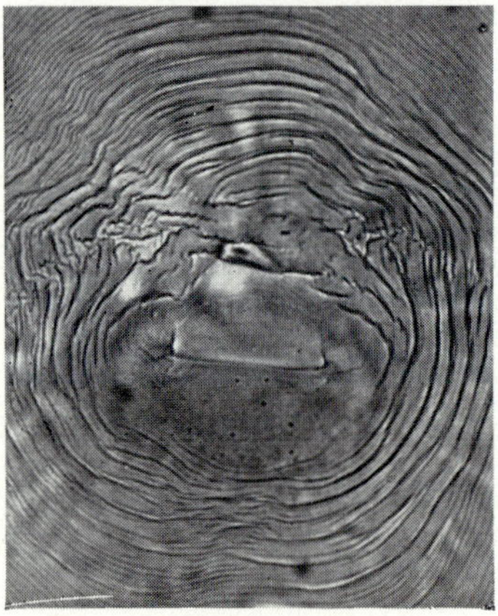

M. arenaria

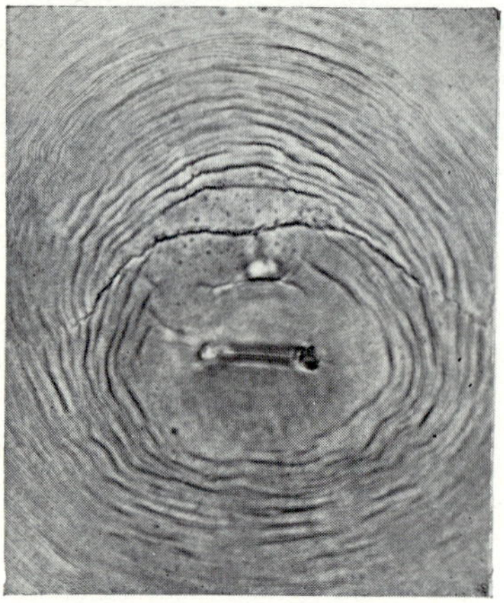

M. hapla

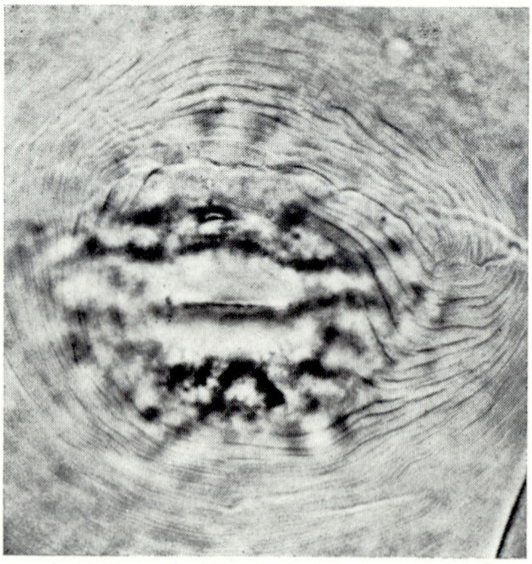

M. hapla (with 'wing')

Photographs of perineal patterns of *Meloidogyne* spp. (Photos: Rothamsted Experimental Station).

PLATE VI

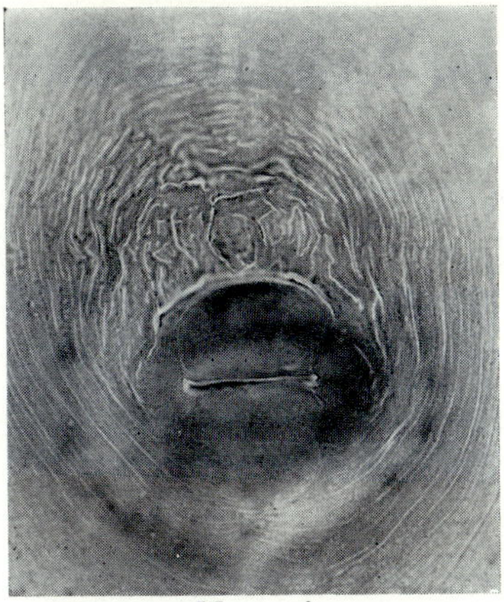

M. naasi

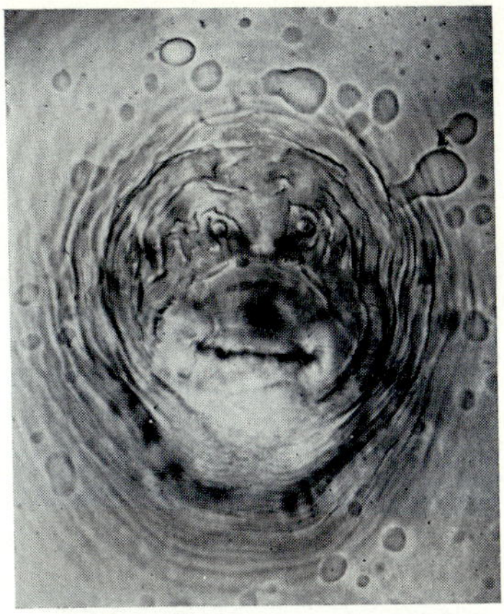

M. naasi low focus showing phasmids

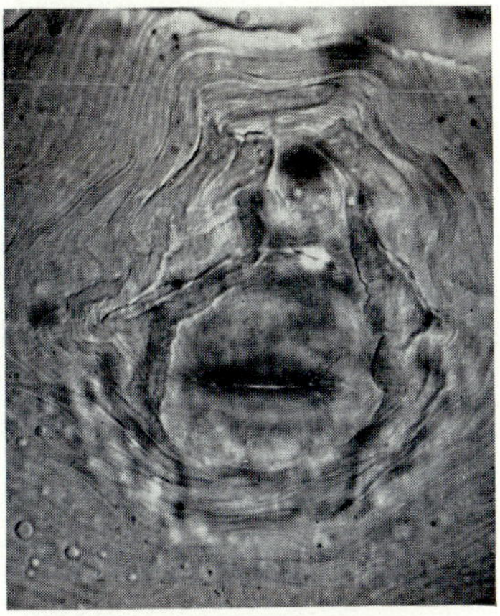

M. artiellia

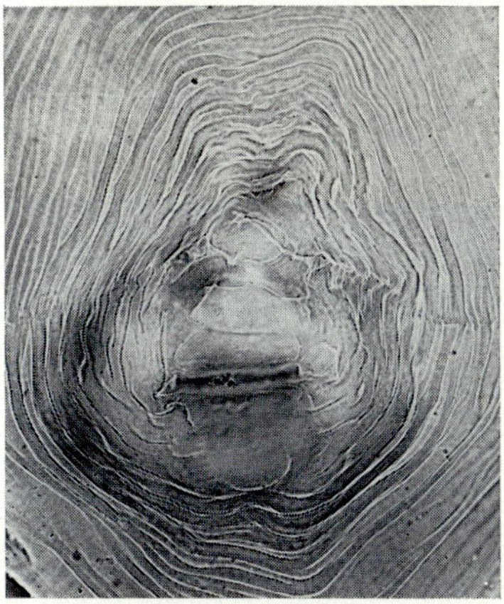

M. ardenensis

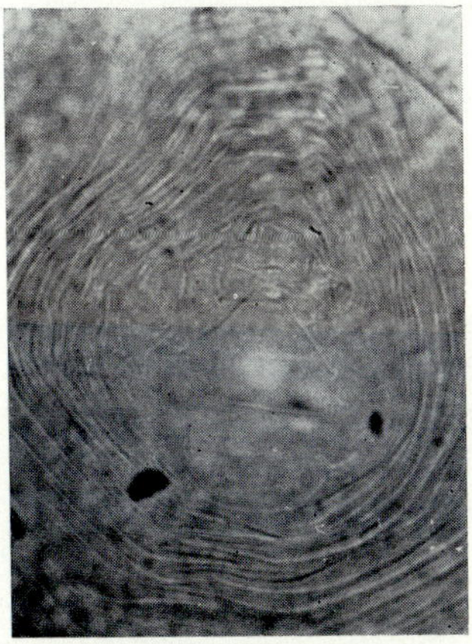

M. deconincki

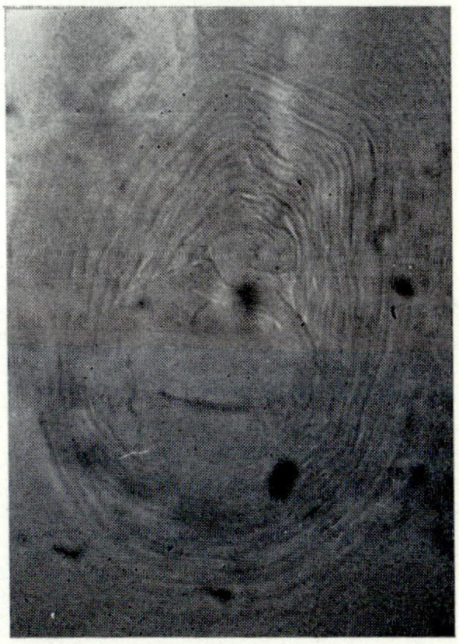

M. litoralis

Photographs of perineal patterns of *Meloidogyne* spp. Two lower photos courtesy Dr. E. Geraert, Laboratorium voor Morfologie en Systematiek, Rijksuniversiteit, Gent, Belgium; remainder, Rothamsted Experimental Station.

PLATE VII

Cyst of *Heterodera trifolii* with contents of a living female of similar size. The female was cut in water to release the contents *in toto*; these were immediately larger than the female body, indicating that there was turgor within it. (Photo: C. C. Doncaster).

Left: Adult females of *Heterodera trifolii* showing, on the nematode surface, a patchy subcrystalline layer of variable thickness and, at the terminal end, the extruded gelatinous matrix or egg sac. Right: Adult female of a *Heterodera* sp. found on sea-shore grasses, showing the thick subcrystalline layer which is typical of this species. (Photos: Glasshouse Crops Research Institute).

PLATE VIII

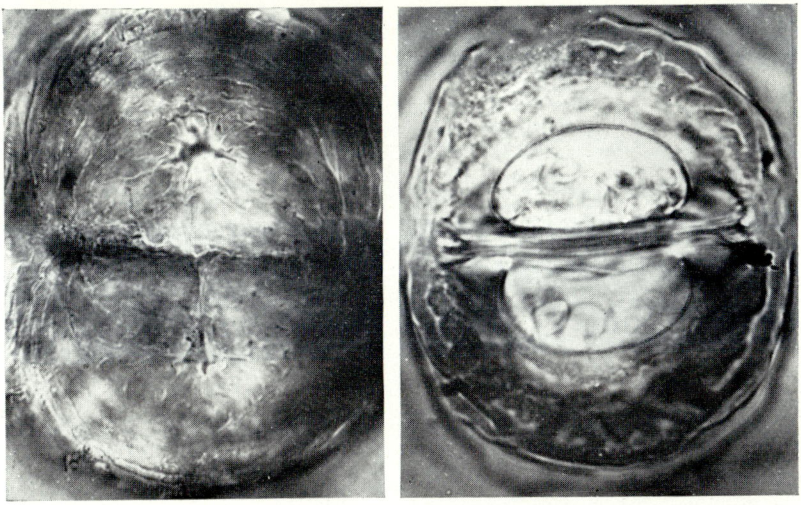

Fenestration in *Heterodera schachtii*: left, unfenestrated female; right fenestrated cyst. This species is ambifenestrate. (From Goffart, 1960a).

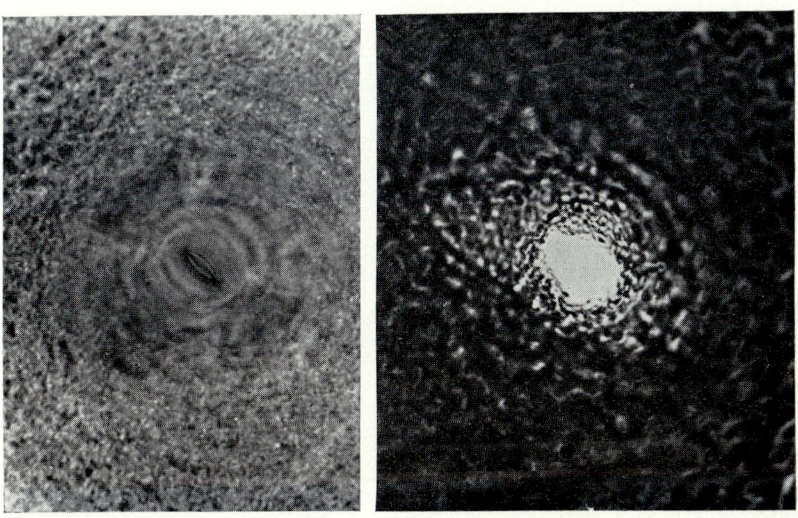

Fenestration in *Globodera rostochiensis*: left, vulval slit on unfenestrated female; right, fenestrated cyst—the vulval slit is lost and a roughly circular hole is formed. This species is circumfenestrate. (From Goffart, 1960a).

PLATE IX

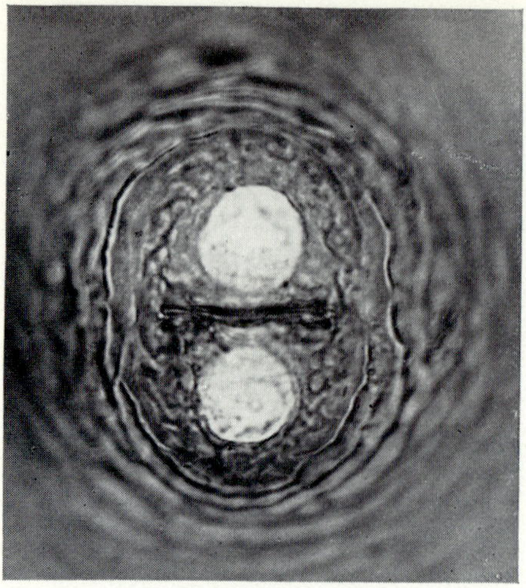

Cone-top in *Heterodera humuli*. The almost circular semifenestrae are separated by a wide vulval bridge. This species is bifenestrate. (Photo: Canada Dept. of Agriculture).

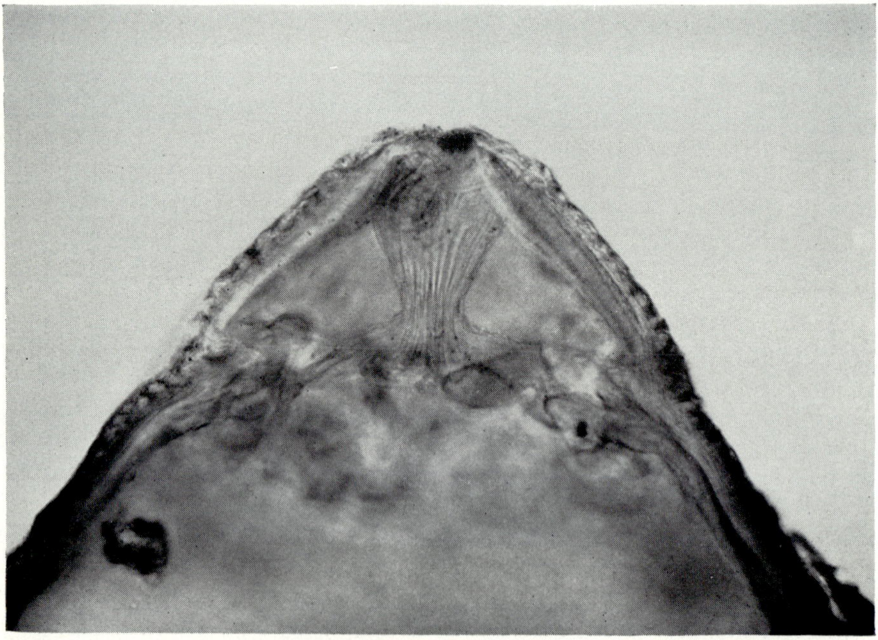

Heterodera schachtii: dorso-ventral view of vulval cone. The bullae are situated a short distance within the cyst, the vagina ('sheaf-like organ') appears wide and the underbridge extends across the cone between the bullae. (Photo: C. C. Doncaster).

PLATE X

Heterodera galeopsidis: lateral view of vulval cone showing the vagina (a flattened tube) in narrow view. The underbridge is seen end-on and appears as a dense structure in the centre between the bullae. (Photo: C. C. Doncaster).

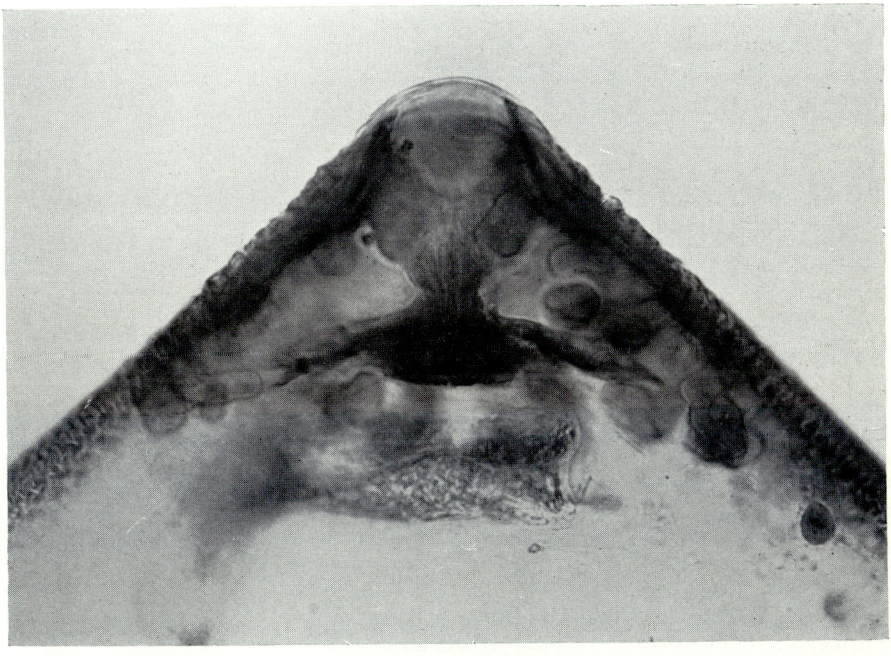

Heterodera galeopsidis: dorso-ventral view of vulval cone. Note 'cervix' and underbridge. (Photo: C. C. Doncaster).

PLATE XI

Heterodera avenae: vulval cone showing bullae close to vulva. (Photo: C. C. Doncaster).

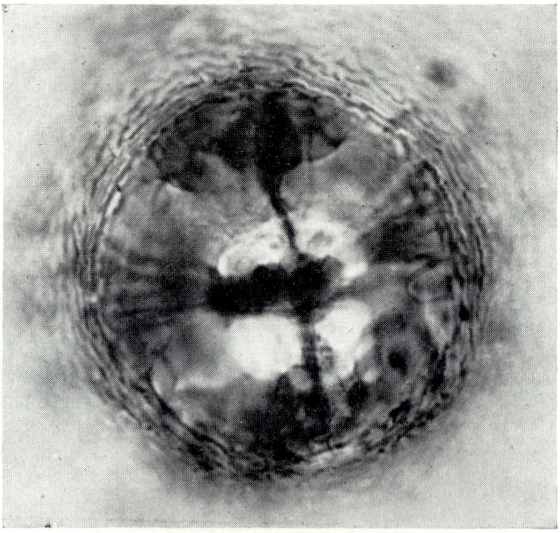

'Mulvey's bridge' in *Heterodera trifolii*—a muscular structure running just beneath the fenestra and at right angles to the underbridge. (From Mulvey, 1959a. Photo: Canada Dept. of Agriculture).

PLATE XII

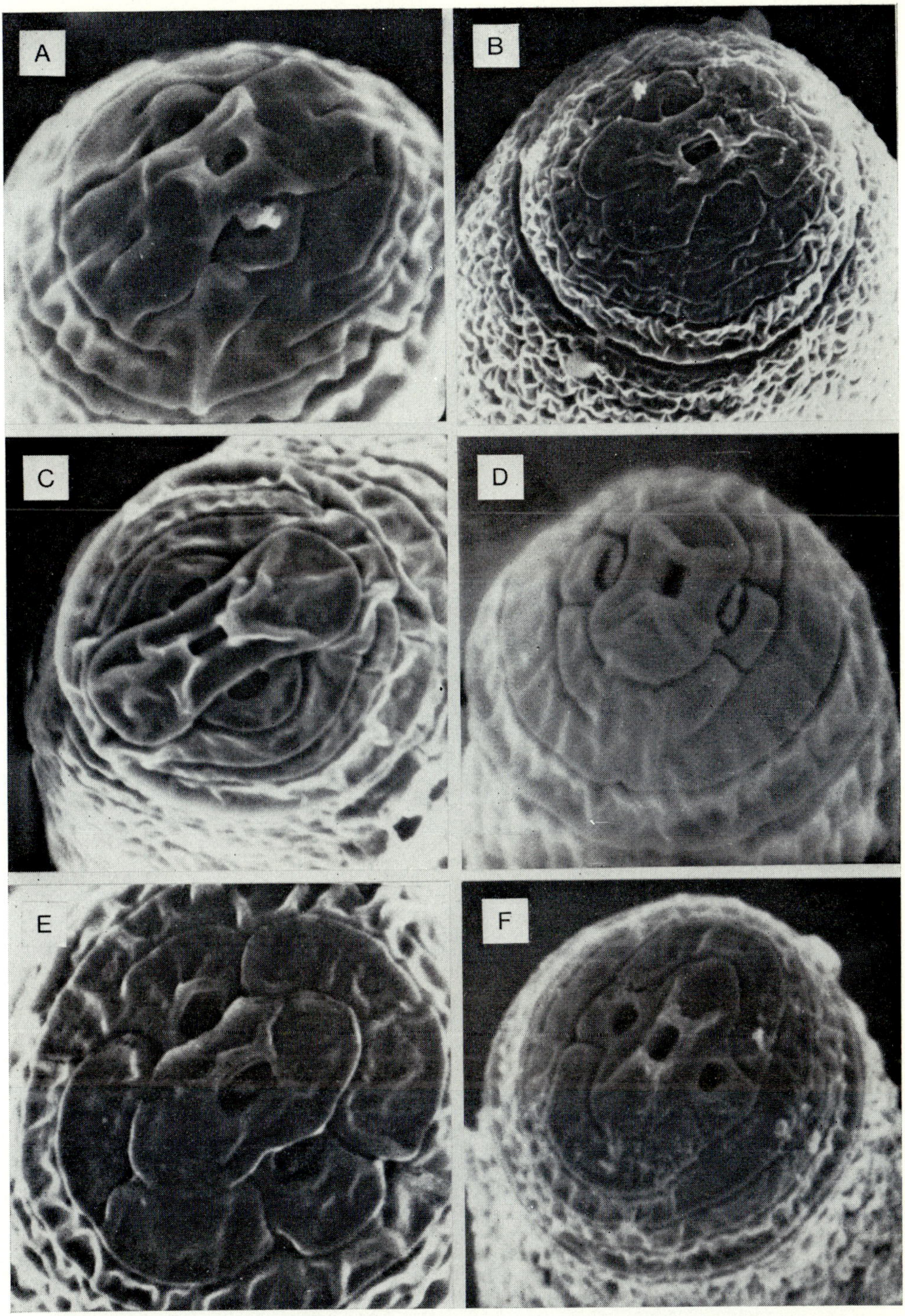

'Stereoscan' photographs of *en face* view of the heads of second-stage larvae of *Heterodera*, *Globodera* and *Punctodera* spp.: **A**, *H. rosii*; **B**, *H. cruciferae*; **C**, *H. avenae*; **D**, *H. betulae*; **E**, *G. tabacum*; **F**, *P. punctata*. (All at the same magnification). (Photos: Rothamsted Experimental Station).

PLATE XIII

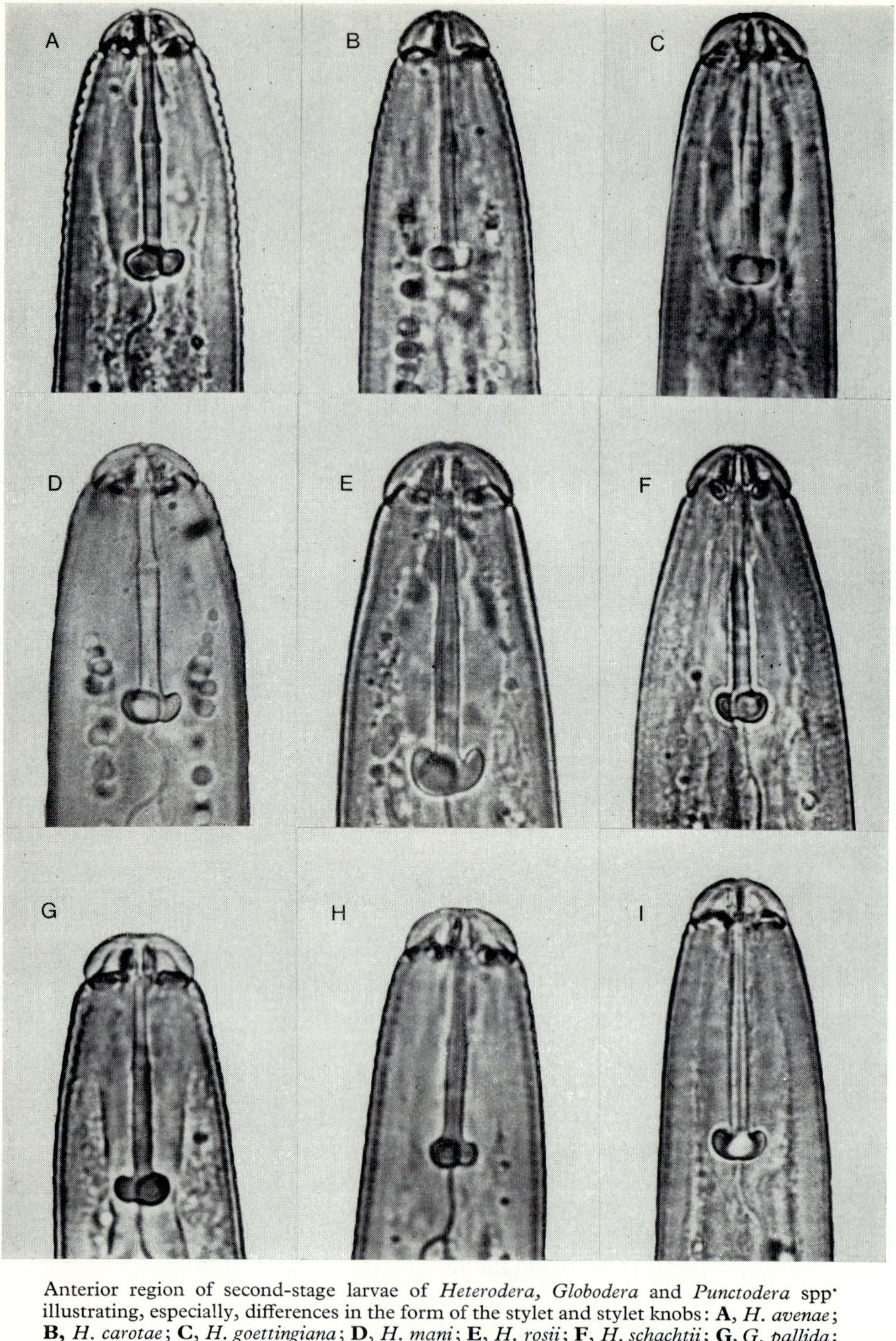

Anterior region of second-stage larvae of *Heterodera, Globodera* and *Punctodera* spp. illustrating, especially, differences in the form of the stylet and stylet knobs: **A**, *H. avenae*; **B**, *H. carotae*; **C**, *H. goettingiana*; **D**, *H. mani*; **E**, *H. rosii*; **F**, *H. schachtii*; **G**, *G. pallida*; **H**, *G. rostochiensis*; **I**, *P. punctata*. (All at the same magnification). (Photos: D. P. Webley).

PLATE XIV

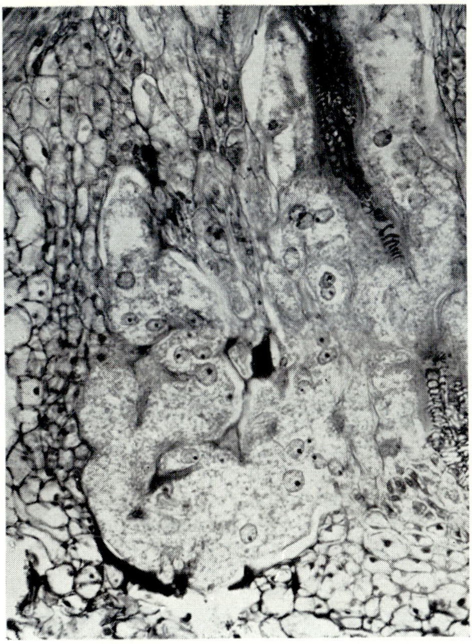

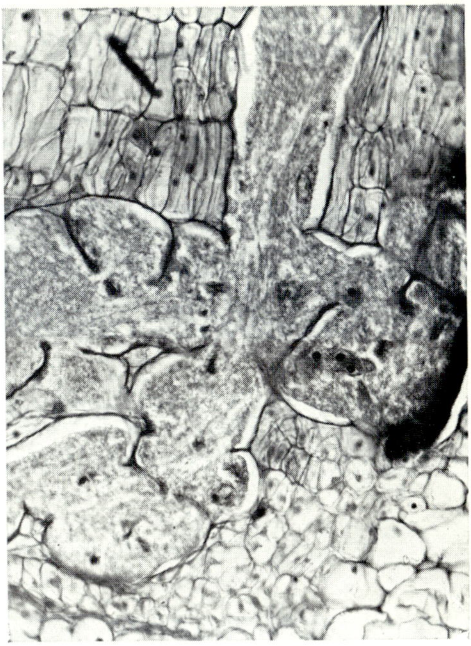

Syncytia in various host roots caused by the feeding of *Heterodera* species: top left, *H. schachtii* on *Beta maritima,* L.S. root at the origin of a secondary root showing multinucleate condition of the syncytium and granular cell contents; top right, *H. schachtii* on *Beta maritima* showing gaps in the cell walls of the syncytium; lower, *H. avenae* on spring wheat (*Triticum vulgare*), T. S. of a seminal root showing active deeply-staining syncytium in stele, 'ageing' syncytium with prominent nuclei in the central xylem vessel and mature *H. avenae* female. (Photos: Rothamsted Experimental Station).

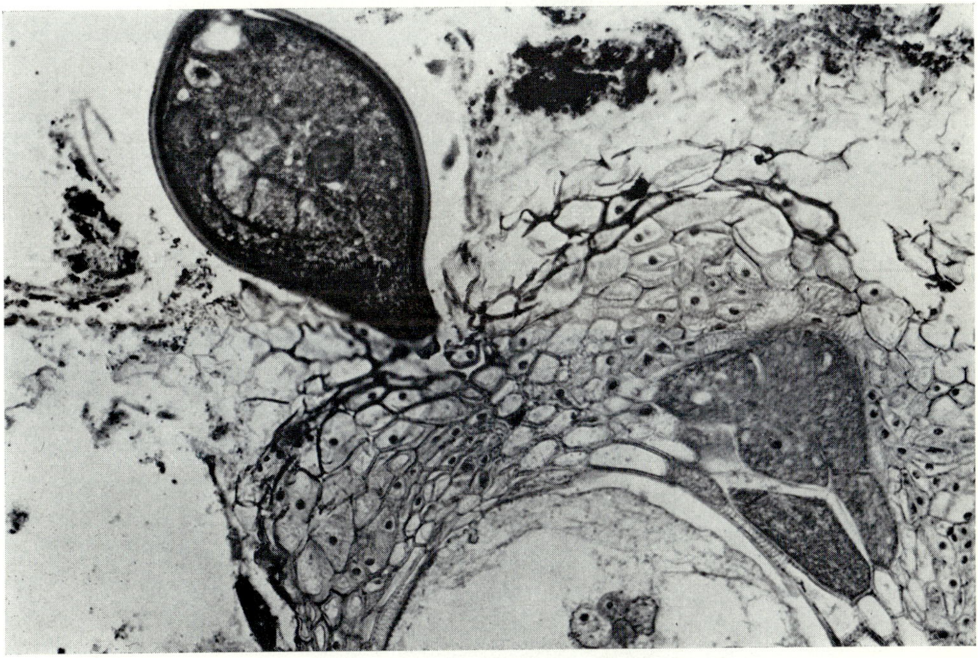

PLATE XV

Symptoms of attack by species of *Aphelenchoides* on leaves of host plants: top, surface damage ('feeding areas') on leaflets of strawberry due to earlier feeding by *A. ritzemabosi* or *A. fragariae* in the bud; lower photo, typical interveinal necrosis of chrysanthemum leaf due to feeding of *A. ritzemabosi* in the mesophyll tissue. (Photos: East Malling Research Station and Rothamsted Experimental Station respectively).

PLATE XVI

Interveinal necrosis due to feeding of *Aphelenchoides* spp. in leaves of host plants: top left, *A. ritzemabosi* in *Primula*; top right, *A. fragariae* in *Saxifraga crassifolia*; lower photo, *A. fragariae* in fern (*Pteris* sp.).

PLATE XVII

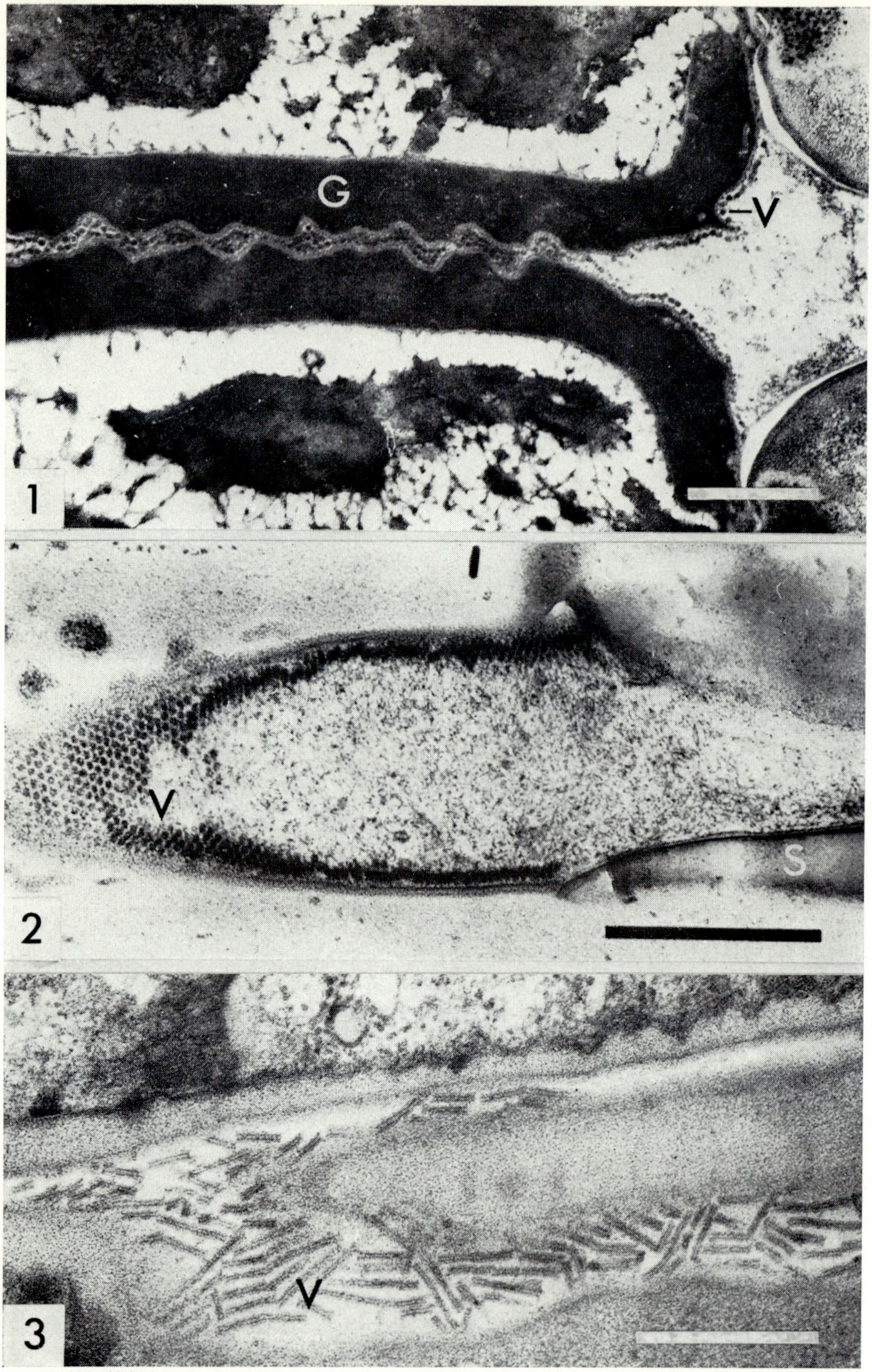

Sections of anterior parts of nematode vectors showing location of virus particles: (1) *Longidorus elongatus*, guide sheath (G) and raspberry ringspot virus particles (V); (2) *Xiphinema diversicaudatum*, junction of odontostyle (S) and odontophore showing arabis mosaic virus particles (V); (3) *Paratrichodorus pachydermus*, tobacco rattle virus particles (V) in oesophagus. (Electron micrographs: Scottish Horticultural Research Institute).

PLATE XVIII

Outbreaks of raspberry ringspot virus in (upper photo) 'Malling Jewel' raspberry and (lower photo) 'Talisman' strawberry at Blairgowrie, Perthshire. (Photos: Scottish Horticultural Research Institute).

PLATE XIX

Upper photo: an outbreak of strawberry latent ringspot virus in 'Malling Jewel' raspberry at Cupar, Fife. Lower photo: part of an experiment with 'Redgauntlet' strawberry two years after planting in soil infested with *Longidorus elongatus* carrying raspberry ringspot and tomato black ring viruses. Untreated (centre right), D–D at 449 kg/ha (centre left), quintozene at 269 kg/ha (foreground left), quintozene at 67 kg/ha (foreground right). (Photos: Scottish Horticultural Research Institute).

PLATE XX

Other species:

X. arcum Khan, 1964
X. attorodorum Luc, 1961
X. australiae McLeod & Khair, 1971
X. bakeri Williams, 1961
X. basilgoodeyi Coomans, 1965
X. basiri Siddiqi, 1959
X. bergeri Luc, 1973
X. brasiliense Lordello, 1951
X. brevicolle Lordello & da Costa, 1961
X. cavenessi Luc, 1973
X. chambersi Thorne, 1939
X. clavatum Heyns, 1965
X. costaricense Lamberti & Tarjan, 1974
X. coxi Tarjan, 1964
X. cubense Razjivin, in Razjivin, O'Relly & Milian, 1973
X. denoudeni Loof & Maas, 1972
X. dimorphicaudatum Heyns, 1966
X. diversicaudatum (Micol., 1923 & 1927) Thorne, 1939
X. douceti Luc, 1973
X. ebriense Luc, 1958
X. elongatum Sch. Stek. & Teun., 1938
X. ensiculiferum (Cobb, 1893) Thorne, 1937
X. filicaudatum Loof & Maas, 1972
X. flagellicaudatum Luc, 1961
X. hallei Luc, 1958
X. hygrophilum Southey & Luc, 1974
X. ifacolum Luc, 1961
X. imambaksi Loof & Maas, 1972
X. imitator Heyns, 1965
X. index Thorne & Allen, 1950
X. ingens Luc & Dalmasso, 1964
X. insigne Loos, 1949
X. italiae Meyl, 1953
X. krugi Lordello, 1955
X. lambertii Bajaj & Jairajpuri, 1976
X. longicaudatum Luc, 1961
X. longidoroides Luc, 1961
X. macrostylum Esser, 1966
X. malagasi Luc, 1973
X. malutiensis Heyns, 1976
X. mammillatum Sch. Stek. & Teun., 1938
X. manubriatum Luc, 1975
X. marsupilami Luc, 1973
X. mediterraneum Martelli & Lamberti, 1967
X. meridianum Heyns, 1971
X. mluci Heyns, 1976
X. monohysterum Brown, 1968
X. neoamericanum Saxena, Chhabra & Joshi, 1973
X. neoelongatum Bajaj & Jairajpuri, 1976
X. neovuittenezi Dalmasso, 1969
X. nigeriense Luc, 1961
X. opisthohysterum Siddiqi, 1961
X. orbum Siddiqi, 1964
X. orthotenum Cohn & Sher, 1972
X. parvistylus Heyns, 1971
X. paulistanum Carvalho, 1965
X. pini Heyns, 1965
X. pyrenaicum Dalmasso, 1969
X. radicicola Goodey, 1936
X. rivesi Dalmasso, 1969
X. rotundatum Sch. Stek. & Teun., 1938
X. sahelense Dalmasso, 1969
X. seredouense Luc, 1975
X. setariae Luc, 1958
X. simillimum Loof & Yassin, 1971
X. spinuterus Luc, 1973
X. surinamense Loof & Maas, 1972
X. tarjani Luc, 1975
X. tropicale Zullini, 1973
X. turcicum Luc & Dalmasso, 1963
X. vanderlindei Heyns, 1962
X. variabile Heyns, 1966
X. vitis Heyns, 1974
X. vuittenezi Luc, Lima, Weischer & Flegg, 1964
X. vulgare Tarjan, 1964
X. yapoense Luc, 1958
X. zulu Heyns, 1965

Besides the above currently accepted species there are about 12 *species inquirendae*. The genus is predominantly tropical and sub-tropical. About 12 species occur in the Mediterranean, Iberian and Pyrennees areas but few extend into central and northern parts of Europe. In the British Isles only one species (*X. diversicaudatum*) is common and three others have been found rarely (Taylor & Brown, 1976).

Keys for Identification

Loof & Maas (1972) gave a standard dichotomous key to species of the genus. Cohn & Sher (1972) gave a dichotomous key to their eight subgenera. Southey (1973b) chose to regard Cohn & Sher's subgeneric divisions as artificial groupings rather than taxa; he modified their key to these groups and provided tabular keys to the species in each group based chiefly on measurements. Esser (1974) gave similar tabular keys for (functionally) monodelphic and didelphic species.

Stegarescu (1966—translated into German by Weischer (1969)) devised an ingenious polytomous key to 35 species of *Xiphinema*; this has been adapted and extended by Luc & Dalmasso (1975).

Characters Useful for Identification

Females are more useful than males for identification and are usually more readily obtained; in about 60 per cent of described species of *Xiphinema* males are rare or unknown. Measurements that are useful in separating species are given on p. 209. It is sometimes useful also, as in *Longidorus*, to have a series of measurements of juvenile stages, especially the first stage. The more useful diagnostic features are discussed below.

Body size and form. When relaxed by gentle heat *Xiphinema* spp. are curved with the ventral surface inwards, the curvature being greater towards the tail. They may be arcuate, or open or more closed C-shaped (Fig. 10.5A–F). Body shape and the proportion 'a' are useful for separating short, stout species with relatively little curvature (e.g., *X. ensiculiferum* and allied species) from the many, more slender and curved ones. Total length (L) varies between species from about 1·5 to 5·5 mm. Thus this feature distinguishes the small species of the *X. americanum* group (subgenus *Xiphinema* of Cohn & Sher, 1972; L = 1·6–2·4 mm) from the larger species.

Tail shape is probably the most useful feature for species identification. The ratio c′ (tail length/anal body width) indicates relative tail length. *Xiphinema* species could be grouped according to tail-shape of the adult female, viz., hemispheroid without peg (e.g., *ensiculiferum* group), hemispheroid pegged (mammillate) (*diversicaudatum, index, vuittenezi*, etc.), conoid (*americanum, mediterraneum*, etc.), conoid-digitate (*coxi* and many other species), elongate-conoid (e.g., *insigne*), filiform (e.g., *flagellicaudatum*). No long-tailed forms (the last two groups) have been found in temperate areas.

Female genital tract and 'V' value. Cohn & Sher (1972) rated the female reproductive system as of prime importance in *Xiphinema* taxonomy and used it in conjunction with tail shape to divide the genus into eight subgenera. From the vulva, the cuticle-lined vagina leads inwards to a muscular, more or less kidney-shaped ovejector from which paired uteri, or single uterus, extend respectively in both directions or posteriorly only. Each complete genital branch or genital tube consists of uterus, oviduct and ovary—the latter always reflexed at its junction with the oviduct so that the 'youngest' oocytes are nearest to the vulva (See Fig. 10.6D).

Strictly, *didelphic* species are those having two uteri (Gk. δελφυς = uterus) but the term is usually applied where there are two *complete* genital branches each with ovary (as in all known British species). The anterior one, though complete, may be reduced in size and in some didelphic species is probably non-functional. *X. cubense* is exceptional for the genus in that, according to Razjivin *et al.* (1973), the posterior branch is reduced and the anterior fully developed. Cohn & Sher (1972) coined the term *pseudomonodelphic* for species in which the anterior branch lacks an ovary; otherwise it may be more or less complete, or reduced to a short undifferentiated pre-vulval sac. *Monodelphic* (more correctly opisthomonodelphic) species are those with no trace of an anterior branch.

The vulva in *Xiphinema* is median or pre-median (V = 28–56). The value of V is related to the structure of the genital tract; it is smallest (vulva furthest forward) in monodelphic species and largest (vulva ± median) in fully didelphic species. V is a useful character of remarkably small intraspecific variability (Southey, 1973b).

Z-organ. In some *Xiphinema* species the uterus is a simple structure, usually with an enlarged distal portion next to the sphincter which separates uterus from oviduct. In other species the uterus is differentiated into several regions; in some of them the 'median differentiation' proximal to the enlarged portion contains a group of more or less angular sclerotized bodies termed 'organe Z' by Luc (1961) and the 'Z-organ' by Coomans (1965). Flegg (1966a) showed that *X. diversicaudatum* possessed a rather atypical Z-organ (Fig. 10.6J) ('globular bodies' of Coomans, 1965; 'Z-pseudo-organ' of Luc & Dalmasso, 1975) which distinguishes it from the closely related *X. index*. Presence or absence of a Z-organ is a useful character.

Odontostyle and odontophore. These do not vary much in structure between species, nor does the position of the associated guiding sheath. However, their lengths vary between species; odontostyle length in particular is very useful in identification and shows small intraspecific variability (Southey, 1973b).

Shape of lip region. The profile of the lips (high, low, or more or less rounded) and whether or not they are offset from the body by a constriction, are useful features, e.g., among known British species *X. mediterraneum* is the only one with a markedly offset lip region (Fig. 10.5K).

Internal structure of tail. The shape of the internal protoplasmic portion of the tail (e.g., rounded, conical or truncated) can be of value, especially in separating round-tailed species. Some more or less round-tailed species show a rudimentary or vestigial peg or 'mucro' in the form of a slight protuberance of the thick cuticle at the tail tip. Associated with this is a so-called 'internal mucro' or 'blind terminal canal' (Southey & Luc, 1974) seen as a clear zone of the cuticle apparently free from the fine radial striations that surround the rest of the tail. This structure, which was well illustrated by Dalmasso (1969), is also characteristic of many, if not all, peg-tailed species. In some of the latter, e.g., *X. diversicaudatum,* pegless individuals occur occasionally but the presence of an internal mucro may betray them as essentially peg-tailed types. The following brief descriptions apply mainly to adult females, usually mentioning merely the existence or frequency of males.

Common European Species

The biology and distribution of these species, and the role of some as virus vectors, are discussed in Chapter 11.

Xiphinema diversicaudatum (Micol., 1923 & 1927) Thorne, 1939 (Figs. 10.5A, G; 10.6)

One of the commonest species of *Xiphinema* in western Europe and the only one that is common and widely distributed in U.K. See Goodey, Peacock & Pitcher (1960) and Pitcher, Siddiqi & Brown (1974) for detailed descriptions.

Principal ♀ measurements (British populations): L = 4·9 (4·0–5·5) mm; a = 74 (57–92); c′ = 1·0; V = 43 (39–46); ost = 143 (130–157) μm; oph = 85 (70–97) μm. One of the largest species in the genus, females ranging to more than 5 mm long, males occasionally over 6 mm. Lip region slightly indented at junction with body (Figs. 10.5G, 10.6B). Spear and guiding sheath typical of the genus. Vulva a transverse slit, forward of mid-body (V = 39–46), opening into vagina with + shaped lumen. Genital tract didelphic with more or less equal

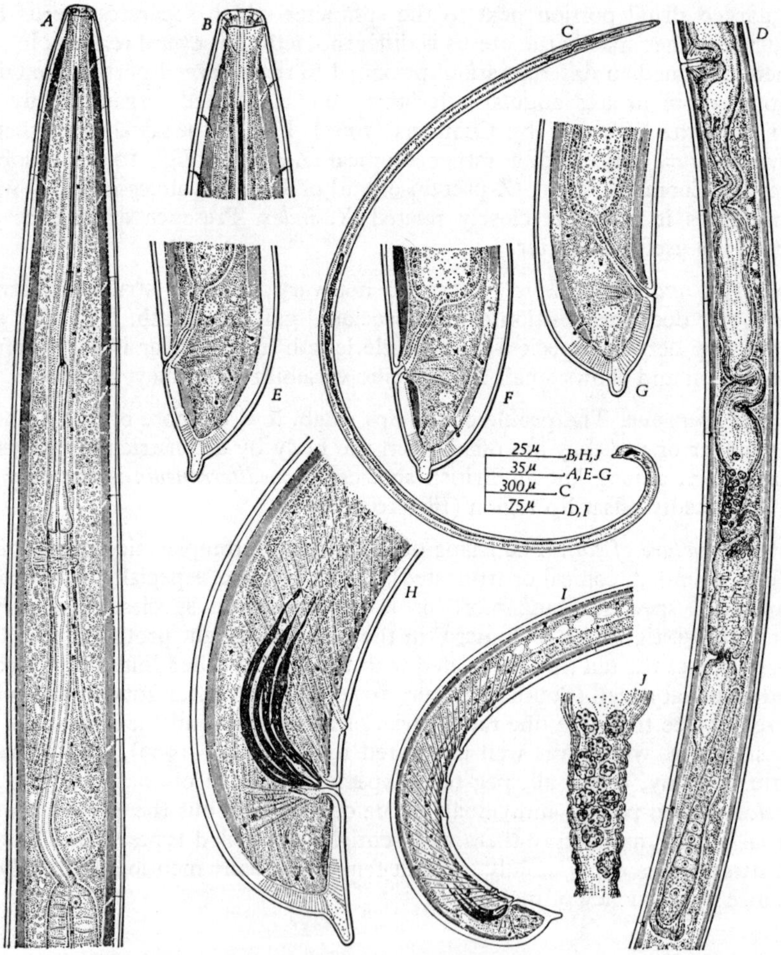

Fig. 10.6. *Xiphinema diversicaudatum,* morphological details. **A, B, C, H** & **I,** lectotype male; remainder, females from hop, England. **A, B,** anterior parts; **C,** entire male showing posture when relaxed by heat; **D,** posterior branch of genital tract; **E–G,** female tails; **H, I,** male tails; **J,** Z-organ containing globular bodies. (From Pitcher, Siddiqi & Brown, 1974, *CIH Descriptions of Plant-parasitic Nematodes*).

anterior and posterior branches (Fig. 10.6D), each occupying about 15 per cent of body length. Each branch consists, progressing distally from the vagina, of a uterus about seven to eight body-widths long, separated by a constricted muscular sphincter from the oviduct that is three or four body-widths long. The ovary, varying in size according to degree of development, is reflexed back towards the vulva at its junction with the oviduct. Both oviduct and uterus show bulb-like swellings next to the sphincter, the uterine one being the longer. Proximal to the uterine swelling is a short constricted part containing the Z-organ (Fig. 10.6D, J) (see p. 221). Proximal portion of uterus a more or less convoluted tube which, just before oviposition, is distended by the presence of eggs. Eggs oval, four to seven times as long as wide, averaging 195 × 44 μm; one to four eggs may be present in the genital tract at one time, maximum two per uterine branch.

Tail of both sexes about one a.b.w. long, rounded, more so dorsally than ventrally, with ventral surface somewhat flattened, terminating in a ventrally placed, mammillate peg which is rounded at the end but scarcely tapered (Figs. 10.5G, 10.6). Pegless specimens are said to occur but seem rare. Male tail showing cuticular ventral 'supplements'; one pair just anterior to anus, then a gap of about two body-widths, then a series of three to five. Spicules stout, curved, averaging 76 μm long.

Table 1 gives body lengths (L) and odontostyle lengths of the four juvenile stages as an aid to their recognition. The J.1. tail is elongate conoid; the remaining stages show progression from conoid-digitate to the typical pegged form of the adult.

Table 1

X. diversicaudatum *Juveniles: Body and Odontostyle (ost) Lengths (after Goodey* et al., 1960)

Stage	L (mm)	ost (μm)	*ost 2 (μm)
J.1	1·1	56	73
J.2	1·9	80	105
	(1·4–2·3)	(71–89)	(85–120)
J.3	2·5	103	127
	(2·2–2·7)	(98–106)	(121–133)
J.4	3·7	123	151
	(3·4–4·1)	(117–130)	(142–158)

*ost 2 = replacement odontostyle.

X. index Thorne & Allen, 1950. (Fig. 10.5B, H)

Principal ♀ measurements: L = 3·1 (2·7–3·5) mm; a = 61 (57–69); c′ = 1·0 (0·9–1·1); V = 40 (38·0–42·5); ost = 128 (119–134) μm; oph = 71 (65–76) μm. Morphologically close to *X. diversicaudatum* from which it differs in the following particulars: Body smaller (average length about 3 mm, maximum 3·5 mm) and other dimensions in proportion. Lip region practically continuous with body contour. Vulva of female on average more forwardly placed though ranges of V overlap. Genital branches usually shorter (about 10–12 per cent of body length). Uteri somewhat simpler in structure; Z-organ absent. Peg on female tail usually more tapered. Males scarce; they show no special features.

X. vuittenezi Luc, Lima, Weischer & Flegg, 1964. (Fig. 10.5C, I)

Principal ♀ measurements (English population, after Luc *et al.*, 1964): L = 3·3 (2·8–3·8) mm; a = 61 (50–70); c′ = 0·9 (0·8–1·2); V = 51 (48–53); ost = 130 (124–134) μm; oph = 75 (66–80) μm. General morphology and anatomy similar to those of *X. diversicaudatum.* Lip region offset from body by a slight constriction. Vulva further back than in the preceding species, almost exactly midway along length of body. Genital branches more or less equal, each occupying about 15 per cent of body length; fully differentiated as in *X. diversicaudatum* but lacking Z-organ. Tail ending in a short rounded peg or protuberance situated more centrally (less ventrally) than the longer peg of *X. diversicaudatum.* Males rare; tail shape similar to that of females.

X. coxi Tarjan, 1964. (Fig. 10.5D, J)

Principal ♀ measurements: L = 3·6 (3–4) mm; a = 75 (66–82); c′ = 1·8 (1·5–2·0); V = 44 (40–46); ost = 122 (113–127) μm; oph = 72 (68–82) μm. General morphology and anatomy similar to those of *X. diversicaudatum.* Lip region slightly offset from body. Vulva forward of mid-body (V = 40–46). Genital tract didelphic; branches ± equal, each occupying about 12 per cent of body length, fully differentiated, Z-organ present. Tail conoid-digitate, i.e., tapering smoothly into terminal peg, usually almost twice as long as body width at anus. Male unknown.

**X. mediterraneum* Martelli & Lamberti, 1967. (Fig. 10.5E, K)

Principal ♀ measurements (Type population, after Lamberti & Martelli, 1971): L = 1·9 (1·7–2·1) mm; a = 64 (59–71); c′ = 1·6 (1·5–2·1); V = 56 (53–58); ost = 86 (77–90) μm; oph = 48 (44–51) μm. A species of the *americanum* group (subgenus *Xiphinema* Cohn & Sher, 1972) separated from most other European species by its small size. First described by Lima (1965) in an unpublished thesis; first published description by Martelli & Lamberti (1967) and taxonomic position further clarified by Lamberti & Martelli (1971) who designated type specimens. Relaxed body-shape spiral or rather closed C compared to the more elongated open C of the foregoing larger species. Body more slender than *X. americanum* ('a' ratio averages around 60). Lip region distinctly offset from body. Vulva slightly behind mid-body (V = 55–61). Tail conoid; tail-tip narrow, rounded, slightly digitate. Males rare.

Less Common European Species

X. brevicolle Lordello & da Costa, 1961. ♀♀ (Italian population, after Martelli & Lamberti, 1967): L = 2·3 (2·2–2·4) mm; a = 47 (41–57); c′ = 0·85 (0·78–1·0); V = 50 (48–52); ost = 91 (84–96) μm; oph = 56 (52–64) μm. A French population had V = 52–55 and ost = 110–115 μm. A species of the *americanum* group (subgenus *Xiphinema* Cohn & Sher) slightly larger than *X. americanum* and *X. mediterraneum* and having a shorter conoid tail with rounded tip;

Added in proof:* Siddiqi & Lamberti (*Nematol. medit.* **5, 133–135, 1977) consider *Xiphinema mediterraneum* Martelli & Lamberti, 1967 to be a junior synonym of *X. pachtaicum* (Tulaganov, 1938) Kir'yanova, 1951 based on study of the holotype and other specimens of the latter.

distinguished from those species also by the longer spear and less offset lip region. Didelphic; Z-organ absent. Males rare. Reported from Italy, southern France, Spain and Czechoslovakia.

X. ingens Luc & Dalmasso, 1964. ♀♀ (Italian and Israeli populations, after Martelli & Lamberti, 1967): L = 5·0 (4·6–5·4) mm; a = 72 (68–80); c′ = 0·72 (0·68–0·85); V = 49 (47–52); ost = 156 (146–168) μm; oph = 91 (84–96) μm. A large species somewhat resembling *X. vuittenezi*. Genital tract didelphic and fully differentiated; Z-organ present. Tail short, hemispheroid with trace of peg. Males plentiful. In Europe, only in Italy and Spain (Arias & Navacerrada, 1973).

X. italiae Meyl, 1953. ♀♀ (Italian populations, after Martelli & Lamberti, 1967): L = 3·0 (2·3–3·5) mm; a = 89 (75–114); c′ = 3·4 (2·1–4·0); V = 45 (43–49); ost = 95 (87–106) μm; oph = 60 (54–67) μm. A medium sized species. Lip-region offset (like that of *X. mediterraneum*). Didelphic; Z-organ absent. Tail rather long, conoid with rounded tip, variable but often tending to be digitate. Males rare. Restricted to southern Europe and the mediterranean region where it is locally common on grapevines and is a virus vector. (See Chapter 11).

X. neovuittenezi Dalmasso, 1969 (q.v.). ♀♀: L = 2·8 (2·6–3·2) mm; a = 50 (44–54); c′ = 0·84 (0·7–0·9); V = 47 (45–48); ost = 129 (124–136) μm; oph = 73 (68–76) μm. Near to *X. vuittenezi* but differs chiefly in the more forward position of the vulva, the tail tip having only a slight prominence to represent the peg or mucro, and in the common occurrence of males. Found on the western mediterranean coast of France and in Spain.

X. pyrenaicum Dalmasso, 1969 (q.v.). ♀♀: L = 4·0 (3·5–4·5) mm; a = 69 (56–79); c′ = 0·81 (0·73–0·93); V = 48 (45–52); ost = 137 (127–149) μm; oph = 82 (76–90) μm. Almost as large as *X. diversicaudatum* but no Z-organ and tail peg almost absent (as in *X. neovuittenezi*). Male unknown. In south-west France and Spain only.

X. rivesi Dalmasso, 1969 (q.v.). ♀♀: L = 2·0 (1·7–2·1) mm; a = 42 (37–49); c′ = 1·38 (1·3–1·5); V = 52 (51–54); ost = 96 (90–101) μm; oph = 51 (48–57) μm. A small species of the *americanum* group (subgenus *Xiphinema* Cohn & Sher) distinguished chiefly by its almost continuous head contour. Z-organ absent. Males rare. Reported from the Bordeaux district of France and from Spain.

X. sahelense Dalmasso, 1969 (q.v.). ♀♀: L = 4·4 (3·8–4·9) mm; a = 78 (69–89); c′ = 1·5 (1·4–1·6); V = 46 (45–48); ost = 130 (124–137) μm; oph = 78 (74–80) μm. A large species with conoid-subdigitate tail and no Z-organ in the female; males plentiful. Originally described from Algeria; also present in Spain according to Arias & Navacerrada (1973).

X. turcicum Luc & Dalmasso, 1963 (q.v. and also Dalmasso (1969) for description) has been reported, in Europe, from Bulgaria (Stoyanov, 1974), Sardinia (Lamberti, Roca & Bleve-Zacheo, 1972) and Spain (Arias & Navacerrada, 1973).

Other Near-temperate Species

X. americanum Cobb, 1913 (Fig. 10.5F, L). The nominal species *X. americanum* is very variable and appears to comprise a collection of geographic variants or closely related species (Tarjan, 1969). The following measurement ranges (♀♀) are those within which most populations fall, according to Tarjan: L = 1·6–1·9 mm; a = 38–55; c′ = 1·4–1·8; V = 48–54; ost = 78 μm (av.); oph = 52 μm (av.). Lima, in his Ph.D. thesis (1965, unpublished) recognized four closely related species in this complex, only one of which (*X. mediterraneum* above) has been formally described. Excluding *X. mediterraneum* and other described species in Cohn & Sher's subgenus *Xiphinema, X. americanum* has not been reported in Europe although almost world-wide in warmer areas and, in North America, extending as far north as Ontario (Potter & Townshend, 1973). Important as a virus vector, (see Chapter 11). *X. americanum* resembles *X. mediterraneum* in its small size, spirally curved body when heat-relaxed, and sharply offset lip region. It is distinguished by its slightly stouter body, median vulva, and tail which is short-conoid with more or less rounded tip. Tarjan (1969) shows a wide variation in shape of tail from broadly rounded at the tip to conical and pointed. Males rare.

X. bakeri Williams, 1961 (q. v., also Sutherland, Dunn & Cousens, 1970). ♀♀: L = 4·1 (3·7–4·6) mm; a = 65 (60–71); c′ = 1·5 (approx.); V = 33 (31–34); ost = 142 μm (av.); oph = 77 μm (av.). A large species near to *X. index* and *X. diversicaudatum* from which it differs chiefly in the more anteriorly placed vulva and the longer, more conoid tail with digitate terminus. Z-organ absent. Males rare. Occurs on raspberry, strawberry and coniferous trees in British Columbia, Canada. (Anon., 1970; McElroy, 1970; Sutherland & Dunn, 1970).

Suborder DIPHTHEROPHORINA Coomans & Loof, 1970 emend Siddiqi, 1974

Superfamily Trichodoroidea Thorne, 1935 (Siddiqi, 1974)
Family Trichodoridae Thorne, 1935 (Clarke, 1961)
Diagnosis as for superfamily see p. 22.
Type genus *Trichodorus* Cobb, 1913

For some species in the genus *Trichodorus* (*sensu lato*) Siddiqi (1974a) erected the genus *Paratrichodorus* which he further subdivided into subgenera *Paratrichodorus, Atlantadorus* and *Nanidorus*. Allen (1957), Siddiqi (1974a), Hooper (1975) and Loof (1975) should be referred to for more detailed information on the morphology and taxonomy of this group.

Trichodorids (stubby root nematodes) are root ectoparasites and vary in size from about 0·5 to just over 1·5 mm long; the females, which usually have a subterminal anus, often appear rather plump and cigar-shaped especially when moribund. The following characters of adults are usually used to distinguish species in the genera *Trichodorus* and *Paratrichodorus*:

General—Body length.
Onchiostyle (stylet) length.
Body straight or curved ventrally when killed by heat, especially males.
Degree of swelling, looseness, of cuticle especially when fixed.
Location of excretory pore.
Type of junction of oesophagus with intestine and position of oesophageal gland nuclei.
Occurrence of males.

Males —Bursa (caudal alae) present, reduced or absent.
Spicule length and shape, with or without transverse striations or bristles (bristles usually only visible when spicules are extruded).
Gubernaculum shape and position relative to spicules.
Number and position of ventromedian cervical pores or papillae.
Lateral cervical pores present or absent.
Number and position of preanal ventromedian papillae (supplements).
Number and position of caudal pores or papillae.

Females—Shape of vulval opening in ventral view—transverse slit, longitudinal slit or pore-shaped.
Shape of refractive (sclerotized) thickenings at vulva-vagina junction when viewed laterally.
General shape and size of vagina in lateral view.
Number and position of lateral body pores.
Number and position of ventral body pores.

GENUS *TRICHODORUS* COBB, 1913 (Fig. 10.7)

The cuticle does not swell abnormally on fixation as it usually does in *Paratrichodorus*. Oesophageal glands abut onto intestine. Females didelphic with one pair of lateral body pores within one body width of vulva. Vagina half a body width or more long with conspicuous musculature and prominent sclerotization at vulva-vagina junction. Males usually abundant; posterior part of body ventrally arcuate when killed by heat, due mainly to the diagonal copulatory muscles which extend from cloacal level to well in front of the spicular region. Bursa absent. Spicules arcuate, with or without transverse striations; spicule suspensor muscles prominent. Three preanal supplementary papillae present two of which are anterior to spicular region. Lateral cervical pores usually close to level of onchiostyle base and anterior to level of excretory pore. Ventromedian cervical papillae usually more than one, usually more prominent than excretory pore, and located opposite onchiostyle.

Type species *Trichodorus primitivus* (de Man, 1880) Micoletzky, 1922. (Fig. 10.7).

♀♀: L = 0·71–0·81 mm; a = 22–27; b = 4·9–5·5; c — anus subterminal. V = $^{22-19}55–57^{19-25}$; onchiostyle 45–52 μm.

♂♂: L = 0·66–0·87 mm; a = 24–29; b = 4·4–5·4; c = 40–58; spicules 34–41 μm; onchiostyle 50–55 μm.

Female. Body only slightly curved ventrally when killed by heat. Lateral body pores present; one slightly posterior to vulva, one about three body widths and the other about seven body widths anterior to vulva. Refractive thickenings at vulva conspicuous, characteristically rod-shaped and appear at right angles to body axis when viewed laterally.

Male. Three ventromedian cervical papillae anterior to excretory pore, two of them usually on a level opposite posterior half of onchiostyle. Three preanal ventromedian supplementary papillae present. Spicules paired, without transverse striations, ventrally curved and narrowing towards distal end. Gubernaculum characteristically shaped, lying between or ventral to spicules for much of its length.

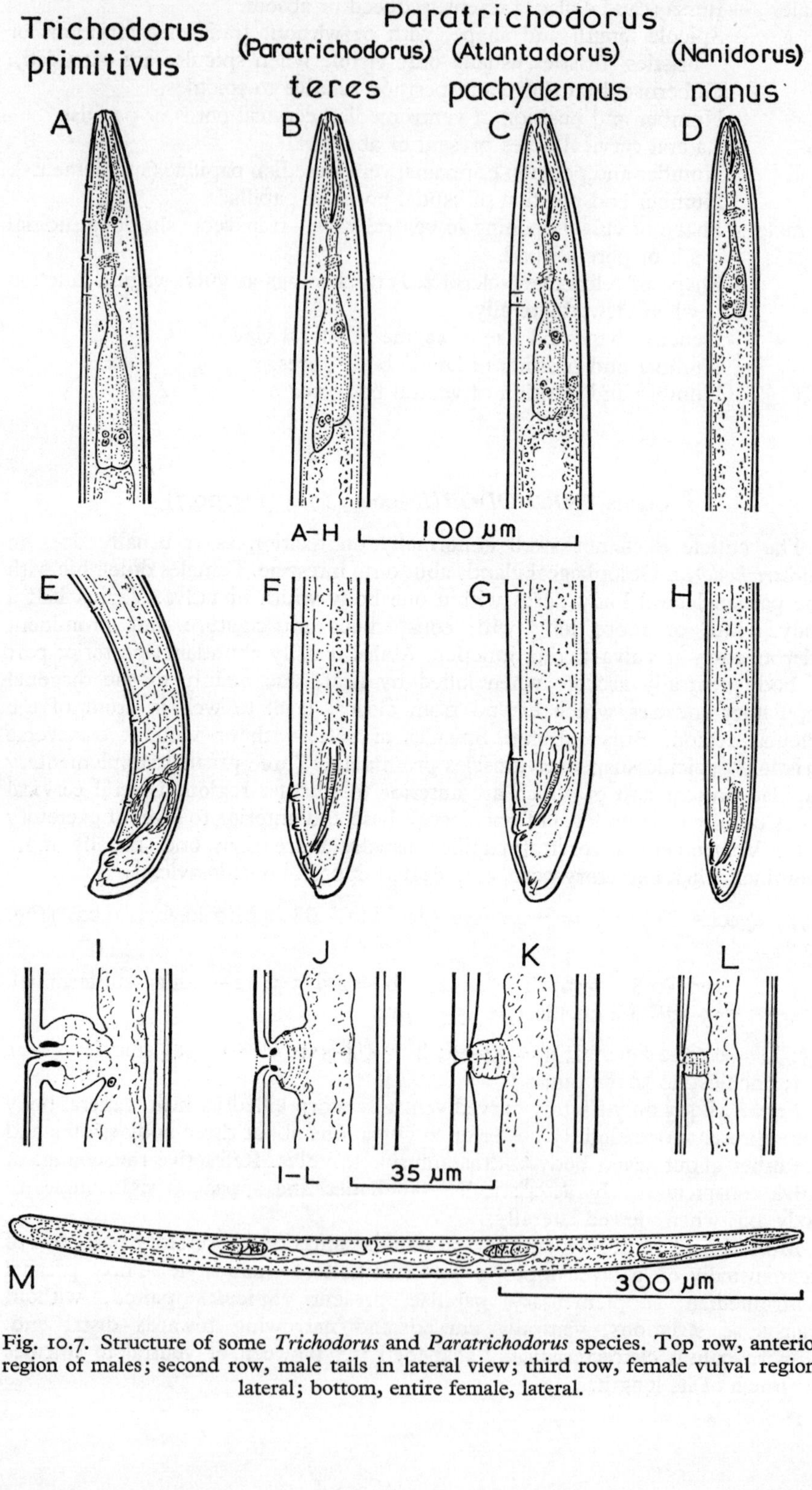

Fig. 10.7. Structure of some *Trichodorus* and *Paratrichodorus* species. Top row, anterior region of males; second row, male tails in lateral view; third row, female vulval region, lateral; bottom, entire female, lateral.

Other species:

T. aequalis Allen, 1957
T. borneoensis Hooper, 1962
T. californicus Allen, 1957
T. castellanensis Arias-Delgado *et al.*, 1965
T. cedarus Yokoo, 1964
T. cottieri Clark, 1963
**T. cylindricus* Hooper, 1962
T. elegans Allen, 1957
**T. hooperi* Loof, 1973
T. kurumeensis Yokoo, 1966
T. longistylus Yokoo, 1964
T. lusitanicus Siddiqi, 1974
T. obscurus Allen, 1957
†*T. obtusus* Cobb, 1913
T. pakistanensis Siddiqi, 1962
T. proximus Allen, 1957
**T. similis* Seinhorst, 1963
**T. sparsus* Szczygiel, 1968
**T. variopapillatus* Hooper, 1972
**T. velatus* Hooper, 1972
**T. viruliferus* Hooper, 1963

†Often regarded as a junior synonym of *T. primitivus* but Seinhorst (1963) regards it as a distinct species.
*Species reported from Europe.

GENUS *PARATRICHODORUS* SIDDIQI, 1974 (Fig. 10.7)

Cuticle usually swells abnormally when specimens are killed by heat, especially when fixed in acid fixatives. Posterior oesophageal glands overlapping intestine, or base of oesophagus flattened to rounded with intestine extending forward dorsally and laterally. Females didelphic, usually with one or two pairs of lateral body pores but rarely within one body width of the vulva (none at all in subgenus *Nanidorus*). Vagina inconspicuous, less than half a body width long with weakly developed musculature and sclerotization. Male usually straight when killed by heat with conoid or rounded tail surrounded by bursa. Copulatory muscles inconspicuous and restricted to bursal region. Spicules transversely striated, suspensor muscles inconspicuous. Usually two ventromedian supplementary papillae within bursal region, or almost so, and well separated from the third which may be reduced or absent; if single, the ventromedian supplement is preanal. Lateral cervical pores, if present, located about on a level with excretory pore and well posterior to base of onchiostyle. Ventromedian cervical papillae similar in appearance to excretory pore and, if present, always well posterior to base of onchiostyle.

SUBGENUS *Paratrichodorus* SIDDIQI, 1974

Posterior subventral oesophageal glands overlap intestine for some distance. Males sometimes present. Females with caudal pores and sometimes with lateral body pores.

Type species *Paratrichodorus* (*Paratrichodorus*) *tunisiensis* (Siddiqi, 1963) Siddiqi, 1974.

P. (*P.*) *teres* (Hooper, 1962) Siddiqi, 1974 (syn. *T. flevensis* Kuiper & Loof, 1962) is described because, unlike the type species, it occurs in Europe (Fig. 10.7).

♀♀: L = 0·80–1·23 mm; a = 20–33; b = 4·5–8·9; c — anus subterminal; V = $^{22-14}$51–59$^{11-16}$; onchiostyle 42–52 μm.
♂♂: L = 0·72–0·86 mm; a = 19–23; b = 4·6–7·8; c — 50–65; onchiostyle 47–54 μm; spicules 45–52 μm.

Female. Body almost straight when killed by heat, rather slender. Subventral oesophageal glands overlap intestine for a distance of about two thirds the body width. Excretory pore opposite anterior half of oesophageal bulb. Lateral body pores present, one on either side, usually within two to five body widths behind vulva; a further pair of pores may be present in this region. Vulva a short longitudinal slit in ventral view; inconspicuous, rather widely spaced oval refractive thickenings visible in lateral view, vagina weakly developed.

Male. Anterior end similar to female without ventromedian or lateral cervical papillae. Spicules almost straight with fine transverse striations. Gubernaculum thickened and characteristically bifurcate distally. Bursa present with two preanal ventromedian supplements within its range and a less conspicuous one three body widths anterior to its second supplement. One pair of large postanal ventrosubmedian papillae present and a pair of subterminal caudal pores. Males extremely rare in many populations but when present even in good numbers they are apparently non-functional (Loof, 1965).

Other species: *P.* (*P.*) *acaudatus* (Siddiqi, 1960) Siddiqi, 1974
P. (*P.*) *alleni* (Andrássy, 1968) Siddiqi, 1974
P. (*P.*) *allius* (Jensen, 1963) Siddiqi, 1974
P. (*P.*) *lobatus* (Colbran, 1965) Siddiqi, 1974
P. (*P.*) *mirzai* (Siddiqi, 1960) Siddiqi, 1974
P. (*P.*) *rhodesiensis* (Siddiqi & Brown, 1965) Siddiqi, 1974
P. (*P.*) *tansaniensis* Siddiqi, 1974

SUBGENUS *Atlantadorus* SIDDIQI, 1974

Intestine usually extending forward a short distance dorsally and laterally over the oesophageal bulb. Males usually common with a ventromedian cervical papilla just anterior to excretory pore; usually with two pairs of post-anal ventrosubmedian papillae. Females usually with lateral body pores.

Type species *Paratrichodorus* (*Atlantadorus*) *atlanticus* (Allen, 1957) Siddiqi, 1974.

P. (*A.*) *pachydermus* (Seinhorst, 1954) Siddiqi, 1974 which is common in Europe is described in preference to the type species which is not recorded there (Fig. 10.7).

♀♀: L = 0·71–0·81 mm; a = 22–27; b = 4·9–5·5; c — anus subterminal; $V^{22-19}55-57^{19-25}$; onchiostyle 45–52 μm.
♂♂: L = 0·73–0·83; a = 22–26; b = 5·2–6·0; c = 55–61; onchiostyle 40–50 μm; spicules 33–40 μm.

Female. Intestine usually distinctly extended forward dorsally and laterally over base of oesophagus. Vulva pore-shaped in ventral view; in lateral view the small elongate refractive thickenings lie parallel to body axis. Usually two or three lateral body pores either side posterior to vulva, one just over a body width posterior and the others following at about two body width intervals.

Male. Excretory pore opening opposite beginning of expanded part of oesophagus with a single ventromedian cervical papilla just in front of it and a lateral cervical pore just anterior to this region. Spicules curved slightly ventrally with transverse striations; gubernaculum thickened at its distal end. Bursa present; three ventromedian supplements, one just forward of adanal, the second at the anterior end of the bursa and the third about two body widths anterior to the second. Two pairs of ventrosubmedian papillae, one post-anal the other near the subterminal caudal pores. Males usually common.

Other species: **P.* (*A.*) *anemones* (Loof, 1965) Siddiqi, 1974
P. (*A.*) *porosus* (Allen, 1957) Siddiqi, 1974

SUBGENUS *Nanidorus* SIDDIQI, 1974

Excretory pore near or posterior to level of base of oesophagus, over one body width posterior to nerve ring. Lateral body and caudal pores absent. Second onchiostyle tip usually inserted in dorsal slit of functional onchiostyle of adults as usually seen in juveniles in Trichodoridae. Males rare, bursa prominent, single preanal ventromedian supplement present and a pair of ventrosubmedian papillae near tail tip; cervical papillae absent.

Type species **Paratrichodorus* (*Nanidorus*) *nanus* (Allen, 1957) Siddiqi, 1974 (Fig. 10.7).

♀♀: L = 0·49–0·60 mm; a = 18–23; b = 6·7–8·1; c—anus subterminal; V = 52–56; onchiostyle 21–23 μm.
♂♂: L = 0·49–0·55 mm; a = 21–25; b = 5·9–7·3; c = 41–46; onchiostyle 21–23 μm; spicules 42–44 μm.

Female. Oesophagus abuts onto intestine. Excretory pore at a level 1·5 to 2·5 body widths posterior to base of oesophagus. Vulva a transverse slit in ventral view; with inconspicuous refractive thickenings in lateral view.

Male. Excretory pore as in female. Spicules almost straight with transverse striations. Gubernaculum thickened at its distal end. Bursa conspicuous, arising opposite middle of spicules and enveloping tail. A single preanal ventromedian supplement and large, paired, subterminal, ventrosubmedian papillae present.

Other species: *P.* (*N.*) *acutus* (Bird, 1967) Siddiqi, 1974
P. (*N.*) *christiei* (Allen, 1957) Siddiqi, 1974
P. (*N.*) *minor* (Colbran, 1956) Siddiqi, 1974
P. (*N.*) *renifer* Siddiqi, 1974

Siddiqi (1974a) gives keys to the subgenera and species of *Paratrichodorus* and to the species of *Trichodorus* but (by oversight) *T. elegans* is omitted from the key. *T. monohystera* Allen, 1957 is deliberately omitted because it has a posterior vulva and a single anterior gonad and 'possibly belongs to a separate genus'†. Siddiqi uses the position of oesophageal gland nuclei to help separate the genera and subgenera but they are not readily seen in many specimens. Loof (1975a) also gives keys to Trichodorid genera and species and synonymizes *P.* (*N.*) *christiei* with *P.* (*N.*) *minor*.

*Species reported from Europe.
†Added in proof: Andrássy (1976) proposed the genus *Monotrichodorus* to contain *T. monohystera*.

CHAPTER 11

Plant-parasitic Dorylaimida: Biology and Virus Transmission

C. E. TAYLOR

Scottish Horticultural Research Institute, Invergowrie

Introduction

Interest in the biology of Dorylaimid nematodes was stimulated by the discovery of their role as vectors of certain plant viruses. Hewitt, Raski & Goheen (1958) demonstrated that *Xiphinema index* transmitted grapevine fanleaf virus and discoveries of other vector species in the genera *Xiphinema, Londigorus, Trichodorus* and *Paratrichodorus* followed (see Table 1). Many of the viruses transmitted by nematodes cause economically important diseases of crop plants. For example, grapevine fanleaf virus occurs in most areas where grapevines are grown and can halve the crop yields; raspberry ringspot virus severely affects raspberry and strawberry (Plate XIX), causes spoon leaf disease of red-currant and also diseases in blackberry and cherry. Tobacco rattle virus, transmitted by many species of *Trichodorus* and *Paratrichodorus,* is prevalent in sandy soils in Western Europe, and also occurs in North and South America, causing internal necrosis or spraing disease of potato tubers and serious diseases of many other crop plants. In general, both the viruses and their nematode vectors have wide host ranges among wild and cultivated plants; tobacco rattle virus is known to infect more than 400 species. Most of the records of the occurrence of nematode-transmitted viruses are from Western Europe and North America, largely stemming from the early research interest and development in those areas, but there are recent records from Australia (Stubbs, 1971), New Zealand (Thomas & Procter, 1972), Japan (Komuro, Yoshino & Ichinohe, 1970; Iwaki & Komuro, 1971; Yagita & Komuro, 1972) and Russia (Romanenko, 1971; Ivanova, 1972), and it seems probable that these viruses have a world-wide distribution in cultivated soils.

Nematode-transmitted viruses fall into two groups called nepoviruses and tobraviruses. Nepoviruses have isometric particles, about 30nm in diameter, and are transmitted by species of *Xiphinema* or *Longidorus.* Most have wide host ranges and cause ringspot diseases, but other symptoms are commonly met with and a single virus may cause different symptoms in different hosts, e.g., arabis mosaic virus causes raspberry yellow dwarf, strawberry yellow crinkle, rhubarb mosaic or lettuce chlorotic stunt. Nepoviruses are transmitted through the seed of many of their host plants, and sometimes through pollen. A close association or specificity is shown between nepoviruses and their nematode vectors. For example *X. diversicaudatum* transmits arabis mosaic or strawberry latent ringspot viruses (Plate XX), but not tomato black ring or raspberry ringspot viruses which are transmitted by *L. elongatus* (Table 1). Where individual viruses exist as distinct strains they may have different, though closely related, species of the same genus as vectors. Thus *L. elongatus* transmits the Scottish strain of tomato black ring virus but the English strain is transmitted by *L. attenuatus*; likewise outbreaks of raspberry ringspot virus (Scottish strain) are associated with *L. elongatus* and the English strain with *L. macrosoma*, although under experimental conditions in the laboratory *L. elongatus* has been shown to be capable of transmitting both strains of the virus (Taylor & Murant, 1969) and *X. diversicaudatum* has been shown to transmit both the hop and type strains of arabis mosaic virus (Valdez, McNamara, Ormerod, Pitcher & Thresh, 1974). The tobraviruses have straight tubular particles and there are only two viruses in this group, tobacco rattle and pea early-browning, which are transmitted by *Trichodorus* and *Paratrichodorus* species. Many isolates of tobacco rattle virus have been identified and these differ somewhat in symptomatology, antigenic constitution and in particle length, but these differences are not associated with any specificity of transmission. Several Trichodorid species can transmit the same isolate, and several isolates can be transmitted by one *Trichodorus* or *Paratrichodorus* species (Table 1). On the other hand, there is some experimental evidence of specificity: *P. anemones* transmitted an English but not a Dutch strain of pea early-browning virus (Harrison, 1967), and *P. pachydermus* transmitted only one of five Dutch isolates of tobacco rattle virus (van Hoof, 1968).

Many species within the four vector genera are known to cause considerable crop losses by the direct effect of their feeding on plant roots. *Xiphinema* and *Longidorus* tend to feed at or near the root tips, causing a characteristic galling and a general stunting of the root system. *X. index* causes severe galling of the roots of grapevine and fig, and *X. diversicaudatum* has been known to damage roses in the U.S.A. (Schindler, 1957), Britain (Brown, 1965) and Belgium (Coolen & D'Herde, 1970). *Longidorus* is morphologically similar to *Xiphinema*, particularly in the structure of the feeding apparatus (see Chapter 10), and several species have been associated with root galling on crops such as sugar beet, strawberry, onions, carrots, celery and peppermint. *Trichodorus* and *Paratrichodorus* species also feed at the root tips, but feeding is probably by repeated, direct thrusts of the solid odontostyle into the epidermal cells, whereas *Xiphinema* and *Longidorus* nematodes insert their hypodermic-like odontostyle into deeper tissues and feed for longer periods at each site than Trichodorids. In 1951, Christie and Perry first demonstrated the pathogenic capabilities of the free-living migratory group of nematodes with a Trichodorid species which later was named *T. christiei* (now *P. christiei*); they found that the nematode caused a stunting of the roots producing a condition they descriptively termed 'stubby-root'. Most other Trichodorids are similarly pathogenic but necrosis of the root cells may also occur to a greater or lesser degree (Pitcher & Flegg, 1965; Wyss, 1973).

Some other genera in the Dorylaimida have the elongated stylet of *Xiphinema* and *Longidorus* and there is evidence that some are plant parasites, but only species of *Xiphinema, Longidorus, Trichodorus* and *Paratrichodorus* have so far been implicated as pathogens of crop plants and it is only from these four genera that virus vectors are known. Research in the field of nematode transmission of plant viruses has progressed rapidly and different aspects have been the subject of several critical reviews (Harrison, 1960, 1964a; Raski & Hewitt, 1963; Cadman, 1963; Pitcher, 1965; Dalmasso, 1967; Weischer, 1968b; Taylor & Cadman, 1969; Taylor, 1971, 1972a,b; Lamberti, Taylor & Seinhorst, 1975).

Table 1

Nematode-transmitted Viruses, Vectors, Crops Infected and Distribution of Diseases

Virus	Vector	References	Examples of crops infected	Geographical occurrence of disease
		Nepoviruses		
Arabis mosaic				
Type strain	*Xiphinema diversicaudatum*	Harrison & Cadman, 1959 Jha & Posnette, 1959	Cherry, cucumber, grape-vine, raspberry, rhubarb	Europe, New Zealand
	X. coxi	Fritzsche, 1964		
Hop strain	*X. diversicaudatum*	Valdez *et al.*, 1974	Hop	Europe, Tasmania
Artichoke Italian latent	*Longidorus apulus*	Roca *et al.*, 1975 (See footnote p. 238)	Artichoke	Southern Italy
Cherry leaf roll	*X. diversicaudatum*, *X. coxi*, *X. vuittenezi*	Fritzsche & Kegler, 1964 Fritzsche, 1964 Flegg, 1969	Cherry, blackberry, elm, rhubarb	Europe and N. America
Grapevine fanleaf	*X. index* *X. italiae*	Hewitt *et al.*, 1958 Cohn *et al.*, 1970	Grapevine	All temperate regions where *Vitis vinifera* is grown
Mulberry ringspot	*L. martini*	Yagita & Komuro, 1972	Mulberry	Japan
Raspberry ringspot Type (Scottish) strain	*L. elongatus*	Taylor, 1962	Blackberry, raspberry, red currant, strawberry	Europe
English strain	*L. macrosoma* *L. elongatus*	Harrison, 1964b Taylor & Murant, 1969	Blackberry, raspberry	

Cherry strain	*L. macrosoma*	Fritzsche & Kegler, 1968	Cherry	
	X. diversicaudatum	Fritzsche & Kegler, 1968		
Strawberry latent ringspot	*X. diversicaudatum*	Lister, 1964	Black currant, cherry, celery, peach, plum, raspberry, rose, strawberry	Europe
	X. coxi	Putz & Stocky, 1970		
Tobacco ringspot	*X. americanum*	Fulton, 1962	Bean, blueberry, gladiolus, grapevine, tobacco	U.S.A., Canada
	X. coxi	van Hoof, 1971		
Tomato black ring			Celery, globe artichoke, lettuce, peach, potato, raspberry, strawberry, sugar beet, tomato	Europe
English strain	*L. attenuatus*	Harrison *et al.*, 1961		
Scottish strain	*L. elongatus*	Harrison *et al.*, 1961		
Tomato ringspot	*X. americanum*	Breece & Hart, 1959	Blackberry, cherry, grapevine, peach, raspberry, tobacco	U.S.A., Canada, Japan
		Iwaki & Komuro, 1971		
Tobraviruses				
Pea early browning				
Dutch isolates	*Paratrichodorus pachydermus*	van Hoof, 1962	Pea, lucerne	Western Europe, particularly The Netherlands and eastern England
	P. teres	van Hoof, 1962		
English isolates	*P. anemones*	Harrison, 1967	Pea, lucerne	
	Trichodorus primitivus	Harrison, 1966		
	T. viruliferus	Gibbs & Harrison, 1964		

Table 1 (*continued*)

Virus	Vector	References	Examples of crops infected	Geographical occurrence of disease
		Tobraviruses (*continued*)		
Tobacco rattle				
European isolates	*P. anemones*	van Hoof, 1968	Potato, bulbous ornamentals, tobacco	Europe, U.S.A., Brazil, Japan
	P. nanus	van Hoof, 1968		
	P. pachydermus	Sol *et al.*, 1960		
	P. teres	van Hoof, 1968		
	P. minor	Komuro *et al.*, 1970		
	T. primitivus	Harrison, 1961; Sänger, 1961		
	T. similis	Cremer & Kooistra, 1964		
	T. viruliferus	van Hoof, 1968		
American isolates	*P. allius*	Jensen & Allen, 1964	Potato, globe artichoke, lettuce, pepper, tobacco	U.S.A., Brazil
	P. christiei	Walkinshaw *et al.*, 1961		
	P. porosus	Ayala & Allen, 1968		
		Other viruses		
Brome mosaic	*X. diversicaudatum*	Schmidt *et al.*, 1963	Grasses, cereals	U.S.A., Europe
	L. macrosoma	Fritzsche, 1970		
Carnation ringspot	*X. diversicaudatum*	Fritzsche & Schmelzer, 1967	Carnation	In temperate regions where carnations are grown.
Prunus necrotic ringspot	*L. macrosoma*	Fritzsche & Kegler, 1968	Plum	World-wide in temperate regions

Occurrence and Life Cycles

Xiphinema

At least 12 *Xiphinema* species occur in Western Europe but in Britain only *X. diversicaudatum* is prevalent. A few isolated populations of *X. coxi* and *X. vuittenezi* have been recorded in England, and *X. mediterraneum** has been found only in Kent and may well be a chance introduction from its more usual mediterranean habitat (Taylor & Brown, 1976). *X. diversicaudatum* appears to be common in a wide range of habitats, including grassland, deciduous woodland and many arable crops, and occurs in moisture-retaining soils ranging from sandy loams to heavy clays and peats. It is widely distributed in England and Wales and is fairly common in the clay soils in the Lothians of southern Scotland, but does not extend further north than the Tay area. Relatively few of the populations encountered are likely to be associated with arabis mosaic or strawberry latent ringspot viruses, which *X. diversicaudatum* transmits. Out of about 2000 soil samples collected throughout Britain during the period 1969–73, *X. diversicaudatum* occurred in 325 but only 18 of these populations were shown to be viruliferous (Taylor & Brown, 1976).

The vectors of grapevine fanleaf virus, *X. index* and *X. italiae,* do not occur in Britain but are common in mediterranean countries. *X. index* is also widespread in California, U.S.A. and in the State of Victoria, Australia, probably having been introduced to these areas with vine rootstocks imported from France. *X. americanum,* the vector of tobacco ringspot and tomato ringspot viruses, is widely distributed in warmer climates such as North and South America, West Africa, India and Japan, but has not been found in Europe, previous records for this region all probably relating to *X. mediterraneum* which it closely resembles. Plant material infected with tobacco ringspot virus has occasionally been found in Britain but this can usually be traced to importations from an infected area, or more rarely to bird-carried infected seeds.

The life-cycles of *Xiphinema* species are long, and in the field *X. diversicaudatum* and *X. americanum* may require at least a year to complete each generation under temperate conditions (Norton, 1963; Taylor & Thomas, 1968; Flegg, 1968). Each of the four larval stages plus adults are likely to be found in soil samples taken at any time during the year but there is often a sharp increase in the relative number of mature females when there is a seasonal flush of new root growth of the host plant, and it is at this time that egg-laying occurs. Local climatic conditions shift the period of reproduction towards summer or winter. For example, in Scotland *X. diversicaudatum* reproduces mainly from April to July (Taylor & Thomas, 1968) and in the west of France from March to June, but in southern Italy gravid females occur as early as January (Dalmasso, 1970). Under most circumstances the eggs hatch fairly soon after laying, but under the severely cold winter conditions of Wisconsin and Iowa, U.S.A., they appear to be the only stage of *X. americanum* to survive (Griffin & Darling, 1964). Under less severe conditions, most stages are capable of surviving for several months without feeding, and *X. mediterraneum* may be capable of aestivating to overcome periods when the soil is dry and hot (Dalmasso, 1970).

Longidorus

Of about 44 species of *Longidorus*, seven have been recorded in Britain, and of these *L. elongatus, L. macrosoma* and *L. attenuatus* are generally recognized as

*See footnote p. 224.

virus vectors. *L. elongatus* occurs in many temperate regions of the world, and has been recorded from most areas of Britain although it is especially prevalent in north-east England and eastern Scotland in sandy to medium loam soils. As with *X. diversicaudatum*, few of the populations are viruliferous and in the soil sampling survey referred to above *L. elongatus* was present in 265 of the samples, but only four of them were viruliferous. *L. macrosoma* is well distributed in southern and central England usually in heavy soils, whereas *L. attenuatus* appears to be restricted to light sandy, well-drained soils and in Britain most records of its occurrence are from East Anglia; it is common in the sandy soils of the south-eastern coast of Italy where it has been found in association with outbreaks of a disease of globe artichoke caused by artichoke Italian latent ringspot virus (Rana & Roca, 1973; Roca, Martelli & Rana, 1975).*

L. goodeyi and *L. leptocephalus* are widely distributed in Britain, usually in heavy soils and often associated with grassland. *L. caespiticola* is common in arable soils and grassland in Wales and southern and central England, but is only occasionally found in Scotland; in France it is common in the coastal north-west and frequently found in grassland, but is much rarer elsewhere (Dalmasso, 1970). *L. profundorum* is widespread in south-east England, especially in association with orchard trees, but is uncommon elsewhere in Britain, although it is well distributed in many parts of France (Dalmasso, 1970).

Most *Longidorus* species have long life cycles and the adults may have a life span of several years, particularly in temperate climates. Flegg (1968) found that *L. profundorum* has an annual egg-laying period between May and August, coinciding with the flush of root growth on orchard trees, and he estimated that the life cycle from egg to adult takes two years or more. *L. elongatus* probably has one generation per year in Britain, egg laying again coinciding with active root growth of the host plant, which in some years and on hosts such as strawberry may also lead to some egg-laying activity in the autumn (Taylor & Murant, 1968; Thomas, 1969). In each egg-laying cycle a single female may produce up to 20 eggs, but mortality at different stages in the life cycle results in a much lower annual rate of multiplication, and for *L. elongatus* there is only a two- to four-fold population increase on favoured hosts (Taylor, 1967; Thomas, 1969; Yassin, 1969).

Trichodorus and *Paratrichodorus*

Eleven species of these genera are known to occur in Britain and these invariably occur in light, open-textured soils. In Scotland, *P. pachydermus* and *T. primitivus* each occurred in about 65 per cent of soil samples from freely drained podsols and about 80 per cent of these carried tobacco rattle virus (Cooper, 1971). In samples taken from arable sandy soils in eastern England, *P. pachydermus* and *T. primitivus* were found in 33 per cent, *T. viruliferus, T. similis* and *T. cylindricus* in about 10 per cent, and *P. teres* and *P. anemones* in about two per cent of the samples (Whitehead & Hooper, 1970). In southern Sweden *P. pachydermus, T. primitivus* and *T. similis* were found commonly in light soils (Persson, 1968) and in California, U.S.A., *P. allius* favoured soil with a large proportion of coarse sand (Ayala, Allen & Noffsinger, 1970).

*These Italian vector populations are now recognized as *Longidorus apulus* Lamberti & Bleve Zacheo (*Nematol. medit.* 5, 73–83, 1977).

The occurrence of Trichodorid species in the soil appears to be much more variable than that of *Xiphinema* or *Longidorus* because many of them are capable of multiplying rapidly if suitable host plants are available, consequently this may give rise to relatively rapid fluctuations in their numbers. Soil moisture appears to be a particularly important factor, and large populations often coincide with the substantial rainfall during the early spring growth of the host plants (Jones, Larbey & Parrott, 1969; Cooper & Harrison, 1973). Rohde & Jenkins (1957a, b) found that *P. christiei* completed its life-cycle on *Zea mays* in about 22 days at 22°C and 17 days at 30°C, with a tenfold population increase in 60 days; Ayala *et al.* (1970) considered 16–24°C to be the optimum temperature range for the reproduction of *P. christiei*, 21–24°C for *P. allius* and 24°C for *P. porosus*. On apple, *T. viruliferus* completed its life-cycle in about 45 days at laboratory temperature (Pitcher & McNamara, 1970).

Ecological Associations

Outbreaks of nematode-transmitted viruses characteristically appear as patches of diseased plants within a crop corresponding to the distribution of vector populations in the soil. Spread of disease from the periphery of such patches is usually slow because of the general lack of mobility of the vectors. In a study of the association of *X. diversicaudatum* with an outbreak of strawberry latent ringspot virus in Malling Jewel raspberry (Plate XX), Taylor & Thomas (1968) plotted the 'isopathological curves' (Scotto la Massese, 1972) of the nematode populations and showed that the edge of the infestation, and the corresponding virus infection in the crop, advanced only about 2m per year (Fig. 11.1).

A further characteristic is the persistence of infection in the soil for long periods even during a fallow or when crops that resist infection are grown. Viruses transmitted by *Xiphinema* or Trichodorids may be retained for several months in their vectors and thus can, for example, easily survive winter fallow between susceptible crops. In contrast, *Longidorus* vectors do not retain infectivity for such relatively long periods. Raspberry ringspot and tomato black ring viruses persist in *L. elongatus* for a maximum period of about eight weeks and infectivity is lost during a winter fallow. Instead, these viruses can survive in the dormant weed seeds which act as sources of infection when they germinate in the spring. Seed transmission also occurs with viruses transmitted by *Xiphinema* and Trichodorids but is less frequent in the field than with viruses transmitted by *Longidorus* spp. (Murant & Lister, 1967; Taylor & Thomas, 1968). Root pieces remaining after the removal of an infected crop may also act as reservoirs of infection and sources of food for vector nematodes; this occurs in the case of grapevine fanleaf virus which seems rarely to infect weeds in the field but the virus and the vector, *X. index*, can survive for several years on root pieces remaining after a crop has been grubbed (Raski, Hewitt, Goheen, Taylor & Taylor, 1965).

Outbreaks of nettlehead disease of hop spread slowly, often recur in successive plantings, and are particularly prevalent alongside hedges, where hedges have been removed and after orchard crops or permanent pasture (Keyworth & Davies, 1946; Keyworth & Hitchcock, 1948). These features are characteristic of a nematode-transmitted virus and an unusual strain of arabis mosaic was

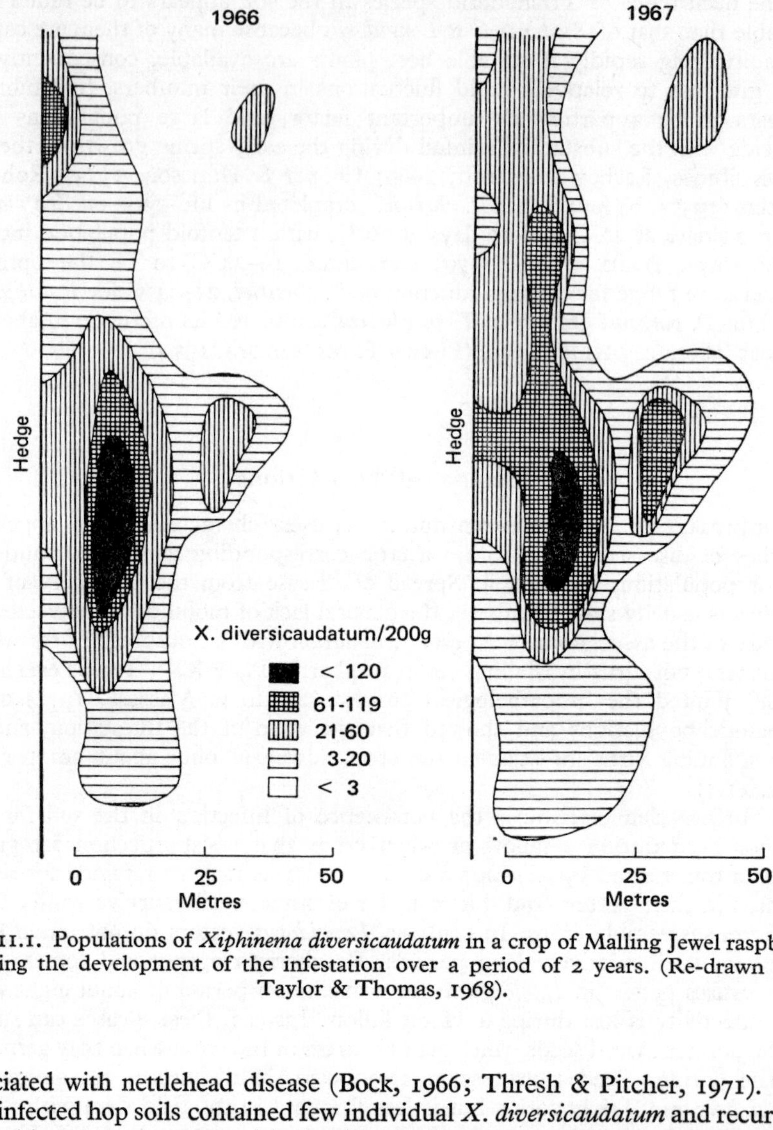

Fig. 11.1. Populations of *Xiphinema diversicaudatum* in a crop of Malling Jewel raspberry showing the development of the infestation over a period of 2 years. (Re-drawn after Taylor & Thomas, 1968).

associated with nettlehead disease (Bock, 1966; Thresh & Pitcher, 1971). The first infected hop soils contained few individual *X. diversicaudatum* and recurring difficulties were encountered in implicating it as a vector (Valdez, McNamara, Ormerod, Pitcher & Thresh, 1974). However, laboratory experiments showed that the hop strain of arabis mosaic virus can be transmitted by *X. diversicaudatum*, although only a proportion of the experimentally infected hop plants developed typical symptoms of nettlehead disease (Valdez *et al.*, 1974).

Hops are large, long-lived perennials and the epidemiology of AMV in the crop poses special problems. Each individual hop root system occupies at least $2m^3$ of soil, and so a site containing only one *X. diversicaudatum* in 200 ml, which is the smallest population detectable in the usual soil sample unit, exposes each hop plant to about 10 000 vector nematodes, each of which is capable of infecting it (Pitcher, 1975). Thus the population of *X. diversicaudatum* that can be tolerated in a hop field, when the hop strain of AMV is present, is very low.

Transmission and Retention

Nematodes acquire and transmit viruses when they feed at the root tip of the host plant. However, the transmission process is a complex sequence of events involving the biological properties of the virus, vector and plant—and the probability and efficiency of transmission depends on successful interactions between the three components.

Adult and juvenile stages appear to be equally efficient in acquiring virus in short feeding periods of an hour or less. Nepoviruses, and probably also tobraviruses, are not retained by nematodes through a moult and do not pass through the egg stage, yet viruses can persist in their vectors for longer periods than the longevity of the virus *in vitro*. This suggests that the virus is retained extracellularly within the nematode and probably after adsorption on some part of the tissue of the mouthparts that is shed during a moult.

The sites of retention of viruses in nematode vectors (Plate XVIII) have been discovered by means of electron microscopy of ultrathin sections of virus-carrying nematodes (Taylor & Robertson, 1969; 1970a, b; 1973; McGuire, Kim & Douthit, 1970; Raski, Maggenti & Jones, 1973). In *L. elongatus*, raspberry ringspot and tomato black ring viruses accumulate as a single layer of particles on the inner surface of the stylet guiding sheath. Viruses transmitted by *Xiphinema* spp. are found only in association with the cuticular lining of the oesophageal lumen, and may be present as a monolayer from the anterior end of the odontophore to the posterior end of the oesophageal bulb. Particles of tobacco rattle virus in *Paratrichodorus pachydermus* have also been found in association with the cuticular lining of the lumen of the pharynx and the oesophagus (Taylor & Robertson, 1970a).

It is thought that the virus particles are selectively and specifically adsorbed on to the cuticular guiding sheath or the lining of the oesophagus as virus-containing plant sap passes from the plant to the nematode's intestine. Other viruses present in the plant sap but not transmitted by the nematode are not retained at these sites. For example, arabis mosaic virus particles were not retained by *L. elongatus* after feeding on infected plants (Taylor & Robertson, 1969). Specificity of transmission therefore seems to reflect the specific association between the protein surface of the virus and the cuticular surface in the nematode where the particles are retained. Although this provides an explanation of specificity of transmission in terms of serological behaviour of the viruses, the precise features of the interface between virus particle and nematode cuticle that are important for transmission are unknown.

For virus to be transmitted, virus particles must associate with, and later dissociate from, the site of retention in the nematode. Dissociation seems most likely to occur when saliva passes from the nematode into the plant cell during the initial phases of feeding; it may be brought about because the pH or ionic conditions then change, or because of some enzymic effect of the saliva either on the surfaces of nematode or virus particles, or on materials which may be involved in binding the virus particles to the nematode cuticular surface. However it occurs, the process of dissociation is likely to take place relatively slowly because viruses can persist for many weeks in their vectors and some can be transmitted to several plants in a series. Inoculation presumably occurs when the virus is carried with the saliva into a plant cell that has not been damaged too severely by the nematode's feeding.

Methods of Control

The distribution and use of infected planting material is probably the most important way by which nematode-transmitted viruses become disseminated to sites that may or may not already be infested with vector species. Viruses associated with grapevine, strawberry, potato and some bulbous ornamentals (e.g., *Narcissus*) have been distributed in commerce in this way. Weed seeds may also play a part in the distribution of certain viruses such as raspberry ringspot and tomato black ring viruses transmitted by *L. elongatus* although there is as yet no direct evidence of this, and the importance of weed seed infection may lie in the perennation of the virus in the soil, as already mentioned.

Populations of nematode vectors remain more or less static and travel only short distances under their own volition. However, they may be carried in the soil adhering to the roots of transplants, or on farm machinery, or perhaps on the feet of birds or other animals. Establishment at a new site is then dependent on suitable environmental conditions persisting in the soil, such as pore space, moisture and temperature and the presence of host plants. Because of the long life cycle of *Xiphinema* and *Longidorus*, undisturbed conditions are preferable and hence grass leys, orchards and woodlands, or hedgerows become important as ecological niches in which nematode populations can become established.

The problem of nematode-transmitted viruses can be avoided by not planting susceptible crops on land where virus-carrying nematodes are present. The presence of the vector can be detected by means of soil sampling, and the extent of virus infection can be determined by growing seedlings of suitable herbaceous hosts in the soil ('bait testing'). The avoidance of infected land is particularly important in respect of high-grade nursery stocks, such as strawberry, black currant, raspberry, cherry, grown in official certification schemes (see Chapter 17, p. 329). Even if the use of virus-infected land cannot be avoided, the results of soil sampling can help to decide whether chemical treatment of the soil or other control measures need be undertaken.

Apart from measures taken to prevent viruses and vectors reaching new sites, such as ensuring the use of virus-free planting material, control measures are mainly dependent on the eradication of virus from land already infected. Crop rotations are largely precluded as a means of control because of the wide host range of virus and vectors, and while periods of fallow result in some decline in nematode populations they are ineffective in controlling infection because of the perennation of the viruses in root pieces and weed seeds. Weed control may, however, have some effect particularly when the crop itself is a poor host for the vector. For example, *L. elongatus* fails to reproduce on raspberry, so weed control from the time of planting eliminates sources of virus infection and in the absence of any host plants, vector populations gradually decline. With perennial crops such as grapevines, stone-fruits and raspberries, the planting of virus-immune cultivars provides a solution to the problem of nematode-transmitted viruses; but although the crop may be immune from virus infection it may nevertheless be a suitable host for the vector and allow it to multiply. This not only increases the chances of infection in subsequent susceptible crops but also increases the damage caused by the results of the nematodes' feeding.

The use of chemical nematicides has proved very satisfactory in dealing with nematode vectors and preventing the infection of newly planted crops with the viruses they transmit (Plate XIX. See also Chapter 14, p. 293). Although their application rarely results in the complete elimination of nematode populations, particularly where the nematodes are present at some depth in the soil, the effect can be long lasting because of the low rate of multiplication of most of the species involved.

CHAPTER 12

Some Free-living Nematodes found in Plant Material and Soil Samples

J. F. SOUTHEY

Plant Pathology Laboratory, Harpenden

Introduction

Soil samples and diseased plant material that are examined for plant-parasitic nematodes frequently contain many nematodes that are not plant parasites or whose feeding habits are imperfectly known; many are microbivorous, some are predatory. This chapter will serve as an introduction to the commoner ones; further details are given in textbooks and other publications. In particular, Nicholas (1975) has provided a useful review of the biology of free-living nematodes, which also covers morphology, taxonomy, ecology and techniques. For identification, Goodey's *Soil and Freshwater Nematodes* (T. Goodey, 1963) remains the best general guide to genera, though now sadly out of date; Heyns (1971) is more up to date and is useful over a much wider geographical area than that for which it was written. For those able to read Russian, Kir'yanova & Krall (1969) give keys to genera including all the groups mentioned in this chapter as well as plant-parasitic genera. To see where the free-living groups fit into the general scheme of nematode classification, reference should be made to Chapter 1.

The numerous soil-inhabiting Tylenchids that are not, or are doubtfully, pathogenic to higher plants are excluded. Some of these feed on fungi, moss rhizoids, root hairs of vascular plants or casually on root surfaces but for the present purpose they are regarded as essentially plant-parasitic, although a few are predatory. Some of the commoner ones are mentioned in Chapters 4, 8 and 9. Some of the many soil Dorylaimoids whose feeding habits are little known may fall into a similar category, but apart from a few specialized genera (see Chapters 10 and 11) the majority of nematodes in this group are probably not parasitic on plants, or at any rate on higher plants, and they are not covered elsewhere in this book. They are therefore given brief treatment here.

This chapter is in part a revision of Chapter 10 by the late Dr. J. B. Goodey in the second edition (1965) of this book.

Order RHABDITIDA

(see page 24)

Family DIPLOGASTERIDAE (Fig. 12.1)

The many genera show considerable morphological diversity but a common basic structure. Goodey (1963) gives a tabular key. The name is derived from the two swellings of the oesophagus which has a cylindrical procorpus, swollen median bulb with crescentic valve plates (like that of Tylenchids), narrower

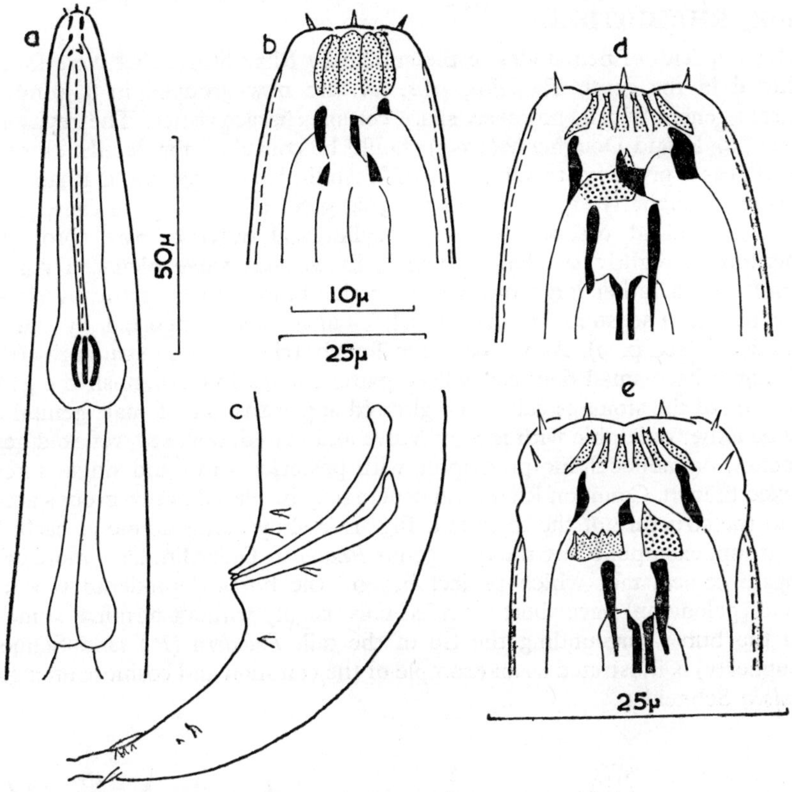

Fig. 12.1. Diplogasterids. *Diplogasteritus consobrinus* (de Man) Paramonov: **a,** oesophageal region; **b,** anterior end; **c,** male tail. *Monochoides striatus* (Bütschli) J. B. Goodey: **d,** female anterior end in left lateral view; **e,** male anterior end in ventral view. (From J. B. Goodey, 1965 after Weingärtner, 1955).

isthmus, and swollen posterior bulb without valve plates. Female genital tract usually has paired, opposed, reflexed branches (didelphic) with median vulva; rarely consisting of an anterior branch only with posterior vulva (monodelphic, prodelphic). The male sometimes has a bursa and 7–10 pairs of caudal papillae arranged in typical pattern. In other genera three of the last four pairs form a latero-ventral group often located in a small rudiment of the bursa and the last pair is dorso-lateral. Stoma very variable, presumably reflecting different feeding habits, long and narrow to short and broad. Those with short broad stomas armed with conspicuous teeth include predatory forms; others are probably microbivorous. The stoma consists basically of the same five parts seen typically in the Rhabditidae (see also p. 9). The fourth part (metarhabdions) bears denticles that are sometimes modified to large movable teeth. Fig. 12.1 shows examples of less and more specialized forms.

Diplogasterids are often abundant in habitats rich in decomposing organic matter, especially where nitrogen is plentiful, e.g., animal droppings, manure heaps, compost, etc. A few genera (e.g., *Micoletzkya* Weingärtner) are associates of bark beetles or other Coleoptera.

Family RHABDITIDAE

Many species of nematodes in the subfamily Rhabditinae that were formerly included in the genus *Rhabditis sens. lat.* are now grouped in a number of separate genera which, however, share certain characteristics. The revisions of Osche (1952) and Dougherty (1955) should be consulted for details. Hopper & Cairns (1959) give keys to subgenera of Rhabditidae and genera of Rhabditinae, following Dougherty (1955). Goodey (1963) gives tabular keys to these groups.

The Rhabditid oesophagus has a cylindrical anterior part (procorpus), sometimes a slightly swollen median bulb without valve plates, a narrower isthmus and a rather large rounded posterior bulb with tripartite overlapping valve plates. The stoma is cylindrical, elongate and supported by cuticular 'rhabdions' (see p. 9). At its base the fourth trio of these structures (meta-rhabdions) have small denticles whose pattern helps to distinguish the genera. This part of the stoma is called the glottoid apparatus. The female genital tract may be either didelphic with median vulva and paired, opposed, reflexed genital branches, or monodelphic-prodelphic with posterior vulva and single anterior, reflexed branch. Common Rhabditid genera may be placed in two groups according to the structure of the male tail. Fig. 12.2 shows an example of each. The first group, exemplified by the type genus *Rhabditis* Dujardin, have more or less elongate conical tails which project beyond the bursa (leptoderan type); the second (peloderan) have short rounded tails usually without terminal spine and with the bursa surrounding the tip of the tail. *Pelodera* (*P.*) *teres* Schneider (Dougherty) is illustrated as an example of the common and cosmopolitan genus *Pelodera* Schneider.

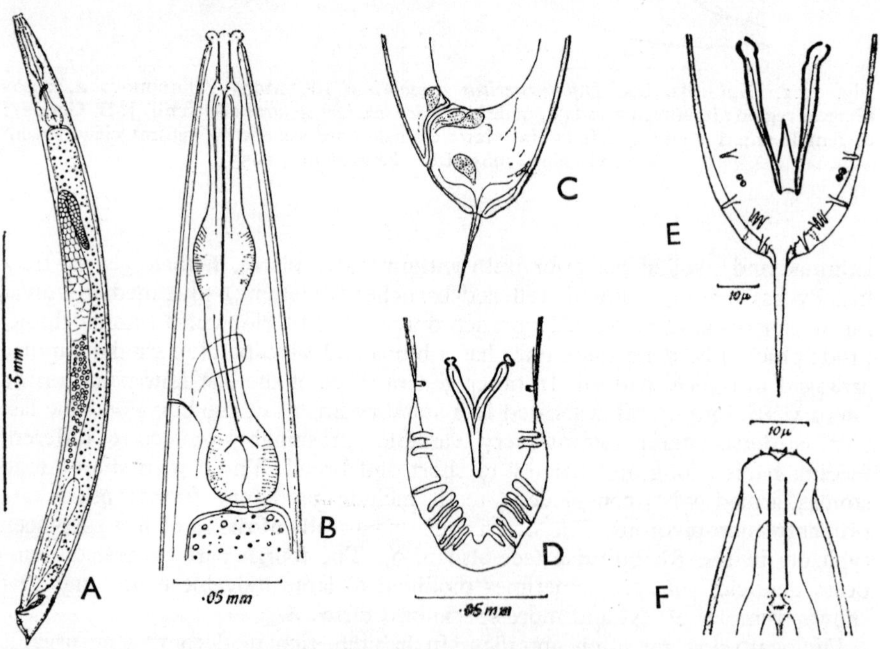

Fig. 12.2. Rhabditids. *Pelodera* (*P.*) *teres* Schneider: **A,** whole male; **B,** oesophageal region; **C,** female tail; **D,** male peloderan tail. *Rhabditis* (*Cephaloboides*) *musicola* Rahm: **E,** male leptoderan tail; **F,** anterior end. (From T. Goodey, 1963).

Rhabditids (subfamily Rhabditinae) are among the commonest associates of soil that is rich in organic matter, and of peat, moss, damaged plant tissue and similar materials; they are found in any situation where bacteria, on which they feed, are plentiful. They sometimes occur in astronomical numbers on mushroom compost (Staniland, 1957) but such excessive multiplication there probably indicates bad growing conditions (Hussey, Read & Hesling, 1969). Some species are common associates of invertebrates, e.g., earthworms, insects and other terrestrial arthropods.

Family PANAGROLAIMIDAE

Generally similar in structure to Rhabditidae but usually of relatively more slender body proportions, with a fairly wide stoma that is shorter than in typical

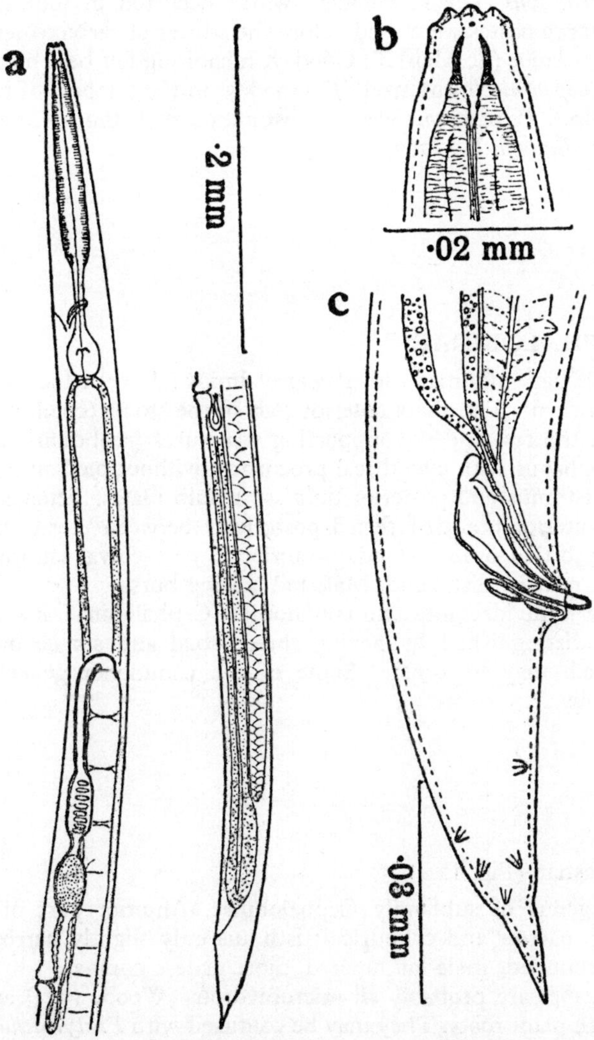

Fig. 12.3. *Panagrolaimus rigidus* (Schneider) Thorne: **a,** female; **b,** anterior end; **c,** male tail. (From T. Goodey, 1951).

Rhabditids and often tapers posteriorly. Oesophagus usually lacking a median bulb; procorpus cylindrical, or occasionally with valveless median bulb, followed by narrow isthmus and swollen posterior bulb with valve plates. Female genital tract with one fully developed anterior branch (prodelphic) with reflexed ovary; either truly monodelphic, or pseudomonodelphic with a short post-vulval sac to represent the posterior branch. Male tail lacking a bursa.

Type genus: *Panagrolaimus* Fuchs. *P. rigidus* (Schneider) Thorne (Fig. 12.3) is a very common non-parasitic associate of plants, e.g., about roots and stem bases of cereals. It is a fairly slender nematode with pointed tail, only slightly stouter in proportions than *Ditylenchus dipsaci* with which it can be confused under low magnification.

The family includes species that provide often-cited examples of curious nematode habitats, e.g., *Turbatrix aceti* (Müller) Peters, the vinegar eelworm; *Panagrellus redivivus* (L.) T. Goodey, which occurred in sour paperhangers' paste when starch pastes were used before the advent of carboxymethyl cellulose adhesives; *P. silusiae* (de Man) T. Goodey, inhabiting felt beer mats in German inns; and *P. nepenthicola* (Menzel) T. Goodey, in the pitchers of pitcher plants (*Nepenthes* spp.). *P. redivivus* has the distinction of being a Linnaean species (Syn. *Chaos redivivum* L., 1767).

Family CEPHALOBIDAE

Cephalobids have affinities with Panagrolaimids (above) which were formerly grouped with them. The short anterior part of the stoma (cheilostom, see p. 9) is wider than the rest and the supporting structures (rhabdions) are distinctly jointed. Oesophagus with cylindrical procorpus, without median bulb, followed by narrower isthmus and posterior bulb with bulb plates. Female genital tract prodelphic; anterior branch reflexed posteriorly between uterus and ovary (at spermatheca) so that ovary extends posterior to post-vulval sac; ovary reflexed once or twice near its distal end. Male tail lacking bursa.

The family is divided into two subfamilies, Cephalobinae and Acrobelinae; the latter is distinguished by having three labial and six cephalic lobes or probolae which may be ornate. Some of the commoner genera are briefly described below.

Cephalobus Bastian (Fig. 12.4)

The type genus of subfamily Cephalobinae. Anterior part of oesophagus (corpus) long, narrow and cylindrical; isthmus only slightly narrower. Female tail blunt or rounded; male tail tapered, more or less pointed.

Cephalobus spp. are probably all microbivorous (Wood, 1973) and are often associated with plant roots. They may be confused with *Pratylenchus* spp. under low magnification, because of their similar proportions and rounded female tail, especially when stained roots of cereals etc. are examined for the latter.

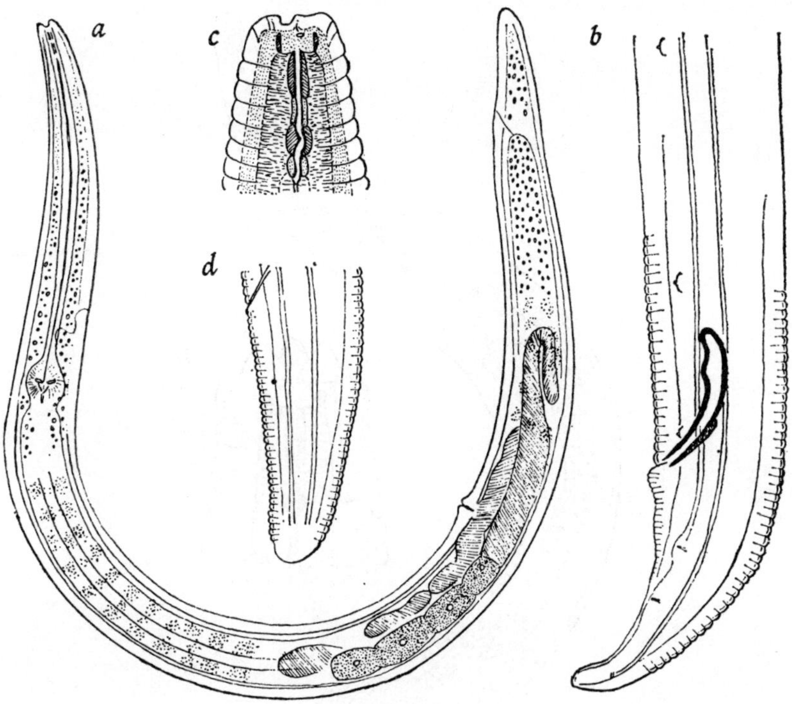

Fig. 12.4. *Cephalobus persegnis* Bastian: **a,** female; **b,** male tail; **c,** anterior end; **d,** female tail. (From T. Goodey, 1963).

Acrobeloides (Cobb) Thorne (Fig. 12.5)

Usually short, stout nematodes with rounded conoid tails. Lip region with small, rounded or conical, forwardly pointing labial probolae. First body annule may also bear lobe- or plate-like projections (cephalic probolae). Corpus of oesophagus usually with elongate, fusiform swelling. Males very rare. For key to species, see Brzeski (1962).

Common in soils rich in organic matter; often associated with roots or rotting plant tissues. Microbivorous (Wood, 1973). Thorne (1928) found *A. buetschlii* in cysts of *Heterodera schachtii*.

Acrobeles von Linstow (Fig. 12.6)

Distinguished by the conspicuous and elaborately fringed probolae, the labial ones being deeply forked and projecting well forward of the cephalic ones. Longer and more slender in proportions than *Acrobeloides*; tails of both sexes pointed. For review of genus, see Thomas (1965a).

Not encountered very frequently in routine sample examinations but when found they are attractive microscopical subjects; occur mainly in sandy soils. Probably microbivorous.

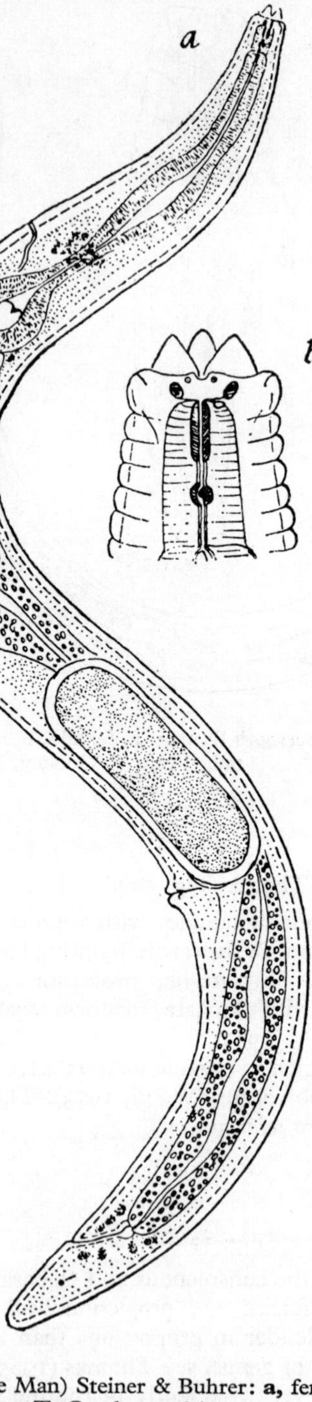

Fig. 12.5. *Acrobeloides buetschlii* (de Man) Steiner & Buhrer: **a,** female; **b,** anterior end. (From T. Goodey, 1963).

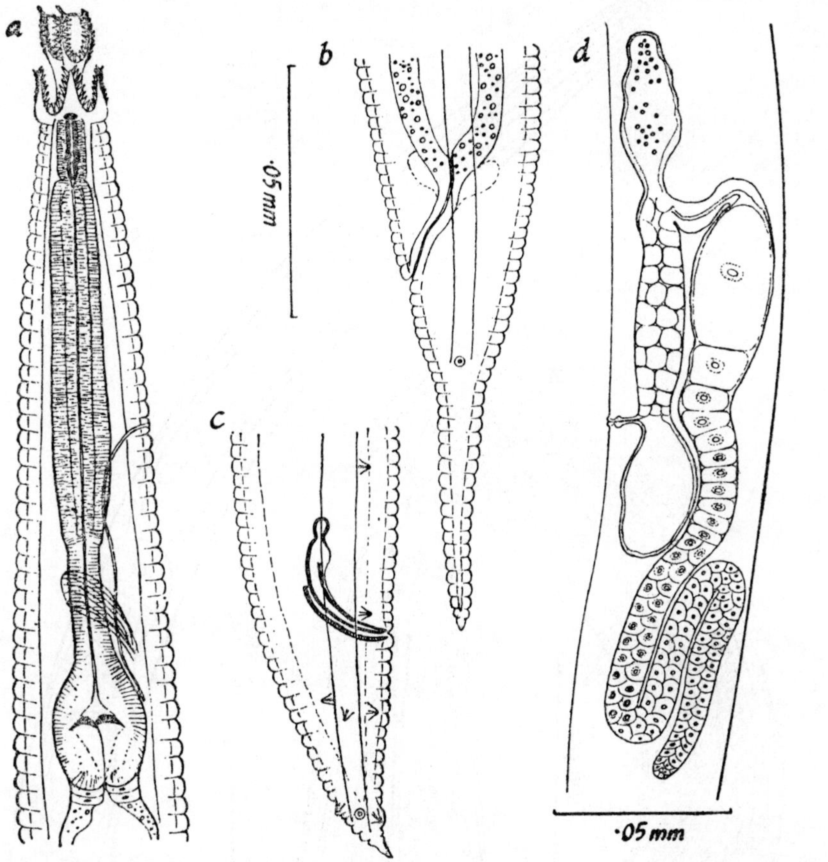

Fig. 12.6. *Acrobeles ciliatus* von Linstow: **a,** oesophageal region; **b,** female tail; **c,** male tail; **d,** female genital tract. (From T. Goodey, 1963).

Order ARAEOLAIMIDA

This order (see p. 20), comprising very active nematodes with adhesive caudal glands and spiral or loop-like amphids, is divided into two superfamilies, PLECTOIDEA and AXONOLAIMOIDEA. The former, especially members of the family PLECTIDAE, are by far the commoner in soil and about plant roots.

Plectus Bastian (Fig. 12.7)

Stoma tubular, lined with rod-like rhabdions appearing different in ventral and lateral views. Four cephalic setae just behind anterior extremity of body. Oesophagus almost cylindrical anteriorly with nerve ring encircling it mid-way; slightly constricted posteriorly before joining the swollen terminal bulb which has conspicuous, triradiate, denticulate, overlapping valve plates; end of terminal bulb projecting into anterior end of intestine which is slightly invaginated as a result. Excretory duct twisted, often conspicuous (generic name derived from

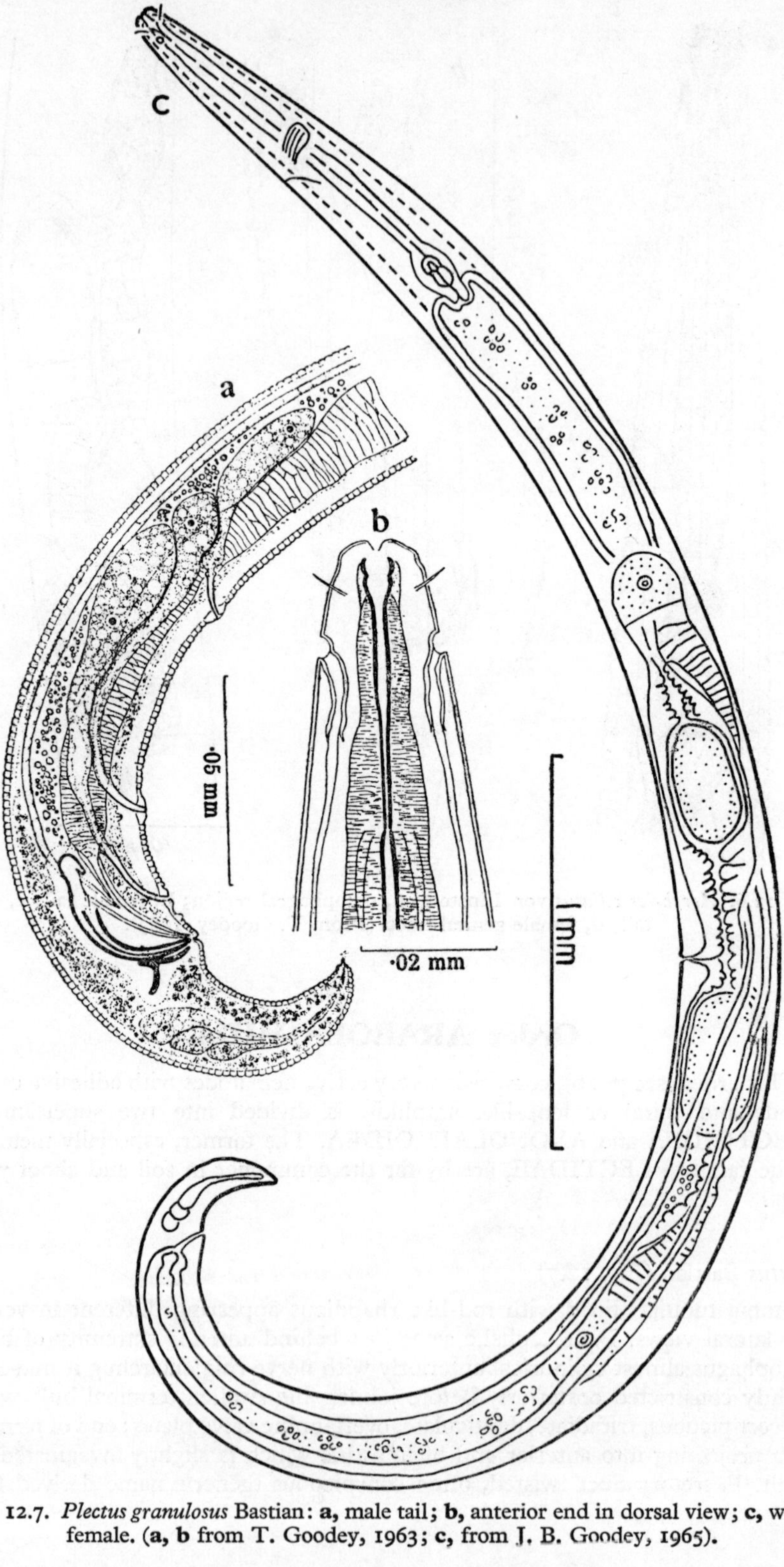

Fig. 12.7. *Plectus granulosus* Bastian: **a,** male tail; **b,** anterior end in dorsal view; **c,** whole female. (**a, b** from T. Goodey, 1963; **c,** from J. B. Goodey, 1965).

this twisted duct). Tail of both sexes conoid, pointed and ventrally curved, containing three gland cells with tube-like terminal duct forming an adhesive organ. Female has paired reflexed genital branches and median vulva. Male has paired opposed testes attached to common vas deferens, paired asymmetrical spicules and characteristic gubernaculum with dorsal projection. Median, tubular ventral supplements usually present on male tail. For review of genus with key, see Maggenti (1961b).

Plectus species are common in soil with plenty of organic matter; they occur also in leaf mould and litter, amongst mosses and lichens and in fresh water. Their rapid whip-like movement and tendency for the tail to adhere to the substrate are characteristic habits. They can withstand periods of desiccation and become mobile on re-moistening. Microbivorous feeders (Wood, 1973).

Wilsonema Cobb (Fig. 12.8)

This Plectid genus is recognized by the transparent wing-like expansions of cuticle extending dorsally and ventrally at the anterior end of the body. These are in reality backward extensions from the middle member of a group of three cirri on the outer edges of the dorsal and ventral lips. Found occasionally in soil and associated with lichens. Wood (1973) found that *W. otophorum* fed on bacteria and multiplied in bacterial cultures.

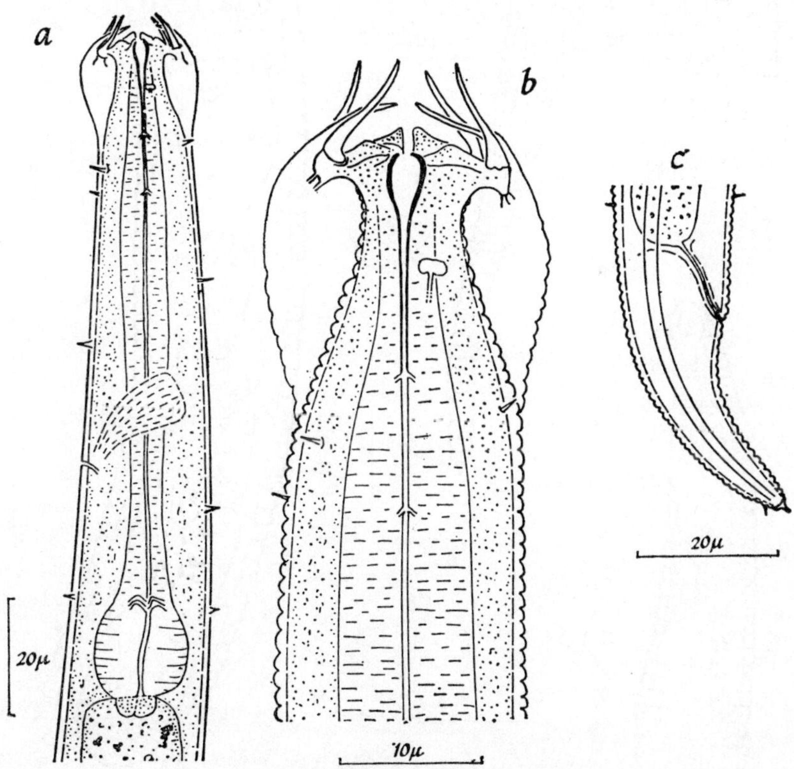

Fig. 12.8. *Wilsonema auriculatum* (Bütschli) Cobb: **a,** oesophageal region; **b,** anterior end; **c,** female tail. (From T. Goodey, 1963).

Aphanolaimus de Man (Fig. 12.9)

An example of the family Camacolaimidae of the Plectoidea. Amphids very large and near circular, occupying practically the whole cephalic region in lateral view. Head with four fine setae. Cuticle coarsely annulated, refractive. Posterior oesophageal bulb absent. Andrássy (1968) gave a key to species.

Most species are fresh- or brackish-water forms but *A. attentus* de Man occurs in wet soils, e.g., in meadows or woodland. The example illustrated was from a Welsh mountain pasture with *Vaccinium,* at over 610 m.

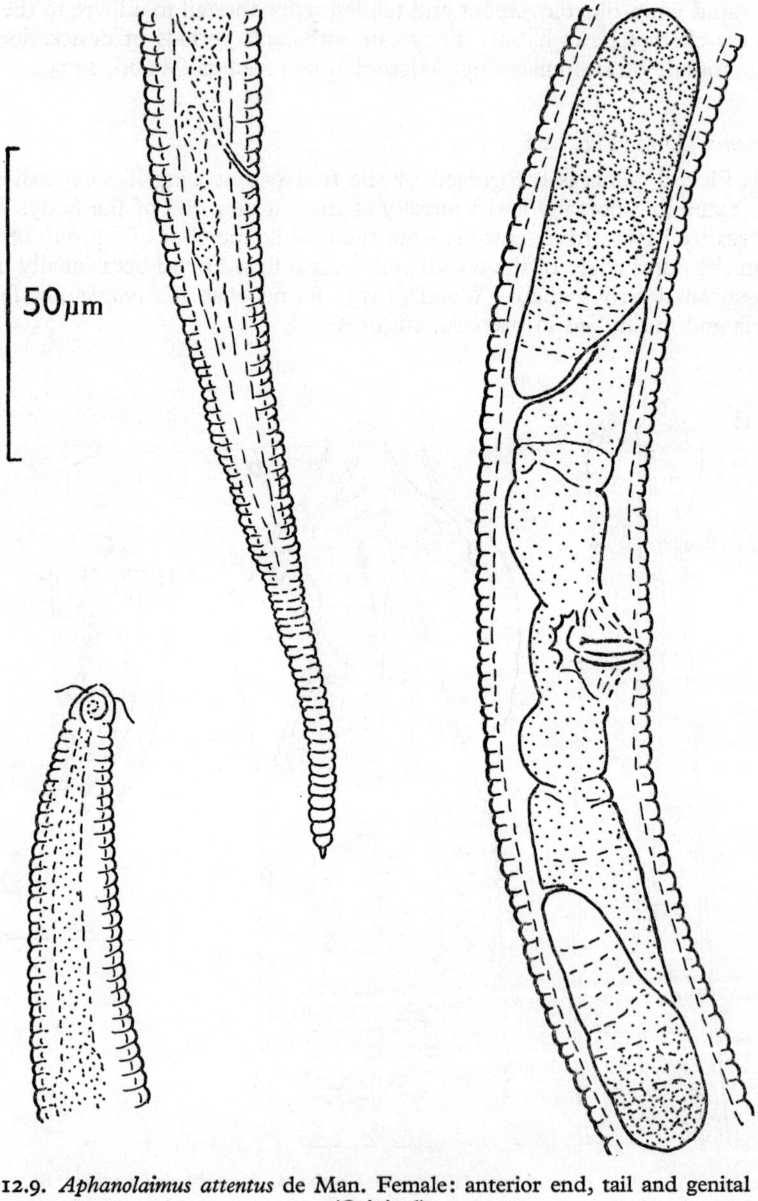

Fig. 12.9. *Aphanolaimus attentus* de Man. Female: anterior end, tail and genital tracts. (Original).

Order MONHYSTERIDA

Monhystera Bastian (Fig. 12.10)

Small to medium-sized (< 1 mm to several mm long); cuticle with fine bristles or setae. Lips indistinct, continuous with body; lip region bearing setae. Amphids usually circular. Stoma small, usually unarmed; oesophagus cylindrical. Ovary usually single, anterior, outstretched; testis always so. Tail tapered with caudal glands and terminal duct.

A large genus; many species fresh-water or marine, some common in soil. Recognizable by their very active wriggling movements. Feeding habits little known; Wood (1973) cultured *Monhystera* sp. in mixed bacterial cultures.

Order ENOPLIDA

This order contains many marine species but the families TRIPYLIDAE and ALAIMIDAE occur in fresh water and moist soils.

Family TRIPYLIDAE, genus *Tripyla* Bastian (Fig. 12.11)

Lip region with three large rounded lips bearing bristle-like papillae. Amphids inconspicuous. Stoma a narrow tube with tooth in dorsal wall. Oesophagus cylindrical. Genital tract didelphic with opposed reflexed gonads. Tail tapered with caudal glands and terminal duct.

Some species are fairly common in lighter loam soils. Predators on small nematodes, rotifers, etc. which may be seen in the gut contents.

Family ALAIMIDAE

Stoma vestigial, without spear, tooth or cuticular support. Oesophagus dorylaimoid. Excretory pore present (absent in Dorylaimoidea). Males with single testis, straight or arcuate spicules and without gubernaculum. Two genera, *Alaimus* with minute pore-like amphid apertures and *Amphidelus* with large crescentic slit-like amphid apertures; the former is by far the commoner in soil. Alaimids were formerly placed in Dorylaimida but were transferred to Enoplida by Gerlach & Riemann (1974).

Alaimus de Man (Fig. 12.12)

Medium sized (1–3 mm) or occasionally large (4 mm), mostly slender nematodes with rounded to conoid lip region continuous with body contour, and pointed tail in both sexes. Stoma inconspicuous, amphid apertures minute and pore-like, oesophagus enlarged in posterior third or less. Genital tract usually lacking anterior branch (opisthomonodelphic). Siddiqi & Husain (1967) gave a key to species.

These superficially almost featureless nematodes are very common members of the soil fauna; their feeding habits are unknown but the small and unspecialized stoma suggests fluid or particulate food. Lack of apparent economic significance and obscurity of structure have tended to discourage their study.

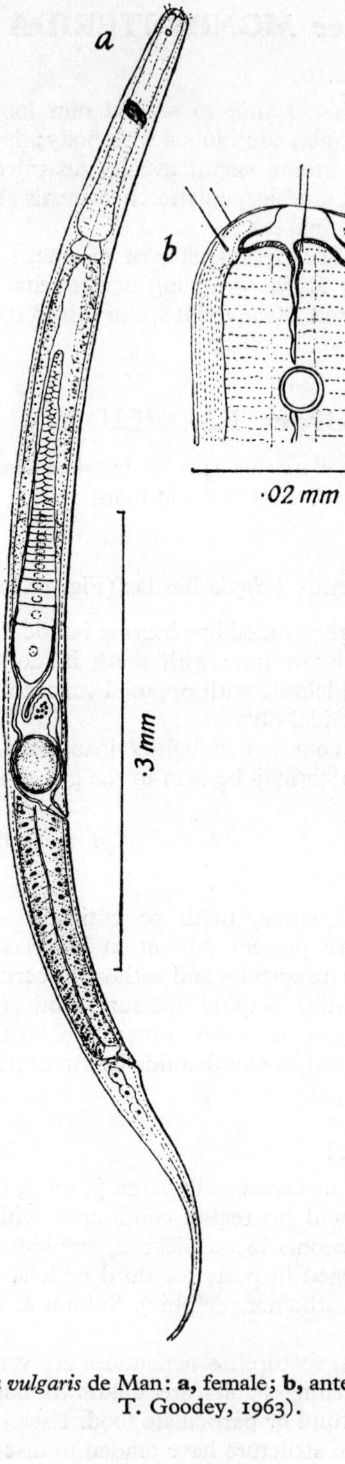

Fig. 12.10. *Monhystera vulgaris* de Man: **a,** female; **b,** anterior end in lateral view. (From T. Goodey, 1963).

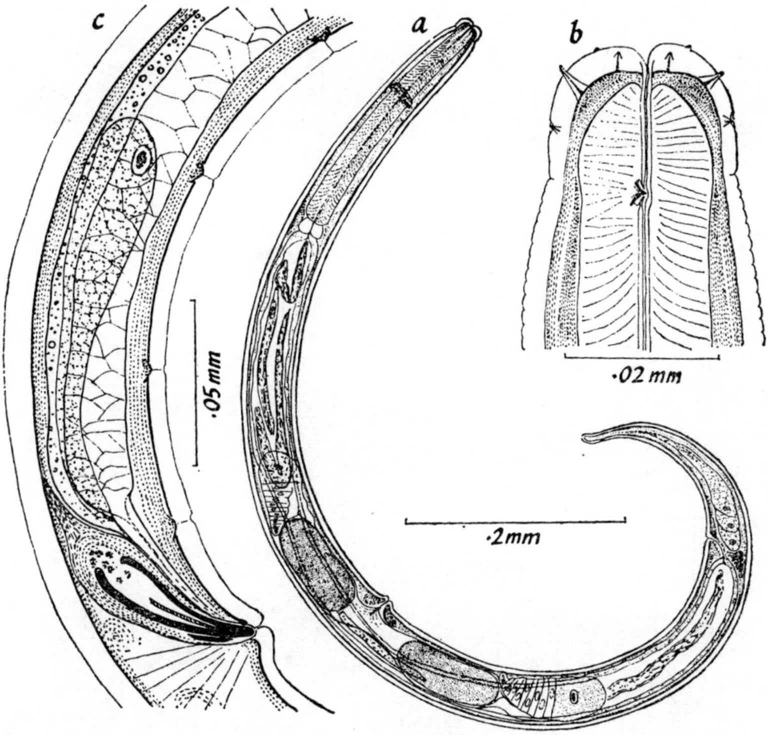

Fig. 12.11. *Tripyla affinis* de Man: **a,** female (note small nematodes in intestine); **b,** anterior end in lateral view; **c,** posterior part of male body. (From T. Goodey, 1963).

Order DORYLAIMIDA

For higher classification of this large Order, see p. 21, also Ferris (1971). Dorylaimid nematodes are abundant and conspicuous members of the soil-nematode fauna. They are generally larger (1·5 mm upwards) and more robust-looking than most Tylenchid and Rhabditid groups; most Dorylaimids are recognizable by their usually thick, smooth, refractive cuticle, a stoma armed with mural tooth or teeth or, in the largest group, the Dorylaimoidea (p. 21), an axial spear with oblique opening, and a cylindrical oesophagus. The latter is usually in two parts, an anterior more slender and posterior wider and more muscular. Many Dorylaimids have conoid or rounded tails surrounded by thicker cuticle than that on the rest of the body, but in some the tail is abruptly tapered to an acute or filiform terminus. The female genital tract is typically didelphic with opposed branches and reflexed ovaries, but one branch, usually the anterior, may be reduced or vestigial (pseudomonodelphic condition) or absent (monodelphic).

Two specialized groups of Dorylaimida are ectoparasites on the roots of higher plants and contain many economically important species (see Chapters 10 & 11), but little is known of the feeding habits of the majority. Some others

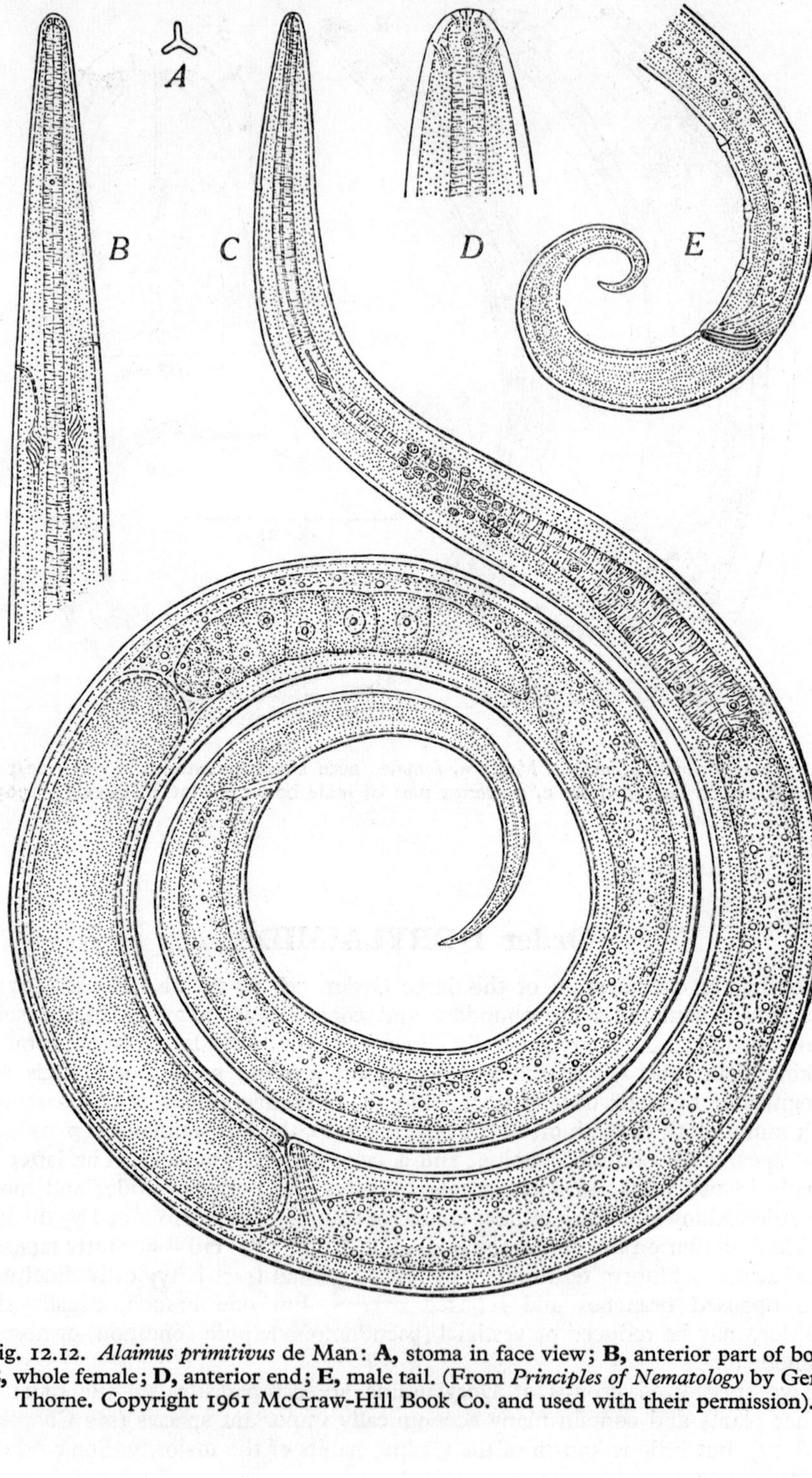

Fig. 12.12. *Alaimus primitivus* de Man: **A,** stoma in face view; **B,** anterior part of body; **C,** whole female; **D,** anterior end; **E,** male tail. (From *Principles of Nematology* by Gerald Thorne. Copyright 1961 McGraw-Hill Book Co. and used with their permission).

may feed on plant roots; some are predatory on other nematodes or other soil animals; some feed on algae, and evidence of this may be seen in green intestinal contents; some may feed on moss rhizoids or protonemata; and some on soil fungi.

Knowledge of the taxonomy of soil Dorylaimids has outstripped other aspects of their study; the biology and ecology of the commoner groups deserve more attention.

Superfamily DORYLAIMOIDEA, family DORYLAIMIDAE

Stoma with a stout axial spear (odontostyle) with oblique dorsal opening (like that of a hypodermic needle) just behind the tip, attached posteriorly to a less well defined spear extension (odontophore) which, unlike the spear, is not shed at moulting. The four juvenile stages each show two spears, the functional odontostyle and a replacement located in the subventral wall of the oesophagus. At moulting the replacement moves forward through the odontophore as the functional spear is shed, and another replacement is formed in the oesophageal wall; occasionally three spears may be seen briefly in a moulting juvenile. Oesophagus cylindrical; in two parts, a more slender anterior and a wider muscular portion containing gland cells.

Three common genera of this large and diverse group are briefly described and illustrated as examples. The type genus *Dorylaimus* Dujardin (*sensu stricto*) comprises mostly fresh-water forms.

Eudorylaimus Andrássy (Fig. 12.13)

Subfamily Dorylaiminae, i.e., having relatively short dorylaimoid spear with simple extension. Guide ring well-developed. Body medium-sized mostly 1–3 mm long. Lip region slightly offset. Tails of both sexes similar, short, conical or bluntly rounded. Genital tract usually didelphic, sometimes monodelphic or pseudomonodelphic. The genus was reviewed by Tjepkema, Ferris & Ferris (1971) who divided it into six species groups.

A large genus, very common in soil and frequently found around roots of cultivated plants. Feeding habits little known but apparently polyphagous; species have been cultured with blue-green algae, green algae, protozoa and fungi. Wood (1973) showed that *Eudorylaimus* sp. fed on algae and moss protonemata. *E. carteri* has been seen feeding on mite eggs (Goodey, 1963).

Aporcelaimellus Heyns (Fig. 12.14)

Dorylaiminae. Body medium-sized (1–3 mm), rather robust (a = 20–30). Cuticle fairly thick, especially towards extremities. Lip region clearly offset. Spear short, stout, with large aperture occupying from slightly less to more than half the spear length. Guide sheath lacking a sclerotized ring. Tails of both sexes similar, short; bluntly rounded, conoid or slightly digitate. Genital tract didelphic. Baqri & Jairajpuri (1968) gave a key to species and Tjepkema, Ferris & Ferris (1971) reviewed the genus.

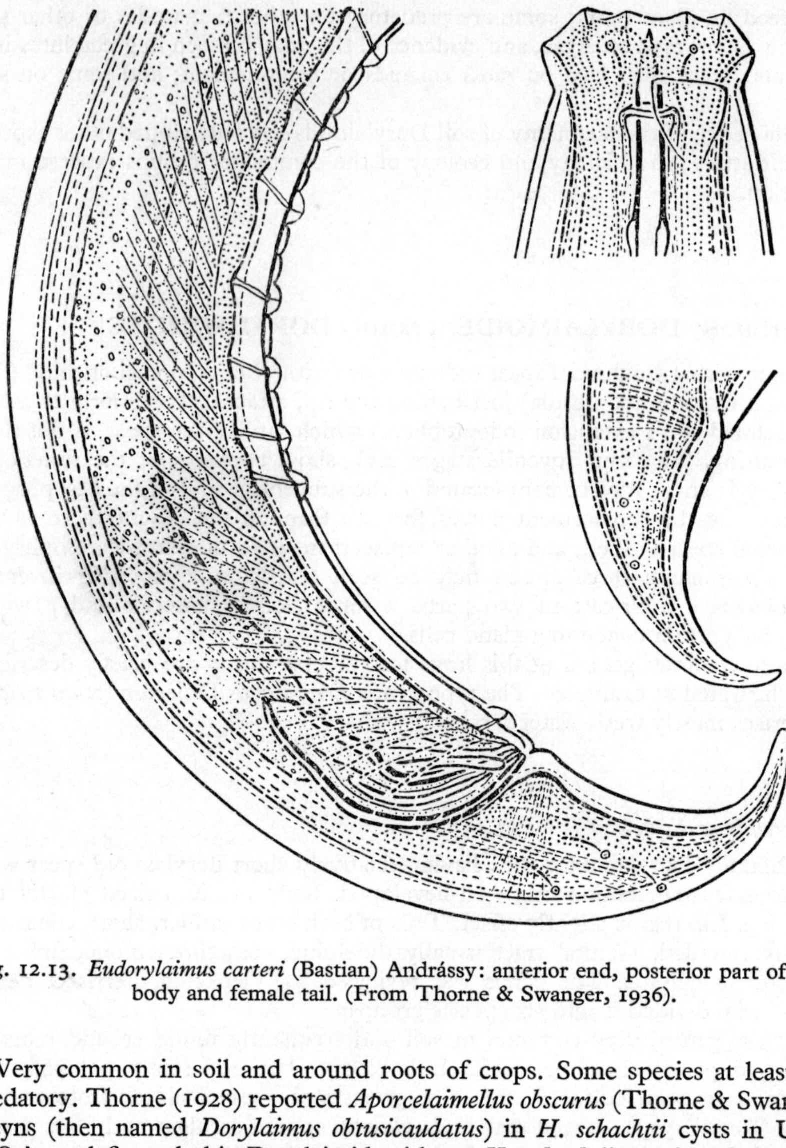

Fig. 12.13. *Eudorylaimus carteri* (Bastian) Andrássy: anterior end, posterior part of male body and female tail. (From Thorne & Swanger, 1936).

Very common in soil and around roots of crops. Some species at least are predatory. Thorne (1928) reported *Aporcelaimellus obscurus* (Thorne & Swanger) Heyns (then named *Dorylaimus obtusicaudatus*) in *H. schachtii* cysts in Utah, U.S.A. and figured this Dorylaimid with an *H. schachtii* egg impaled on its spear. *Aporcelaimellus* sp. (mostly juveniles) were found in 2–5 per cent of cysts of *Globodera rostochiensis* in an English garden soil (1–7 nematodes per cyst). The invaded cysts usually contained few eggs, occasionally they were empty. Wood (1973) found that *Aporcelaimellus* sp. fed and multiplied on mixed algal cultures and moss protonemata; he observed feeding also on Enchytraeids and rotifers.

Pungentus Thorne & Swanger (Fig. 12.15)

Dorylaiminae. Similar in size to the preceding genera but characterized by the presence of four sclerotized platelets surrounding the oral opening (two

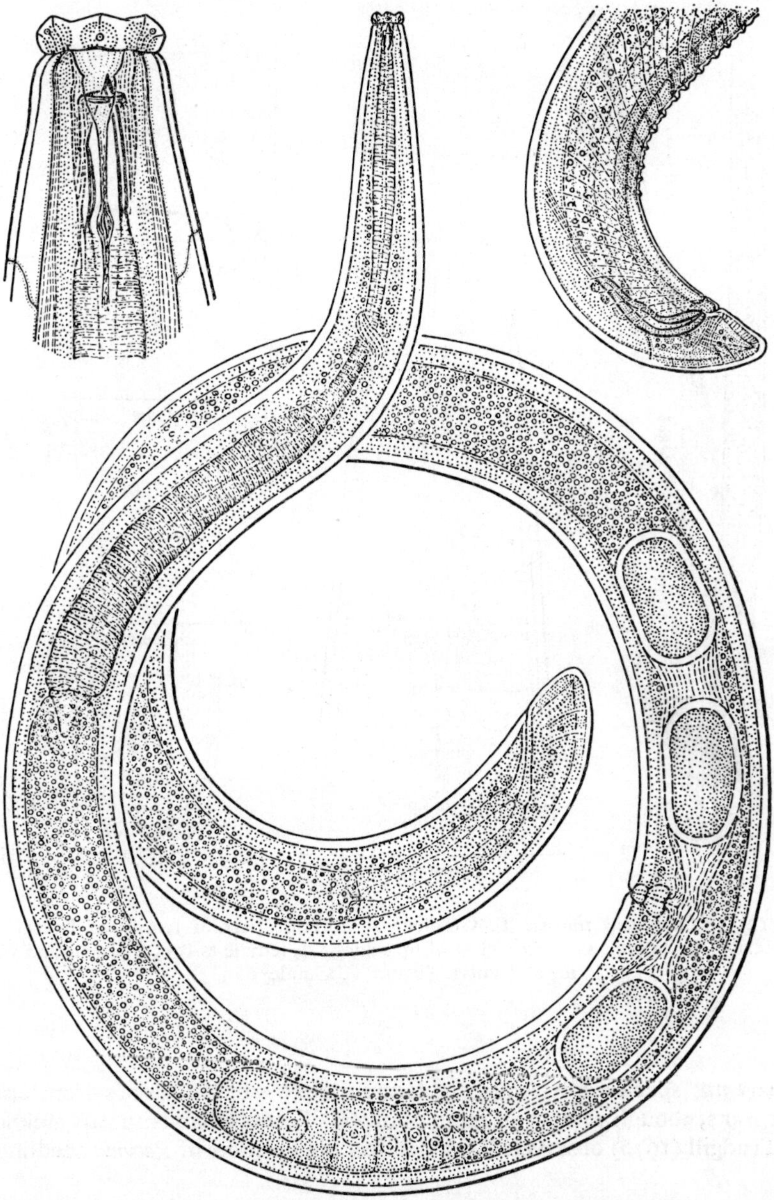

Fig. 12.14. *Aporcelaimellus obscurus* (Thorne & Swanger) Heyns: anterior end, whole female and male tail. (From *Principles of Nematology* by Gerald Thorne. Copyright 1961 McGraw-Hill Book Co. and used with their permission).

usually seen in lateral profile). Lip region offset. Spear typically relatively long and slightly curved (convex ventrally), sometimes short and straight. Guide ring usually appears double. Junction of spear extension with oesophageal lumen surrounded by conspicuous muscle sheath. Tails of both sexes short, conoid or rounded. Genital tract didelphic or with posterior branch only (opisthomonodelphic). Andrássy (1963) gave a key to species.

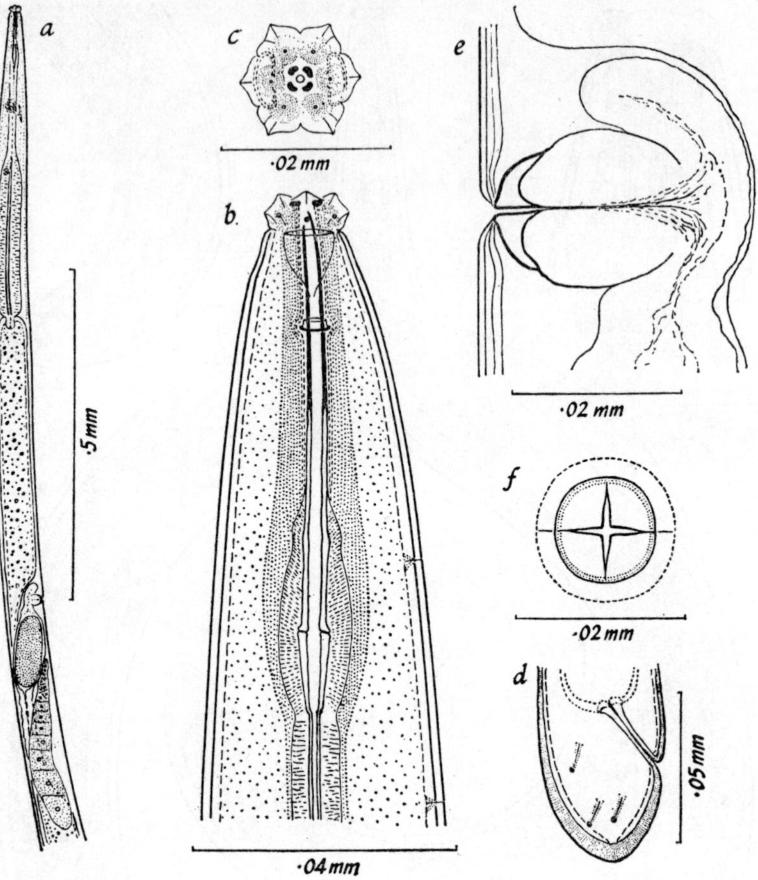

Fig. 12.15. *Pungentus thornei* T. Goodey: **a,** anterior part of female in lateral view; **b,** anterior end lateral; **c,** *en face* view of lip region; **d,** female tail; **e, f,** lateral and ventral views of vulva. (From T. Goodey, 1963).

Pungentus species are frequently found among soil nematodes from around plant roots, though seldom in large numbers. Feeding habits mostly unknown, but Trudgill (1976) observed *P. thornei* feeding on roots of *Petunia* seedlings.

Other Dorylaimoid families whose members may be found in soil samples include TYLENCHOLAIMIDAE, in which the genus *Tylencholaimus* has an odontophore (spear extension) with basal knobs, giving it an appearance similar to the spear of Tylenchids; and NYGOLAIMIDAE (Ref. Heyns, 1968), which differ from other Dorylaimoids in having an eversible pharynx with a mural tooth in its subventral wall instead of a typical Dorylaim spear. Loof & Jairajpuri (1968) gave a key to the species of *Tylencholaimus*. According to Wood (1973) *Tylencholaimus* feeds on fungi and *Nygolaimus* is predatory, feeding on small oligochaets, nematodes and nematode eggs.

Three other superfamilies of Dorylaimida deserve mention:

Superfamily LEPTONCHOIDEA

Like the Tylencholaimidae (above) this group contains genera (e.g., *Tylencholaimellus*) whose spear extensions have conspicuous basal knobs or flanges, sometimes causing them to be mistaken at low magnification for Tylenchids. Spears vary in other genera, being sometimes very slender. However Leptonchoids are characterized by an oesophagus with long slender anterior portion and short ovoid posterior expansion.

Genera common in soil include *Leptonchus, Tylencholaimellus* and *Dorylaimoides*. See Ferris (1971) for discussion and classification of the group.

Superfamily BELONDIROIDEA

This group is at once distinguished from other Dorylaimida by the thick refractive sheath of muscles surrounding the basal part of the oesophagus; the sheath muscles are often spirally arranged but sometimes more or less parallel to the long axis. Anterior part of body often tapers to a narrow lip region so that Belondiroids appear to have pointed heads; spear usually small. For further information, see Ferris (1971). Nair (1973a, b) and Nair & Coomans (1973, 1974) have reviewed the common genus *Axonchium*.

Order MONONCHIDA

(see page 22)

Suborder MONONCHINA

Medium to large nematodes with large, strongly sclerotized, barrel-shaped stoma bearing a large, fixed dorsal tooth which may or may not be opposed by subventral teeth or denticles. Oesophagus more or less cylindrical with thickened lumen. Excretory pore degenerate. Males with ventral supplements, paired spicules, well developed gubernaculum and prominent lateral accessory pieces. Tail sometimes blunt and rounded, more usually tapering and curved when killed by heat. In most genera caudal glands are present and the tail terminates in an adhesive organ or spinneret.

These common predacious nematodes are found in most soils and occur also in fresh water. They are conspicuous members of the soil fauna because of their vigorous movements, usually robust build and large well-armed stoma. They often have a tendency to float on the surface of water until pushed under and those with a spinneret tend to adhere to the substrate by the tail tip. They multiply particularly under conditions that favour build-up of Rhabditids and other bacterial feeders, and may then play a part in the natural control of plant nematodes. (See also Chapter 13).

In contrast to the Alaimoids (see above), the Mononchs have been popular subjects for taxonomic study because of the convenience of their stomal structures for classification. Thus, since 1959, the group has been elevated from single genus (*Mononchus* Bastian) to suborder. Cobb (1917) first showed that the species could be grouped according to the number, direction and arrangement of stomal teeth or denticles, and erected 6 subgenera. There the situation rested until Andrássy (1959a) gave Cobb's subgenera generic status and made five new

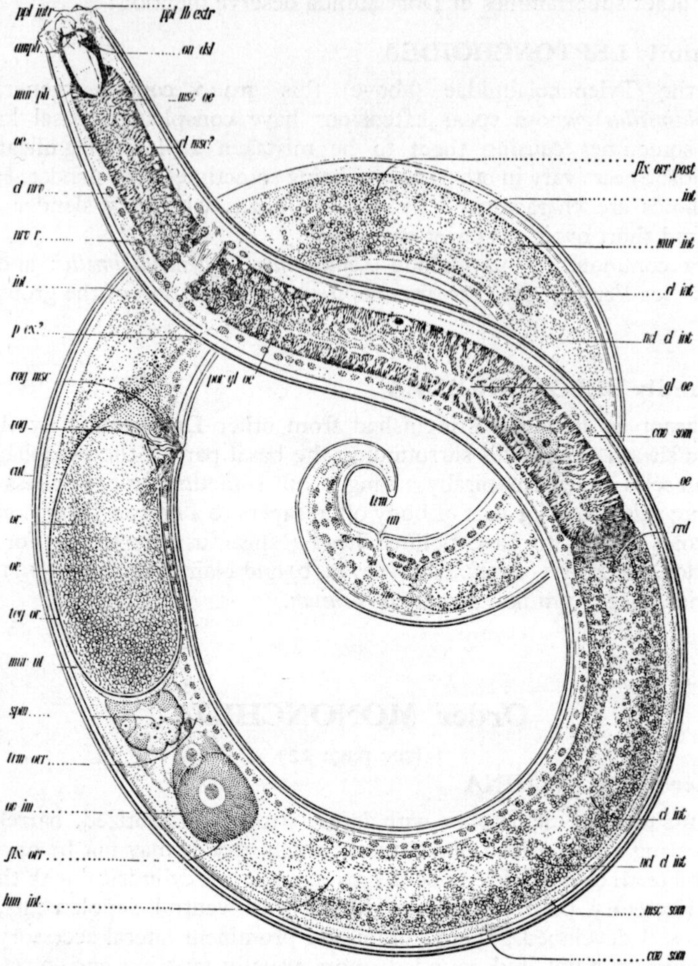

Fig. 12.16. *Mononchus papillatus* Bastian: female. (From Cobb, 1917).

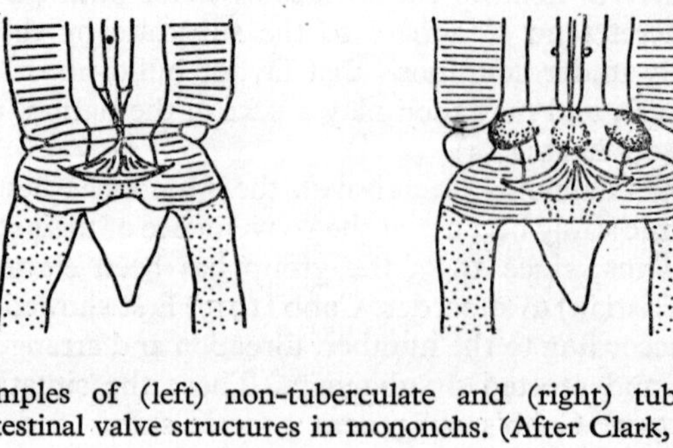

Fig. 12.17. Examples of (left) non-tuberculate and (right) tuberculate oesophago-intestinal valve structures in mononchs. (After Clark, 1960).

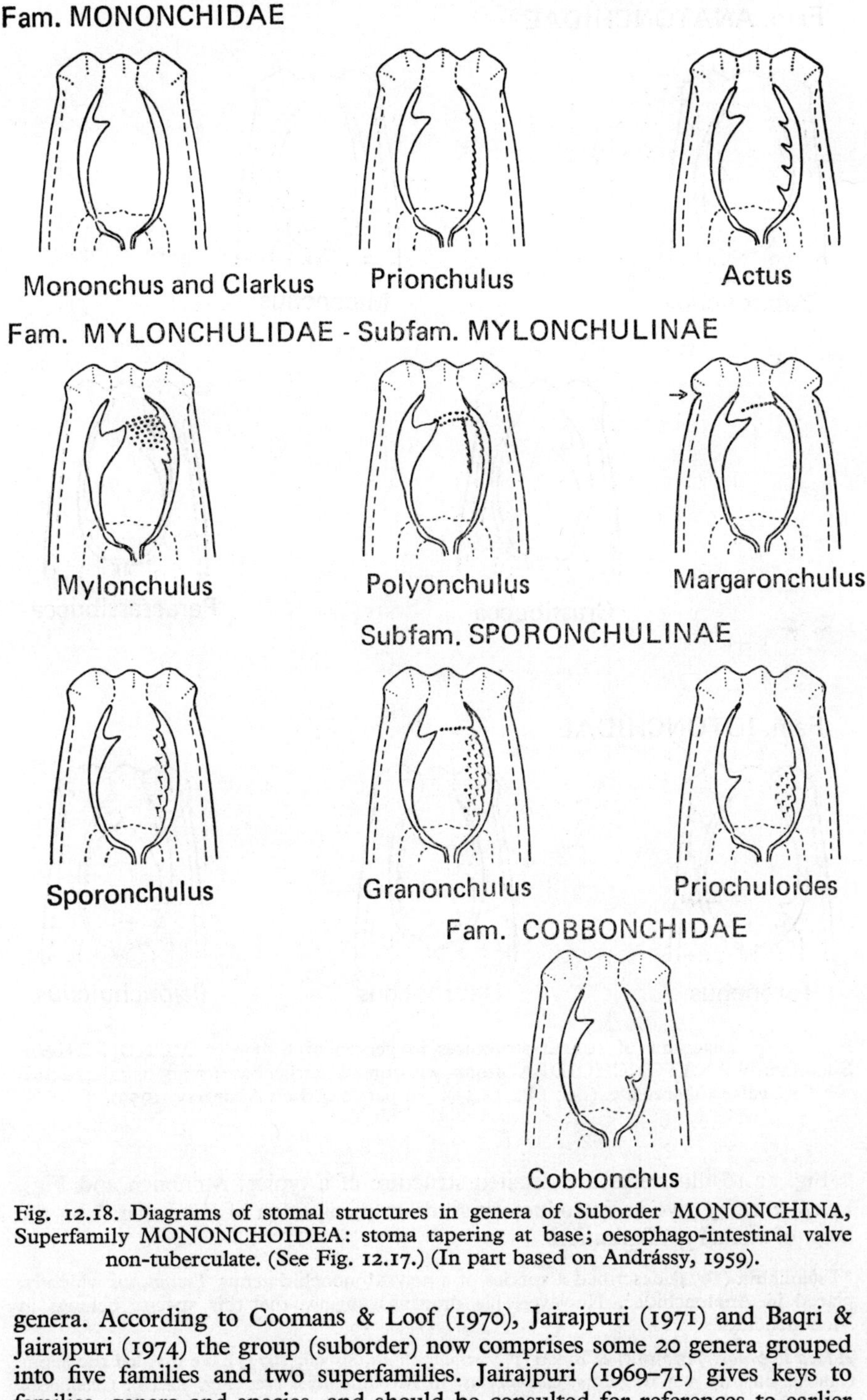

Fig. 12.18. Diagrams of stomal structures in genera of Suborder MONONCHINA, Superfamily MONONCHOIDEA: stoma tapering at base; oesophago-intestinal valve non-tuberculate. (See Fig. 12.17.) (In part based on Andrássy, 1959).

genera. According to Coomans & Loof (1970), Jairajpuri (1971) and Baqri & Jairajpuri (1974) the group (suborder) now comprises some 20 genera grouped into five families and two superfamilies. Jairajpuri (1969–71) gives keys to families, genera and species, and should be consulted for references to earlier important studies of the Mononchs by Clark and Mulvey.

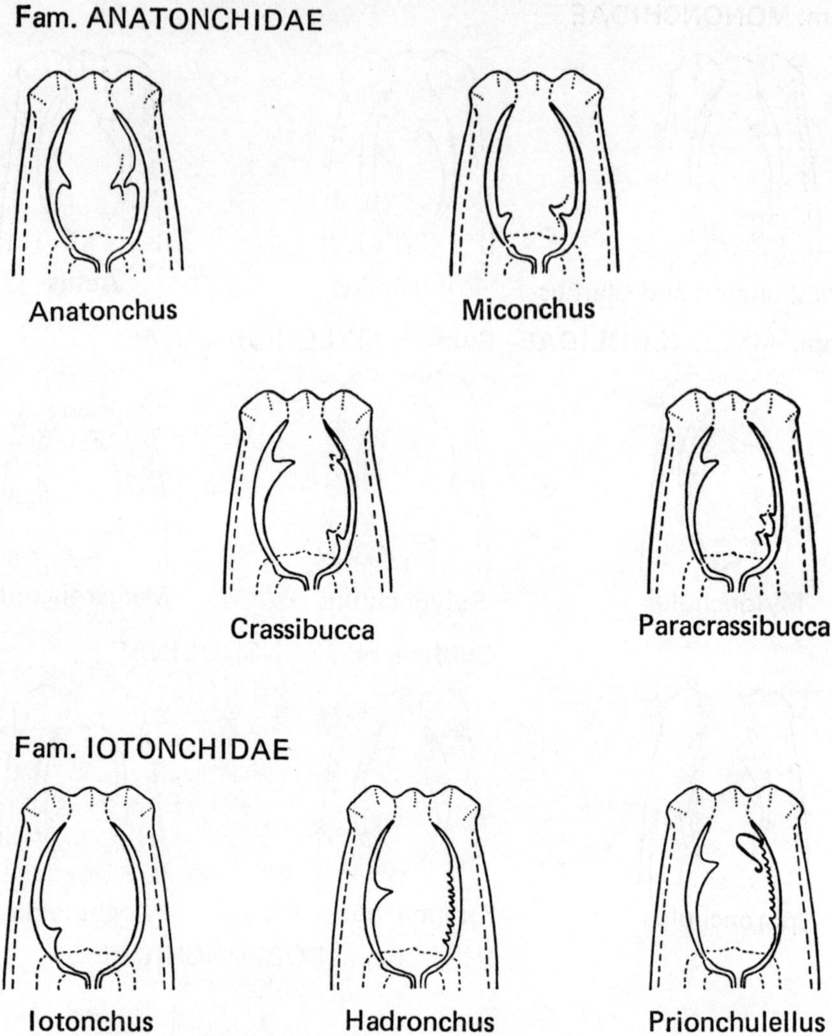

Fig. 12.19. Diagrams of stomal structures in genera of Suborder MONONCHINA, Superfamily ANATONCHOIDEA: stoma with broad flattish base; oesophago-intestinal valve tuberculate. (See Fig. 12.17). (In part based on Andrássy, 1959).

Fig. 12.16 illustrates the detailed structure of a typical Mononch and Figs. 12.18 & 12.19 give an introductory guide to recognition of the genera* on the basis of stomal configuration.

*Tsalolikhin (1974) described a species of a new Mononchid genus *Tectonchus* which he placed in Anatonchidae. However, his drawings suggest that this species belongs in Mononchoidea.

Added in proof: Jairajpuri & Khan (*Nematologica* **23,** 89–96, 1977) have divided the genus *Clarkus* making a new genus, *Coomansus,* in Mononchidae (see Fig. 12.18). *Coomansus* differs from *Mononchus* in having a short tail and lacking a spinneret; *Clarkus* (*sensu stricto*) differs from both in that the stoma has a non-denticulate ventral ridge.

Part **II**
Principles of Control

CHAPTER 13

Cultural and Biological Control Methods

E. B. BROWN

A.D.A.S., Woodthorne, Wolverhampton

Cultural control

Progress is being made towards the control of nematodes by chemical means (Chapter 14) but cultural methods still play a valuable part. The degree of control exercised by these methods can vary widely. Cultural methods are attempts to adapt husbandry practices so as to minimize crop losses due to nematodes. Before deciding on the best means of combating a nematode, its life history, population dynamics, host range, and the efficiency and susceptibility of its main hosts must all be considered. Usually the operation will be a normal one carried out at little or no extra expense: occasionally it may be an operation aimed specifically against the nematode, when the whole cost must be considered.

With chemical control, the actual species is not often of great importance, but before much progress can be made in suggesting suitable cultural measures it is essential to know which species of nematode is present. Advances in taxonomy are often followed sooner or later by advances in knowledge of cultural control. Occasionally, as happened with the root-knot nematodes (*Meloidogyne* spp.), knowledge of host-parasite relationships has led to advances in taxonomy (Chitwood, 1949). There are known to be several races within some species (e.g., *Ditylenchus dipsaci*) and new races or pathotypes may be encountered at any time, leading to variation in cultural advice.

Cultural control can be divided into three main categories.

(1) *Crop rotation*—the attempt to keep a balance on individual fields between the population of the nematode and the frequency of growing the host crop.

(2) *Prevention of spread*—the means of minimizing the spread of a nematode from field to field, from farm to farm, from one country to another or from onc variety to another. This is more important with nematodes which multiply rapidly.

(3) *Selection of healthy propagating material*—with strict attention to hygiene—chiefly horticultural crops.

In addition, the *influence of manuring* on the general problem of nematode control is discussed.

CROP ROTATION

Some kind of rotational cropping has been practised in agriculture for hundreds of years. In early times it was common practice to clear a piece of land, crop it until it was exhausted of nutrients and then to abandon it for another site. Nematode attacks may have contributed to yield loss even in those days, but this can only be conjecture. In modern times rotational cropping has been recognized as a means of controlling weeds, pests and diseases. But with the development of and reliance on modern weedkillers there has been less emphasis on weeds. Even before this there was a tendency to forget pests and diseases and to grow the more remunerative crops more frequently. This encouraged the build-up of nematode pests which often are not noticed until they are causing considerable yield losses. Nusbaum & Ferris (1973) have reviewed in detail the role of cropping systems in nematode control.

The aim of rotation for nematode control is to allow sufficient interval after each host crop to allow the nematode population to return to a safe level before the next host crop. To be able to make recommendations it is essential to know the host range of the nematode, to have some idea of the efficiency and susceptibility of the various hosts, to have information on population dynamics, and knowledge of the relation between population density and yield loss. There is a danger that if one crop is precluded from a rotation because of one nematode species, another crop may be grown more frequently, resulting in the build-up of another species. For example, in some potato-growing areas of England where fields stopped growing potatoes because of potato cyst nematodes (*Globodera rostochiensis* and/or *G. pallida*), this resulted in *Heterodera cruciferae* becoming important due to the greater frequency of brassica crops. Occasionally, however, a nematode may have such a wide host range or be so persistent that rotational methods are virtually useless.

The less persistent nematodes in the soil often have a wide host range and, conversely, the more persistent ones may have only one or few hosts. For example, the potato cyst nematodes, with potatoes as the only agricultural host, have a relatively slow rate of decline. Grainger (1951) found that in Scotland the annual rate of population decline was about 30 per cent whereas in England it is generally considered to be 40–50 per cent. This means that the decline rate may vary according to latitude or conditions, and that in Scotland a longer interval between potato crops may be required than in England, to prevent populations from increasing. But, in England, there is evidence that the rate of decline varies with area or soil type. A four- to five-year rotation may be adequate on peats and medium soils, but on light silts a longer period, up to seven years, may be necessary (Anon, 1977). The reason for this is not fully understood. In some areas there may have been selection for longevity or the difference may be due to biological differences between the two species, *G. rostochiensis* and *G. pallida*.

In England and Wales the Agricultural Development and Advisory Service and some private consultants operate a soil examination service to inform farmers which fields are infested and, ideally, to help them to plan a good rotation for potatoes. These services can also be used to inform farmers of the likely yield losses due to potato cyst nematodes on individual fields. According to the degree of infestation field samples are placed by A.D.A.S. advisers in three main categories:

(i) samples in which it has been possible to detect few if any cysts, indicating that potatoes can be grown with little risk of yield loss;
(ii) samples with a moderate population which will probably result in some yield loss;
(iii) samples with a high population, on which yield loss may be severe.

It is recognized, however, that failures may occur at low cyst densities when other factors are involved. (See Chapter 3).

As a result of such advice there may be a tendency for farmers, by avoiding fields with high populations, to grow potatoes more frequently on fields with low populations. This could lead to a gradual increase in the average population level on the farm, so that it becomes progressively more difficult to find suitable fields. However, the introduction of potato cultivars resistant to *G. rostochiensis* and of effective nematicides may lessen the importance of rotations in future.

Cereal cyst nematode (*Heterodera avenae*) has a wide host range in the family Gramineae and an annual decline rate of over 60 per cent (Hesling, 1958a). The various cereals differ widely in their susceptibility and host-efficiency, and grasses can usually be ignored as hosts. Such information enables cereals to be grown intensively, even on the light soils where *H. avenae* most commonly occurs (Southey, 1956). The relatively rapid population decline enables a high population to be brought down to a safe level quickly by growing crops that are either non-hosts or very poor hosts (Stone, 1960; Hesling, 1958b). Wheat and barley are less efficient hosts when sown in autumn than when spring-sown. Time of drilling can, therefore, be important when planning a sequence of cropping.

In many areas populations of *H. avenae* tend to stabilize at a low density (less than 10 eggs/g), even under the most efficient hosts, oats and barley. It appears that drenching the soil with formalin (Williams, 1969) removes an inhibiting factor, possibly biological in nature, allowing populations to increase. Kerry (1974, 1975) has suggested that this factor is an *Entomophthora*-type fungus. Another possible factor may be that some light sandy soils do not retain enough moisture in dry years to allow the nematodes to invade the roots or the females to be fertilized.

Heterodera goettingiana has a decline rate similar to that of *Globodera* spp. (potato cyst nematodes) but its main host, pea, appears to be both efficient and very susceptible to quite low populations. In England, pea failures occur on fields which have not been cropped frequently with host crops. There is evidence (Brown, 1958) that these populations are of long standing and have been maintained by infrequent crops of field beans at a level high enough to cause pea failures. On such fields rotation appears to be of little value.

With cyst-nematode species having an average rate of multiplication of ten times and an annual decline rate of about 50 per cent, growing an efficient host not more than once every four or five years is usually regarded as sufficient to prevent populations increasing rapidly.

Little progress was made with the cultural control of *Meloidogyne* species until 1949 when Chitwood described several new species, all of which had previously been included in the one species, *Heterodera marioni*, with some 1300 recorded hosts. This host list was thus rendered useless but progress has been made in compiling new ones for the more important of the species now known. (See Chapter 5).

Daulton & Nusbaum (1961) investigated the survival in the soil of three populations of *M. javanica* and one of *M. hapla*. They showed that many *M. hapla* eggs were still alive after 250 days, but, although there were differences among the three populations of *M. javanica*, egg viability had fallen considerably by 100 days and few eggs were alive after 140 days. Thorne (1961) gave details of an area in California where a mixed population of *M. incognita* and *M. arenaria* occurred. After three years, lucerne had to be ploughed because of damage by *M. arenaria*. This was followed by cotton for two years, when the other species multiplied, and lucerne was then sown again.

Several species of *Meloidogyne* survive the winter only in glasshouses in

Great Britain but, as crop rotation is not usually practicable under glass, their longevity in the absence of a host is of little importance. *M. hapla* occasionally occurs as a pest outdoors in Great Britain (Brown, 1955) and experience here and in The Netherlands (s'Jacob, 1960) has shown that after a one-year cereal break the population is depressed to a low level.

These examples show that *Meloidogyne* species are not persistent and decline rapidly under fallow or non-host crops. However, most species probably have wide host ranges and are maintained at low level in the field on weed hosts.

With *Ditylenchus dipsaci,* the part that rotation plays in control is less important than with many other species. Seinhorst (1956) has shown that in The Netherlands *D. dipsaci* is a persistent menace on the heavier soils even when host crops are not grown frequently, whereas on light soils it only becomes important when there has been overcropping with host plants. Similarly in Great Britain this nematode is largely confined to the heavier soils, where severe attacks occur sporadically on red clover and oats on fields which have not grown a known host for several years. Such attacks almost certainly originate mainly from the soil although seed-borne infestation may be more important on red clover than has been recognized. On light sandy soils *D. dipsaci* is rare and only becomes a problem, for example on lucerne, when it is introduced with seed and spread in the crop over a period of years (Brown, 1957).

There are several references in the literature to the interval between host crops necessary for the control of *D. dipsaci* (Lewis & Mai, 1960). This varies between two and six years and may be related to particular conditions, or sometimes may be little more than guess-work. On heavy soils it is usually impracticable to suggest a suitable interval between host crops. On uninfested heavy silt land, Cooper (1953) suggested that it is better to grow onions every year on the same land as long as it remains economic to do so. Use of nematode-free seed is advised but the nematode is likely to be introduced sooner or later, and once it is introduced it spreads at about the same rate whether or not onions are grown. Thus rapid contamination of other fields may be avoided.

Hesling & Wallace (1961a) showed that chrysanthemum eelworm (*Aphelenchoides ritzemabosi*), which has a wide host range, will not overwinter in the soil provided the land is kept weed-free. Most of the nematodes have disappeared after two months, although a few may remain for up to four months. Nothing is known about the survival in the soil of other leaf and bud infesting *Aphelenchoides* species but they probably behave similarly to *A. ritzemabosi*. Rotation can, therefore, be largely disregarded in controlling *A. ritzemabosi*.

In The Netherlands, crops such as narcissus, strawberry and lilies are damaged by low soil populations of *Pratylenchus penetrans* (Seinhorst, 1957). Unlike *D. dipsaci, P. penetrans* is common on light soils, especially those high in organic matter, and degree of persistence appears to be independent of cropping practices. Oostenbrink (1956) gave figures for the changes in populations of *P. penetrans* and *P. pratensis* under different crops. He suggested that, with *P. penetrans,* there is a direct relation between susceptibility and size of root system: the bigger the root system, the smaller the amount of damage and *vice versa*. Later (1961) he demonstrated a significant linear relationship between initial population density (log scale) and effect on the host plant. Beet and mangolds were found to suppress populations of *P. penetrans* and Oostenbrink advises that one of these crops should be grown on infested land before potatoes and red clover, which are susceptible.

Thus, rotation has a big part to play in the control of some nematodes (e.g.,

Heterodera, Globodera, Meloidogyne, Ditylenchus dipsaci on light land) but is of little importance with others either because the nematode is non-persistent (*Aphelenchoides*) or because it is extremely persistent (*D. dipsaci* on heavy land).

Importance of self-sown host crops and weed hosts. Volunteer potatoes are often abundant, particularly in the year following potatoes. Den Ouden (1967) has shown that volunteer potatoes are of little or no importance at higher densities of potato cyst nematode but may encourage increase at low densities (below about 10 eggs/g).

Weed control is very important for the elimination of soil populations of *Aphelenchoides ritzemabosi* and for reduction of *Meloidogyne* populations. The importance of weed hosts for the maintenance of *Ditylenchus dipsaci* and *D. destructor* populations on heavy land is unknown. In Great Britain *D. destructor* appears to be associated on the lighter silt soils with the presence of *Mentha arvensis,* a known host weed. Linford (1937a) observed *D. dipsaci* from *Hypochaeris radicata* feeding on soil fungi and *D. destructor* has also been reared successfully on several common soil fungi (Baker, Brown & James, 1954; Darling, Faulkner & Wallendal, 1957). Therefore these species may sometimes be able to subsist on fungi in the complete absence of host plants.

PREVENTION OF SPREAD

Nematodes do not move more than a few inches a year of their own accord and most spread is passive. They may be spread in soil or in the roots of a host plant (e.g., *Heterodera, Globodera, Meloidogyne, Pratylenchus*), or in the aerial parts of plants including the seed (*Aphelenchoides, Ditylenchus dipsaci*). They may also be spread by such agencies as rain, flood-water and wind.

With modern national and international traffic in plants and seeds the spread of plant-parasitic nematodes must be regarded as inevitable. Some will establish themselves in a new area or country if conditions are suitable and host plants are found. Several *Meloidogyne* species occurring in Great Britain probably originated in tropical countries and have been able to survive here under glass but not out of doors. Oostenbrink, s'Jacob & Kuiper (1956) have shown that in newly reclaimed polder soil it is not many years before plant-parasitic nematodes appear and cause damage.

It is impossible to prevent local spread from one field to another, from one farm to a neighbouring one or from one variety of plant to another. But there are often worthwhile steps which will help to lessen both this local spread and spread over greater distances. Many countries have introduced regulations designed to prevent the introduction of plant nematodes (Chapter 17).

Spread is more important with species that have several generations a year and a high rate of multiplication, e.g., *Ditylenchus dipsaci* and *Aphelenchoides* spp. than it is with species with a relatively slow multiplication rate, e.g., cyst nematodes. The introduction of a *Heterodera* or *Globodera* species to a field will not lead to trouble until the host crop has been grown several times. But a low infestation of *D. dipsaci* in narcissus bulbs or in lucerne seed can easily lead to serious damage during the life of that crop.

Spread in soil. Any means whereby soil is moved will result in soil-inhabiting nematodes being spread and some, e.g., *Heterodera* and *Globodera* spp., remain viable even if the soil becomes dry. These nematodes can be spread in soil adhering to farm implements, workers' boots, seed potatoes, brassica transplants, sugar-beet stecklings, black-currant bushes, etc., and how far the nematodes are moved depends solely on how far the soil is carried.

There is no practicable means of preventing spread on farm machinery to all fields on which the machinery operates, as it is impossible to avoid soil contamination, especially in wet conditions, and complete removal by cleaning is impracticable. There is more point in trying to prevent the introduction of particular species to areas which grow the host plant intensively and where the nematode is not known to occur. Such introductions are likely to occur mainly on seed potatoes or transplants. In Great Britain and most other European countries some safeguard against potato cyst nematodes is provided by regulations which require that seed potatoes for marketing can only be grown on fields on which no cysts of these species have been found in official soil samples taken before planting. (See Chapter 17).

Spread with seed. Several nematodes are known to be seed-borne, e.g., *Ditylenchus dipsaci* on onions, lucerne, red clover and teasel, *Aphelenchoides* on rice and *Callistephus,* and *Anguina* species on wheat and grasses. With infestations of *D. dipsaci* on clover and lucerne, and those of *Aphelenchoides* on *Callistephus,* seed is contaminated by nematode-infested debris from aerial parts of the plant, whereas with onion the nematodes are chiefly attached to or beneath the seed-coat (Southey, 1965).

Seed-borne infestation is an important means of spread, especially with *D. dipsaci*. In onion seed, infestations of up to 235 nematodes per g of seed have been recorded (Southey *in litt.*) but such heavy infestations are rare. The work of Seinhorst & Koert (1969) suggests that infestation of onion seed is seldom likely to cause serious damage to a crop grown from the seed. They found up to 20 *D. dipsaci* per g of seed in commercial samples in The Netherlands. In plot experiments, sowing seed with up to 200 *D. dipsaci* per g produced no field symptoms and a maximum of three per cent of harvested bulbs infested. However, as Seinhorst & Koert point out, small numbers of nematodes introduced into previously clean fields could multiply to serious proportions on subsequent crops and, since the nematodes tend to be aggregated on individual seeds, they are favourably placed for multiplication.

In lucerne, recorded infestations in Great Britain are very much lower, of the order of 10 nematodes per 1000 g of seed, probably giving rise to a few small patches per acre (Brown, 1957). Thorne (1961) quoted data from an infested field in Utah, U.S.A., where about 2000 nematodes per 1000 g were found in uncleaned seed and nearly 500 in cleaned seed. Such infestations would probably result in a very large number of nematode-damaged patches per acre, much higher than encountered so far in Great Britain. T. Goodey examined 504 *small* samples of red clover seed and found *D. dipsaci* in 32 of them (J. B. Goodey, 1949). This would have detected only the higher infestations and it is probable that many were missed because of the small sample size. Oostenbrink (1960a) gives an example of 93 nematodes extracted from 20 g of clover seed.

Efficient seed cleaning methods can considerably reduce the amount of nematode-infested debris with the seed. With *Anguina tritici* modern seed cleaning methods eliminate the galls from the seed and this species is now rare except in regions of the world with backward agricultural methods. Marshall (1960) showed that between 1918 and 1934 the percentage of gall-containing wheat samples submitted to the Official Seed Testing Station, Cambridge generally varied between 2 and 4, with 9 per cent in 1921. After that the percentage fell rapidly and between 1949 and 1957 it varied between 0 and 0·02. However, *D. dipsaci* and *Aphelenchoides* are probably not so easily eliminated from seed as *Anguina*.

Little effort has so far been made to produce nematode-free seed. Onions are often grown on one site and then moved to another for seed production; hence they have two chances of picking up *D. dipsaci* from the soil. In a certification scheme for England and Wales, administered by the Ministry of Agriculture, Fisheries and Food, red clover seed crops are inspected in July or August and if *D. dipsaci* infestation is found the seed must be fumigated by a recognized method. Ideally there are three steps which should be taken to ensure the production of 'clean' seed:

(a) only nematode-free seed should be used for growing seed crops;
(b) this should be sown on land where the nematode is not known to exist;
(c) the crop should be inspected for the presence of the nematode, the best time usually being in the early spring after growth commences.

Seed may occasionally be taken from crops which are being grown primarily for another purpose (e.g., teasel, *Callistephus*) and it may be possible then to select uninfested plants for seed collection.

Spread in bulbs and aerial parts of plants. Ditylenchus dipsaci in narcissus and tulip is spread mainly by the sale of infested bulbs. The only way to avoid introducing infested stocks is to buy from reputable sources and, if possible, to see the bulbs in advance during the growing season. Roguing of infested stocks may be worth considering, particularly if the stock is to be hot-water treated. If there are numerous slightly infested plants all over the field, roguing will be impracticable, but it may be feasible if there are only a few scattered patches. The roguing of dry bulbs is unlikely to do any good as only those obviously infested are removed.

There is likely to be spread of *D. dipsaci* within and between fields on harvesting machinery, particularly with clover and lucerne crops. After cutting an infested clover or lucerne field, it may be worth cleaning the machinery to get rid of as much plant debris as possible before entering an uninfested crop.

Spread by wind and water. Agencies such as wind, flood-water and rain-water normally give rise to local spread and little can be done to obviate this. White (1953) showed that large numbers of potato cyst nematode cysts can be windborne from fields such as those on light peat soils which blow readily. The addition of clay, partly to prevent wind erosion, helps to reduce spread by this means.

Rain or irrigation water is responsible for the localized spread of, for example, *Ditylenchus dipsaci* in bulb crops and *Aphelenchoides* in chrysanthemums. With ridge-planting of bulbs, spread is mainly along the ridge in contrast to flat planting where it usually occurs in all directions or down a slope. With the ridge method there is, therefore, less spread between varieties.

It is good practice, particularly with tulip and chrysanthemum where a large number of varieties are often grown in close proximity, to plant the stocks in the same order each time. This ensures that the same plants will always be in contact with an infested stock and so lessens the risk of spread into other stocks. The spread of *Aphelenchoides* in chrysanthemums is reduced if overhead watering is kept to a minimum.

SELECTION OF HEALTHY PROPAGATING MATERIAL

With plants that are propagated vegetatively it is often possible to eliminate nematodes by selecting uninfested plants or parts of plants for propagation. For example, by rooting stem cuttings of shrubs in nematode-free soil, infestations of *Meloidogyne hapla* and any other root parasites can be eliminated. Similarly, *Heterodera cacti* and *Meloidogyne* spp. attacking cacti can be avoided by cutting off existing roots or preferably by taking aerial divisions and rooting in nematode-free compost. Conversely, nematodes attacking aerial parts of the plant can sometimes be avoided by taking root cuttings. In this way *Ditylenchus dipsaci* and *Aphelenchoides* spp. attacking *Phlox* can be eliminated.

Some pot-grown, glasshouse ornamental plants (e.g., *Peperomia, Saintpaulia*) are propagated from leaf cuttings. Healthy plants should be isolated from those infested by *Aphelenchoides* and inspected regularly to make certain they are uninfested. This ensures that healthy material only is used for propagation. The same principle can be applied to scabious infested by *A. blastophthorus*. During the growing season healthy plants can be marked and used for propagation by division in the spring.

The Ministry of Agriculture administers voluntary certification schemes for black-currants and strawberries, designed to ensure the production of plants reaching a high standard of health. Thus strawberry runners which hold an 'S.S.' certificate can be regarded as virtually free from *Aphelenchoides* and *D. dipsaci*. There is not the same assurance with certified black-currants since, with the stool-bed method of propagation, the stocks cannot easily be inspected for *Aphelenchoides ritzemabosi*. However, this nematode is troublesome on two varieties only. (See also p. 182).

General hygiene. Although the importance of hygiene has already been made evident, one aspect warrants special mention. This concerns avoiding the contamination of clean planting material or compost, which would defeat the object of propagation methods for nematode control, and also has a bearing on prevention of spread and on persistence. For example, in *Saintpaulia* propagating beds infested with *Aphelenchoides* or in mushroom beds infested with various nematodes, it is important to clean out the old compost thoroughly and not to allow it to contaminate the new material which is to be introduced. Much of the *Aphelenchoides* infestation of chrysanthemum stools can be removed at lifting time by a thorough washing to clean off the soil, and removing the new shoots.

Hygiene is also important in the dry-bulb store. Narcissus bulb stocks known to be infested with *Ditylenchus dipsaci* should be separated from healthy bulbs as the desiccated nematodes can drift on air currents. Hygiene is equally important in hot-water treatment sheds (Chapter 15) and it is vital that treated stocks should not be allowed near untreated bulbs.

Under this heading might also be considered the isolation or quarantining of any plant whose state of health is unknown—a procedure particularly applicable to horticultural plants.

INFLUENCE OF MANURING

It has been known for several years that organic manuring influences development of cyst nematodes. O'Brien & Prentice (1932) suggested that attack by *Globodera* sp was delayed by organic manuring until the potato plant was well established. Investigations on *Heterodera schachtii* in Germany led to the conclusion that an extra application of organic manure increased the tolerance of the plant. Similarly Thompson (1949), Goffart (1949), Oostenbrink (1950) and Schmidt (1954) mention the beneficial effect of farmyard manure on potato crops in the presence of *Globodera*. Recent work has shown that yield loss is independent of potential yield so that manuring may be increasing the potential without affecting loss.

Van der Laan (1956) investigated the influence of organic manuring on the development of *G. rostochiensis*. He found that larvae in the roots of potatoes treated with organic manure developed significantly more slowly than in other plants, and suggested that this was due to physiological changes in the plant, resulting in slight resistance.

The application of organic manure can provide several additional benefits:

(a) it can result in better soil structure, supplying the plants with a more suitable medium in which to grow; such improved soil will have a higher water-holding capacity;
(b) in a soil deficient in plant food it can supply nutrients;
(c) it may result in the build-up of organisms which feed on nematodes (see Biological Control section p. 280 *seq.*).

Under the conditions of (a) and (b) it is logical to expect, with a more vigorous plant, that there would be a corresponding increase in nematode population. That this is not always so strongly supports van der Laan's suggestions.

Oostenbrink (1954) showed that the number of *Pratylenchus* was significantly lower in farmyard-manured plots than in plots to which artificial fertilizer was applied. Kirkpatrick *et al.* (1959a) showed that populations of *Xiphinema* and *Pratylenchus* were significantly lower under cherry trees receiving high rates of potash (5 and 15 lb K_2SO_4 per tree) but were not affected by nitrogen and phosphate. They also showed (1959b) that the numbers of *Xiphinema* and *Paratylenchus* were correlated negatively with leaf potash and positively with leaf nitrogen. These results suggest that both organic and inorganic manuring may affect nematode populations, and further studies on these lines may enable manuring to be used more effectively in cultural control.

Biological control

In the following section, higher plants, fungi, viruses and animals known or suspected to be antagonistic to nematodes are reviewed. This includes the simple observation of an organism attacking a nematode in addition to efforts made to use organisms for biological control of nematode pests. At present, few if any of these organisms have been used for control commercially and there seems little likelihood of rapid progress, chiefly because of the high percentage kill needed to give more than a temporary benefit. It is possible that natural enemies may explain some unexpected nematode population declines.

ENEMY PLANTS

Gramineae. Triffitt (1934) and Franklin (1937) studied the influence of the root diffusates from various Gramineae on the cyst contents of potato cyst nematode (*Globodera* sp.). In field experiments Triffitt found that meadow grasses (*Poa pratensis* and *P. trivialis*) decreased the cyst contents compared with controls, and Franklin obtained a reduction with maize. Later workers (Oostenbrink, 1950; Stelter, 1958; Hesling, Pawelska & Shepherd, 1961) have been unable to confirm these results. It is difficult to explain this lack of agreement but none of the decreases obtained by Triffitt and Franklin was greater than the 40–50 per cent which might be expected under a fallow or non-host crop. It must therefore be concluded that there is no reliable evidence that any Graminaceous plant brings the eelworm population below the level obtained by fallowing.

Mustard. Morgan (1925) accidentally discovered that potatoes grown with white mustard in a pot of infested soil were less heavily attacked by cyst nematodes than potatoes growing alone. This work was extended by Triffitt (1929a, 1930) who showed that potato root diffusate was ineffective in the presence of leachings from the roots of mustard seedlings. Later Ellenby (1945a) confirmed these results but showed that larval emergence became normal on return to potato root diffusate alone. Ellenby (1945b, 1951) showed that, in field trials, mustard oils increased the yield of potatoes by reducing the severity of nematode attack.

Tagetes. Oostenbrink, Kuiper & s'Jacob (1957) showed that, by growing *Tagetes*, *Pratylenchus* populations could be reduced by 90 per cent. This effect has been shown to be due to a nematicidal action of the growing *Tagetes* plants (Uhlenbroek & Bijloo, 1958). Sixteen varieties of *Tagetes patula* and *T. erecta* were found to be effective but a growing period of three to four months was necessary. Later Meijneke & Oostenbrink (1958) showed that populations of *Paratylenchus*, *Tylenchorhynchus* and *Rotylenchus* could also be considerably reduced by this treatment.

It has been shown (Hesling, Pawelska & Shepherd, 1961; Omidvar, 1961) that *Tagetes* root diffusates do not stimulate or inhibit hatching of potato cyst nematodes. However, Omidvar (1962) obtained results which suggested that by growing *Tagetes* for four months in soil infested with *Globodera* sp. a slight reduction in nematode population could be obtained, possibly due to weak nematicidal action. He concluded that this reduction would be of negligible value and would not be worth considering as a control measure.

Hijink & Suatmadji (1967) have shown that several other Compositae (*Helenium* hybrids, *Gaillardia* hybrid 'Burgunder', *Eriophyllum caespitosum*) very effectively suppress populations of *Pratylenchus penetrans*.

Asparagus. Rohde & Jenkins (1958) showed that *Asparagus officinalis* would not support populations of *Paratrichodorus christiei* for more than 40–50 days. Decline of populations on young plants was at the same rate as in fallow soil but was more rapid after fleshy storage roots had formed. Tomato, normally a good host of this nematode, supported only a low population when asparagus was growing in the same pots. The juice from asparagus roots was found to be toxic to *P. christiei* and to several other nematode species.

Crotalaria. Ochse & Brewton (1954) have suggested that *Crotalaria* root systems exert a toxic effect on *Meloidogyne* populations. Barrons (1940) demonstrated that *Meloidogyne* larvae freely entered roots of *Crotalaria* but failed to survive. Although the possibility remains that this was due to toxic action, the main effect is probably that of a non-host crop (van der Linde, 1956; McBeth & Taylor, 1944).

Taylor & Murant (1967) have shown that populations of *Longidorus elongatus* and other plant parasitic nematodes were much reduced when raspberry canes were incorporated into the soil. This was thought to be due to a toxic substance in the canes.

TRAP CROPPING

Trap cropping as a means of controlling nematodes, particularly *Heterodera* and *Globodera* spp., has been investigated by a number of workers. A crop which causes the nematodes to hatch is planted; if the crop is a host plant it must be destroyed before the nematodes mature, but if the larvae do not develop, the crop can be allowed to mature normally.

Kühn (1891) investigated the use of trap crops for control of *H. schachtii.* He used rape at a high seed rate and grew several crops in one season. Carroll & McMahon (1937, 1939) investigated the use of potatoes for trapping *Globodera* sp. They obtained encouraging results by removing an April-planted crop five weeks after planting. It was not necessary to remove the complete root system but the best results were obtained by digging up the plants with as many stolons as possible.

More recently Stone (1961) concluded that oats are an efficient trap crop for *H. avenae.* He obtained a large reduction in infestation in strips ploughed up in early June compared with strips in which the oats were allowed to mature. However, it is generally accepted that *H. avenae* is not stimulated to hatch by oat root diffusate and hence the reduction is likely to be similar to that obtained by growing non-host crops or by fallow.

Winslow (1955b) investigated the effect on *G. rostochiensis* of trap cropping with *Solanum nigrum.* The roots of this plant stimulate hatching and are invaded but cysts are not normally formed. Winslow concluded that trap cropping with *S. nigrum* would not be an economic proposition with the reduction he obtained.

The roots of *Hesperis matronalis* (Cruciferae) stimulate *H. schachtii* to hatch but cysts are not formed. Moriarty (1961) gives evidence which suggests that populations of *H. schachtii* are decreased slightly more by this plant than by fallow or non-host crops, but this is insufficient to render it a useful method of control.

There are many difficulties associated with the use of trap crops which generally preclude their use. If a crop has to be destroyed before it matures or if the crop is useless (e.g., *S. nigrum*), the whole cost of the operation, including loss of cropping time, must be balanced against the degree of control obtained. This depends on how much of the infested soil mass is reached by root diffusates and what proportion of the larvae are hatchable; it is unlikely that a sufficiently high degree of control would be obtained. An additional hazard is that a host crop may not be destroyed at the correct time, due perhaps to unfavourable weather conditions. The nematodes may then mature and the population, instead of being reduced, will increase.

The trap cropping principle operates in some early potato growing areas where the crop may be lifted before the nematodes mature. This may allow potatoes to be grown annually on the same land.

NEMATOPHAGOUS FUNGI

The existence of fungi that prey on nematodes has long been known (Zopf, 1888) but it was not until Drechsler (1937) published an account of seventeen species that it was realized that they were fairly common. Until recently, most recorded species came from decaying organic matter but Duddington (1954) showed that they were also common in soils.

These fungi probably exist in the soil mainly as spores. When the spores germinate they have two possible phases of activity—saprophytic and predatory. Mankau (1962) showed that soil may often contain a substance which inhibits spore germination. The addition of vegetable organic matter suspended this fungistasis at least temporarily, but animal manure did not do so. Mankau suggested that predacious fungi are poor competitors within the soil and that their ability to grow in the saprophytic phase may be limited. In his experiments, germinated spores in highly antagonistic soils generally formed traps immediately and then depended on capturing nematodes for survival, but some soils appear to be too antagonistic for these fungi. Concurrently with this work Cooke (1961, 1962) investigated the relationship between nematophagous fungi, nematodes and organic matter. He postulated that, when organic amendments are added to the soil, the nematode-trapping fungi are stimulated first into saprophytic competition with other organisms; only after competition has become intense, and the nematode population has built up, do the predacious fungi begin to feed on nematodes. Decomposing cabbage induced a larger nematode population than a comparable amount of sucrose and yet sucrose stimulated a longer and more intense period of predacious activity than did cabbage. But with both sucrose and cabbage activity declined while the nematode population was still very high. Cooke suggests that there are three requirements if these fungi are to be used successfully—(1) the consistent stimulation of the fungi, (2) an increase in their period of activity and (3) an intensification of their activity.

Various attempts have been made in pots and in field trials to use predacious fungi for nematode control. The first was in Hawaii where Linford (1937b) claimed a degree of control by adding chopped pineapple tops to soil infested with *Meloidogyne*. Later, Linford & Yap (1939) investigated the possibility of inoculating soil with predacious fungi for *Meloidogyne* control in pineapples, but the results were disappointing. Mankau (1961) obtained similar results when attempting to reduce *Meloidogyne* damage to tomato and okra.

Work by Duddington, Jones & Williams (1956), and Duddington, Jones & Moriarty (1956) gave inconclusive results when predacious fungi were tried against *Globodera* spp. infesting potatoes and *Heterodera schachtii*. Later work (Duddington & Duthoit, 1960; Duddington, Everard & Duthoit, 1961) gave significant differences in the number of *H. avenae* larvae entering oat roots from control and treated plots. Hams & Wilkin (1961) suggested that congenial conditions for growth of predacious fungi in soil can be produced in the laboratory but, in the field, only sparse growth can be obtained, insufficient to affect nematode attacks.

It therefore appears that the problem of the correct use of predacious fungi for the control of nematodes in the field has not been solved. It seems to be important to add organic matter as well as making sure that the desired fungi are present. A rich energy source (e.g., sucrose) may be of greater value than green manure. However, with *Heterodera, Globodera* and *Meloidogyne* spp. for example, a very high degree of control is needed to give anything more than a temporary benefit and it seems doubtful if predacious fungi will be of use for control of these nematodes.

BACTERIA

Several reports of bacterial infection of plant-parasitic nematodes (by *Pseudomonas* spp.) are cited by Webster (1972) but there seems to be little information on the status of bacteria as natural enemies.

VIRUSES

Loewenberg, Sullivan & Schuster (1959) have demonstrated a virus disease of *Meloidogyne incognita*. In their experiments the infected nematode larvae became sluggish and no galls were formed on the plant roots. This is so far the only recorded instance of a virus disease of nematodes.

PREDACIOUS AND PARASITIC INVERTEBRATES

Nematodes have many natural enemies among invertebrate animals, including Protozoa, predacious nematodes, Enchytraeids, tardigrades and Collembola. At least some of these occur in most soils but it is not known how important they are as natural controlling agents. They are almost certainly more active than predacious fungi in normal soils and may provide the explanation for the occasional unexpected decline in a nematode population. However, there have been few attempts to assess their importance. It is doubtful if these organisms are likely to prove of practical value in biological control.

Amoeboid organisms have been found attacking larvae of potato cyst nematodes in The Netherlands and England, and of *Heterodera schachtii* in Ontario (Weber, Zwillenberg & van der Laan, 1952; Winslow & Williams, 1957). Canning (1973) has reviewed the literature on protozoan parasites of nematodes. She concluded that some of these organisms deserve investigation as possible biological-control agents, especially *Duboscquia penetrans* Thorne which produces lethal infections of various nematodes including plant parasites; however she questioned the protozoan nature of *D. penetrans*. Mankau (1975) believed that *D. penetrans* was a spore-forming bacterium but Birchfield & Antonopoulos (1976) disagree; at the time of writing the true nature of this parasite remains in doubt.

Many predacious nematodes (e.g., Mononchids, Dorylaimids—see Chapter 12) are common in soil. Thorne (1927) investigated the importance of several species of Mononchids (then all included in one genus, *Mononchus*) in controlling *H. schachtii* and concluded that it was doubtful if they were of economic importance.

Schaerffenberg (1950) claimed that, when Enchytraeids were added to pots containing sugar-beet plants attacked by *H. schachtii*, the plants were able to recover provided that the nematodes had not matured. He stated that the Enchytraeids penetrated the plant roots to feed on the nematode larvae and suggested that dunging or green manuring would help to control *H. schachtii* by creating favourable conditions for the Enchytraeids.

Tardigrades have been observed to feed on nematodes in U.S.A., Ceylon and Great Britain (Hutchinson & Streu, 1960; Doncaster & Hooper, 1961).

Murphy & Doncaster (1957) found that the number of *H. cruciferae* reaching maturity on rape in a plot appeared to be rather low; sometimes up to 30 per cent of the females had large holes in them and the body contents had usually disappeared. This was due to predation by Collembola, chiefly *Onychiurus armatus*. In a pot culture an estimated seven per cent of cysts were damaged by this species.

CHAPTER 14

Chemical Control
(a) Soil Treatment

A. G. WHITEHEAD

Rothamsted Experimental Station, Harpenden

Populations of plant-parasitic nematodes can be controlled by applying nematicidal or nematostatic chemicals to the soil. All such chemicals are referred to as 'nematicides'. Disinfesting soils chemically, although cheaper than steam sterilization, is still costly, especially if the bulk of the soil must be treated, and the cost of the nematicides is usually much more than the cost of applying them. The amount of nematicide needed and whether the whole of the topsoil or only a part of it is to be treated depends on the nematodes to be controlled, the kill required, the distribution of the nematodes in the soil, and the effectiveness of the nematicide.

Control and its Assessment

Damage to field crops may be minimized by killing or immobilizing three quarters of the nematodes in the soil. For example, killing or immobilizing about 75 per cent of the Trichodorids or *Longidorus* in sugar-beet rows prevents injury to sugar-beet seedlings in sandy soils in eastern England (Whitehead, Dunning & Cooke, 1971). Damage caused by stem nematodes (*Ditylenchus dipsaci*), by heavy infestations of cyst nematodes (*Heterodera* and *Globodera* spp.) or by nematode-transmitted viruses is prevented only by 'killing' more than 90 per cent of the nematodes. If a susceptible crop is to be grown without increasing the numbers of a nematode in the soil after harvest, more than 75 per cent of the nematodes must usually be 'killed'. PARTIAL CONTROL is achieved if the nematode increases less than expected following the growing of a susceptible crop. FULL CONTROL is achieved when the percentage of nematodes 'killed' equals K; $K = 100\frac{(x-1)}{x}$, where x is the number of nematodes left in the soil after a susceptible crop has been grown, divided by the number of nematodes in the soil before it was treated with a nematicide. For example, K is 90 per cent for *Trichodorus, Paratrichodorus, Longidorus* or *Xiphinema* (up to 10-fold increase on host plants), 98 per cent for *Heterodera* or *Globodera* (up to 50-fold increase), and 99·9 per cent for *Meloidogyne* or *Ditylenchus dipsaci* (up to 1000-fold increase). COMPLETE OR TOTAL CONTROL is achieved only when every nematode is killed or immobilized; this is rarely if ever possible.

Which of these degrees of control is needed depends on the frequency with which susceptible crop plants are grown, on the nematode species to be controlled and on whether further control measures can be applied. Full control is needed, for example, if tomatoes or potatoes susceptible to potato cyst nematodes (*G. rostochiensis, G. pallida*) are to be grown every year in infested soil, but partial control is adequate if they are grown in rotation with non-host crop plants. Controlling nematodes which harm perennial crop plants is usually more difficult than controlling nematodes on annual crop plants. No treatment is known which can eliminate nematodes from soil, i.e., achieve total control and, after a perennial crop is planted, only nematicides non-toxic to the plants can be applied and these must be injected or drenched into the soil to avoid damage to established roots. Killing 99 per cent of a population of one *Xiphinema diversicaudatum*/200 ml soil in a hop garden still leaves 120 surviving nematodes per hop plant, some of which can remain infective with arabis mosaic virus-H for 9–12 months (Pitcher & McNamara, 1973). If transmission of this virus to hop is to be prevented by a nematicide, total control is needed. As already stated this is impractical, if not impossible to achieve.

Methods of assessing chemical control of cyst nematodes were reviewed by Peachey, Rao & Chapman (1963) and by Whitehead (1973). Methods suitable for the principal nematode pests of temperate soils are in Table 1. Control cannot be assessed reliably by the growth of susceptible crop plants in nematicide-treated soils because nematicides may affect other soil micro-organisms and the supply of soil nutrients, especially nitrogen.

Table 1

Methods of Assessing Nematode Control by Nematicides

Assessment	Assay	Relevant genera
1. *Kill*	Dead eggs in cysts (Shepherd, 1962c)	*Heterodera, Globodera*
2. *Kill or nematostasis*	(i) Hatched larvae/g soil	*Heterodera, Globodera, Meloidogyne*
	(ii) Nematodes/g root or shoot	*Heterodera, Globodera, Meloidogyne, Pratylenchus, Ditylenchus dipsaci*
	(iii) Galls or stubby root tips/g root or /root system	*Heterodera* or *Globodera* (rarely), *Meloidogyne, Xiphinema, Longidorus, Trichodorus, Paratrichodorus*
	(iv) Cysts on roots Egg-sacs on roots	*Heterodera, Globodera* *Meloidogyne*
3. *Population control*	(i) Cysts/100g soil, before treatment and after susceptible crop	*Heterodera, Globodera*
	(ii) Eggs in cysts/g soil, before treatment and after susceptible crop*	*Heterodera, Globodera*

*Not very reliable if many eggs are also laid in egg-sacs or if some eggs hatch and the larvae have left cysts before pretreatment soil samples are collected.

Types of Nematicide

Four groups of pesticides include nematicides suitable for temperate soils: (i) halogenated aliphatic hydrocarbons—1,3-dichloropropene mixtures (e.g.,

D–D mixture, Telone) in field soils and methyl bromide, ethylene dibromide and 1,2-dibromo-3-chloropropane in glasshouse soils, (ii) methyl isothiocyanate mixtures, including solutions in xylol or in dichloropropene mixtures, with or without added chloropicrin, and methyl isothiocyanate precursor compounds (metham-sodium, dazomet), (iii) organophosphates, especially diethyl phosphorothioates and (iv) oximecarbamates, especially aldicarb and oxamyl. Compounds in groups (i) and (ii) act as soil fumigants, compounds in groups (iii) and (iv) have much smaller vapour pressures, so they exert little fumigant action on the soil. None of these nematicides is very persistent in soil. Various other chemicals are mentioned in this chapter which do not fit in any of the above categories but have been reported to control plant-parasitic nematodes. Nematicides suitable for temperate soils are characterized in Table 2. Chloropicrin has been omitted because although it is a good nematicide and fungicide its irritating and harmful vapour makes it unpleasant and inconvenient to apply. There is also the possibility that large amounts will produce gas clouds that may settle around houses and annoy the inhabitants. Similarly, no organophosphate is included, because none is known which works well in all soils. Ethylene dibromide and dibromochloropropane, much used in tropical soils, are suitable only for heated glasshouse or warm outdoor soils.

SOIL FUMIGANTS

Soil fumigants kill many different soil animals simultaneously, including nematodes and their eggs. Large amounts are needed to kill most of the nematodes in soil, i.e., 100–600 or more kg/ha. The relationship between log dosage and probit mortality is linear (Peters, 1954; Seinhorst, 1973a), probit mortality increasing by about one half of one probit unit with each doubling of the dosage of fumigant. Large amounts of soil fumigants are phytotoxic and must be applied to the soil several days, weeks or months before planting, but dibromochloropropane can be applied around the roots of woody perennials. Detailed accounts of the diffusion of fumigants in soil are given by Youngson & Goring (1962) and Leistra (1972).

Soil fumigants can kill nematodes deep in the soil (Harrison, Peachey & Winslow, 1963; Taconis & Kuiper, 1964; Whitehead, Fraser & Greet, 1970). Except for methyl bromide and chloropicrin, soil fumigants are only moderately toxic to man and other vertebrates. Methyl bromide, chloropicrin, metham-sodium and dazomet also kill or inhibit soil fungi and may kill weed seeds or delay their germination. Dichloropropene mixtures are usually weakly toxic to fungi. Soil fumigants also inhibit or kill some soil bacteria. Surviving ammonifying bacteria, which convert organic nitrogen in soil to ammonium nitrogen, recolonize the soil more quickly after fumigation than the nitrifying bacteria, which convert ammonium nitrogen to nitrate nitrogen. Ammonium nitrogen, therefore, accumulates in the soil (Gasser & Peachey, 1964). This may be an advantage to crops in sandy soils from which nitrate but not ammonium nitrogen is rapidly leached from the surface by spring rainfall. Soil fumigants may also alter the availability of other soil nutrients, e.g., phosphorus, manganese, iron, aluminium and silicon (Smith, 1963).

Harvested crops may contain soil fumigant residues, e.g., bromine after fumigation with methyl bromide (Kempton & Maw, 1972) and 3-chloroallyl methyl sulphide (Dekker, 1972), 2,2-dichloropropane or 1,2,3-trichloropropane (Shepherd, 1952) after fumigation with dichloropropene mixtures.

Ta

Properties and Applica

		HALOGENATED HYDROCARB		
COMPOUND	Common names	methyl bromide	dichloropropene (Telone) dichloropropane-dichloropropene mixture (D–D)	ethylene dibromide
	UK trade names	Embafume C Dowfume MC-2	D–D Vidden D Telone Telone II	—
	Chemical name	bromomethane	1,3-dichloropropene	1,2-dibromo-ethane (ethylene dibromide)
	Formulation	+ 2% chloropicrin	1,3-dichloropropene and other chlorinated hydrocarbons	concentration required in solvent
ACTIVE INGREDIENT	Boiling point (°C)	4	104–112	131–132
	Vapour pressure (mm Hg)	1420 (20°C)	about 55 (20°C)	8 (20°C)
	% water solubility	1·75 (20°C)	0·1 (20°C)	0·3 (20°C)
MAMMALIAN TOXICITY	Acute oral LD_{50} or LC_{50}	(rat) 514 ppm	(rat) 140–500 mg/kg	(male rat) 146 mg/kg
APPLICATION TECHNIQUE	Suggested minimum soil temperature (°C)	5	7	16
	Dose	400–900 kg/ha	50–900 kg/ha	up to 250 kg/ha
	Method of application	by pipeline direct from cylinder to soil under gas-tight cover	(i) injected 15 cm deep in rows 25–30 cm apart. (ii) injected 15 cm deep in crop row positions	injected 15 cm deep in rows 25–30 cm apart

ome Useful Nematicides

MIGANTS	CARBAMATE FUMIGANTS		OXIME *N*-METHYLCARBAMATES	
romo- loropropane	dazomet	metham-sodium	aldicarb	oxamyl
magon, mazone	Basamid	Vapam, Vitafume, Unifume, Sistan, Trimaton, Maposol	Temik	Vydate
-dibromo-3- loropropane	3,5 dimethyl- tetrahydro-1,3, 5-2H- thiadiazine-2- thione	sodium *N*-methyl dithiocarbamate	2-methyl-2- (methylthio) propionaldehyde -*O*-(methyl- carbamoyl) oxime	2-dimethyl- amino-1- (methylthio) glyoxal *O*- methylcarbamoyl -monoxime
ulsifiable ncentrates or anules	fine granules 98–99% a.i. (prill)	32–35% aqueous solution	5, 7½, 10% granules; in U.K. 10% coal based granules only	10% granules
6	Principal active breakdown product is methyl isothiocyanate 119		Decomposes above 100°C	—
3 (21°C)	21 (20°C)		1×10^{-4} (25°C)	$2{\cdot}3 \times 10^{-4}$ (25°C)
(20°C)	0·8 (20°C)		0·6 (25°C)	28 (25°C)
ale rat) 73 mg/kg	(rat) 500 mg/kg	(rat) 820 mg/kg	(rat) 0·9 mg/kg	(rat) 5·4 mg/kg
	10	10	—	—
to 120 kg/ha	200–500 kg/ha	1200 l soln./ha	up to 11 kg/ha	up to 11 kg/ha
ected around ots of woody rennials or in ws 25–30 cm art 15 cm deep	spread on soil surface and incorporated by rotary cultiva- tion	diluted and rotary cultivated in or injected 6 ml every 25 cm	broadcast to soil and rotary-cultivated in or applied in seed furrows	

ORGANOPHOSPHATES AND OXIMECARBAMATES

Much smaller amounts of organophosphates or oxime *N*-methylcarbamate nematicides [*O*-(methylcarbamoyl) oximes] (0·5–10 kg a.i./ha) than of soil fumigants are needed to control nematodes in soil and, as such amounts are not usually toxic to plants, they can be applied at or just before planting—a big advantage. Diethyl phosphorothioates are more effective nematicides than dimethyl phosphates (Whitehead, Fraser & Storey, 1972). Other organophosphates effective against potato cyst nematode (*Globodera rostochiensis*) in sandy loam are Dowco 275 [*O,O*—diethyl *O*–(6–fluoro–2–pyridyl phosphorothioate)], phoxim, CGA 12223, [*O,O*–diethyl *O*–(1–isopropyl–5–chloro–1,2,4–triazolyl–(3)) phosphorothioate], cyanthoate, chlorpyrifos and azinphos-ethyl (Whitehead, Tite, Finch, Fraser, French & Wright, 1974).

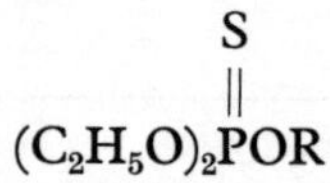

diethyl phosphorothioates

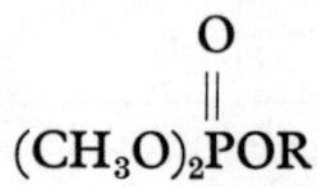

dimethyl phosphates

Oxime *N*–methylcarbamates have the general formula:

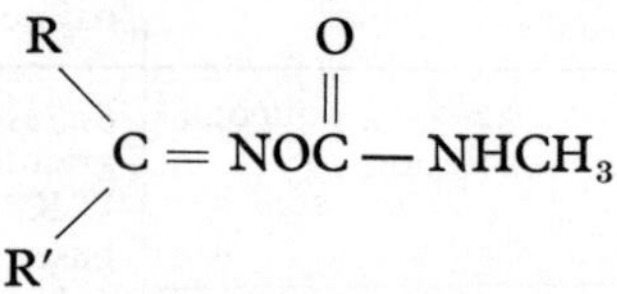

and four are very effective nematicides—aldicarb, oxamyl, methomyl and Tirpate (2,4-dimethyl-2-formyl 1–1,3-dithiolane oxime *N*-methylcarbamate). Methomyl can be phytotoxic and Tirpate is not available commercially. Aldicarb and oxamyl, unlike several organophosphate nematicides, are not affected by the nature of the soil (Moss, Crump & Whitehead, 1975).

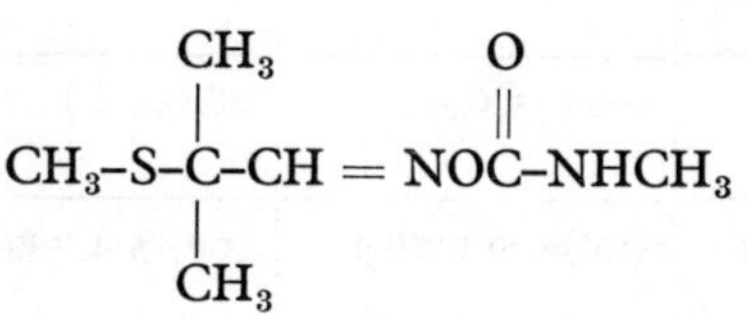

aldicarb

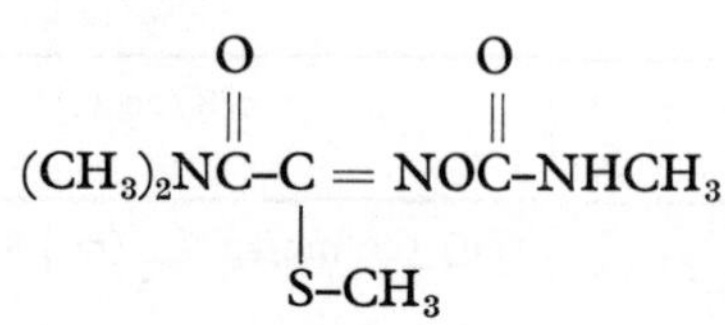

oxamyl

Factors Affecting Control by Nematicides

Many factors affect the control of nematodes by nematicides, especially fumigant nematicides. The most important are the distribution of the nematodes in the soil, soil pore space, soil moisture and temperature, nematicide concentration $\times$ time product and whether the nematicide is sufficiently well distributed in the soil.

DISTRIBUTION OF NEMATODES IN SOIL

Choice of nematicide treatment and the control that can be obtained are influenced by the distribution of the nematodes in the soil. Because it is a shoot parasite, stem nematode (*Ditylenchus dipsaci*) may be controlled by nematicides mixed shallowly into the soil in the crop rows. In contrast, *Longidorus attenuatus* and *L. leptocephalus* are often more abundant below 30 cm deep in sandy soil but can be controlled fully by injecting the soil in winter with dichloropropene mixtures or with chloropicrin (Whitehead, Fraser & Greet, 1970; Evans & Pandé, 1972). *Xiphinema diversicaudatum* may also be plentiful in soil to 60 cm or deeper, whereas *L. elongatus* is found mostly in soil to 30 cm deep (D'Herde & van den Brande, 1964). Potato cyst nematodes (*Globodera rostochiensis*, *G. pallida*) may be as numerous in soil 25–40 cm deep (below ploughing depth) as in soil to 25 cm deep. Bird (1969) found the root-knot nematode, *Meloidogyne incognita*, to depths of 120 cm in glasshouses growing tomatoes or cucumbers; at such depths it was beyond the reach of normal steam or chemical soil sterilization techniques.

SOIL TYPE AND MOISTURE

Most soil fumigants are less efficient in soils containing much organic matter or clay than in soils containing little of either (Martin, 1947; Peters & Fenwick, 1949; Fenwick, Peters & Libbey, 1953; Allen & Raski, 1950; Whitehead, Tite, Fraser & French, 1972, 1973b, e). The gases are largely adsorbed on the clay and organic matter fractions. Soils with too little air space (pores $< 30\ \mu m$) or containing too much water which blocks too great a part of the pore system, prevent the gases from diffusing readily, for diffusion in water is many times slower than in air. Also the concentration of the material in the greater amount of water is much smaller. Stark (1948) found that more chloropicrin was adsorbed the greater the clay content of the soil but that dry organic matter adsorbed little. He also found that the greater the boiling point of the fumigant the greater its potential for adsorption. Adsorption is reduced by raising the temperature and decreasing the moisture content of the soil and Lewis & Mai (1963) found that 1,3-dichloropropene mixture was more effective against *Ditylenchus dipsaci* in organic soils kept dry and hot than moist and cold. Soil fumigants are effective in moist but well drained soils (pF 2–3). Soil moisture has a big effect on the control of potato cyst nematode with D–D (van den Brande, Kips & D'Herde, 1954a, b, 1956) and it affects the fumigation of soil with D–D, Telone, dazomet, chloropicrin or methyl isothiocyanate much more than does soil temperature.

Nematodes close to the soil surface are often poorly controlled by soil fumigants (Peters, 1955), for the surface soil is looser and frequently dries out and cracks. Moreover, it is next to the atmosphere where the concentration of vapour is rapidly attenuated, so gases escape rapidly from it. Covering the soil surface with gas-tight sheeting, thoroughly wetting the surface to seal it, incorporating dazomet or oxamyl in the topsoil (Kaai & Windrich, 1971; Seinhorst, 1973b; Whitehead, Tite, Fraser & French, 1973e) or drenching the soil surface with solubilized para- and meta-cresol mixture (Peters, 1955; Stone, 1956, 1957) increases nematode control near the surface. Large amounts of water are needed for sealing or drenching the soil.

Several organophosphate nematicides are also affected by soil composition. For example, thionazin controlled *G. rostochiensis* better in a peaty loam than in a silt loam (Whitehead, Tite, Fraser & French, 1973a), whereas fenamiphos controlled the nematode better in a sandy loam than in a peaty loam or a silt loam (Whitehead, Tite, Fraser & French, 1973d). In two other soils, one a peaty loam, the other a silty loam, fenamiphos fully controlled the nematodes

(Whitehead, Tite, Fraser & French, 1973a). Three oxime *N*-methylcarbamate nematicides (aldicarb, oxamyl and Tirpate) were little affected by the clay or organic matter content of the soil (Whitehead, Fraser & Storey, 1972; Whitehead, Tite, Fraser & French, 1973d; Moss, Crump & Whitehead, 1975).

How well nematodes are controlled by nematicides with small vapour pressures also depends on how well they are distributed initially and how long they persist in the soil. Bromilow (1973) showed that the distribution of a non-volatile nematicide in soil is determined by the method of application and incorporation, and by subsequent redistribution by diffusion and leaching. He also showed that leaching of compounds through soil columns was related to the ratio Q, where $Q =$ (chemical concentration in soil organic matter) $\times$ (chemical concentration in soil water)$^{-1}$.

Aldicarb, aldicarb sulphoxide, aldicarb sulphone and oxamyl have small Q values and so leach further down columns of soil than fenamiphos, which has a large Q value. The size of Q is determined by the polarity of the compound, i.e., the degree of charge separation in the molecule. The greater the organic matter content of the soil and the greater the Q value, the less the nematicide will move and the greater the need to mix it thoroughly. In finely structured soils the organophosphate and oximecarbamate nematicides will move less than in coarsely structured soils. The amount of water flowing through the soil profile also affects the redistribution of the nematicide.

Prolonged persistence of nematicides in soil is undesirable for environmental reasons but they must last long enough to give a satisfactory 'kill'. With the exception of methyl benzimidazole carbamate, the principal breakdown product of benomyl, which may persist in soil for several years, most nematicides are of short persistence, i.e., they have a half life of one month or less. Harvested crops may contain minute traces of oximecarbamates, organophosphates or their derivatives (usually less than 0·1 ppm).

SOIL TEMPERATURE

Fumigants which have small vapour pressures, such as ethylene dibromide or dibromochloropropane, take longer at low temperatures to reach lethal concentrations, kill nematodes and escape from the soil into the atmosphere than fumigants with large vapour pressures, such as D–D, Telone, dazomet, chloropicrin or methyl isothiocyanate. Dibromochloropropane is slow acting, even in warm soils (Ichikawa, Gilpatrick & McBeth, 1955). In moist, sandy or peaty loam at 5°C or 10°C oxamyl remained effective against *G. rostochiensis* for several months (Whitehead, Tite, Fraser, French & Smith, 1975). In moist soils, organophosphate and oximecarbamate nematicides are probably degraded faster the higher the temperature.

NEMATICIDE CONCENTRATION × TIME PRODUCT

Except at extremes of time and concentration (Hague & Sood, 1963) the number of *G. rostochiensis* larvae killed by methyl bromide (and presumably other soil fumigants) is related to the concentration of gas multiplied by the time over which it operates (CTP). At the same CTP more nematodes are killed in soil by small concentrations of the gas for long periods than by large concentrations for short periods (Hague, Lubatti & Page, 1964). The CTP achieved determines nematode control and this is not necessarily the same as the applied

CTP. For example, the smaller the porosity of the soil, the longer it took a dose of ethylene dibromide to kill *G. rostochiensis*, i.e., to reach the required CTP (Call & Hague, 1962). The CTP achieved with a soil fumigant is greatly affected by the rate at which the gas escapes from the soil surface into the air. In such a leaky system the CTP equals the area under the concentration × time curve. Organophosphate and oximecarbamate nematicides are, in time, degraded in soil to substances not toxic to nematodes. It is therefore important, for this and other reasons, to know what CTPs are needed for nematode control with these compounds.

Applying Nematicides

Efficient use of a nematicide requires it to be thoroughly mixed with the layer of soil in which it is to control nematodes. Adding a wetting agent to soil doubled the toxicity of dazomet, dichloroisobutylene, fensulphothion and aldicarb to several plant-parasitic nematodes (Dern, 1971).

SOIL PREPARATION

For pre-plant fumigation the residues of any previous crop should be ploughed in and allowed to rot. Then, when the soil is moist but well drained (pF 2–3) it should be worked down to a fine seedbed and the fumigant applied. The soil temperature after injection should exceed 16°C for several weeks (ethylene dibromide) or several months (dibromochloropropane), but lower temperatures (down to 7°C or even 5°C) are adequate for other soil fumigants (dichloropropene mixtures, dazomet, metham-sodium, chloropicrin, methyl isothiocyanate mixtures) provided the soil remains well drained and several weeks or months elapse before planting. Methyl bromide (b.p. 4°C) kills nematodes rapidly at soil temperatures down to 5°C. After fumigation the soil should be aerated by harrowing, to rid it of any remaining phytotoxic gases.

LIQUID NEMATICIDES

In glasshouses and nursery beds the topsoil can be fumigated with methyl bromide applied under polyethylene sheeting spread over the soil, the gas being fed in through delivery pipes from small, disposable canisters of liquified methyl bromide or under pressure from large cylinders. Methyl bromide formulations include 2 per cent of chloropicrin as a warning agent, as methyl bromide itself is extremely poisonous but odourless. Other liquid fumigants (1,3-dichloropropene mixtures, methyl isothiocyanate formulations, ethylene dibromide, chloropicrin, dibromochloropropane) can be injected into the soil at points 20–30 cm apart and 15–25 cm deep by hand-held injectors (e.g., Shell HI injectors), under the ploughsole during ploughing, or through blade coulters mounted 25–30 cm apart on a tractor-drawn toolbar. The fumigant can be delivered under the ploughsole or to the bottom of the trailing edges of the coulters by a constant-head gravity feed, but variation in forward tractor speed results in uneven dosage. Even dosage is ensured by a land-wheel driven pump (e.g., 'John Blue' fumigant pump). The slits left in the soil by blade coulters must be sealed immediately and the soil surface harrowed to a fine tilth to prevent rapid loss of fumigant to the atmosphere. The toxicity of many organosphosphate and oximecarbamate nematicides makes them generally unsuitable for spray applications to the soil.

GRANULAR NEMATICIDES

Some soil fumigants are formulated as solids. Dazomet is formulated as dust, granules or microgranules (prill); dibromochloropropane, and in the United States D–D and methyl bromide, have also been formulated as granules. Nematicide granules should be as small as practicable, flow easily, be non-dusty and should be made unattractive to wild animals. Granular nematicides can be applied evenly to the soil surface by a 'Sisag-Lospred' fertilizer distributor, provided it is shielded against the wind. 'Horstine Farmery' microband granule applicators fitted with modified fish-tail spreaders or with air-flow distributors are also suitable. Granules are best incorporated in soil by rotary cultivation (rotavation) 10–15 cm deep. Rotavators with L-shaped or spike tines incorporate granules in peaty (organic) soils better than a reciprocating harrow or a modified power harrow (Roterra) (Whitehead, Bromilow, Lord, Moss & Smith, 1975). Rotavation ensures fairly even distribution of granules to the depth of rotavation (Whitehead, Tite, Fraser, French, Bromilow, Lord & Smith, 1974) and in peaty loam to 20 cm deep *G. rostochiensis* was better controlled by rotavating granules of oxamyl or Dowco 275 to 10 cm deep than to 20 cm deep (Whitehead, Tite, Fraser, French & Smith, 1975). To ensure that all granules applied to the soil surface are incorporated in the soil the granule applicator should be mounted on the tractor in front of the rotavator. The seed drill or automatic potato planter could be attached to the rotavator so that granule application, incorporation, seedbed preparation and planting are all accomplished in one operation. Covering the soil with polyethylene sheeting after rotavating in dazomet prill did not increase control of *G. rostochiensis* in peat soil (Whitehead, Tite, Fraser, French & Smith, 1975) or *G. pallida* in sandy clay loam (Whitehead, Fraser, French & Wright, 1975).

COMBINED NEMATICIDE TREATMENTS

Sealing the soil surface with polyethylene sheeting or by drenching it with water or solubilized chemicals is impractical in large fields but more nematodes can be killed or immobilized near the soil surface as well as deep in the soil if dazomet or oxamyl is incorporated in the topsoil and a 1,3-dichloropropene mixture or Di-Trapex CP is injected 15–25 cm deep in the soil. Such combined treatments may be cheaper and more effective than applying very large amounts of one nematicide. Alternatively, dazomet can be rotovated in in two equal amounts, one before and one after ploughing, so the granules are incorporated more deeply in the soil and the nematodes controlled more efficiently than with the same amount applied in a single dose after ploughing (Whitehead, Tite, Fraser & French, 1973e). Dazomet incorporated in sandy loam 15 cm deep controlled potato cyst nematode (*G. rostochiensis*) to 40 cm deep (Whitehead *et al.*, 1973e) so split application is not always advantageous.

Nematode damage is more harmful to seedlings than to older plants. For example, Efremova (1966) found that the younger cucumber plants were when infested with root-knot nematodes, the greater was the yield lost. Treatment with nematicides applied only to the planting holes or to the crop rows before planting seeds or setts may, therefore, lessen nematode damage as well as, and more cheaply than, treating all the topsoil (Walter & Kelsheimer, 1949; Winfield, Murdoch & John, 1971; Whitehead, Tite & Fraser, 1970, 1973a). In wide-row crops, such as potatoes or sugar beet, narrow-band row treatments with oxime-carbamate or organophosphate nematicides will not usually control cyst nematodes effectively, because these nematicides are not spread far laterally by leaching nor are they redistributed by the plant to roots growing outside the treated soil. Treating potato ridge soil with effective nematicides can control potato cyst nematodes effectively (Whitehead, Tite & Fraser, 1973b; Whitehead, Tite, Fraser & French, 1973a–e).

Effective Treatments

ECTOPARASITIC NEMATODES

Large amounts of methyl bromide, D–D or chloropicrin (450–900 kg or more/ha) injected into the soil may kill nearly all Trichodorids, *Longidorus* or *Xiphinema* (e.g., Harrison, Peachey & Winslow, 1963; Boubals, Pistre & Dalmasso, 1964; Cooke & Draycott, 1969; Whitehead, Fraser & Greet, 1970). Such treatments are expensive and can only be justified to protect valuable crops from nematode-transmitted viruses, e.g., arabis mosaic virus in strawberries or fanleaf virus in vines, where the aim is to kill all or nearly all viruliferous nematodes. Thus, Taylor & Murant (1968), found that 400 lb D–D/acre (448 kg/ha) or 60 lb quintozene a.i./acre (67 kg/ha) controlled *L. elongatus* and largely prevented transmission of raspberry ringspot and tomato black ring viruses to strawberry.

The incidence of 'spraing' in potatoes, caused by tobacco rattle virus and transmitted by *Trichodorus* and *Paratrichodorus* spp. has been greatly reduced by treating the soil with D–D, dazomet or methomyl (Cooper & Thomas, 1971) or by applying small amounts of aldicarb, ethoprophos or oxamyl in the seed furrows at planting time (Brown & Sykes, 1973). Steudel (1969) also found aldicarb and fenamiphos controlled spraing in potatoes. As complete control of a nematode is practically impossible to achieve, the control of nematode-transmitted viruses of perennial crops is unlikely to last more than a few years.

The damage inflicted by Trichodorids or *Longidorus* spp., feeding on roots of sugar beet seedlings in sandy soil in England (Docking disorder), is severe and results in much yield loss in affected fields only in years with frequent abundant rain in May and early June (Jones, Larbey & Parrott, 1969). The cost of treatment may not be repaid by increased yields in years when Docking disorder does not develop, so the annual cost of controlling these nematodes must be kept small. When 45 l Telone or 67 l D–D/ha are injected 15 cm deep in the rows in which sugar beet seed is sown one to ten days later, about 85 per cent of the nematodes in the rows are killed. Similarly, aldicarb granules at about 0·6–1·2 kg a.i./ha applied in the seed furrow during sowing prevent the nematodes feeding on the roots of the seedlings and also control other seedling pests. These row treatments enable the seedlings to establish effective root systems and to grow away from the nematode attack thus avoiding serious yield losses (Whitehead, Dunning & Cooke, 1971; Cooke, Dunning & Winder, 1974).

ROOT-LESION NEMATODES

Full or partial control of root-lesion nematodes (*Pratylenchus* spp.) has been obtained by fumigating infested soils with dichloropropene mixtures (D–D,

Telone, Telone PBC) (e.g., d'Herde, van den Brande & Gillard, 1960; Whitehead, Tite & Fraser, 1970), metham-sodium (Klinkenberg & Seinhorst, 1956; Peachey & Winslow, 1962), chloropicrin (Apt & Gould, 1961; Miller & Hawkins, 1969; Whitehead, Fraser & Greet, 1970), methyl bromide alone (Peachey & Winslow, 1962) or with chloropicrin (Peachey & Winslow, 1962; Taylor, Canter-Vischer, Newhook & Grandison, 1970), dazomet (Thompson & Willis, 1970; Peachey & Winslow, 1962), methyl isothiocyanate in D–D ('Vorlex') (Miller & Hawkins, 1969) or ethylene dibromide (Peachey & Winslow, 1962; Whitehead, Tite & Fraser, 1970). Several non-fumigant nematicides have also given good control of *Pratylenchus* spp. Thompson & Willis (1970) obtained full control of *Pratylenchus* spp. on lucerne, red clover and birdsfoot trefoil with large amounts of fensulphothion, aldicarb, methomyl, carbofuran or ethoprophos. Krusberg (1971) found that aldicarb or fenamiphos, applied before planting antirrhinums, controlled *P. penetrans* fully in heavy glasshouse soil that was not amenable to fumigation with chloropicrin or dibromochloropropane. Miller & Kring (1970) have reported good control of *P. penetrans* on potatoes by applying aldicarb, carbofuran, fensulphothion or disulphoton as side dressings to the rows. They also found that carbofuran applied to one side of the potato row protected roots on the other, untreated, side of the row from *Pratylenchus*, though they did not establish how this occurred.

CYST NEMATODES

Cyst nematodes have been fully or partially controlled by treating infested soil with (i) large amounts (100–1500 kg/ha) of halogenated aliphatic hydrocarbons (dichloropropene mixtures, chloropicrin, methyl bromide, ethylene dibromide, 1,2-dibromo-3-chloropropane, dichlorobutene mixtures), various isothiocyanates, especially methyl isothiocyanate and precursor compounds (metham-sodium, dazomet), (ii) smaller amounts (10–100 kg a.i./ha) of many diethyl phosphorothioates or diethyl phosphates, (iii) small amounts (2–11 kg a.i./ha) of several oxime *N*-methylcarbamates, especially aldicarb and oxamyl. Two benzimidazoles (benomyl and thiabendazole) at 22 kg a.i./ha have controlled *Globodera rostochiensis* (Hide & Corbett, 1973; Whitehead, Tite, Finch, Fraser, French & Wright, 1974).

Using large amounts of expensive nematicides can only be justified for high value cash crops and as a cheaper alternative to steam sterilization of soil in glasshouses. Murdoch & Murdoch (1963) found that steam sterilization alternating in successive years with D–D gave a greater profit margin, and as good control of potato cyst nematodes, as steaming every year. Treating the soil with oxamyl and Di-Trapex CP, dazomet and Telone, or dazomet and Di-Trapex CP, fully controlled *G. pallida* and allowed good crops of tomatoes to be grown in heavily infested soil under glass (Whitehead, Fraser, French & Wright, 1975).

H. goettingiana has been fully controlled by incorporating aldicarb or oxamyl in the seed-bed before sowing peas (Whitehead, Tite, Fraser & French, 1974). Similarly *G. rostochiensis* and *G. pallida* have been fully controlled by incorporating aldicarb or oxamyl in the ridge soil before planting susceptible potatoes and *Heterodera schachtii* has been fully controlled by incorporating either nematicide in the topsoil before sowing sugar beet seeds (Whitehead, 1974). *G. rostochiensis, G. pallida* and *H. schachtii* are not controlled if the nematicides are placed in the seed furrows at sowing time because the rows are wide apart and the nematicides are not translocated to roots growing beyond the treated soil.

Row treatments with oximecarbamate nematicides are likely to be more effective if the crop rows are closely spaced, for Brown (1973) found 2·2 kg aldicarb a.i./ha drilled with wheat seed increased yield as much and controlled *H. avenae* as well as did 9 kg a.i./ha incorporated to 10 cm deep in the topsoil. Similarly oxamyl drilled with pea seed controlled *H. goettingiana* as well as did oxamyl mixed in the seed-bed before sowing (Whitehead, Bromilow, Lord, Moss & Smith, 1975).

STEM NEMATODES

Stem nematodes are difficult to control with nematicides because they can multiply up to 1000-fold on a susceptible plant, so they can only be fully controlled if up to 99·9 per cent of them are killed or immobilized.

Large amounts of soil fumigants have sometimes given good control of stem nematodes. Chitwood & Newhall (1942) prevented 'bloat' of onions by injecting a 1 to 9 mixture of chloropicrin in ethylene chloride into holes 6 in. (15 cm) deep and $9 \times 10\frac{1}{2}$ in. (23×27 cm) apart at 10 ml/hole before onions were sown. Lewis & Mai (1958) found that large amounts of D–D injected into organic soils before sowing were similarly effective in controlling damage caused by stem nematodes. Thomas, Wiggell & Moyse (1958) found that 100 or 200 gallons of D–D or ethylene dibromide/acre (1123–2246 l/ha) ploughed into the soil in August, and sealed in by rolling, harrowing and drenching the soil with 2 gallons of 1 per cent solubilized xylenol or cresylic acid/sq. yard (10·8 l/m^2), prevented injury to narcissus by stem nematodes for at least one year. Although large amounts of dazomet incorporated in the topsoil can kill nearly all the nematodes, this is not enough to prevent injury to onions in heavily infested soil (Duggan, 1971). Soil fumigants, therefore, may prevent serious injury to crops by stem nematodes but large amounts are needed and population control is not achieved.

Small amounts of aldicarb, oxamyl, fenamiphos or thionazin in the crop rows can protect seedlings of sensitive crops from injury (e.g., onions) and greatly increase yields (Graf, Joseph, Keller, Liechti & Savary, 1964; Brzeski, 1970; Winfield, Murdoch & John, 1971; Whitehead, 1972; Whitehead, Tite & Fraser, 1973a). Treating onion rows with small amounts of oxime *N*–methylcarbamates during sowing is better than incorporating the same amounts throughout the topsoil, because stem nematodes affect only the shoots, before and after emergence from the soil, and row treatment concentrates the nematicides close to the plant. Surviving nematodes may multiply rapidly on the crop later in the growing season, when temperatures decrease and moisture returns (Whitehead, Tite & Fraser, 1973a). To control this damaging phase of the attack a second application of systemic nematicide is needed or, for ornamental crops, a more persistent nematicide could be used.

ROOT-KNOT NEMATODES

Like stem nematodes, root-knot nematodes are difficult to control fully, for they can multiply 1000-fold or more on a susceptible plant in the conditions of a heated glasshouse.

Most soil fumigants are unable to kill root-knot nematodes inside roots, so infested roots of a previous crop must be removed from the soil or allowed to rot before treatment with dibromochloropropane or ethylene dibromide (Lear &

Raski, 1956, 1962). D–D and methyl bromide penetrate unrotted root galls better than chloropicrin (Stark & Lear, 1947).

Several soil fumigants will control damage to annual crops by root-knot nematodes. Flushing hydroponic gravel beds repeatedly with 2 per cent formaldehyde prevented injury to tomatoes by *M. incognita* (Schuster & Wagner, 1972). Injecting infested soil with D–D (Heeley, 1951; Marien, 1956; Wuyts, 1971), ethylene dibromide (Harranger, 1956; Wilson & Hedden, 1964), methyl isothiocyanate mixtures (McKeen & Sayre, 1964), metham-sodium (Stettmeier, 1957) or dibromochloropropane (Fenili & Marinari, 1959; Kantzes, Jenkins & Davis, 1959; Kantzes & Morgan, 1962; Potter & Morgan, 1956; Wilson & Hedden, 1964) has also prevented galling of the roots of susceptible crop plants.

Several non-fumigant nematicides, especially oxime *N*–methylcarbamates have given good protection against root-knot nematodes. Panayi & Estey (1971) found a drench of methomyl (32 ppm) controlled *M. hapla* better on tomatoes than dibromochloropropane, trichloronate, fensulphothion or fenthion. Jarnevic & Coffee (1965) found that fensulphothion, fenthion or trichloronate prevented galling of tomato roots by *M. incognita* at rates as low as 12·5 ppm in the soil, dibromochloropropane did so at rates down to 6 ppm but D–D was not nematicidal below 50 ppm. Two alkane sulphonates, 3,5–dichloro–2–pyridyl methane sulphonate at $4\frac{1}{2}$ lb/acre (5 kg/ha) (Hamer, Huffman & Ristich, 1970) and 2–4–dichlorophenyl methane sulphonate at 8–20 lb/acre (9–22 kg/ha) (McBeth, White & Ichikawa, 1964), Nellite (phenyl *N,N'*–dimethyl phosphorodiamidate) at 1·25–2·5 ppm in soil (Youngson & Goring, 1963), diphenyl sulphide (Torgeson & Haynes, 1961) and 3,4–dichlorotetrahydrothiophene–1,1–dioxide (Schuldt & Bluestone, 1957) have also been reported to be effective for protecting plants from root-knot attack.

(b) Treatment of Plants, Planting Material and Seeds

A. L. WINFIELD

Agricultural Development and Advisory Service, Wye

Treatment of growing plants against nematodes is a relatively recent development because until the advent of the organophosphorus and oxime carbamate pesticides most known nematicides were very phytotoxic. Such compounds could not be used chemotherapeutically, i.e., for treating plant tissues to prevent attack or to kill nematodes already present (Nelmes, Trudgill & Corbett, 1973). Many plant-parasitic nematodes feed on roots and other underground parts of plants. The search for non-phytotoxic chemicals that are translocated downwards from leaves and stems to roots and underground shoots has so far been pursued with the empiricism, and not a little of the frustration, that marked the ancient alchemists' search for the philosopher's stone! Peacock (1966) reviewed progress made up to that date in screening compounds for plant chemotherapy and Nelmes *et al.* (1973) have summarized the latest successful developments.

So far there has been little success in applying chemicals to growing plants to control *Heterodera, Globodera, Meloidogyne* and other root endoparasites; there has been even less progress against the root ectoparasites *Xiphinema, Longidorus, Trichodorus, Paratrichodorus* and various Tylenchid species although recent experiments with aldicarb and oxamyl on sugar beet and potatoes have given promising results. However, there have been notable advances in the chemical control of the endoparasites of stems, buds and leaves, e.g., *Ditylenchus dipsaci* (stem nematode) and *Aphelenchoides ritzemabosi* (a leaf and bud nematode). Especial progress has been achieved in the treatment of bulbs, corms and seeds against stem nematodes, and in spray or drench treatment of chrysanthemums and other ornamental plants against leaf nematodes.

Plant Chemotherapy

NARCISSUS

Narcissus bulbs remain in the soil for at least two years and stem nematode infestations are usually more obvious in the second and subsequent years than in the first. Early work on chemical control was concentrated on narcissi, mainly to find alternatives to hot-water treatment (see p. 305) which was very effective in killing the nematodes but difficult to accomplish correctly on a large scale; incorrect timing or excessive water temperatures often damaged the bulbs. Chitwood, Haasis & Blanton (1941) added formaldehyde solution to the hot water and this is now standard practice not only in America but also in Europe. Staniland (1950b) tried various phenolic additives but they have proved to be unpredictably phytotoxic in practice, though both chlorphenol and formaldehyde greatly improve kill of nematodes. Newton, Hastings & Bosher (1933) forced silver nitrate/potassium cyanide mixture into bulbs under vacuum and achieved 96 per cent kill of nematodes without phytotoxicity. However, the treatment was too expensive and it was impractical to treat large bulks of bulbs in this way. Thomas (1965b) and Purnell & Hague (1965a) found that fumigation with methyl bromide was phytotoxic and not completely effective as some nematodes survived treatment.

Feder (1952) and Bergeson (1955) drenched growing narcissi with suspensions of demeton (Systox) and diethyl-2 chlorovinyl phosphate (Shell OS 1836) and Bergeson also used octamethyl phosphoramide, but effective doses of all three chemicals were phytotoxic. Oliff (1966a) successfully drenched narcissi with thionazin; three years afterwards plants still showed no symptoms and populations of nematodes in bulbs and soil were still small; unfortunately this result has not been repeated, either experimentally or in the field. Purnell & Hague (1965a) and Oliff (1966b) found that dipping bulbs in thionazin solution was as effective as hot-water treatment but the chemical was useless as an additive to the latter; all bulbs died when thionazin at 1800 ppm was added to water held at 100°F (37·8°C). Experience in the field with thionazin dips for narcissus bulbs showed that the degree of nematode kill was not sufficiently great for treated bulbs to be planted out for more than one year. Survivors caused severe disease in the second and subsequent years after planting. At present, therefore, there is still no reliable chemical alternative to hot-water/formaldehyde treatment for pre-planting bulb treatments, except possibly for forced narcissi; a thionazin dip might be considered if stocks of bulbs for forcing are found to be infested with stem nematode just before they are boxed in late autumn, usually much too late for hot-water treatment to be feasible (Winfield, 1970a, 1971). Granules containing aldicarb or oxamyl applied at various rates to transplanted narcissus cv. King Alfred, infested with *Ditylenchus dipsaci*, greatly reduced numbers of nematodes in shoots and bulbs at harvest time; carbofuran was moderately effective but phorate was no better than untreated controls (Maggenti & Hart, 1975b).

TULIP AND MUSCARI

Tulips are an annual crop, the bulbs being replanted each year. Most of the nematological problems associated with narcissus culture are therefore absent and chemical control of stem nematode is a more practical proposition. Slootweg (1963) and Oliff (1966a) tested drenches of thionazin, Shell OS 1836 and parathion on growing tulips. None of the treatments was good enough to be adopted commercially but Slootweg (1963) found that thionazin drenches worked well on *Muscari*. Windrich (1970) found that uninfested tulips were protected for up to four months from invasion of nematodes from infested soil by dipping the bulbs in thionazin solution or by incorporating granular aldicarb at planting time.

During the 1960's, Hague and his colleagues at Imperial College Field Station (Hague, 1966; Oliff, 1966a, 1966b; Purnell, 1964; Purnell & Hague, 1965b) established the value of dipping tulip bulbs in thionazin solution before planting and this has been confirmed both in practice and by further experiments at Kirton Experimental Horticulture Station (Winfield, 1969, 1972, 1973). At first it was thought that the mode of action of thionazin was as follows: healthy tissue in the dipped bulbs absorbed the chemical and became poisonous to nematodes that fed upon it in much the same way as plants become systemically toxic to aphids and other sap-sucking insects. It is now believed that thionazin and similar organophosphates and also oxime *N*–methylcarbamates cause nematostasis, an arresting of the normal behaviour and metabolism of the nematodes inside the plants. Such eelworms do not feed and eventually some of them may recover. However, a proportion die in the meantime, the exact number depending on how long the stasis is exerted by the chemical.

After lifting and cleaning, tulip bulbs are immersed for 2½ hours in a cool (about 65°F (18°C)) solution of 0·23 per cent thionazin. The dip is prepared from the commercial 46 per cent emulsifiable concentrate at the rate of four pints per 100 gal of water (5 ml/l). For topping-up, as successive loads of bulbs remove the active ingredient and contaminate the water with soil, two pints of concentrate are added for every 25 gal (10 ml/l) of water added. The chemical deteriorates in the bath, especially if the bulbs carry a lot of soil, but it is not known for how long the dip can be used before it needs to be replaced completely. The solution should be continually agitated during treatment, preferably by using a circulating pump, to prevent the active ingredient settling to the bottom of the tank. Winfield (1969, 1972) showed that tulip bulbs were not damaged by dipping in thionazin during July–September, even at double the recommended strength and that a high degree of nematode kill could be expected from the recommended treatment.

OTHER BULBS, CORMS, TUBERS AND ROOTED PLANTS

Nematodes cause about 5 per cent loss in woody ornamental shrubs and as these crops in Britain were valued at over £12 million in 1972 (Hague, 1972) chemical means of control may often be well worthwhile. For example, of the 20 or so species of nematodes known to damage azaleas and rhododendrons most are controlled by dibromochloropropane treatment either before or after planting (Esser, 1967; Barker, Worf and Epstein, 1965). Other decorative foliar plants such as ferns and caladiums are attacked by *Aphelenchoides fragariae* and *Pratylenchus penetrans*; the latter is controlled by treatment of the soil in which the plants are growing with a mixture of thionazin and phorate granules at 4·5–9·0 kg/ha (Rhoades, 1968). *Meloidogyne incognita* on caladium is killed by hot-water treatment of 50°C for 30 minutes or by dipping the tubers in 1000 ppm thionazin or ethoprop; *A. fragariae* on begonias and gloxinias is controlled by a dip or drench of demeton-S-methyl or thionazin. Root-knot nematodes (*Meloidogyne* spp.) often attack container-grown plants; several authors have tried bare-root dips and drenches to control these pests, and also *Heterodera cacti* and citrus nematodes, with varying success (Perry & Miller, 1964; Miller & Perry 1965; O'Bannon & Esser, 1970; Rhoades, 1970; Johnson, Ratcliffe & Freeman, 1970; Johnson & Gill, 1972; Malek, 1974; Harlan & Jenkins, 1967; O'Bannon & Taylor, 1967).

The potato tuber nematode (*Ditylenchus destructor*) on bulbous iris has been controlled by dipping the bulbs in thionazin solution but the timing of treatment is important; sometimes the dip is phytotoxic (Anon., 1971). Granular aldicarb, carbofuran, oxamyl or phorate applied at various rates below the surface of the soil, before planting bulbous iris, prevented death of bulbs and decreased the number of *D. destructor* per bulb. Oxamyl as a drench (900 ppm) worked as well at 8 or 12 lb/acre (9 or 13 kg/ha) a.i. applied as granules (Maggenti & Hart, 1975a). Overman (1970) found that root-knot nematodes which survived cold storage in infested gladiolus corms were controlled by dipping the corms, before planting, in thionazin, fensulfothion or fenamiphos (Nemacur) at the rate of 0·5 lb/100 U.S. gal water (60 g/100 l). Greater doses of these three chemicals were phytotoxic but three other experimental nematicides at 0·75 lb/100 U.S. gal water (90 g/100 l) were also effective and did not damage the plants. Parathion improved plant stand, flower production and yield of new corms but some root-knot nematodes survived rates of up to 1·0 lb/100 U.S. gal water (120 g/100 l).

CHRYSANTHEMUM

The leaf nematode (*Aphelenchoides ritzemabosi*) is partially controlled on mid-season and late-flowering chrysanthemums by three applications of parathion at 8–10 fl. oz of 20 per cent emulsifiable concentrate per 100 gal (50–63 ml/100 l) water (Bryden & Hodson, 1957). The first drench is applied to the unrooted cuttings when they are set out, the second immediately the cuttings have been rooted and the last drench one month later. Additional protection against the nematode may be obtained by drenching the stool beds with double-strength parathion just before taking the cuttings. Unfortunately, local spread of eelworm is favoured in wet seasons or whenever the soil is very moist and in these circumstances parathion might not be effective. The nematode is more easily controlled with thionazin drenches applied to the rooted cuttings (Anon., 1972). The plants are drenched twice, at fortnightly intervals, with 10 fl. oz of the 46 per cent emulsifiable concentrate per 100 gal of water (63 ml/100 l). Cuttings should be well rooted and the soil thoroughly moistened by the drench because the chemical seems to be taken up by the roots. Thionazin is not phytotoxic and sometimes it seems slightly to stimulate growth of chrysanthemums.

Seed Fumigation and Seed-Potato Treatment

Wallace & Wood (1943) showed that stem nematode (*Ditylenchus dipsaci*) is carried on onion seed and Brown (1957) concluded that infested seed was responsible for infestations found in lucerne crops in East Anglia. Clover seed crops are carefully inspected and seed is fumigated from those in which plants showing symptoms are found. T. Goodey (1945) was the first to show that all the nematodes were killed when onion seed was fumigated with methyl bromide and J. B. Goodey (1949) successfully continued his father's fumigation experiments with infested seed of teazle and red clover. Page, Hague, Jakabsons & Goldsmith (1959) showed that methyl bromide fumigation killed all nematodes on lucerne seed and did not reduce germination, and that this treatment was practicable for 6 ton lots of seed beneath a gas-proof sheet. Fumigation of seed with methyl bromide at a concentration-time product (CTP) of 650 and 1250 oz hours/1000 cubic feet* completely disinfests onion and lucerne seed respectively. Shallot seed can similarly be treated against stem nematode (Hague, 1968). The bent-grass nematode *Anguina agrostis* was killed by CTPs of 600–800 mg h/l*; provided the moisture content of the seed was kept below 12 per cent there was little adverse effect on germination (Hague, 1963). Field-bean seed is a serious means of spread of stem nematode (Hooper, 1971), and tests by Powell (1974) indicate that good, though not necessarily complete, control is possible by methyl bromide fumigation at a CTP of 1000 mg h/l.

Wood & Foot (1975) immersed 300-kg lots of potato tubers contaminated with *Globodera rostochiensis* in dilute sodium hypochlorite solution (1 per cent available chlorine) which destroyed the cysts and their contents. The appearance of the tubers was enhanced by treatment; there seemed to be no residue problem and no adverse effect on culinary quality, on tubers in storage or on sprouting.

Mushroom Compost

Hesling (1972) has reviewed the nematode pest problems of mushrooms. Both *Ditylenchus myceliophagus* and *Aphelenchoides composticola* can seriously reduce yield and both are basically problems of hygiene in the composting, pasteurizing and casing processes, or during 'cooking-out' after cropping has finished.

Mushrooms appear in 'flushes', the first about five weeks after 'spawn' has been placed in the compost, or about three weeks after 'casing' which is done after the spawn has run. Any chemical treatment of the growing crop must be non-toxic to the mushrooms and non-persistent, because cropping continues for six to eight weeks and most mushrooms are either canned or eaten very soon after picking. Once compost has been spawned it should not be disturbed, so mixing nematicides with the compost after spawning is very difficult. Mushrooms are sensitive to many chemicals, and the sporophore on the surface is sometimes more sensitive than the mycelium so that chemical treatment of mushroom beds should be restricted to the compost (Hussey, 1967). The nematicide thionazin is effective at very low concentration (e.g., 4 ppm) against *D. myceliophagus* and *A. composticola*. No detectable residues were found in mushroom sporophores grown in correctly treated compost and thionazin was

*mg. h/l is approximately equivalent to oz. h/1000 cu. ft and g.h/m^3.

non-fungitoxic at concentrations of up to 80 ppm (Oliff, 1965; Hesling, 1966). The emulsifiable concentrate is diluted in water and applied as a drench to the compost as it passes through the spawning machine. Provided care is taken and suitable protective clothing is worn at this stage there is little or no toxic hazard to workers afterwards because the 'half-life' of thionazin in compost is about 14 days (Hesling & Kempton, 1969). McLeod (1973) showed that thiabendazole at 2·5–20 ppm and benomyl at 5–20 ppm inhibited multiplication *in vitro* of both species of nematode on *Agaricus bisporus* although the nematodes were not killed. Efficient pasteurization of compost before spawning and post-cropping 'cooking-out' should effectively control nematodes but occasionally these operations are not entirely satisfactory and some nematodes survive. An alternative to pasteurization is to fumigate the compost with methyl bromide under a gas-proof sheet and, instead of cooking-out, the same fumigant can sometimes be used to fumigate the houses which must be effectively sealed for the treatment to work; details of concentration-time products and safety precautions to be observed are given by Hesling (1972). Before methyl bromide is used it is wise to consult a contractor or a representative of the Ministry of Agriculture; in some countries, including Britain, the use of this dangerous gas is restricted.

Future Developments

New approaches to the control of parasitic nematodes (Nelmes, 1971) are the treatment of plants with substances that subtly alter their physiology, cell division processes or otherwise manipulate their internal environment so that plants become repellant to nematodes or prevent the endoparasites from developing normally. Some plant growth regulators and steroids inhibit the development of nematodes once they have entered the plants; the antimitotic agent morphactin (methyl-2-chloro-9-hydroxyfluorene-(9)-carboxylate) was found to suppress giant-cell formation in tomato and consequently reduced the rate at which *Meloidogyne javanica* developed (Orion & Minz, 1971). These approaches are still in the experimental stage but a better understanding of antimetabolites, antimitotic agents, sex-attractants, egg-hatching agents and the inherited resistance mechanisms of plants may suggest new lines to follow in controlling nematode pests of crop plants in future.

CHAPTER 15

Physical Methods of Control

J. F. SOUTHEY

Plant Pathology Laboratory, Harpenden

It is difficult to draw sharp distinctions between methods used for nematode control; for example temperature and other physical factors affect dispersion and toxicity of chemicals, while cultural practices used for control may depend on physical effects. This chapter is concerned with control treatments that depend primarily on application of energy, especially as heat, at levels lethal to the nematodes or on separating them from contaminated material, e.g., by washing. Chemical factors naturally present in the nematodes' immediate environment may play a part or chemicals may sometimes be added to improve control. Physical treatments may involve hazards for plants or operators but, except possibly with radiation treatments, there are unlikely to be residual effects.

Heat Treatments

STEAM STERILIZATION

Partial sterilization of soil by steam is a long-established practice in glasshouses and for the preparation of seed and potting composts; it has occasionally been used in the open but cost limits its use there to small seed or propagating beds. Insects, earthworms, nematodes, weed seeds, and many bacteria (including nitrifying bacteria) are killed by steam sterilization. Many of the ammonifying bacteria, however, form spores that can survive at 100°C for several hours. These bacteria multiply rapidly after steaming, leading to liberation of quantities of ammonia that would otherwise be converted by nitrifying bacteria into nitrates (Anon., 1959). Seedlings may be injured by free ammonia in seed and potting composts but, in soil borders and with more mature plants, the problem is avoided by balancing the extra nitrogen with potash in the base fertilizer and, if necessary, allowing time for dispersal of free ammonia. Many mineral nutrients become more available as a result of steaming but this may be a drawback rather than a benefit, e.g., where manuring has previously been excessive or where a small increase of micronutrients such as manganese may lead to phytotoxicity.

The most important nematode pests controlled by glasshouse steaming in Britain are potato cyst nematodes attacking tomatoes and root-knot nematodes on tomatoes, cucumbers, lettuces and some ornamentals. For potato cyst nematode under glass, thorough steaming is an efficient and lasting method, but some of the chemical sterilants, particularly methyl bromide, are rapidly replacing steam on both cost and efficiency grounds. Root-knot nematodes are generally easier to kill than potato cyst nematodes, nevertheless in cucumber houses it has sometimes proved surprisingly difficult to control them by steaming because of very rapid build-up in the warm conditions of these houses from galls on deep roots below the level reached by steaming. Methyl bromide with its power of deep penetration into soil has advantages over steam in this respect.

The process of soil heating by steam, the practical problems and the equipment used were described in the Ministry of Agriculture's Bulletin on Soil Sterilization (Anon., 1959). The following is an outline of the more important features. When glasshouse soil is sterilized by steam its final temperature is theoretically that of boiling water, 100°C; it cannot be greater and should be very little less. Steam is generally introduced through perforated pipes buried in the soil and heating takes place by condensation on the soil particles. When the layer of soil immediately around the steam outlets reaches 100°C, the steam passes through this layer to the cold soil beyond. Thus heating occurs by progressive advance of a thin layer or 'heat front' only about $\frac{1}{4}$–$\frac{1}{2}$ in. thick, which forms the boundary between hot and cold soil. If steaming is not continued until treatment is complete, part of the soil will be at 100°C and the rest will be cold. The soil surface is the last part to be treated and those parts of the surface directly above the pipes become hot before the remaining parts. It is essential that the soil surface should be covered during steaming; usually plastic sheets are employed and the formation of a bubble of steam under them indicates that treatment is complete.

In *loose* soil, heat generally travels about six inches downwards below the pipes for every foot above them, but there is virtually no penetration into consolidated soil. In practice, unevenness of soil texture causes deviations from this ideal pattern of heat flow. It is important to prepare the soil thoroughly to the required depth before steaming and to ensure that any large clods are broken up. The heat capacity of soil increases rapidly with moisture content. Soil should be as dry as possible before treatment, so as to economize in fuel and minimize the risk of puddling due to the increased moisture content from condensation of steam.

Steam may be injected into soil through perforated steel pipes buried below the surface or by downward penetration from a tented plastic sheet placed on the surface. Due to the laborious nature of the work and the high cost of labour, the older methods in which pipes were buried in trenches or spiked grids pushed into the soil have largely been superseded. The process has now been mechanized by the use of plough-type grids. These consist of a series of perforated pipes coupled together to form a rigid frame. The frame is buried at one end of a glasshouse and hauled through the soil by means of a winch, depth and direction being maintained by rollers and shares fitted to the frame. Steam is supplied through a flexible hose trailed behind the moving grid. Modern mobile grids will work up to 15 in. deep and sterilize 18 in. of loose soil.

Steaming from the surface downwards is attractive on grounds of low capital and low labour cost but the depth of penetration of steam is limited. Penetration rarely exceeds 10 in. and this only after passing steam for up to seven hours. The method has its main application in beds used for chrysanthemum and carnation production but it is of limited use for nematode control in soil borders.

ELECTRICAL SOIL HEATING

Electric soil sterilizers that make use either of soil resistance or of immersion heaters deal only with small quantities of soil but are useful for the amateur gardener (Anon., 1959). It is most important, however, that they should conform with safety requirements; in particular home-made sterilizers of the resistance type can be dangerous unless built by an expert.

Gillard & van den Brande (1959) investigated the control of root-knot nematodes (*Meloidogyne* spp.) by heating glasshouse soil electrically with wire netting (chicken wire) as a resistance; current was supplied by a transformer with an output of 16–20 volts. This system of soil warming was already in operation in many Belgian glasshouses for propagating frames and witloof-chicory forcing beds. Root-knot nematodes were controlled by maintaining a temperature of 50°C for one hour. This temperature was reached with a minimum of cost when soil temperatures were naturally at their highest, so treatment in late August or early September is recommended. The treatment also controlled soil insects and was cheaper than sterilization.

SHORT-WAVE DIATHERMY

Mai (1958) described experiments on killing encysted larvae of potato cyst nematode in large di-electric ovens used for the destruction of parasites of farm animals contaminating burlap bags (hessian sacks). He found that the treatment gave 100 per cent kill of the contents of cysts placed in small bags inside the bales of sacks and considered that, where such equipment were already available, it might be used to disinfest potato bags. (In this instance the ovens had been set up by a cooperative organization at an initial cost of $90 000 each!). Dielectric heating, or short-wave diathermy (heating by short waves of radio frequency) has obvious attractions for destroying parasites because of its penetrating power and the rapid heating obtained even at the centre of large masses of material. Conventional 'cooking' methods often result in overheating at the surface before the desired temperature is reached inside. In hot-water treatments, for example, the thickness of the plant material that can be treated is often limited by the time taken for heat to penetrate to the parasites within.

PEAK-HEATING AND 'COOKING-OUT' OF MUSHROOM COMPOSTS

Nematodes that destroy mushroom mycelium, such as *Ditylenchus myceliophagus* (see p. 87) and *Aphelenchoides composticola* (see p. 186) are controlled along with insect pests and fungi by the processes of peak heating of mushroom compost before spawning and 'cooking-out' after cropping (Anon., 1960; Hussey, Read & Hesling, 1969). Peak heating means raising the temperature of the prepared compost to about 54–60°C for at least a day; some growers peak-heat for several days. The process is as much a cultural as a pest-control one. Mushroom mycelium grows better and yields more in compost that has been produced at about 54°C; in practice it is difficult to maintain even temperature during composting but peak heating before spawning has a similar effect and results in better yields, though the exact reasons for this are not clear. Cooking-out is a similar process whose aim is purely hygienic, in which the mushroom houses are heated to 71°C (160°F) by injecting live steam for 12 hours after cropping and before removal of spent compost. In peak heating the temperature is maintained partly by the continuing fermentation of the compost, with assistance as necessary from the injection of live steam and by the control of ventilation. All mycelium-destroying nematodes should be killed provided that all parts of the compost are sufficiently heated. Moreton & John (1962) made temperature measurements in a fairly simple commercial peak-heat room and recorded considerable variations. The design of modern peak-heat rooms has decreased this variability and there should be no difficulty in killing nematodes.

HOT-WATER TREATMENT

Fortunately many nematodes are less tolerant of heat than their host plants, especially when the plants are at a dormant stage, as with bulbs, tubers, perennial root-stocks, etc. Water has usually proved the most suitable medium for heat treatment, partly because it is cheap and readily available and partly because dry heat tends to damage plants by causing water-loss, while sometimes making the infesting nematodes more resistant. The principles, methods and applications of hot-water treatment are fully described in the Ministry of Agriculture, Fisheries and Food's Bulletin on the subject (Anon., 1972). Therefore only the more important uses are reviewed below.

Historical Review with Special Reference to Bulb Treatments. The first published record of hot-water treatment of living plant material for nematode control appears to be that of Marcinowski (1909) who treated ferns and begonias for five minutes at 50°C to kill *Aphelenchoides fragariae* (then known as *Aphelenchus ormerodis*). In Ireland, Hewitt (1914) attempted to control *Ditylenchus dipsaci* in narcissus bulbs by soaking them in various liquids and chemical solutions and he appears to have been one of the first to try hot-water treatment of bulbs for nematode control. He immersed them in water heated to 120°F (49°C) for one, three and six hours; the nematodes were all killed, but so were the bulbs!

In 1917, in a lecture to the Royal Horticultural Society, Ramsbottom described his experiments on control of *D. dipsaci* in narcissus bulbs by immersion in water and various chemical solutions at different temperatures, and recommended hot-water treatment at 110°F (43·3°C) for two to four hours (Anon., 1917; Ramsbottom, 1918). Two other workers had already observed that *D. dipsaci* in narcissus bulbs was controlled by hot-water treatment at 110°F, namely an English bulb grower, J. W. Barr, and J. C. F. Fryer, then entomologist to the Board of Agriculture and later Director of the Plant Pathology Laboratory, Harpenden; both mentioned their findings in a discussion following Ramsbottom's lecture (Anon., 1917). Fryer's experiments had been concerned primarily with controlling narcissus fly larvae. In The Netherlands, van Slogteren (1920, 1931) investigated hot-water treatment for control of *D. dipsaci* in bulbs independently at about the same time as Ramsbottom. He recommended treatment of narcissi at 43·5°C for three to four hours and showed that similar treatment also controlled *D. dipsaci* in hyacinths.

In Britain, there is little doubt that the introduction of hot-water treatment saved the narcissus bulb industry, so serious was the nematode threat. However, many difficulties were encountered in efficient commercial application of the process and Staniland's work in the 1930s did much to put it on a sound practical footing. He investigated the killing time for different development-stages of *D. dipsaci* from bulbs, the penetration of heat into bulbs and temperature variations found in various commercial baths (Staniland, 1933; Staniland & Barber, 1937). His work drew attention, for example, to the greater resistance of the desiccated 'wool' stage of the nematode to treatment, the importance of ensuring that all parts of the bulbs were wetted, the effect of bath design on temperature distribution and the importance of using open-mesh containers for the bulbs. Staniland also explored the possibilities of hot-water treatment for controlling nematodes and other pests in a variety of other horticultural plants, e.g., violets (Staniland & Goodey, 1934), muscari, gladiolus and potatoes (Staniland, 1936, 1937, 1939). The chief aim of much of this work was to test the tolerance of the host-plant to hot water.

In Canada and the United States it was found that the three-hour treatment at 110°F, as used in Britain, was insufficient for fully-effective control of *D. dipsaci* in narcissus bulbs. Hastings (1933) quoted Weiss of the U.S. Department of Agriculture who had shown that active stages of *D. dipsaci* escaping into the water of the bath could survive even four hours at 110°F, and had proposed the addition of one or two per cent formalin to the water. Hastings' own experiments confirmed that some nematodes escaped the effects of the treatment and he concluded that the survivors were probably pre-adult juveniles (fourth stage) that occurred as 'eelworm wool', especially on the base plates of diseased bulbs. Hastings & Newton (1934) considered that bulbs should be treated for three hours at 110–112°F (43·3–44·5°C) as soon as possible after lifting so as not to allow time for eelworm wool to form. They drew attention to the fact that lethal time/temperature combinations for nematodes in water are likely to be different from those in bulb tissue.

Chitwood & Blanton (1941) considered that a practical treatment should have an efficacy of better than 90 per cent, preferably 95 per cent, and, after extensive tests over nine years, recommended treatment at 110°F (43·3°C) for four hours in water containing one pint commercial formalin (40 per cent formaldehyde) per 25 gal (1:200); this has since been standard practice in U.S.A. and has proved very effective there (Courtney, 1960). Formaldehyde is not very nematicidal at low temperatures but becomes much more so as temperature rises. Both Canadian and U.S. workers recommend pre-soaking of bulbs in water, usually at 70–80°F for two hours just before hot-water treatment. Hot-water formalin treatments at 110°F (43·3°C) have also been used in U.S.A. against *Ditylenchus destructor* in iris bulbs (three or four hours; Courtney & Gould, 1951) and *Aphelenchoides fragariae* in lily bulbs (one-hour treatment; Courtney & Breakey, 1945).

In Britain, the introduction of hot-water treatment for narcissus bulbs led to such a spectacular improvement in their health that for many years little attention seems to have been given to the precise degree of control achieved. Later, both here and in The Netherlands, it became apparent that the standard three-hour treatment at 110°F (or 43·5°C) frequently gave incomplete control. Some suggested that the nematodes had developed resistance to heat treatment, but a more likely explanation was that demands for higher standards of health, particularly for export, directed attention more critically to the residue not killed by conventional treatment.

Some instances of unsatisfactory control of nematodes in bulbs were undoubtedly due to inefficient equipment or incorrect use of it; other alleged failures may be traced to reinfestation as a result of poor hygiene or planting in infested land. Weaving (1960 and *in litt.*) investigated the efficiency and operation of commercial hot-water treatment plants; he confirmed many of Staniland and Barber's (1937) recommendations but modified others. He concluded, for example, that all hot-water baths should have efficient forced circulation, that, provided this was so, square tanks were as good as round ones, and that all tanks should be well insulated and fitted with insulated lids. Staniland relied on convection to maintain movement of water in smaller baths, the top being left open. But Weaving pointed out that convection is itself an indication of temperature variation which may be considerable. The popularity of open-topped baths in Britain was one reason why formalin was not usually added to the water, as in U.S.A.

There has been remarkable variation in the figures quoted by different workers for *in vitro* killing times of *D. dipsaci* from narcissus at various temperatures and the following table of average killing times at 110–111°F (43·3–44°C) illustrates this:

Killing Time	*Reference*
17 min (active stages)	Staniland, 1933
78 min ('wool' stage on outside of bulb)	Staniland & Barber, 1937
45 min (active stages)	Ramsbottom, 1918
at least 120 min (pre-adult juveniles)	Hastings & Newton, 1934
more than 150 min at 43·3°C (active stages)	Woodville & Morgan, 1961
246 min at 43·3°C (active stages) 122 min at 44°C (active stages)	Woodville, 1964

However, such variation is not surprising when one considers the different techniques used for *in vitro* treatments, the fact that determination of viability has usually been based on observation of movement after varying periods to allow for recovery, the different numbers of nematodes used in the tests, the varying proportions of different development stages, and the fact that all figures have been based on attempts to determine 100 per cent mortality, a notoriously difficult point to establish precisely (Finney, 1962). Figures for 50 per cent mortality would have enabled more valid comparisons to be made between determinations at different times and by different workers.

The estimates of time taken for complete kill *in vitro* make it seem surprising that hot-water treatment has been so generally successful. However, determinations of the time taken to kill nematodes suspended in water can be misleading. As Hastings (1933) and Hastings & Bosher (1934) implied, eelworm wool on the outsides of bulbs presents a problem, not so much because it is more resistant but because it is treated in water rather than in bulb tissue (Green, 1963). The addition of a suitable chemical, such as formaldehyde, is a sound way of dealing with these external nematodes.

Several refinements have improved the efficiency of hot-water treatment for control of *D. dipsaci* in narcissus bulbs and lessened the risk of damage to the bulbs. These include early lifting, storage at 25–35°C followed by presoaking, with a wetting agent, immediately before hot-water treatment, use of a wetter and formaldehyde in the treatment tank, use of anti-foaming agents, and increasing the temperature to 45–47·7°C (Anon., 1972a, 1975b). Use of these higher temperatures, both in England and The Netherlands, has been made possible by the warm-storage treatment. However this advantage has to be balanced against the fact that warm storage increases the resistance of the nematodes to heat (Winfield, 1970b) so that it is often better for known infested stocks to be lifted and hot-water treated early at no more than about 45°C.

Improvements in equipment and handling include use of front-loading tanks (often double—one for loading and one for holding water) and bulk handling systems. With these large-scale, semi-mechanized methods, water circulation, temperature control and post-treatment drying need extra attention; with front-loading tanks in particular, operation of a 'one-way traffic' system is no longer feasible so hygiene becomes doubly important and wetters can cause acute foaming problems unless anti-foamants are used.

Shallot and Onion Sets. In The Netherlands and Denmark treatment of shallot sets at 43·5°C for two hours was recommended to control *D. dipsaci* (Anon., 1972a). Southey & Jones (1963) tested this treatment on different cultivars of onion sets with and without the heat storage that is used to prevent bolting. They found that neither spring nor autumn treatment had adverse effect on the onions. Meijneke & Scheele (1970) found that treatment of onion sets at 45°C for three hours eliminated *D. dipsaci* without affecting yield of bulbs.

Treatment of Chrysanthemum and Strawberry. A further application of hot-water treatment is for control of *Aphelenchoides ritzemabosi* in chrysanthemums by treating the dormant stools. Here the problem is simpler than with bulbs, since one is dealing with nematodes that are either external contaminants or occur within shoot remains and leaf fragments, etc. attached to the stools—the heat does not have to penetrate to worms deep-seated within plant tissues. Staniland (1950a) showed that good control could be obtained by treating stools at 115°F (46°C) for 5 min with less damage to the plants than from the previously recommended treatment of 110°F (43·5°C) for 20–30 min (Hodson, 1933).

Hesling (1961) showed that it was virtually impossible to treat at constant temperature for so short a time as 5 min by automatic means like those used for the longer bulb treatments. However, Bryden (*in litt.* and Anon., 1976) developed a simple modification of the five-minute treatment in which sufficient stools were added to a bath of water at about 48°C to bring the temperature down to about 46·5°C. Usually no further heating is necessary and the stools are removed 5½ min later when the temperature should be about 46°C. Although the short treatment is usually safer and more convenient, Bryden showed that a few varieties were less harmed by the older treatment of 110°F (43·3°C) for 20–30 min and he found it possible to classify chrysanthemum varieties according to their tolerance of both systems of treatment.

Staniland (1953) reviewed early attempts to control nematodes in strawberry runners by hot-water treatment—usually at 110°F (43·3°C) for 20 min. This treatment never became regular practice because it tended to check the plants too severely. Following success with the 46°C treatment for chrysanthemums, he treated strawberry runners at that temperature and obtained good control of *Ditylenchus dipsaci* without damage to the plants, by treating for 7 min. Slightly longer treatment appeared to be necessary to kill *Aphelenchoides* spp. in strawberries, and for these he suggested 10 min. Carroll, McKay & Hannigan (1957) found that a more severe treatment (49°C for 20 min) was necessary to produce stocks of nematode free runners for a strawberry certification scheme, potted plants being treated upside down with only the crown and leaves immersed. However, Staniland's 10 min treatment at 46°C remained the standard recommendation in Britain, though it is now scarcely used commercially because most growers prefer to obtain healthy runners from specialists (Anon., 1972a).

VAPOUR-HEAT AND HOT-AIR TREATMENTS

Vapour-heat treatments, i.e., exposure to heated, moisture-laden air, have been used to control various plant pests, and particularly bulb flies. Air is passed through a mixing chamber with steam and cold-water sprays, balanced to give the required temperature, and the mixture circulated through an insulated treatment chamber containing the bulbs or other plants (Latta, 1932). The method has some possible advantages over hot-water treatment in reducing risk of fungus infection and facilitating subsequent handling and drying, but Chitwood & Blanton (1941) considered that it had no particular advantage from the point of view of bulb nematode (*Ditylenchus dipsaci*) control. Their experiments indicated that some of the nematodes survived plain vapour-heat treatments, just as they did hot-water treatment without formalin, unless the treatment was severe enough to damage the bulbs. *D. dipsaci* survived vapour-heat treatments for three, four and five hours at 114, 116 and 118°F (45·6–47·8°C). Presumably, as with hot-water treatment, the problem is to kill the quiescent pre-adult juveniles or 'wool' on the outsides, or between dry scales, of the bulbs. Chitwood & Blanton suggested less severe vapour-heat treatments preceded by a formalin pre-soak, but low concentrations of formaldehyde or other toxic gases introduced with the water vapour in the treatment chamber might be an alternative.

Sherman (1936) similarly compared hot-water and vapour-heat treatments for control of *Meloidogyne* sp. in *Polianthes tuberosa*; she found that, whereas hot-water treatment for 30 min was effective at 48°C and above, with vapour-heat 30 min at 51°C was necessary for complete control. However, in this instance the plants were said to withstand all treatments without injury.

Dry heat would be expected to be less successful than moist heat for treating nematode-infested plants (see p. 305). Klingler & Lengweiler-Rey (1969) tested active and desiccated specimens of *D. dipsaci, D. myceliophagus* and *Anguina tritici* for survival at 50°C in water and at 100 or 50 per cent humidity. All three species were most susceptible when active in water and most resistant when in the desiccated state at 50 per cent relative humidity. Nevertheless hot-air treatments at 45°C for 30 h and 50°C for 4–8 h have successfully eliminated root-knot nematodes (*Meloidogyne* spp.) from sweet-potato roots used for propagation, and in some instances proved less injurious to the plants than hot-water treatments (Thomason, van Gundy & McKinney, 1960; Martin, 1962).

Electrical Treatments

Several attempts have been made to control nematode pests by electric currents, usually applied to soil, though it has not always been easy to distinguish between true electric shock effects and heat effects arising from electrical resistance. Electrical soil heating as a nematode control method has already been discussed.

In 1952, Daulton & Stokes reported that root-knot nematodes were destroyed in dry soil, damp soil and water on exposure to a pulsating electrostatic field. In tests in Rhodesian tobacco fields they claimed good results with a machine that treated soil to a depth of 24 inches (0·6 m) over a track width of 36 inches (0·9 m). However, more critical laboratory experiments (Stokes & Martin, 1954) showed that about 800 volts per inch (32 v/mm) were needed to kill larvae in water films and that it would be impossible to produce sufficiently high voltages in the field with the equipment previously used.

Turligina & Vershinskiy (1957) have also described small-scale laboratory experiments in which soil samples containing *Meloidogyne* larvae were exposed to high-voltage pulses in glass tubes. They reported considerable control with 30–100 v/cm and complete kill with 500–600 v/cm. Caveness & Caveness (1970—two papers) found that *Meloidogyne* larvae in tap water were killed by potentials of 5 v/mm and greater but 60 v/mm were needed to prevent larval hatching from treated egg masses; with 1, 10 and 20 v/mm hatching was actually increased relative to the controls.

Lear & Jacob (1955) attempted to repeat Daulton & Stokes' field tests, without success. In more critical laboratory experiments, using root-knot infested soil samples, they failed to demonstrate a significant reduction in the number of galls produced on tomato seedlings grown on the treated soil, even with repeated application of considerably higher voltages, and concluded that high-voltage, non-thermal electrical soil treatments were impracticable. Thus, although it appears possible under some conditions to kill nematodes by electric shock, the very high energy applications that would be needed to have useful effect in soil treatments make it unlikely that the method will have any practical application.

Radiation Treatments

Green & Webster (1966) examined the possibility of using ultra-violet radiation to treat recirculated bulb washing water that had been contaminated with *Ditylenchus dipsaci*. They found that *D. dipsaci* failed to reproduce after doses of 100 μw/min/cm^2, were immobilized after 10 000 μw/min/cm^2 and ruptured after 80 000 μw/min/cm^2. Green & Plumb (1967) found that UV dosages needed to inhibit activity of potato cyst nematode (*Globodera rostochiensis*) and beet cyst nematode (*Heterodera schachtii*) were similar to those for *D. dipsaci* but that much lower dosages (500 μw/min/cm^2) would prevent establishment of *G. rostochiensis* larvae in host-plant roots. However it is difficult to see how the latter result could be put to practical use.

The fact that X-ray treatments were used to inhibit sprouting of potatoes in storage led Fassuliotis & Sparrow (1955) to examine the effect of such treatments on encysted larvae of *G. rostochiensis*. They found that X-ray doses of 20 000 r and above (r = roentgen) prevented formation of embryonated eggs in females that developed from the irradiated larvae and considered that treatments to inhibit sprouting of tubers might, therefore, also remove the risk of spreading the nematode. However, more extensive tests with both X- and γ-rays (Fassuliotis, 1958—abstract) showed that at least 80 000 r were required to produce sterility. Weischer (1957) tested the effect of radium treatment on cysts of *G. rostochiensis* and *H. schachtii* and observed no apparent effect after less than 72 hours at 400 r/h (total dose 28 800 r). After longer exposure, up to 196 h, hatching of *H. schachtii* was reduced, but that of *G. rostochiensis* increased. Development and reproduction rate of both species was retarded. Weischer noted that these nematodes showed remarkable resistance to ionizing radiation compared with other animals. Evans (1970a) studied the effects of γ-rays from a Co^{60} source on second-stage larvae, adults and cysts of potato cyst nematode and Townshend (1968) similarly tested beet cyst nematode. Both found that high dosages were needed to decrease hatching from cysts but that irradiation of white females could markedly affect their fertility. Evans (1970a) noted that larvae from irradiated cysts competed poorly with those from untreated cysts when potato plants were inoculated with mixtures of the two.

Wood & Goodey (1957) studied the effect of γ-rays on samples of mushroom compost infested with various nematodes. They found that the inactivation dose for *Ditylenchus myceliophagus* and *Rhabditis* sp. lay between 48 000 and 96 000 r.

Most of the above findings are in general agreement with the results of Myers & Dropkin (1959) who made a thorough study of the effects of γ-rays from a Co^{60} source on eleven species of plant-parasitic nematodes. The dose for complete sterilization was at least 40 000 r and usually above 160 000 r; however, reproduction of most species was considerably reduced by 80 000 r. They concluded that soil treatment to sufficient depth at such doses was impracticable, and treatment of plants was out of the question because their roots were much more sensitive to radiation than were nematodes.

Krnjaić (1973) irradiated galls of *Anguina tritici* using a Co^{60} source and used them for inoculation tests on wheat. At 200 krad (absorbed dose) symptoms were produced but no galls; at 300 krad and above both infection and gall formation were apparently prevented. But before such treatment could be used for seed grain, effects on germination would need to be checked carefully.

Bingefors (1971) has drawn attention to a possible indirect use of radiation treatments in nematode control, namely the use of radiation-induced mutations in breeding for resistance.

Ultrasonics

Kämpfe (1962b) has examined the effect of high-frequency vibrations in the ultrasonic range on potato cyst nematode, beet cyst nematode and several genera of freeliving nematodes (Rhabditida). Free larvae of the cyst-forming species were more resistant to the treatment than were the other genera and intact cysts were scarcely affected. He concluded that the possibility of the treatment being put to practical use for nematode control was slight, especially against cyst-forming species.

Washing Processes

Plant-parasitic nematodes are often spread in soil adhering to potato tubers, bulbs and other planting material, and such spread is especially important with cyst-forming species. Careful washing can considerably reduce the risk but, in practice, it is difficult to wash with sufficient thoroughness to remove all soil. Commercial vegetable or sugar-beet washing equipment, for example, has not proved efficient enough and causes too much damage to potato tubers. However, a potato-tuber washing apparatus was developed in Scotland that cleans seed potatoes to a remarkably high standard (Mabbott, 1956, 1960). The tubers are first passed through a modified vegetable-washing plant in which they are washed with high-pressure water jets while being turned and scrubbed by rotating rubber rollers with surface projections. They are next soaked in a tank containing a fungicide to prevent rot diseases, then passed under a second battery of water jets, and finally excess moisture is removed by passage between sponge-rubber rollers. Where conditions are suitable this process can greatly reduce the risk of transporting nematode cysts with seed potatoes. (See Chapter 14, p. 300 for chemical treatment of potato tubers).

It is well-known that when nematode-infested plant material is suspended in water, nematodes emerge from the tissue and settle in the bottom of the container. This principle is made use of in the Baermann funnel method for extracting nematodes for examination, and it is at least theoretically possible that it might be applied as a control measure for removing them from plants. The method appears to have been used with some success in the U.S.S.R., e.g., Danilov (1952) stated that the number of stem nematodes (*Ditylenchus* sp.) in strawberry plants could be considerably reduced by placing the plants on a sieve immersed in water at 14–17°C for five days, the water being replaced daily. However, the method might be expected to have greater chance of success with ectoparasitic nematodes, such as *Aphelenchoides* spp. on strawberry or some root-infesting genera, than with endoparasites like *Ditylenchus*. Sveshnikova (1956) reported that soaking garlic and onion seed in water at 15–18°C for three or four days, with changes of water, removed *Ditylenchus allii* (probably = an 'onion race' of *D. dipsaci*) and produced 85–100 per cent healthy bulbs. Petrova (1961) described similar work with onion sets: soaking the sets in water for three or four days at 18°C significantly reduced infestation by *D. allii*. She did not consider that under the local conditions, true seed played any important part in spreading the nematode.

Seed Cleaning

Anguina tritici, which causes 'ear cockles' of wheat (see p. 91) is no longer found in Britain though formerly quite a common and serious pest. Although it has been recorded from most parts of the world where wheat is grown, it is probably only of consequence now in less developed countries, and this fact can largely be attributed to the development of mechanical seed-cleaning methods which remove the galls or cockles along with weed seeds and other foreign matter. More refined seed-cleaning methods might be expected to assist in the control of other seed-borne nematodes, e.g., *Ditylenchus dipsaci* occurring in small particles of debris associated with clover and lucerne seed (see also p. 300).

Acknowledgments

I am indebted to Mr. G. F. Sheard and his colleagues at the Glasshouse Crops Research Institute, Littlehampton for help in the revision of this chapter.

CHAPTER 16

Nematode-resistant Crop Plants

H. W. HOWARD

Plant Breeding Institute, Cambridge

J. COTTEN

M.A.F.F., Plant Pathology Laboratory, Harpenden

Introduction

For certain crops the use of nematode-resistant plant cultivars may be the only method of control available, or economic on a field scale. Provided the resistant cultivar is equivalent in other characters, for example yield and quality, to the best susceptible cultivars, its use will entail no extra cost. Additionally, control will not depend, as it does with some nematicides, on having suitable soil and climatic conditions, or on awaiting a suitable stage in the growth of the crop.

It seems likely that the use of nematicides will become commoner, particularly with high value crops, as new nematicides are developed and the cost of using them is reduced. For field crops of low value, such as cereals, and where nematodes are limiting yields only to a small extent, it is unlikely that nematode control by chemical means will become an economic proposition in the forseeable future, particularly where the crop is grown in monoculture.

Integrated control, using both nematicide and resistant cultivar, has advantages over either method alone (see also Chapter 18). Nematicides do not usually produce a complete kill but may, additionally, stimulate growth of a following crop; thus surviving nematodes are able to reproduce rapidly so that the post-cropping population may be as large as or even larger than the initial. On the other hand, most nematode-resistant cultivars inhibit growth and development of the nematode, so that the post-cropping population is decreased.

The occurrence of resistance to nematodes varies with the type of parasitism. Some ectoparasitic nematodes are able to feed, develop and reproduce on a wide range of plants, their feeding apparently producing little reaction in the host tissue. Nematode resistance has been found most frequently where there is a more complex host-parasite interaction, and results from changes in the balance of reactions between parasite and host. Resistance occurs for example to migratory endoparasites like *Aphelenchoides, Ditylenchus* and *Pratylenchus,* but especially to the sedentary endoparasites such as *Heterodera, Globodera* and *Meloidogyne*.

Resistance as measured by the ability of the nematode to feed, develop and reproduce on its host could operate at any time during its life-cycle. Hatching factors for cyst nematodes do not play a part in resistance mechanisms, neither is there any clear evidence that nematodes are less attracted to roots of resistant plants. There are some cases where larvae fail to penetrate resistant plants or do so in reduced numbers and secretions from the roots of some plants e.g., *Tagetes* and *Asparagus* (see Chapter 13) are toxic to nematodes. Mostly though, resistance operates after invasion and results in the larvae being unable to complete their development and reproduce. There may be damage to plants resulting from the initial invasion, but this can be minimized by the use of integrated control measures. Additionally there may be pathotypes of the nematode which can develop normally in a resistant plant; such pathotypes occur commonly in the cyst nematodes.

Tolerance to nematode attack, that is, the ability of the host to withstand nematode invasion without affecting its performance, has been little studied although factors for both resistance and tolerance to *Ditylenchus dipsaci* have been reported in oats.

The mechanism of resistance has been investigated for a number of host/nematode interactions. Resistance is usually associated with the inability of the nematode to induce the correct response in the host. This may result in a hypersensitive reaction as in *Chrysanthemum* cvs. resistant to *Aphelenchoides,* a failure to induce adequate cell separation as in oats resistant to *D. dipsaci* or a failure to induce adequate giant-cell formation as exemplified by many plants resistant to *Meloidogyne, Heterodera* and *Globodera.*

The inheritance of resistance may be simple, due to single dominant or recessive genes, or polygenic. Resistance to the same nematode may be controlled by widely differing genetic systems; for example, resistance in the genus *Solanum* to *G. rostochiensis* is due either to single dominant genes or to polygenes. Resistance controlled by single genes is easy to use in breeding programmes, but is more likely to prove ineffective against resistance-breaking pathotypes than that controlled by polygenes. However, because of their slow rate of reproduction and dissemination, such pathotypes of nematodes do not present as serious a problem as do pathotypes of fungi producing millions of air-borne spores.

Resistance to nematodes has been known for many years. In 1889 Neal pointed out the possibility of solving certain nematode problems by the use of resistant varieties or 'stocks' and Zimmermann in 1897 advocated grafting on resistant Liberica stock, as a means of improving yields of Arabica coffee on nematode-infested land. Early genetic studies of nematode resistance clearly recognized the possibility of producing nematode-resistant cultivars (Nilsson-Ehle, 1920). Breeding has been carried out on a wide range of crops for many years; one of the earliest successes being Merkur, a red clover variety resistant to *D. dipsaci* developed by the Swedish Seed Association at Svalöf and released in 1937.

In this chapter, mostly crops of agricultural or horticultural importance in north western Europe are considered. Thus resistance to *Heterodera glycines* is not included, although several resistant cultivars of soya bean have been developed in the U.S.A. where this nematode is economically important. (See also Chapter 7, p. 171). Reference to work on nematode resistance in other crops can be found in *Helminthological Abstracts*. In addition, useful summaries of information on nematode resistance are provided by Hare (1965), Kehr (1966), Malo (1964), Rohde (1965, 1972) and Webster (1969).

Potatoes

YELLOW POTATO CYST NEMATODE

Work on breeding potato cultivars resistant to the yellow potato cyst nematode, *Globodera rostochiensis* (Woll.), has been extensive during the past two decades following the discovery by Ellenby (1954b) of resistance in the wild

Argentina potato, *Solanum vernei*, and in five clones of the cultivated Andigena potato, *S. tuberosum* ssp. *andigena*. Ellenby's criterion of resistance was the absence of new cysts when plants were grown in infested soil and this has also been the criterion used by breeders.

Initially, breeders concentrated on the use of the Andigena source and in particular clone CPC 1673 which originated from material collected at La Paz, Bolivia. Many resistant cultivars have been bred from this source and they include the British varieties Maris Piper, Maris Anchor, Pentland Javelin, Pentland Lustre and Ulster Glade. By far the most widely grown of these is Maris Piper with some 25 000 hectares in 1976.

The resistance of CPC 1673 is due to a single dominant gene, H_1, and, as the Andigena potatoes are cultivated clones which are similar to those brought to Europe at the end of the 16th century and which were the ancestors of the Tuberosum potatoes (*S. tuberosum* ssp. *tuberosum*), it was comparatively easy to produce cultivars with gene H_1 which were commercially acceptable in long-day conditions (the Andigena potatoes are adapted to form tubers in short days).

Cultivars with gene H_1 produce a root diffusate which has a hatching power similar to that of susceptible cultivars and their roots are invaded by larvae in numbers comparable to those invading susceptibles (Williams, 1958). Although mature males may be produced, only a few larvae develop into mature females. The growing of a resistant cultivar such as Maris Piper therefore produces a marked decrease in the nematode population (Table 1).

Table 1

Decreases in Nematode Populations (Eggs) from Growing Resistant Potatoes and from Fallow or Non-host Crops (From Cole & Howard, 1962)

(a) *Resistant potatoes*

Author[1]	Egg population as percentage of initial after 1 year	2 years	3 years	4 years
Cole & Howard (a)[2]	48	23	14	15
Cole & Howard (b)	19	4	2	1
Huijsman (a)	30	15	4	1
Huijsman (b)	34	5	1	—
Goffart	8	1	—	—

(b) *Non-host crops or fallow*

Author[1]	1 year	2 years	3 years	4 years
Cole & Howard (a)	76	—	37	32
Cole & Howard (b)	35	28	27	25
Huijsman (a)	50	41	25	21
Huijsman (b)	67	44	36	—

[1] The references are given in Cole & Howard (1962).
[2] This population contained a small proportion of *G. pallida* which increased markedly after 4 years, leading to no decrease in the *G. rostochiensis* plus *pallida* total egg population after the third year.

The first soil with a cyst-nematode population to which cultivars with gene H_1 were not resistant was found by Dunnett (1957) at Duddingston, Midlothian. Subsequent survey work, e.g., Jones (1957, 1958) and Brown (1970), has shown that such populations are common in England and Wales. They are uncommon in Scotland (Mabbott, 1974). Although these populations were originally thought to be pathotypes of *G. rostochiensis*, it is now known that they are of a distinct species (see below). In Europe there are, however, several pathotypes of *G. rostochiensis* (Tables 2 and 3). Cultivars resistant to more than one pathotype of *G. rostochiensis* have been bred, e.g., the Dutch cultivars Mara and Proton which are resistant to pathotypes Ro1, Ro2 and Ro3. There has not yet been introduced a cultivar with resistance to the other species, *G. pallida.*

WHITE POTATO CYST NEMATODE

When it was found that there were populations of potato cyst nematode which could produce many cysts on the roots of resistant cultivars with gene H_1, such populations were described as being different pathotypes of *G. rostochiensis*. However, Guile (1966) showed that the pathotypes could be divided into two groups on cyst chromogenesis, one having a long yellow stage in the development of females into cysts and the other a long creamy-white stage. Subsequent work showed other differences and two species are now recognized, *G. rostochiensis* (yellow potato cyst nematode) and *G. pallida* (white potato cyst nematode) (Stone, 1973). Three pathotypes of *G. pallida* have so far been recognized in Europe (Tables 2 and 3).

Table 2

Scheme for Recognition of Pathotypes of G. rostochiensis *and* G. pallida *Proposed by Kort, Ross, Rumpenhorst & Stone* (1977)

Clone	Resistance code	Pathotypes* Ro 1	Ro 2	Ro 3	Ro 4	Ro 5	Pa 1	Pa 2	Pa 3
Susceptible Tuberosum	—	+	+	+	+	+	+	+	+
Andigena CPC 1673 (gene H_1)	Ro 1, 4	−	+	+	−	+	+	+	+
S. kurtzianum hybrid—KTT 60-21-19	Ro 1, 2	−	−	+	+	+	+	+	+
S. vernei hybrid G-LKS 58.1642/4	Ro 1, 2, 3	−	−	−	+	+	+	+	+
S. vernei hybrid $(VT^n)^2$ 62.33/3	Ro 1, 2, 3, 4 Pa 1, 2	−	−	−	−	±	−	−	+
S. vernei hybrid 65.346/19	Ro 1, 2, 3, 4, 5	−	−	−	−	−	+	+	+
S. multidissectum hybrid P55/7 (gene H_2)	Pa 1	+	+	+	+	+	−	+	+
S. vernei hybrid 69. 1377/94	Ro 1, 2, 3, 4, 5 Pa 1, 2, 3	−	−	−	−	−	−	−	−

*+ = many cysts; − = few or no cysts; see also Table 3.

Table 3

Relationship between Pathotypes of Kort, Ross, Rumpenhorst & Stone (see Table 2) and Previous Designations

Scheme of Table 2	Previous designation		
	British	Dutch	German
Ro 1	A	A	Hiltrup
Ro 2	—	B	Obersteinbach
Ro 3	—	C	—
Ro 4	—	F	—
Ro 5	—	—	Harmerz
Pa 1	B	—	—
Pa 2	—	D	—
Pa 3	E	E	Frenswegen, Charvorney

Several sources of resistance to *G. pallida* are known and breeding to produce resistant cultivars is in progress (Howard, 1972a, b). The first source used was the wild, diploid Argentina species, *S. vernei*. Hybrids from different stocks of this potato show differences in resistance to the several pathotypes of *G. pallida* and *G. rostochiensis* (Table 2). Resistance is thought to be due to polygenes; this makes breeding programmes difficult because they must consist of a series of back-crosses to standard Tuberosum cultivars. There are also difficulties in obtaining good cooking quality when a wild species is used as the source of resistance compared with using cultivated potatoes only (Howard, 1970). In addition the resistance of progenies bred from *S. vernei* is not easy to determine, being less than that for gene H_1 against pathotype Ro1 of *G. rostochiensis* and there being many plants with intermediate numbers of cysts on their roots (Kort, Jaspers & Dijkstra, 1952).

Another source of resistance is the wild species, *S. multidissectum*, which has a gene H_2 giving a very high degree of resistance to Pa1 but not to Pa2 and Pa3 (Table 2). However, some resistant clones with genes H_1 and H_2 have a high degree of resistance to Pa3 populations (Parrott & Trudgill, 1972). Other sources of resistance being used include the wild species *S. oplocense* and *S. spegazzinii* (Bouwman & Ross, 1972) and some clones of Andigena potatoes (Howard, Cole & Fuller, 1970). A clone bred from one of these Andigena potatoes was found to be resistant to a large number of *G. pallida* populations (Howard & Fuller, 1971). The resistance was at first thought to be due to a single gene, H_3, but later results (Fuller & Howard, 1974) do not support this hypothesis. Also, as with breeding from *S. vernei*, the progenies showed variable levels of resistance and there was usually no clear segregation into resistant and susceptible categories. No cultivar of commercial importance has so far been bred which has resistance to pathotypes of *G. pallida*.

OTHER NEMATODES ATTACKING POTATOES

Potato tubers can be attacked by the tuber nematode, *Ditylenchus destructor*. All cultivars appear to be susceptible but extensive investigations have not been made as the nematode appears to be of little importance in Western Europe (Moore, 1971). (See also Chapter 4). In semi-tropical conditions potatoes may be attacked by root-knot nematodes. Resistance to *Meloidogyne incognita* has been found in a number of wild potato species and breeding for resistance is being investigated in India (Nirula, Khusha & Raj, 1969).

Migratory nematodes of the genera *Trichodorus* and *Paratrichodorus* are important pests of the potato because they transmit tobacco rattle virus which may cause spraing in tubers (see Chapter 11). The cultivars Arran Pilot and Bintje do not show symptoms of rattle-produced spraing in their tubers (Harrison, 1968) but this is due almost certainly to a resistance of the potato cultivar to the virus and not to the nematode vector.

Sugar Beet

BEET CYST NEMATODE

It is possible to find plants of sugar beet with some resistance and this resistance may be controlled by polygenes (Curtis, 1970). Very high resistance is found in some wild species of *Beta* such as *B. patellaris*, *B. procumbens* and *B. webbiana*. The resistance of *B. patellaris* appears to be of a similar type to that conferred by gene H_1 against pathotype Ro1 of *G. rostochiensis*; mature males may develop but only occasional females (Shepherd, 1958). It has been found difficult to use these wild species as sources of resistance in breeding programmes, largely because of sterility barriers. Most progress appears to have been made by Savitsky (1975) using triploid hybrids between tetraploid sugar beet and diploid *B. procumbens*. She claims to have transferred a segment of the *B. procumbens* chromosome bearing the dominant gene for nematode-resistance to a sugar beet chromosome.

Tomato

POTATO CYST NEMATODES

Resistance to both species of potato cyst nematode, *G. rostochiensis* and *G. pallida*, has been found in two accessions of the wild tomatoes, *Lycopersicon pimpinellifolium* and *L. peruvianum* (Ellis & Smith, 1971). Resistance in one of these accessions, B6173 of *L. pimpinellifolium*, to one *G. rostochiensis* isolate was found to be due to a single dominant gene (*Hero*) and breeding work to transfer this gene to cultivated tomatoes has begun.

ROOT-KNOT NEMATODES

Resistance to root-knot nematodes (*Meloidogyne*) species has also been found in tomatoes and resistant cultivars have been bred. Resistance may not be effective against all species; for example, the cv. Nemared has resistance to *Meloidogyne incognita incognita*, *M. incognita acrita*, *M. javanica* and *M. arenaria* but is an excellent host for *M. hapla* (Dropkin & Boone, 1966).

Cereal crops

CEREAL CYST NEMATODE

Resistance to the cereal cyst nematode (*Heterodera avenae*) has been found in barley, wheat and oats. Nilsson-Ehle (1903, 1920), probably the first to describe resistance in barley, found that several spring cultivars, including Abed Rex,

Brage, Chevalier, Primus and Svanhals, were resistant. Wålstedt (ref. Fröier *et al.*, 1959) found races of cereal cyst nematode able to overcome the resistance of all the cultivars tested by Nilsson-Ehle.

Interest in resistance to cereal cyst nematode in Western Europe was renewed with the work of Andersen (1959, 1961) who found additional sources of resistance in barley and, perhaps more importantly, sources of resistance in oats. Andersen was also able to identify two pathotypes of cereal cyst nematode in Denmark. Only two barley genotypes were fully resistant to both pathotypes; these were a six-row barley No. 191, *Hordeum pallidum*, and barley No. 14, a Canadian six-row cultivar. Pajbjerg Kron, a European barley cultivar was partially resistant to both races. A number of West European barley cultivars had resistance to Pathotype 1 only; these included Fero, Pajbjerg Drost and Alfa and the cultivars Brage, Chevalier, Primus and Svanhals previously recorded as resistant by Nilsson-Ehle.

Resistance to cereal cyst nematode in the oat cultivars Gidgee, Guyra, Mortgage Lifter and Mulga was recorded by Millikan (1938) and in Heines Silber and Adler by Goffart (1941). Andersen (1961) was again largely responsible for the discovery of resistance in a range of oat genotypes, many of which were unadapted but nevertheless provided useful sources of resistance for the plant breeder. These included I 652 *Avena strigosa*; CI 194897 *A. abyssinica*; CI 2863, CI 3444, CI 3445, CI 5194, PI 175021 *A. byzantina* and I 376 *A. sterilis*. Most of the plant breeding programmes in Western Europe have subsequently used Andersen's material. Usable resistance was found in the spring wheat Loros CI 3779 (Nielsen, 1967).

The development of resistant cultivars may involve ten to twelve years of work and the screening of several thousand breeding lines if resistance has to be transferred initially from an unadapted or wild source. The facility with which this can be done depends on the genetic mechanism controlling resistance. Genetic studies have indicated that resistance to cereal cyst nematode is usually simply inherited. Resistance in six barley genotypes, Fero, Drost, No. 191, No. 14, CI 8334 (Marocaine 079) and CI 3902 (Morocco) was investigated by Cotten & Hayes (1969) and Hayes & Cotten (1971). No. 191, CI 8334, and CI 3902 have single dominant genes for resistance to pathotypes 1 and 2 (British nomenclature). No. 14 has a single incompletely dominant gene for resistance to pathotypes 1 and 2. Fero and Drost have dominant genes for resistance to pathotype 1 only. The gene in No. 191, designated Rha_2 (= Ha), is located on chromosome 2 (Cotten & Hayes, 1969; Andersen & Andersen, 1973). The genes for resistance in No. 14, CI 3902 and CI 8334 are closely linked to the Rha_2 locus on chromosome 2. The genes in Fero and Drost are probably at the same locus, but are independent of or only loosely linked to Rha_2.

Resistance in oats is probably more complex. Cotten & Hayes (1972) found that resistance in I 376 *Avena sterilis* is probably controlled by two dominant genes and in cv. Mortgage Lifter by two recessive genes. Resistance in PI 175021 *A. byzantina* is monogenic and dominant, and is probably at a locus distinct from the resistance loci in I 376 and Mortgage Lifter.

Resistance in the wheat genotype Loros is controlled by a single dominant gene (Nielsen, 1967) located on chromosome 2B (Slootmaker, Lange, Jochemsen & Schepers, 1974).

The existence of pathotypes of the cereal cyst nematode was clearly demonstrated by Andersen (1959, 1961). Further work in Britain (Cotten, 1963, 1967),

Netherlands (Kort, Dantuma & van Essen, 1964), Germany (Neubert, 1967; Lücke, 1969), Sweden (Wålstedt, 1967) and Norway (Støen, 1971) has confirmed the widespread occurrence of races of the cereal cyst nematode differing in pathogenicity. Over Europe as a whole, the pathotype situation seems complex, but this may be due in part to the non-standardization of testing methods; the subject has recently been reviewed by Cook & Williams (1972).

Essentially, populations can be divided into three types. Those to which a number of West European cultivars, including Fero, Drost and Alfa, are resistant have been designated pathotype 1 (Netherlands A). Populations able to break the resistance of Fero, Drost and Alfa, but unable to reproduce on No. 191 have been designated pathotype 2 (Netherlands C). Barley No. 14, CI 8334 and CI 3902 also have resistance to British pathotype 2. Stone & Williams (1974) were unable to find morphological differences or differences in soluble proteins between pathotypes 1 and 2 and concluded that they are not different species. Populations able to break the resistance of No. 191 have been reported from various countries and may be of more than one type. In S.E. Sweden populations producing cysts on No. 191 are distinguishable morphologically from *H. avenae* populations hitherto described (Andersson, 1973; Videgård, 1973), but resemble British pathotype 3 in cyst morphology (Cook, 1975a). In Britain pathotype 3 can reproduce on No. 191, CI 8334 and CI 3902. In The Netherlands, Race B can also reproduce on No. 191 but not on CI 8334 or CI 3902. Resistance to British pathotype 3 occurs in Harlan 43 (Hayes & Cotten, 1971).

In Britain, pathotype 2 predominates, while pathotype 1 occurs in pure populations and in mixtures with pathotype 2, so that the resistance of the cultivars Fero, Drost and Alfa is of little practical importance. Pathotype 3 occurs only rarely (Saynor, 1975). In Denmark, similarly, pathotypes 1 and 2 occur most commonly. In The Netherlands pathotype A (≡ pathotype 1) is most common, C (≡ 2) is less common, B is rare. Pathotypes 1 and 2 have also been reported from Sweden and Germany. Populations able to reproduce on No. 191 have been reported from both countries and also from Norway, Australia (Victoria) and India (Rajasthan).

The mechanism of resistance to cereal cyst nematode has not been fully investigated. Barley genotypes deriving their resistance from No. 191 are as readily invaded by the larvae of the nematode as susceptible genotypes, but development of the nematode within resistant plants is slower. Few adult females are produced on resistant plants, but males are produced on resistant and susceptible plants in equal numbers (Cotten, 1969, 1970a). Resistant and susceptible oats are also equally invaded by the nematode initially, but subsequently fewer larvae are found within resistant plants, either because larvae which fail to establish feeding sites migrate from the roots or die within the roots and become unidentifiable (Cook *et al.*, 1974).

The first cereal cultivars to be developed with resistance to the most prevalent pathotypes of cereal cyst nematode are now becoming available to growers. The spring barleys Ansgar, Sabarlis and Tyra have resistance to pathotypes 1 and 2 derived from No. 191. The spring oat Nelson has resistance derived from CI 3444. It is too soon to evaluate fully the use of these resistant cultivars, but field trials with near isogenic resistant and susceptible sister lines of Sabarlis and later trials with Sabarlis itself, have given clear indications of the value of resistant barley cultivars in farm practice (Cotten, 1970b; Hayes, 1970; Graham & Stone, 1975). In 18 field trials over a period of eight years carried out on sites with cereal cyst nematode infestations ranging from 3 to 150 eggs/g of soil, resistant lines outyielded comparable susceptible lines by an average of 8·6 per cent (Table 4). In addition, the resistant lines were effective in reducing cereal cyst nematode populations, thus producing higher yields in subsequent crops (Cotten, 1969; Hayes, 1970; Williams, 1970; Hayes & Cotten, 1971) (Table 5).

Unfortunately, resistant oat cultivars, including Nelson and selections deriving their resistance from I 376 *Avena sterilis*, demonstrate sensitivity to cereal cyst nematode invasion and do not yield well on infested land. Further work is needed to produce oat cultivars which in addition to controlling the nematode are able to give acceptable yields on infested land (Cook, 1975b).

Table 4

Comparison of Grain Yields of Resistant and Susceptible Barley Genotypes on Cereal Cyst Nematode Infested Soils (A.D.A.S./Welsh Plant Breeding Station Trials)

R = Resistant; S = Susceptible

	Site	Year	Grain yield t/ha		% R/S
			R	S	
ADAS Bangor[1]	Ruthin (Ty'n Llanfair), Clwyd	1967	4·72	4·52c	104·4
	Ruthin (Glan Hespin), Clwyd	1969	3·74	3·69c	101·4
ADAS Bristol[1]	West Lavington, Wilts.	1967	5·38*	5·11c	105·3
	Kingston Deverill, Wilts.	1967	2·90	2·91c	99·7
	Kingston Deverill, Wilts.	1968a	3·99*	3·37c	118·4
ADAS Reading[2]	Damerham, Hants.	1967	2·52*	2·22c	113·5
	Eaton, Berks.	1969	4·60*	4·25d	108·2
	Overton, Hants.	1969	5·40*	4·80d	112·5
	Steventon, Hants.	1970	2·76*	2·59c	106·6
	Abingdon, Berks.	1971	5·10*	4·20e	121·4
	Pitstone, Bucks.	1972	4·10*	3·20e	128·1
	Abingdon, Berks.	1972	7·00*	6·50e	107·7
ADAS Shardlow[1]	Southwell, Notts.	1969	5·22	5·62c	92·9
Welsh Plant Breeding Station[1]	Ruthin (Glan Hespin), Clwyd	1965	5·90*	4·93c	119·7
	Ruthin (Glan Hespin), Clwyd	1966	4·86*	4·43c	109·7
	Ruthin (Glan Hespin), Clwyd	1967	5·68	5·51c	103·1
	Ruthin (Glan Hespin), Clwyd	1968	4·67*	4·49c	104·0
	Ruthin (Glan Hespin), Clwyd	1969b	3·07	2·72c	112·9
Mean of all trials			4·53	4·17	108·6

* Difference significant at P = 0·05.
a Resown on 1967 plots.
b After four years of continuous resistant and susceptible lines.
c Comparison of near-isogenic resistant and susceptible lines.
d Comparison of resistant line with susceptible cv. Zephyr.
e Comparison of resistant cv. Sabarlis with susceptible cv. Proctor.
[1] From the Jubilee Report of the Welsh Plant Breeding Station 1919–1969, with the permission of the Director.
[2] From Graham and Stone (1975) Annals of Applied Biology **80**, 61–73, with the permission of the authors.

Table 5

The Population Density of Cereal Cyst Nematode and Yield of Grain after Cropping with Resistant and Susceptible Genotypes of Spring Barley[1]
(R. Gair, pers. comm.)

Nematode reaction of barley genotypes	Cereal cyst nematode eggs/g of soil		Grain yield 1968 t/ha
	Post-cropping 1967	Post-cropping 1968	
Resistant	15·4	6·8	3·99
Susceptible	49·9	21·2	3·37

[1] From the Jubilee Report of the Welsh Plant Breeding Station 1919–1969, with the permission of the Director.

ROOT-KNOT NEMATODES

Of several species of root-knot nematode (*Meloidogyne*) recorded in the field in U.K. (see Chapter 5) only *M. naasi* seems to be prevalent on cereals and this species has now been reported from several European countries and from the U.S.A. (d'Herde, 1965; Schneider, 1967; Grujičić, 1967; Golden & Taylor, 1967). The economic importance of *M. naasi* has yet to be demonstrated in the U.K., but field plot trials in California using cropping sequences that included non-host crops and resistant oat varieties have demonstrated that compared with continuous barley cropping, such treatment can double the yield of subsequent susceptible barley crops (Allen, Hart & Baghott, 1970).

Tests of 47 cultivars of *Avena sativa* and other *Avena* species by Siddiqui & Taylor (1970b) against an Illinois isolate of the nematode gave clear indications of differences of reaction between oat genotypes. No galling or reproduction was found on 10 genotypes, namely, Red Rustproof, Chinese (Hull-less), Swedish Select, Algerian, Pusa Hybrid J, La Prevision 13, Canuck, Suregrain, Avoine-Crena and Punjab Algerian. A further six genotypes showed only traces of galling and no reproduction. Tests with two Welsh isolates of *M. naasi* (Cook, 1972) confirmed differences in susceptibility between oat genotypes.

Tests of five isolates of *M. naasi* from England, California, Illinois, Kentucky and Kansas clearly indicated differences in host range (Michell *et al.*, 1973a, b, c). Any breeding programme for resistance to *M. naasi* would have to take such variability into account.

STEM NEMATODE

Damage by the stem nematode (*Ditylenchus dipsaci*) to barley, wheat, oats, rye and maize has been reported from various European countries. In Britain damage is confined to oats, and Goodey & Hooper (1958) found that isolates from oats were unable to attack wheat and barley. The stem nematode is known to occur in a variety of forms or races which differ in their host range, and this variability possibly accounts for the reports of damage to different cereal crops.

Goodey (1937) showed that the land-race oat Grey Winter and the bred cultivar S81 were resistant to the stem nematode. Although neither was immune to invasion, they did not exhibit 'tulip-root' symptoms as expressed in susceptible cultivars such as S82. In the U.S.A. Thomas (1939) tested 130 oat cultivars and found Victoria, Capa and Pampa to be resistant. Resistance in the spring oat Milford (S225) was recorded by McLeod, Golightly & Price (1954). The same cultivar has been recorded as resistant in Sweden (Andersson, 1966). Resistance in Grey Winter, S81 and S225 was confirmed by Jones, Griffiths & Holden (1955). In addition, resistance was found in Early Grey Winter, Unique and S172. Milford, S81, S172 and Unique almost certainly derive their resistance from Grey Winter. Griffiths, Holden & Jones (1957) tested some 250 cultivars and forms of *Avena,* representing several species, for resistance. Most cultivars of *A. sativa* were found to be susceptible. Resistance was found in Avoine d'hiver, Ladbreg, Osijek br. 191, Novi Sad br. 2, Pezega and Fluirse, all winter types. Resistance was also recorded in several unnamed types of *A. byzantina* and in two forms of *A. ludoviciana,* Cc4346 and Cc4347. Resistance in all the forms of *A. sativa* was probably derived from land-race types similar to Grey Winter.

In laboratory tests, Cameron & Speed (1958) found both resistance and tolerance factors to stem nematode in oats, Milford being described as resistant, Early Miller as tolerant and *A. ludoviciana* as both resistant and tolerant. Resistance in the spring-oat cultivar Manod (S235) was reported by Davies (1960) and French & Cairns (1960). The source of resistance in Manod is unknown but is probably not Grey Winter.

Resistance to the stem nematode in the U.S.A. was reported in the oat cultivar Curt (Brendler *et al.,* 1971) and in Belgium in the cultivar Greta which derives its resistance from S172 (Clamot, 1968).

Resistance to the stem nematode in wheat was reported by Belloni (1954) who found the genotypes Funo 31 and R37 to be resistant. Grujičić (1962) in tests with four wheat cultivars, Acciaio, Marimp 8, Leone and Mara, showed that Leone was most resistant though not immune. Resistance to stem nematode in maize was reported by Küthe & Dern (1970) in the varieities Inrakorn and Inrafrüh. The Dutch rye cultivar Heertvelder has resistance to stem nematode derived from the cultivar Ottersumse (Kort, 1972).

Genetic studies of resistance by Griffiths *et al.* (1957) showed that resistance in Grey Winter and in *Avena ludoviciana* Cc4347 was dominant and in Grey Winter was due to a single gene. Studies by Jones *et al.* (1955) of the mechanism of resistance showed that, in the field, resistant and susceptible genotypes were equally invaded and that it is only with subsequent reproduction of the nematode that differences between resistant and susceptible types emerge. This was confirmed in laboratory tests by Blake (1962a) who found that the numbers of larvae invading the susceptible cultivars Sun II and Powys and the resistant cultivars Manod and Pennant did not differ, although massive invasions of seedlings killed fewer resistant plants than susceptible ones. Nematode development and egg-laying was slower in resistant cultivars and the life cycle took twice as long. Histopathological studies (Blake, 1962b) showed that whereas the nematode caused cells to enlarge and separate in tissues of the susceptible cultivar Sun II, in the resistant cultivar Manod, separation and hypertrophy were slight.

Field trials with stem nematode resistant cultivars of cereals on infested land have repeatedly emphasized their superiority over susceptible cultivars, often giving acceptable yields of grain when susceptible crops are almost a total failure (McLeod, Golightly & Price, 1954; Jones *et al.,* 1955; French & Cairns,

1960; Brendler *et al.*, 1971; Küthe & Dern, 1970). Because the stem nematode has several generations in one growing season, the superiority of a resistant cultivar over its susceptible counterpart is very pronounced. In Britain, oat cultivars such as Milford, Manod, Pennant, Peniarth, and Maris Quest, most of them deriving their resistance from Grey Winter, have proved effective in controlling stem nematode in the field. Despite the great variability in the host range of *D. dipsaci,* there have been no authenticated cases where these cultivars have succumbed to attack.

WHEAT GALL NEMATODE

Wheat and rye are the principle crops attacked by the wheat gall nematode (*Anguina tritici*). Modern seed-cleaning techniques have eradicated this pest in the developed countries of western Europe (Marshall, 1960). (See also Chapter 4).

In parts of eastern Europe, the nematode is still a problem and a search for resistance among established cultivars and non-adapted forms of wheat is currently being undertaken. Of the 319 forms of wheat tested by Tešić (1969) seven were classified as resistant and of these the variety Acoude and a form of *Triticum vulgare* var. *lutescens* were recommended for further breeding work. Resistance has also been reported in the winter wheat cultivars Funone, Bison and Concho, and in the spring wheats Dica, Marquis and Măgurele 7 (Romesçu, 1969).

Forage legumes

STEM NEMATODE

Breeding for resistance to stem nematode (*D. dipsaci*) in red clover was initiated in Sweden and was followed by later work in England and The Netherlands. In Sweden, the resistant cultivars Merkur and Resistenta were produced by mass selection for conditions in the south and further work has produced the winter-hardy types Ulva, Disa and Reko (Bingefors, 1969). In England, resistance was found by chance in the broad red clover stock Pearses (Brown, 1959). Further tests of 21 red clover selections by Fiddian & Aldrich (1964) showed a high degree of resistance in the broad red clovers Pearses and Dorset Marl, and in the late-flowering types, S123 and Montgomery. Two more recent resistant cultivars, both included in the N.I.A.B. Recommended List of Red Clover Varieties for 1975/76, are Maris Leda, a tetraploid broad red clover bred at the Plant Breeding Institute, Cambridge, and deriving its resistance from the Swedish cultivar Silo and from Pearses, and Sabtoron, a diploid broad red clover developed at the Welsh Plant Breeding Station.

A comparison of the red clover selections Merkur and Ultuna showed that the nematode invades both resistant and susceptible forms, but fewer resistant plants are invaded and those that are contain fewer nematodes (Bingefors, 1970). Dijkstra (1957) investigated plants with varying degrees of symptoms and found that resistant selections had more plants with only slight swelling or none, no nematode multiplication occurring in non-swollen plants. Resistance is probably due to several genes with additive effect (Bingefors, 1970, 1971).

Stem nematode attacks on white clover were first reported from England but are also known from New Zealand and Denmark (Bingefors, 1969); most damage is probably caused by a distinct white clover race. Tests in New Zealand

(Williams, 1972) have shown differences in reaction between white clover selections and Bingefors (1971) lists eight resistant white clover selections: Milka, Milkanova, Lodi, Mira, Trifo, Zero, Mega and Adefa.

Damage to lucerne by stem nematodes has been reported and extensive resistance breeding work has been carried out both in South America and U.S.A. and latterly in Europe (Bingefors, 1969; Hunt, Faulkner & Peaden, 1972). Several resistant cultivars have been developed in the U.S.A.; probably the best known is Lahontan which derives its resistance from a Turkestan introduction P.I. 19304 (Hunt *et al.*, 1972). In tests in Scandinavia and the U.K., Lahontan was resistant but not agronomically adapted. Bingefors (1971) listed 11 resistant lucerne cultivars including three which originated from Europe, Alfa II, Vertus and Nematol.

Resistance is genetically complex (Bingefors, 1971), in one or two sources being controlled by one gene with tetrasomic inheritance and in others by minor genes. In Lahontan it is not possible to explain resistance on a factorial basis.

Hanna & Hawn (1965) compared Lahontan with the susceptible cultivar Grimm and found Lahontan had fewer invaded plants and fewer nematodes per plant. Electron microscope studies have failed to reveal the mechanism of resistance (Chang, Campbell & Griffin, 1973).

In Europe there is little evidence of variation in pathogenicity between populations of stem nematode attacking red clover, and resistant selections of red clover seem to be universally resistant; however, in the U.S.A. such differences have been demonstrated in the lucerne race (Bingefors, 1969).

ROOT-KNOT NEMATODES

Root-knot nematodes are prevalent on lucerne in the south-west and southern U.S.A. Resistance has been found to *Meloidogyne hapla, M. javanica* and *M. incognita acrita* and it is considered possible to combine resistance in one cultivar (Bingefors, 1969). Only *M. hapla* normally occurs in the field in cool temperate regions; resistance to this species occurs in the American cultivars Vernal and Hilmar (Bingefors, 1971).

Resistance to *M. hapla* in Vernal selection M-4 is due to the failure of larvae to invade (Goplen & Stanford, 1959) and is in contrast to most instances of resistance to root-knot, cyst and stem nematodes where it normally operates after larval invasion. Resistance to *M. javanica* in Vernal selection M-9 is of the latter type. In Vernal selection M-9, two separate but linked single dominant genes with tetrasomic inheritance determine resistance to *M. hapla* and *M. javanica* (Bingefors, 1971). Variation in pathogenicity has been demonstrated for different isolates of each of the three species *M. hapla, M. incognita acrita* and *M. javanica* attacking lucerne (Goplen, Stanford & Allen, 1959).

In the southern states of the U.S.A., red and white clovers are damaged by *Meloidogyne* spp. and some selection work for resistance has been carried out (Bain, 1959).

CLOVER CYST NEMATODE

Clover cyst nematode (*Heterodera trifolii*) has been reported on red and white clovers from various parts of the world but there is conflicting evidence as to its practical importance (Eriksson, 1972). Kuiper (1960) tested 100 clones of white clover and found that eight were resistant, including the British selections S100 and Kersey.

CHAPTER 17

Regulatory Controls

J. F. SOUTHEY
Plant Pathology Laboratory, Harpenden

In most countries the prevention and control of dangerous pests and diseases of plants is reinforced by statutory regulations when the organisms or diseases caused by them are judged to be sufficiently important to warrant such action. In Great Britain 'enabling legislation' is necessary to give Ministers the powers to make regulations (Orders) for specific purposes; thus the Plant Health Act 1967 is at present the basis of all British plant health regulations. Such regulations must be backed by an efficient inspection service, part of whose task is also to inspect exported plant material in order to satisfy the import regulations of other countries.

Cobb (1914), in a paper on citrus nematode, anticipated much regulatory activity that was to follow when he wrote, '*Tylenchulus semipenetrans* is distributed from place to place on nursery stock . . . and it follows that any method of inspection by which infested nursery stock can be readily distinguished from non-infested will make it possible to establish at least one efficient check on the spread of the disease.' The main aim of phytosanitary or 'quarantine' regulations is, indeed, to prevent or limit the spread of the more dangerous pests or pathogens. A secondary purpose is to improve the general health of planting material moving in national or international trade by controlling levels of other pests and diseases. Measures in the latter category are aspects of quality rather than quarantine controls and here it is useful to include domestic certification schemes, whether compulsory or voluntary, one of whose purposes is to make available healthy stocks or seed of particular plants.

McCubbin (1954) may be referred to for a detailed account of plant quarantine measures and an extensive bibliography on the subject; Morschel (1971) is also useful. Bingefors (1967) discussed factors affecting international dispersal of nematodes and means of reducing spread. Moore (1953) gave valuable guidelines on the principles that should govern plant quarantine work. He stressed particularly the need for a spirit of international co-operation in efforts to improve the health of planting material, rather than a too-nationalistic and protective approach by individual countries. He emphasized also the need for regulations to be soundly based scientifically. From both points of view the European and Mediterranean Plant Protection Organization (E.P.P.O.) plays a valuable role. To quote Cohen (1971), 'It strives to advise member countries on the scientific, administrative and legislative measures needed to prevent the introduction and spread of dangerous pests and diseases of plants and plant products so as to facilitate rather than hinder international trade.'

On a world-wide scale, the most significant step forward was the signing of the International Plant Protection Convention at an F.A.O. conference in Rome in 1951 (Anon., 1954; see also Moore, 1953). Under this Convention each contracting government undertakes, *inter alia,* to collect information on the incidence of pests and diseases in its territories, to run an efficient plant health inspection service and to issue phytosanitary certificates conforming to the model F.A.O. pattern.

Nematode pests often pose special problems in plant-health inspection work because of their small size, the ease with which they are spread especially in even small amounts of soil, and the difficulties of their detection and identification. Examples of the use of regulatory measures for nematode control are discussed below under three main headings: domestic, imports and exports.

Domestic Regulations and Certification Schemes

'QUARANTINE' CONTROLS

Domestic plant health regulations may be introduced when it is vital in the national interest to eradicate or contain a recently discovered or introduced pest or pathogen, or to control an established one that is threatening production of the crop(s) which it attacks. This is the quarantine rather than the quality role discussed above; good examples are provided by internal regulations against cyst-nematode species in various countries.

With soil-borne nematodes eradication can seldom be more than a pious hope, or may be attainable only at great cost, especially if host plants are grown regularly and ecological conditions favour establishment. Following the first North American record of *Globodera* (formerly *Heterodera*) *rostochiensis* (potato cyst eelworm, 'golden nematode') on Long Island, New York State, in 1941 (Chitwood *et al.*, 1942), an exhaustive eradication campaign was instituted by the U.S. Department of Agriculture, which involved massive soil-fumigation treatments and was assisted by building development over large areas of former potato-growing land. These measures resulted in virtual eradication of the pest from much of the area. Spears (1968a) gives full details of the official survey and quarantine procedures. All movement of plants, soil, etc. from land still under cultivation on Long Island is rigidly controlled. Nevertheless, the expenditure of many millions of dollars on the U.S.D.A. Golden Nematode Control Project, including most stringent import controls could not prevent the eventual appearance of *G. rostochiensis* in several other parts of North America near both the Atlantic and Pacific seaboards (Anon., 1962; Orchard, 1965; Sewell, 1967; Spears, 1968b, 1969).

North American outbreaks of potato cyst nematode probably originated via Europe rather than directly from the original source of this nematode group in the Andean region of South America (Jones, 1970). Potato cyst nematodes (*G. rostochiensis* and *G. pallida*—see Chapters 6 & 7) have been too long established in the traditional potato-growing areas of Europe for eradication to be considered with present control methods. Nevertheless most European countries legislate against them because of their serious economic effects and the aim is to prevent spread to apparently clean areas and to prevent multiplication to damaging levels. Usually crop rotation is prescribed, sometimes a minimum one-in-three rotation for potatoes on 'clean' land with longer intervals or total prohibition on infested land. Sometimes (e.g., in The Netherlands) some shortening of the rotation from, say, one in four years unconditional is allowed if other control measures—resistant potato cultivars, nematicide treatments—are used in specified integrated regimes. (See Hijink, 1972). In most instances movement from infested land of plants and soil is prohibited or restricted and regular soil sampling of potato land and nurseries is compulsory. Special attention is generally paid to seed potatoes to ensure that they are produced only on 'eelworm-free' land.

The Potato Root Eelworm Order (Northern Ireland) 1968, which revised and consolidated earlier orders going back to 1945, follows the general lines outlined above. The potato cyst nematode situation in Northern Ireland is more favourable than that in ware-potato growing areas of England, for example; cysts are fairly widely distributed but infestations are generally slight and do not occur near the important seed-potato producing areas. This situation is similar to that in western parts of England and much of Scotland, and is related chiefly to the prevailing type of agriculture, usually grassland and hill-farming with emphasis on livestock rather than ware potatoes and other arable crops.

The European Economic Community (E.E.C.) has issued a Directive on the control of potato cyst nematode. Like other E.E.C. directives this lays down minimum provisions that member countries must undertake to implement through appropriate national legislation. Thus, The Potato Cyst Eelworm (Great Britain) Order 1973 came into operation in that year as a result of the U.K. entry into the E.E.C. (The Northern Ireland Order, mentioned above, already met the Directive's requirements). The Great Britain Order requires that seed potatoes intended for marketing may be produced only on land on which no potato cyst nematode was detected as a result of official soil examination. It also prohibits the planting, except under licence, of any potatoes on, or removal of plants for transplanting from, any land declared by notice to be infested. Action under the Order is limited to land sampled by specially authorized officers, normally Plant Health and Seeds Inspectors of the Ministry of Agriculture, Fisheries and Food.

The British Potato Cyst Eelworm Order, though a 'domestic' regulation, was introduced, not because of a domestic plant-health need, but to meet an international (E.E.C.) obligation which concerns the interests of the Community as a whole. In that way it differs from, for example, the Northern Ireland Order which has both functions. In England and Wales, The Beet Eelworm Orders of 1943 and later revisions provide examples of purely domestic regulations. They placed restrictions on the growing of sugar beet and other host crops of *Heterodera schachtii* on infested land in order to protect the sugar beet crop from losses due to this pest. (See Jones & Dunning, 1972). When a Beet Eelworm Order was first introduced (1943) the distribution of *H. schachtii* was thought to be more circumscribed and emphasis was placed on prevention of spread as well as build-up to damaging levels. Population densities of this pest are usually low and it remains more a potential than an actual threat to beet sugar production, but it is now believed that its distribution more or less coincides with that of the sugar beet crop. The latest revision of the Order, The Beet Cyst Nematode Order 1977, provides powers to restrict growing of host crops on infested land, except under licence, if and when official surveys or other evidence show that controls are needed.

In The Netherlands *Ditylenchus dipsaci* in flower bulbs ranks high in importance among quarantine pests. The Dutch plant health regulations require inspection of all bulb crops and nematode-infested stocks must be given a prescribed hot-water treatment or destroyed. Imported narcissus bulbs are similarly hot-water treated, whether infested or not. The treatment is a severe one (four hours at not less than 45°C) being geared more to destruction of *D. dipsaci* than to ensuring satisfactory growth of the bulbs.

QUALITY CONTROLS, INCLUDING CERTIFICATION SCHEMES

Certification schemes aim to make available to growers stocks of planting material or seeds that are of a specified health and purity standard. Although these schemes are organized by Government Departments or semi-official bodies, they are often supported or even initiated by growers' organizations. They protect the interests of producers (by safeguarding the reputation of their products) and of consumers, and only secondarily may have a 'quarantine' function in lessening spread of pests and diseases. Where the national importance of the crop, and risks from pests and diseases, warrant it such schemes are made mandatory by being linked with the appropriate crop marketing (e.g., seed-potato) regulations.

Tuber infesting nematodes, especially *Ditylenchus destructor* (Chapter 4), are usually covered by tuber inspections in seed-potato certification schemes. In most European countries seed potatoes are not eligible for certification unless the land on which they were produced was soil-sampled before planting and the samples found free from potato cyst nematodes. These pests do not affect the quality of the tubers as planting material so this is an example of use of certification schemes for quarantine or phytosanitary purposes, in this case to lessen risks of spread of cysts with seed tubers.

In England and Wales, certification schemes for the highest grades of planting material of strawberry, fruit trees, black-currant and hop require prior approval of the planting sites, on the basis of soil sampling, as apparently free from vectors of nepo-viruses (Chapter 11). Nuclear or other top-grade stocks are generally produced well away from likely sources of virus infection and many indigenous populations of potential nematode vectors appear to be virus-free. Nevertheless the additional safeguard provided by soil sampling can be justified where it is important to avoid even the slightest risk of infection in the early stages of propagation of initially virus-free clonal material, e.g., by transmission via nematodes from weed seedlings.

In The Netherlands, good examples of extensive use of 'quality' controls are provided by the semi-official N.A.K. and bulb-inspection schemes which are sponsored by growers' interests but backed by the Dutch Ministry of Agriculture and Fisheries. There are separate schemes for different crop groups each of which has its own inspection organization, e.g., N.A.K.-A, seed potatoes and agricultural seeds; N.A.K.-B, fruit stocks; N.A.K.-S, specialist flower crops (other than bulbs). In view of the particular importance of flower bulbs in the Dutch economy, they have their own inspection service, the B.K.D., which co-operates with the Dutch Plant Protection Service (P.D.) in making field inspections for quality and quarantine purposes of all commercial bulb crops in The Netherlands.

Import Regulations

Almost all countries have regulations which govern the health of imported plant material and, usually, prohibit the importation of dangerous plant pests and pathogens. Such regulations could be abused, e.g., for commercial or political purposes, but it is generally the aim of responsible authorities to ensure that the requirements are determined only by phytosanitary considerations. To this end, the International Plant Protection Convention and, so far as Europe and neighbouring territories are concerned, the European and Mediterranean Plant Protection Organization give sound guidance (see p. 326). Some countries have comprehensive plant health regulations covering in a single statute both domestic and import requirements. Others, e.g., Great Britain, have separate importation orders. Within the European Economic Community, the recently adopted Directive on protective measures against the introduction of harmful organisms of plants or plant products sets the pattern for Plant Health frontier legislation for all Member States. Pests whose importation they are required to prohibit include potato cyst nematodes, *Ditylenchus dipsaci* on *Allium* spp. flower bulbs and lucerne seed, and *D. destructor* on flower bulbs and potato tubers.

As with domestic regulations, importation controls may have the two aspects which are termed broadly 'quality' and 'quarantine'. The aim of the former is to ensure a reasonable standard of general health, especially as regards common pests and diseases, in imported planting material, whereas the latter seeks to prevent introduction of the more dangerous pests and pathogens, especially those not known to occur in the importing country. Often these organisms are listed as dangerous, prohibited or restricted, in appendices or annexes to regulations.

Plant-health certificates are issued by exporting countries to confirm that the importing country's requirements have been complied with. The International Plant Protection Convention stipulates that contracting governments shall issue certificates conforming to the Model Phytosanitary Certificate that is appended to the Convention and which runs as follows—'This is to certify that the plants, parts of plants or plant products described below or representative samples of them were thoroughly examined on (date) by (name) an authorised officer of the (Plant Inspection Service) and were found . . . to be substantially free from injurious diseases and pests; and that the consignment is believed to conform with the current phytosanitary regulations of the importing country . . .'. Then follows space for additional declarations regarding fumigation or other treatments if required, and freedom from specified organisms. For example, the current U.K. import regulations require lucerne seed to be fumigated with methyl bromide in the exporting country, according to a specified schedule, to control *D. dipsaci*, while several countries (such as Canada and U.S.A.) require a statement to the effect that soil samples taken by approved methods from the place of growth were found free from potato cyst nematodes.

To enable valid health certificates to be issued an official inspection must be made of each consignment immediately before export. This covers most of the general-health requirements for common pests and diseases (e.g., root-knot nematodes) indicated by the words 'substantially free' in the model certificate. However, for some specified organisms, and for pests and diseases not, or not readily, detectable on dormant planting material at pre-export inspection, it is usual to require inspection during active growth, i.e., a 'growing-season inspection' (G.S.I.). A G.S.I. is obviously necessary for virus diseases, and at least desirable (depending on the host) for stem- and leaf-infesting nematodes (*D. dipsaci*, *Aphelenchoides* spp.) on herbaceous plants, bulbs etc. These nematodes are mostly common and widely distributed, so 'substantial freedom' (i.e., a small—say 2 per cent—tolerance at inspection) may suffice, though some species are specifically prohibited by some countries, sometimes when associated with particular host plants (e.g., *D. dipsaci* on flower bulbs by The Netherlands).

Total prohibition of host-plant material can be justified if risks of introducing dangerous new pests or diseases are judged to be too great to depend solely on inspection methods or quarantine treatments, and such extreme measures are provided for in the International Plant Protection Convention. Thus importation of potatoes from certain countries is prohibited by Canada and U.S.A., chiefly because of the risk of introducing potato cyst nematodes and in line with their generally stringent attitude to these pests (see p. 327). This measure has probably been a significant factor in delaying the appearance of these nematodes on any scale in their potato-producing areas.

Export Regulations

Specific export regulations are rare because health and inspection standards for exported plant material are generally determined by the regulations of *importing* countries (see above), or may be covered by domestic regulations or certification schemes (e.g., for bulbs in The Netherlands and seed-potatoes there and in other E.E.C. member countries). However, The Export of Bulbs Ordinance, 1953, of the island of Guernsey provides an unusual example of a regulation covering exports to all countries whether or not they have equivalent import controls. It was introduced as a safeguard to the health standard of exported Guernsey bulbs in the interests of the island's bulb industry. All bulbs for export must be inspected and special attention is paid to *Ditylenchus dipsaci* for which a small tolerance is allowed only if the bulbs are subsequently hot-water treated by officially approved methods before export.

Discussion

Nematodes, especially soil-inhabiting species, present special problems for effective regulatory control, e.g., the ubiquity of many species, the virtual impossibility of detecting them by normal inspection procedures, difficulties of identification, the existence sometimes of marked intraspecific differences in host ranges (see Dalmasso*, 1973), the inevitability of spread short of stopping all movement of plant material and soil and, in many cases, the absence of economic or non-phytotoxic control treatments.

Jones (1969a) was frankly pessimistic about the outlook for plant-quarantine measures when he wrote, '. . . quarantine services everywhere are fighting losing battles against known and unknown means of entry and . . ., at best, they can delay but cannot prevent indefinitely the introduction and establishment of noxious alien species able to multiply on suitable host crops in new territory.' He usefully draws attention to the limitations of quarantine activity, as Peters (1951) did in respect of so-called certificates of freedom from potato cyst nematode based on soil sampling. Nevertheless to delay or to lessen spread is better than doing nothing, and to remove all restrictions would be irresponsible. Although commercial and other pressures would provide some natural sanctions against opening of the flood-gates to obviously infested or diseased material, many pests (especially nematodes) and pathogens, not readily detectable by the recipient of imported plants, could gain much more rapid foothold. At least,

*See especially Dalmasso's remarks concerning potential risks of introduction with import commodities of a new gene pool of nematodes such as *Meloidogyne*, species of which show broad genetical heterogenicity.

valuable time can be gained, for example for plant breeders to develop resistant cultivars of host crops. Thus, the twin functions of regulatory activity, which may be broadly termed quarantine and quality, still appear worthwhile—firstly to minimize spread of the more dangerous organisms, or at least to delay their introduction to new territories, and secondly to ensure a reasonable standard of general health for planting material moving in international trade.

Statutory control measures should always be applied rationally and be sound technically. In attempting to assess risks of importing non-indigenous species, nematologists who have to advise quarantine officials are often hampered by lack of data on the ability of foreign nematodes to survive and reproduce in a new environment. Thus research is called for on lines comparable with those suggested by Baker (1972) with reference to insects. Southey (1974a) reviews principles that should be applied to soil sampling for potato cyst nematodes to improve chances of detection at low population densities and to ensure consistent results. However it is also suggested in the same paper that to have effective systems of control on potato growing land at source so as to keep populations low, and to minimize soil accompanying exported plants and seed potatoes, are probably much more effective measures than soil sampling alone, in checking spread.

There has been a tendency for country to follow country in listing dangerous pests and diseases in their regulations. Potato cyst nematodes (*Globodera rostochiensis* and *G. pallida*), for example, are included in lists of standard quarantine organisms to be regulated against, sometimes regardless of the extent to which they actually threaten crop production in particular circumstances. It is possible to lose sight of the fact that these nematodes are primarily important only as pests of their host crops (chiefly potato and tomato). It is possible also to have the situation where regulations pose a greater threat to crop production than the pest does. When that happens, clearly something is wrong!

CHAPTER 18

Population Dynamics, Population Models and Integrated Control

F. G. W. JONES and R. A. KEMPTON*
Rothamsted Experimental Station, Harpenden

General

Compared with mammals, birds and insects, nematodes are relatively immobile and the mobility of species that live wholly in the soil is further curtailed by the constraints of the soil pore system. Consequently nematode populations large enough to inflict economic injury on crops are often slow to build up, much inbred and difficult to dislodge.

The life strategies of plant-feeding nematodes vary greatly. Some, which may complete several generations a year, have a large potential for increase and may multiply more than a thousand times within a growing season. Nevertheless, this is a far smaller potential than that of viral, bacterial or fungal pathogens and of some Hemipterous and other insects. The populations of this kind of nematode, which may be described as exploiters, fluctuate greatly, e.g., *Aphelenchoides ritzemabosi,* some Trichodorid species, the races of *Ditylenchus dipsaci* and many soil species feeding on bacteria and fungi. Plant-feeding 'exploiters' tend to survive the interval between population outbursts upon weeds and other marginal hosts but some species seem able to survive in soil for long periods without hosts.

Other nematode species, e.g., cyst nematodes in Northern Europe, pass only one or two generations a year and are often limited to one generation by the short vegetative life of their host crops from planting to harvest (Jones, 1950). The maximum observed multiplication rate is usually much less than 100 and population fluctuations are limited largely because not all eggs hatch when a host crop is grown and even fewer hatch when it is withheld. Yet other nematode species have long generation times which may exceed 12 months. These are thought to lay relatively few eggs, e.g., *Xiphinema, Longidorus* (Flegg, 1968; Weischer, 1975) and have effective multiplication rates of less than ten times per year as have species of *Trichodorus* and *Paratrichodorus* common in British soils.

Distribution and Spread

With few exceptions, our knowledge of the distribution and spread of harmful species of nematodes is woefully inadequate and is largely an impression gained from the incidence of economic damage to crops and the activities of a few industrious nematologists (Jones, 1969a). The zoological distribution of nematode pests is wider than indicated by crop injury and the origins of most are obscure. As with viral, fungal and bacterial pathogens, arthropod and other pests, and noxious weeds, many nematode pests are aliens that have entered our fields

*Now at Plant Breeding Institute, Cambridge.

to fill vacant ecological niches created by agriculture and horticulture (Jones & Jones, 1974). Less fit indigenous species have become relicts in hedgerows, copses and other uncultivated ground whilst, where agriculture is long established, the alien species are now naturalized residents.

The fastest spread into unoccupied territory occurs when host plant and nematode parasite are distributed together on or within rootstocks or other planting material. Examples are the wide and discontinuous distribution of the citrus nematode (*Tylenchulus semipenetrans*), the burrowing nematode (*Radopholus similis*) on banana and the dagger nematodes (*Xiphinema* spp.) on vines. Rapid spread also occurs when nematodes are seedborne as is *Ditylenchus dipsaci* on *Vicia* beans, clover, onion and teasel, or occur in galls (*Anguina* spp.) or pellets of soil contaminating seed, e.g., *Heterodera schachtii* and *H. glycines* with unprocessed sugar-beet seeds and soya-bean seeds respectively. Irrigation water, especially that re-used (drainage or return-water) can spread great numbers of nematodes locally (Faulkner & Bolander, 1966) as can floods in low-lying areas and wind in areas of dry, light soil or peat. Casual introductions between areas separated by effective geographic barriers (seas, wide rivers, deserts, high mountains) occur by many means in commerce and are difficult to counter by quarantine measures (Bingefors, 1967; Jones, 1969a). Two well-documented examples of the spread of noxious nematodes in areas previously uninfested illustrate the colonization of new territory.

Potato cyst nematodes appear to have originated in the Andes region of South America at altitudes above 2000 metres (see also Chapter 7). The epicentre for *Globodera pallida* includes Southern Colombia, Ecuador, Peru and Northern Bolivia. That of *G. rostochiensis* appears more southerly extending from Southern Peru around lake Titicaca and Bolivia southwards. Infestation in Venezuela and Panama seem to be of recent European origin (Evans, Franco & de Scurrah, 1975). From their epicentres both species have followed the cultivation of potatoes to other parts of the world (Fig. 18.1). The main vehicle of spread has probably been in soil adhering to seed potato tubers. Spread to Japan and possibly to Europe may also have been in guano, of which great quantities were formerly used as fertilizer. As this is dug on offshore islands, cysts in guano are almost certainly from contamination with soil in sacks, vessel holds or produce from the South American mainland (Inagaki & Kegasawa, 1973).

When potato cyst nematodes were first introduced into Europe is unknown but was possibly after 1850. Females (cysts) were first observed on potato roots in Germany in 1881. In Britain the pest may date from before 1900 but it was not positively identified until 1913. Records of infestations became more numerous in the 1920's and 1930's and by 1950 it can be said that the zoological distribution was virtually coincident with the distribution of potatoes in fields and gardens (Jones, 1970). Race Ro1 of *G. rostochiensis* and Pa3 of *G. pallida* (formerly known as pathotypes A and E respectively—Kort, Ross, Rumpenhorst & Stone, 1977) are the dominant populations in Britain. Their distribution (Fig. 18.2) probably represents initial introductions into different areas, where one or other became firmly established, followed by haphazard spread to other areas with seed potatoes and by other means. Here whole farms, single fields or even part fields on farms may be colonized by one species, whereas neighbouring farms, fields, or part fields are colonized by the other. This suggests that the two species cannot co-exist and that, according to circumstances one or the other becomes dominant (see p. 354 *seq.*).

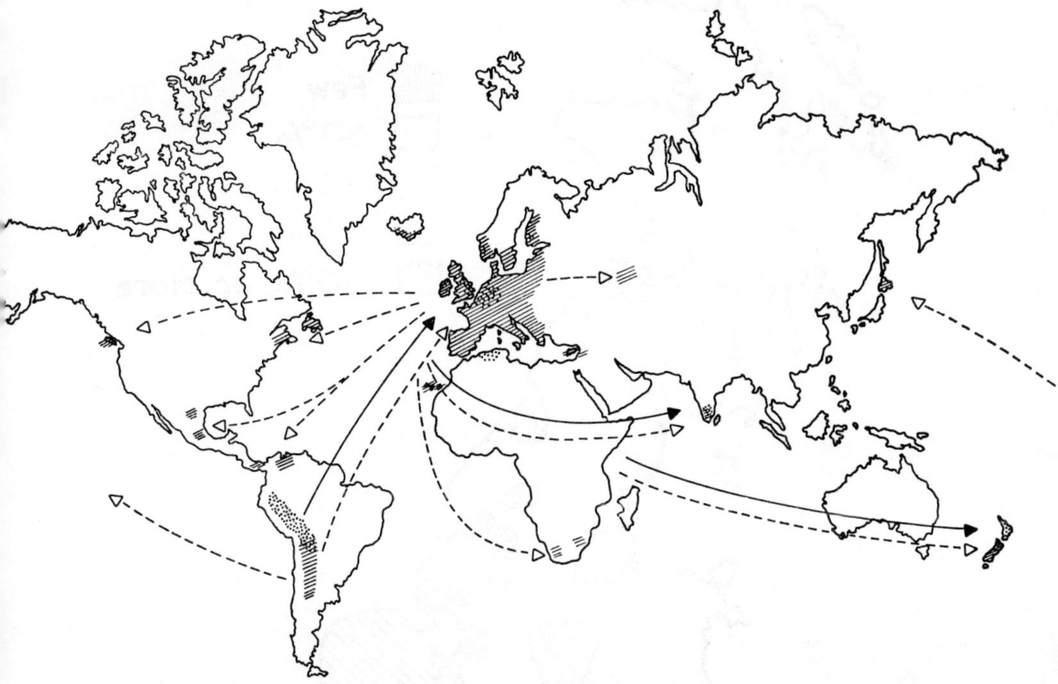

Fig. 18.1. Map of world dispersal of the potato cyst nematodes *Globodera rostochiensis* and *G. pallida* from their epicentres in the South American Andes. (After Evans & Trudgill, 1978 in press). Shaded, *G. rostochiensis*; stippled, *G. pallida*.

The time scale for the colonization of the British Isles has been less than 100 years. The time for the colonization of Europe, 1850 to the present, compares favourably with that of some large, winged insect pests, e.g., the Colorado beetle (*Leptinotarsa decemlineata*). The potato cyst nematodes are still extending their range, notably to Eastern Europe and Russia and, more slowly, in the U.S.A. and Canada.

The inland part of the fenlands of East Anglia consists of peat soils reclaimed from marsh between 1650 and 1850. Although *H. schachtii* is possibly indigenous or long naturalized in Britain it was not observed until 1928 and was absent from the fens before drainage. First records of crop injury which can be attributed to the pest date from 1920 and positive identification dates from 1934 (Petherbridge & Jones, 1944; Jones, 1951a). Despite enforced rotations of sugar beet and all other susceptible crops (Jones, 1951b), the nematode has spread widely until almost all the black peat soil has been colonized and infestations have begun to appear on high land around the fen basin during a time scale of about 60 years. For comparison, colonization of the Holland division of Lincolnshire, mainly by *G. pallida Pa*3 between 1880, when potatoes were first grown in fields, and 1940 when infestations had become widespread, also took about 60 years.

Bedi (1968) studied the spread of *G. rostochiensis* within Scottish potato fields and found a pattern of the type in Fig. 18.3. In fields where *G. rostochiensis* or *H. schachtii* are so scarce as to appear absent, at least five host crops are required

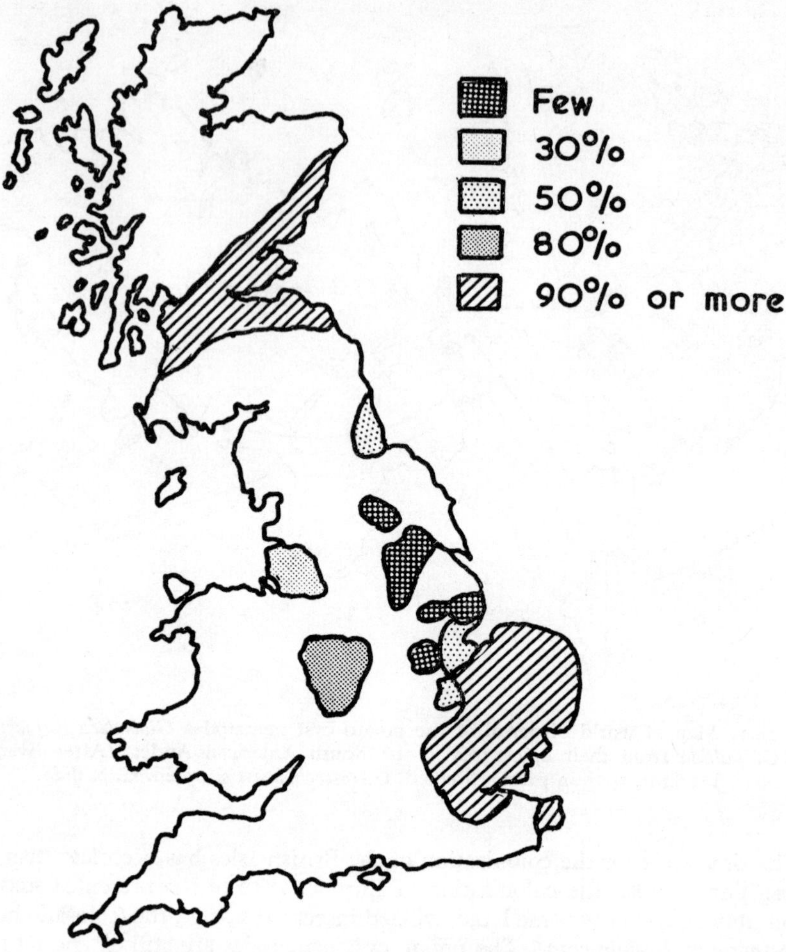

Fig. 18.2. The distribution of fields suitable for growing potato cultivars with nematode resistance gene H_1 derived from *Solanum tuberosum* ssp. *andigena*. Areas cross-hatched are those where 90 per cent of the fields are infested with *G. rostochiensis* Ro1 and areas in black where most of the fields are infested with *G. pallida* Pa3. Other areas contain fields infested predominantly with one or the other in the proportions indicated in the key (After Jones, 1970). *G. pallida* Pa1 is widely but sparsely scattered.

to produce injurious populations and these appear first in patches. Little is known of the origin or the dynamics of population dispersal of most other nematode species although Oostenbrink, s'Jacob & Kuiper (1956) found that the appearance of common soil nematodes in newly reclaimed polder soils in The Netherlands took only a few years after the salt had been leached and cropping commenced.

Once an infestation is established in favourable new territory, spread begins from the first focus, speeds up as new foci are formed and gathers pace until virtually all suitable territory is occupied. The rate of colonization of territory follows the same course at first but slows down as spread leads to the recolonization of territory already occupied. The colonization curve is therefore sigmoid as

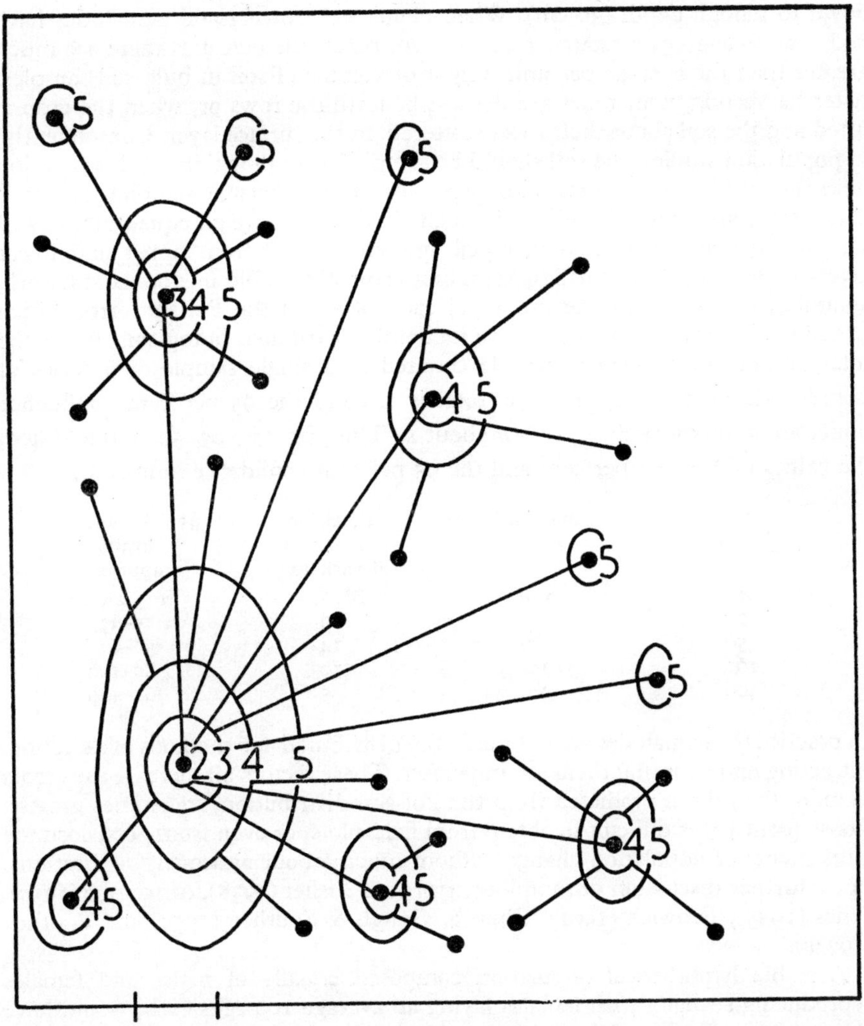

Fig. 18.3. Foundation effects discernible in some Scottish potato fields from the density of cysts of the potato cyst nematode *Globodera rostochiensis*. The numbers 2, 3, 4, 5 represent new foci of infestation established after the harvesting of successive potato crops. (Diagrammatic, after Bedi, 1968).

has been demonstrated for other animals, e.g., the grey squirrel in England and Wales (Jones & Jones, 1974). Once nematodes become established, whether they form cysts or not, spread continues until the area under cultivation by its host crops is colonized and there is little that is effective to prevent extension of the infestation.

Population Size and Structure

The population densities of nematodes are often large. For example populations of cyst nematodes range from less than one to more than 100 eggs/g soil, (i.e., from $2{\cdot}5 \times 10^5$, to $2{\cdot}5 \times 10^7/\text{m}^2$ or from $2{\cdot}5 \times 10^9$, to $2{\cdot}5 \times 10^{11}/\text{ha}$)

down to plough depth (30 cm). When plants are growing and nematodes have had time to aggregate around roots or stem bases, the densities there are much greater than the average per unit weight or volume of soil in bulk soil samples. After harvesting, nematodes are densest beneath the rows or, when the crop is lifted and the soil about their roots scattered, in the surface layer. Consequently, in population studies, the soil should be sampled after ploughing and cultivation have re-distributed the nematodes more randomly. Otherwise sampling methods and sample sizes must be adopted which eliminate bias (e.g., equal numbers of samples in and between rows, equal amounts of soil from upper and lower layers of topsoil). The smallest sampling error obtainable in population work, assuming the organisms do not repel each other, is the Poisson error which would apply only if nematodes were randomly distributed in top soil. If n is the total number of a given nematode counted in a single sample or a series of samples, the standard error of the total is $\sqrt{n}$ and the 95 per cent confidence limits are approximately $\pm\ 2\sqrt{n}$ about n. Thus for $n = 25$, 50, 100 and 400, the values of $\sqrt{n}$, SE per cent and the 95 per cent confidence limits are:

Total	Standard error	Precision (coefficient of variation)	95% confidence limits (approx.)
n	$\sqrt{n}$	SE% n	$n \pm 2\sqrt{n}$
25	5	20	15–35
50	7	14	36–64
100	10	10	80–120
400	20	5	360–440

In practice the nematodes are not randomly distributed and methods of sampling, extracting and counting them are imperfect. These factors all increase the errors to more than those predicted from the Poisson distribution, sometimes greatly. Consequently it is difficult to obtain from field plots, or even from pots, accurate assessments of population change without much labour and many replications. For a further discussion of sampling errors see Fisher (1938), Anscombe (1950), Jones (1955), Fenwick (1961), Church, Gough & Southey (1959) and Southey (1974a).

A stable hypothetical population composed equally of males and females reproducing sexually with females laying an average 100 eggs each, would have the structure illustrated in Fig. 18.4. Mortality of the pre-adult stages would total 98 per cent, leaving two adults, one male and one female, to reproduce. Although in real populations the number of stages and the per cent mortality in different stages might differ from that shown, it is usual for populations to have a pyramidal age structure with an excess of juveniles. This is often obscured because all stages may not be present at any one time (Fig. 18.5).

Our knowledge of the population structure of most plant-parasitic nematodes is imperfect in several respects. Most sampling methods are subject to large errors and not all nematodes are extracted. Only in those species in which all or most of the eggs are retained in cysts or in those with recoverable egg masses, are reasonably accurate estimates of egg population densities obtainable with present-day techniques. For scarcely any species do we have accurate assessments of age-specific mortality and its causes. Yet these might provide clues indicating why some species are serious pests and others not. Nor are we always able to predict the consequences of control measures, especially those of applying different kinds of nematicide, or other crop-protection chemicals.

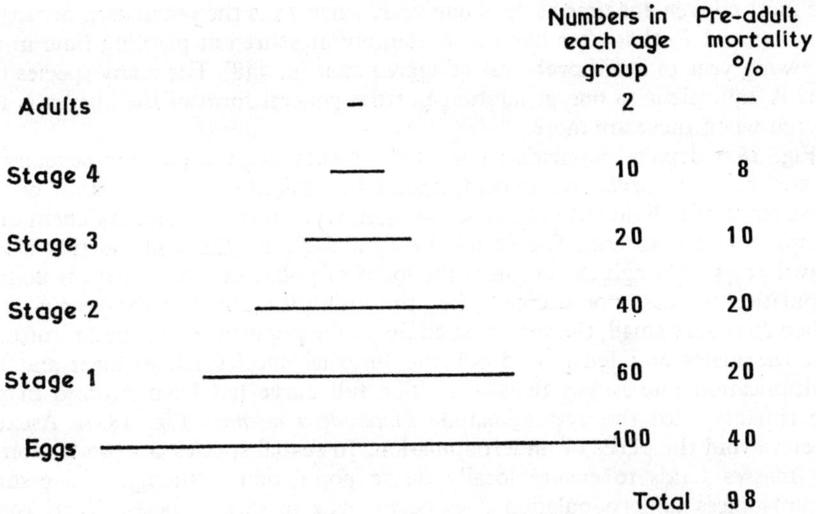

Fig. 18.4. Age structure of a hypothetical population in real equilibrium. Note the excess of juvenile stages and the small proportion of adults. (After Jones & Jones, 1974).

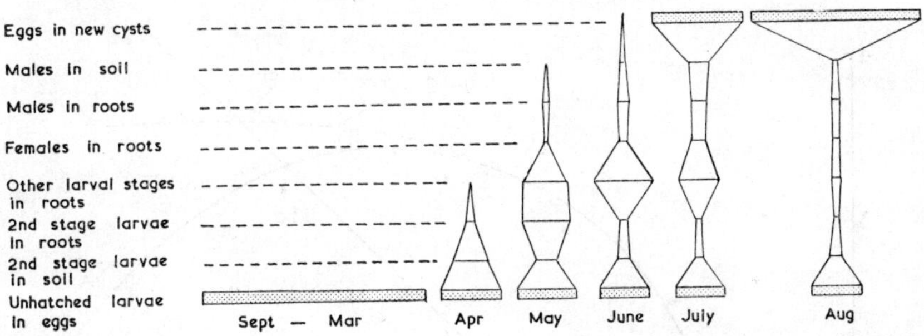

Fig. 18.5. The time structure of a *Globodera rostochiensis* population, diagrammatic. (After Jones & Jones, 1974).

Population Curves

Field experiments on larger, mobile pests (insects, mammals, birds) are complicated by large rates of emigration and immigration or by sampling problems associated with excessively small population densities and great variations from one season to the next. Consequently many fundamental studies can be done only in artificially contrived, closed systems or in very large plots. With nematodes, meaningful population studies are possible in pots and small field plots outdoors. Large plots are inappropriate because populations are usually patchy and the multiplication rate decreases as the population density increases. The average rate for the field as a whole may give no indication what is happening in its parts. If, for example, the infestation overall is slight, but made up of a few patches of high density, the average may suggest that the multiplication rate is small, which is misleading (Jones, 1945, 1956). Agriculturally, the most convenient time scale is one year, when P_i is the population density at planting and P_f that after harvest, preferably measured at planting time in the following year to avoid problems of aggregation (p. 338). For many species one year is equivalent to one generation but the general form of the curves is unaltered when there are more.

Fig. 18.6 depicts population curves for nematodes reproducing sexually or asexually. The curves are uncomplicated by emigration, immigration or the persistence of individuals not participating in reproduction. Effective enemies or competitors are absent. The scales for P_f and P_i are the same so that a line drawn at 45° through the origin is the locus of points at which there is neither population increase nor decrease, i.e., the multiplication over the year is × 1. When P_i is very small, the surviving adults in the population may be so scattered that the males and females of obligate bisexual species fail to meet and the multiplication rate is less than × 1. The full curve has been realized in pot experiments with the cyst nematode *Heterodera avenae* (Fig. 18.6). Asexual species avoid the perils of underpopulation. In sexual species the production of egg masses tends to ensure locally dense populations although under some circumstances underpopulation does occur even in these species (Kort, 1962; Seinhorst, 1968b). As P_i increases the multiplication rate increases to its maximum and then decreases to × 1 or less, largely because of increasing competition between individuals and decreasing food supply.

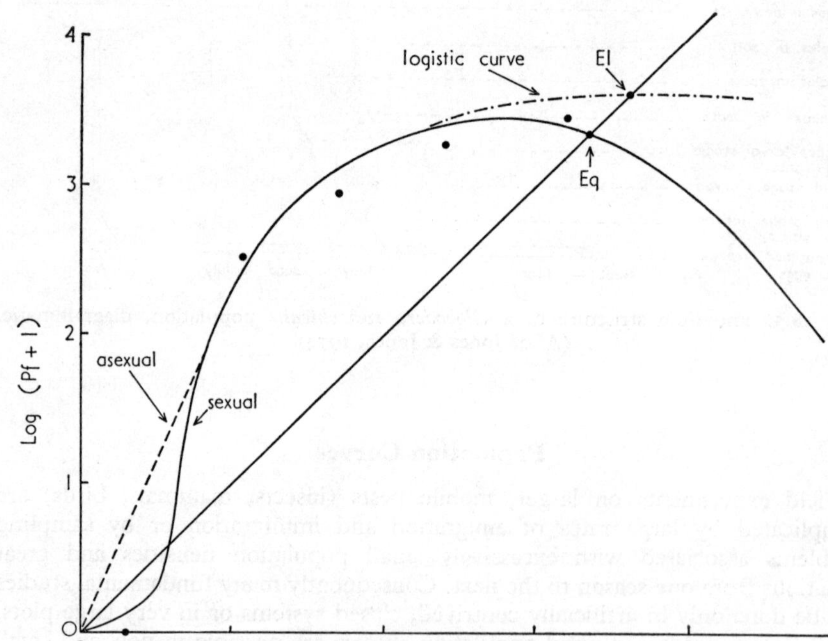

Fig. 18.6. The relationship between preplanting (P_i) and post-harvest population densities (P_f) of asexual and sexual species without effective enemies. The upper, logistic curve is the one that would hold if the food supply remained constant. E_l, logistic equilibrium point; E_q, observed equilibrium point. ●, data for *H. avenae* in pots. (After Mukhopadhaya *et al.*, 1973).

The observed multiplication rate per generation or per year is usually far less than the reproductive potential of the species because eggs fail to hatch, larvae fail to find their hosts and enemies, parasites and competitors take their toll at each stage (Fig. 18.4). Because of the pyramidal age structure of populations, enemies that kill or competitors that eliminate a proportion of juvenile stages tend merely to remove individuals surplus to the carrying capacity of their hosts. In contrast, enemies and competitors that kill the few surviving females or seriously decrease the number of eggs they produce have, relatively, a much greater impact and lead to a much diminished number of progeny at the end of the growing season.

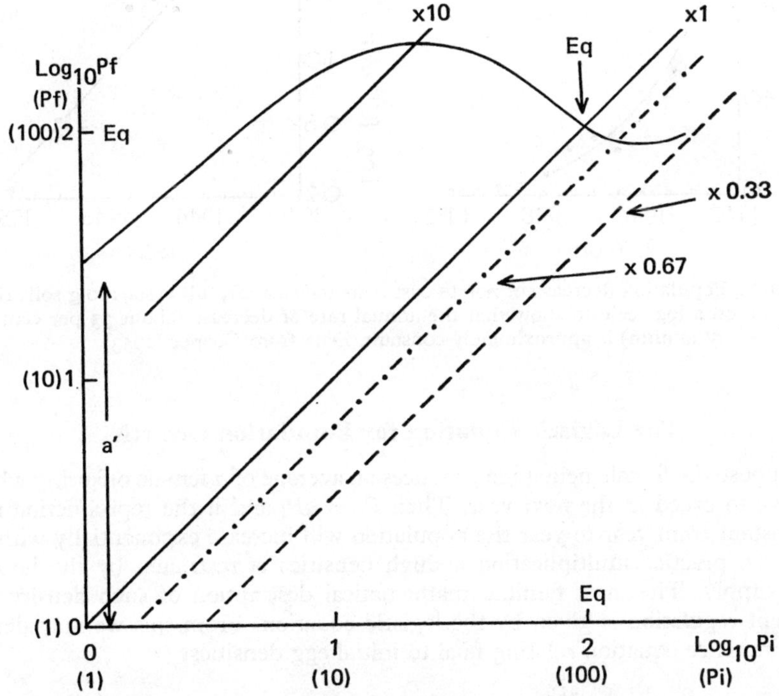

Fig. 18.7. Typical population curves for potato cyst nematodes. P_i and P_f log eggs/g soil. ———, old and new eggs; – – – –, eggs unhatched by resistant or susceptible potato cultivars; – · – · – · –, unhatched eggs when crops other than potatoes are grown or the land is left fallow; *a′*, observed maximum multiplication rate. *E*, observed equilibrium population density. *Note:* for *a′* on graph read a′ $(1-C_p)$.

Since population densities usually change in a multiplicative rather than an additive way it is helpful to plot P_i and P_f on logarithmic scales (Fig. 18.7). Distances measured in a positive direction relative to the Y-axis then indicate rates of population increase (multiplication) and those measured in a negative direction population decrease. The × 1 line can be drawn as in Fig. 18.6. Lines representing × 10 multiplication or a population decrease of say two thirds (× 0·33) do not pass through the origin but lie parallel to the × 1 line (Fig. 18·7). A constant annual rate of population decrease shows up as a straight line when plotted on a log-scale, whereas, on a linear scale, the same data would produce an exponential curve (Fig. 18.8) from which the constant rate is not immediately apparent. Logarithms are also useful when population densities fluctuate widely. Since we cannot take the logarithm of a zero density we use $\log (x + c)$, where c is a constant usually 1. The taking of logarithms may falsely exaggerate differences between very small densities but this may be avoided by choosing c to be suitably large.

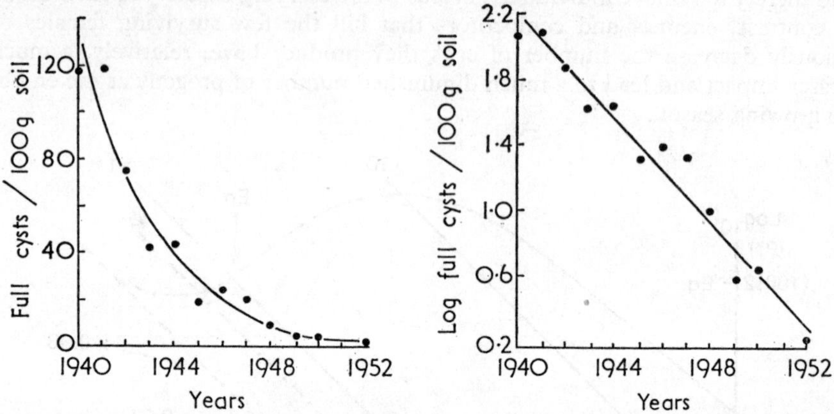

Fig. 18.8. Population decrease of potato cyst nematodes. Left, full cysts/100 g soil; right, the same on a log scale to show that the annual rate of decrease (about 33 per cent per annum) is approximately constant. Data from Cooper (1953).

The Logistic Equation for Population Growth

Suppose the female nematode produces an average of a female offspring which survive to breed in the next year. Then $P_f = aP_i$ and if the reproduction rate is constant from year to year the population will increase exponentially without limit. In practice multiplication at high densities is restricted by the limited food supply. The most familiar mathematical description of such density dependent regulation is given by the logistic equation. From this we may derive the difference equation relating final to initial egg densities,

$$P_f = \frac{aP_i}{1 + (a-1)P_i/E_l} = amP_i \qquad \ldots\ldots(1)$$

(Jones, Parrott & Ross, 1967; Seinhorst, 1966a), where $m = 1/[1 + (a-1)P_i/E_l]$ and E_l is the equilibrium population density. In effect the maximum reproduction rate a is diminished by a fraction m which varies with P_i, the initial population density. When P_i is small m approaches the value 1 and the maximum rate a operates. When $P_i = E_l$, $m = a^{-1}$ and the effective multiplication rate $am = 1$ so that $P_f = P_i$. When P_i is greater than E_l, am is less than 1, i.e., P_f is smaller than P_i, though still slightly greater than E_l (Fig. 18.6). In practice, the food supply, which in a growing plant is an expanding one, is not equal for all values of P_i but diminishes as P_i increases, and the observed equilibrium density may be as small as a half or a fifth of the logistic equilibrium E_l. This necessitates extending the logistic model as shown later (p. 346).

Alternative relationships suggested for describing population growth with no plant damage (e.g., Seinhorst, 1968a, 1970) appear to give no better fit to field data and suffer from unfamiliarity and increased mathematical complexity.

Relationship Between Numbers and Yield

Most nematodes attack roots and their main effect is to decrease the size of the root system. Only in crops like sugar beet is yield related directly to root size. In other crops the yield of leaf tissue, grain or tubers is affected indirectly or is the result of climatic and other stresses, mostly operating after root size is determined. Similarly, for nematodes attacking plant parts other than roots, yield is often only indirectly related to the severity of attack. Seinhorst (1965) argued that the yield of plants was unaffected when nematode density at planting or sowing was below a certain threshold, but above this threshold decreased exponentially with egg density:

$$\frac{y - y\text{min.}}{y\text{max.} - y\text{min.}} = e^{-k(P-T)} = z^{P-T} \qquad \ldots\ldots(2)$$

where y is the expected yield, $y$max. and $y$min. the maximum and minimum yields and $P \geqslant T$ the initial egg density. $k = -\log z$ where z is a constant slightly less than 1 and is notionally the expected yield when nematode density exceeds the threshold T by unity. z varies with the nematode species and the plant attacked. $z^{-T} = c$ is a constant ranging from 1·05 to 1·15 and can be regarded as the tendency of the plant to compensate for damage done to the extent of 5 to 15 per cent.

The yield approaches its maximum asymptotically as nematode density decreases to its threshold (Fig. 18.9). Whether this threshold always exists and whether there is always a minimum yield is debatable. Oostenbrink (1966) has objected in detail. Sometimes in pot cultures the yield or plant growth curve suggests that small numbers of nematodes actually cause an increase. Further, the minimum yield found at high nematode densities is likely to consist of parts of the plant which cannot be utilized by the nematode. In our modelling which concerns the size of root systems rather than that of yields, minimum yield is of no interest and we shall take the minimum usable yield of roots to be zero.

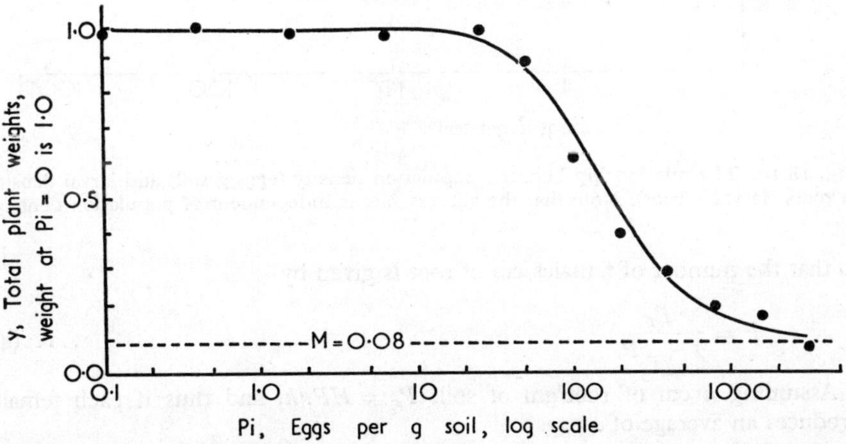

Fig. 18.9. The relationship between total plant weight and P_i for the potato cv. Libertas supplied with extra nitrogen. The tolerance threshold, i.e., the value of P_i below which no yield is lost, lies between 10 and 20 eggs/g soil. Total plant weight in uninfested soil = 1·0 (not shown), M = minimum yield. (Redrawn from Seinhorst & den Ouden, 1971).

Population Equations for Cyst Nematodes

To build up a mathematical model of population changes in nematode egg density from first principles we need additional quantitative information about the life cycle. Of the eggs present in the soil we suppose a proportion H of the larvae they contain eventually establish themselves in the roots. This proportion appeared to be independent of density in plots of Pentland Dell potatoes infested with *G. rostochiensis* (Fig. 18.10). A proportion of the larvae on the root become female and the number of eggs found in females (cysts) appears again to be independent of density (Hesling, 1961a; Jones, 1966; Seinhorst, 1968a). In the absence of any serious predators or parasites, the main density-dependent mechanism seems to act by regulating the sex ratio. Trudgill (1967) found that the proportion of larvae becoming female decreased markedly with density. His data (Fig. 18.11) suggest that the sex ratio (males/females) increases proportionately with numbers of larvae per cm of root (P_r). Thus

$$M/F = bP_r, \qquad \ldots .(3)$$

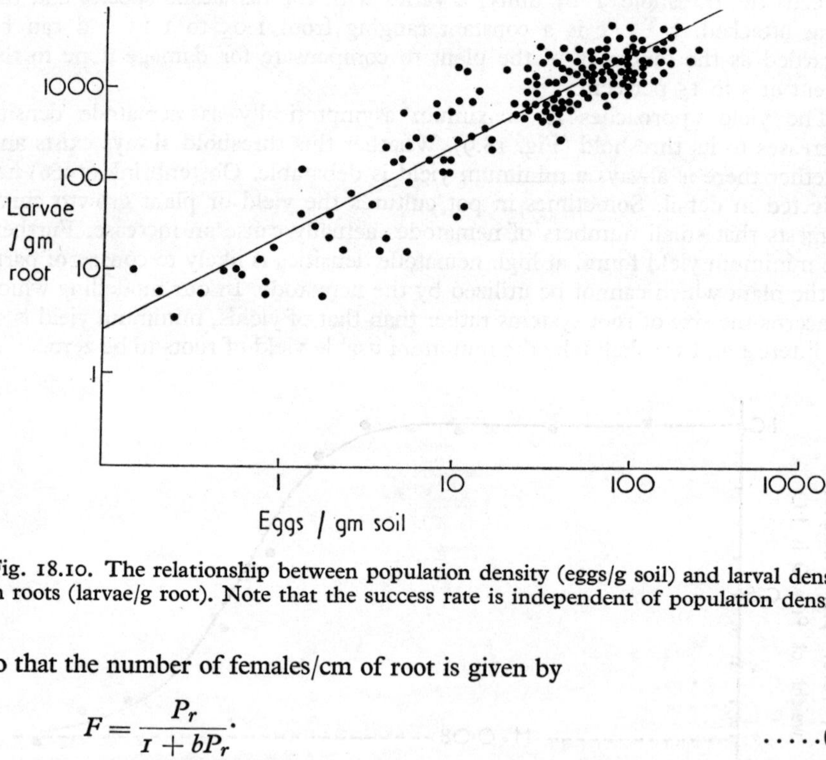

Fig. 18.10. The relationship between population density (eggs/g soil) and larval density in roots (larvae/g root). Note that the success rate is independent of population density.

so that the number of females/cm of root is given by

$$F = \frac{P_r}{1 + bP_r}. \qquad \ldots .(4)$$

Assuming h cm of root/gm of soil, $P_r = HP_i/h$, and thus if each female produces an average of a eggs

$$P_f = \frac{aHP_i}{1 + \frac{bH}{h}P_i}. \qquad \ldots .(5)$$

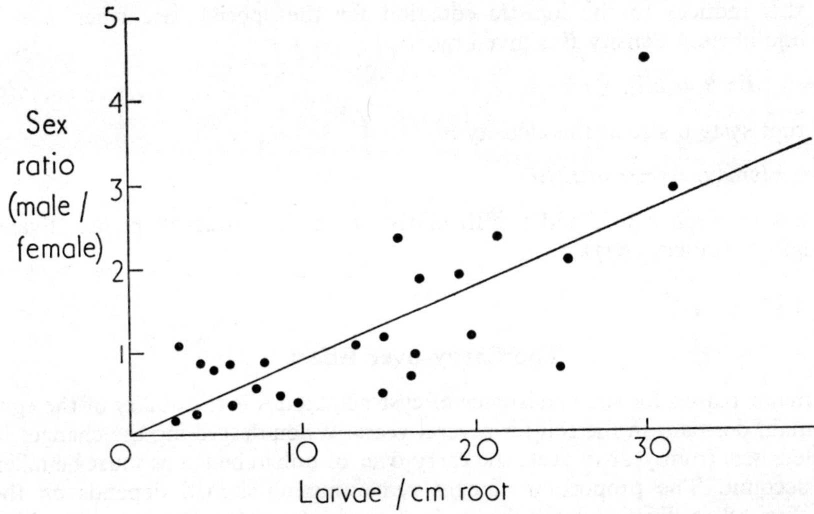

Fig. 18.11. The relationship between larval density in roots and the sex ratio, male/female. The proportion of females diminishes with population density, i.e., is density dependent.

The equilibrium density

$$E_l = \frac{h}{bH}(a' - 1) \qquad \ldots\ldots(6)$$

where $a' = aH$ is the maximum multiplication rate achieved at low nematode densities. Equation (5) can thus be written

$$P_f = \frac{a'P_i}{1 + (a' - 1)\dfrac{P_i}{E_l}} \qquad \ldots\ldots(7)$$

which is identical with the familiar logistic equation (1).

As previously noted the above equation assumes that the food supply, h, is not affected by nematode attack, i.e., P_i is less than the threshold density T. For values of $P_i > T$ we should replace h in equation (5) by hz^{Pi-T}. This comes from equation (2), where h denotes the maximum root length/g soil when no nematodes are present, and the minimum usable root length at very high densities is assumed to be zero.

Equation (5) now becomes for $P_i > T$

$$P_f = \frac{aHP_i}{1 + \dfrac{bHP_i}{hz^{Pi-T}}},$$

that is

$$P_f = \frac{a'P_i}{1 + (a' - 1)\dfrac{P_i z^{-Pi}}{cE_l}} \qquad \ldots\ldots(8)$$

and this reduces to the logistic equation for the special case when $z = 1$. The equilibrium density E is given by

$$Ez^{-E} = cE_l. \qquad \ldots\ldots(9)$$

The root system size at this density is

$$y/y\text{max.} = cz^E = E/E_l$$

and lies between a half and a fifth of that of lightly attacked plants (Evans, Trudgill & Brown, 1977).

The Carry-over Effect

A major reason for the persistence of cyst nematodes is the ability of the eggs to remain dormant in the soil for several years. When describing the changes in egg densities from year to year, the carry-over of unhatched eggs must be taken into account. The proportion of eggs remaining unhatched depends on the crop. Fig. 18.12 illustrates the change in densities for three crop years in which the land was successively fallow (or growing a non-host crop), growing a resistant crop and growing a susceptible crop. In Fig. 18.12 the nematode is *Globodera rostochiensis* pathotype Ro1 and the resistant and susceptible crops are potatoes.

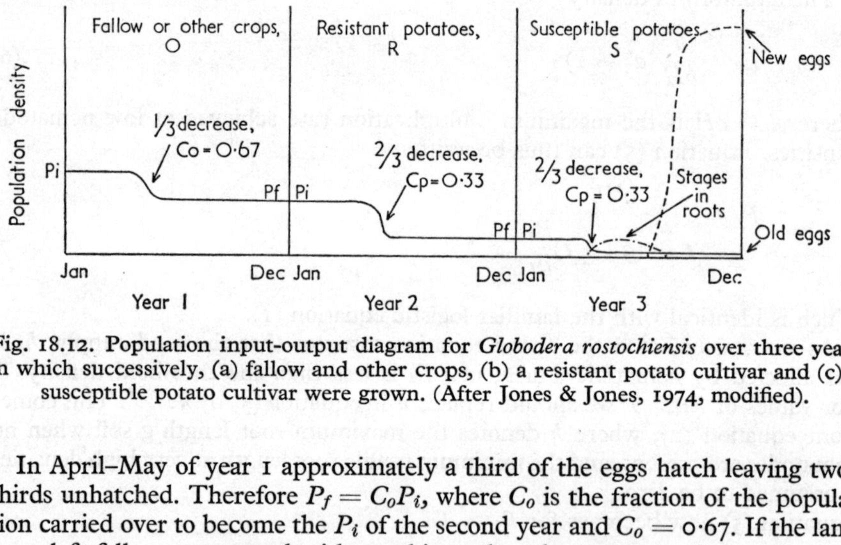

Fig. 18.12. Population input-output diagram for *Globodera rostochiensis* over three years in which successively, (a) fallow and other crops, (b) a resistant potato cultivar and (c) a susceptible potato cultivar were grown. (After Jones & Jones, 1974, modified).

In April–May of year 1 approximately a third of the eggs hatch leaving two-thirds unhatched. Therefore $P_f = C_o P_i$, where C_o is the fraction of the population carried over to become the P_i of the second year and $C_o = 0{\cdot}67$. If the land were left fallow or cropped with anything other than potatoes for n years, we could write $P_f = C_o{}^n P_i$ where P_f would be the population density at the end of n years. In the second year approximately two thirds of the population hatches under the influence of the resistant potato crop. Hence we can write $P_f = C_p P_i$ where C_p is the fraction of the population carried over to become the P_i of the next year and $C_p = 0{\cdot}33$. Also if resistant potatoes were grown for n years we could write $P_f = C_p{}^n P_i$.

In the third year when susceptible potatoes were grown a proportion H of the eggs hatch and the larvae establish themselves in the root to produce a new generation of eggs for the next year (equation 8). Adding to these the proportion C_p carried over unhatched we have

$$P_f = \frac{a'(1 - C_p)P_i}{1 + (a' - 1)\dfrac{P_i z^{-P_i}}{cE_l}} + C_pP_i, \qquad \ldots\ldots(10)$$

where

$$a' = \frac{aH}{1-C_p}$$

is now the maximum multiplication rate including those eggs carried over which hatch in subsequent years.

Population Oscillations and Population Equilibria

The way in which populations behave in the neighbourhood of the equilibrium (*Eq* in Figs. 18.6 and 18.7) depends on the rate of population increase and the degree to which the host is damaged. If increase is moderate and little or no damage is done, the logistic situation applies and, in successive years of continuous cultivation of the same host, the equilibrium population density is approached gradually (Fig. 18.13A). However, when increase is excessive and damage to the root system great, the population tends to overshoot the equilibrium and large oscillations occur (Fig. 18.13C). Examples of the first kind are depicted by Seinhorst (1967) for *Tylenchorhynchus dubius, Pratylenchus crenatus* and *Rotylenchus uniformis*. The large population fluctuations of races of *Ditylenchus dipsaci* on onions and narcissus and of *Aphelenchoides ritzemabosi* on chrysanthemums are examples of the second. Cyst nematodes provide

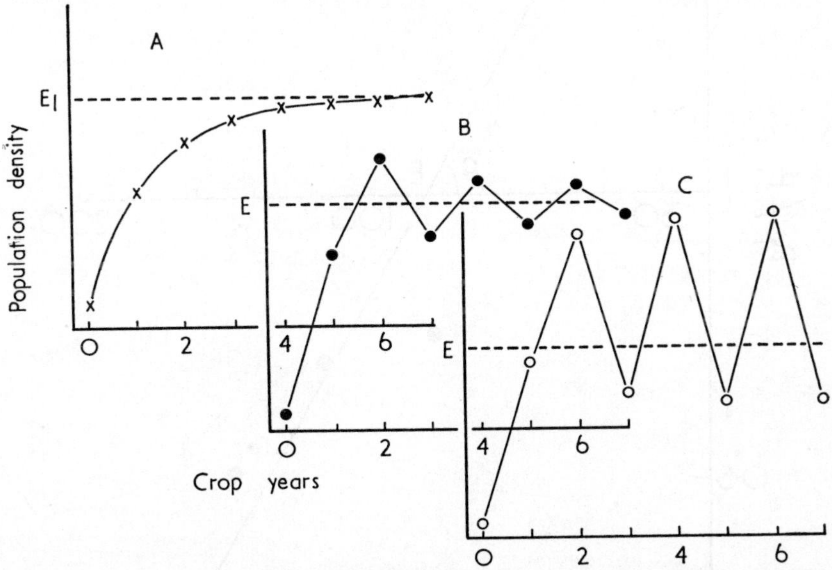

Fig. 18.13. Types of population oscillation when a host crop is grown in consecutive years. A, gradual approach to equilibrium (logistic); B, approach by damped oscillations; C, large fluctuations about the equilibrium.

examples of an intermediate type in which fluctuations about the equilibrium are small and sometimes damped (Fig. 18.13B, see also Jones & Parrott, 1969; Jones, 1974). An important factor restricting oscillations for cyst-nematode populations is the fraction of the population which fails to hatch even when the host crop is grown. This conservative element also slows genetic changes (p. 357).

A simple way of predicting the type of population oscillation, provided the data are not too scattered, is to plot the relative increase or decrease in population density at the end of a crop season against the initial density, i.e., P_f/P_i against P_i. P_f and P_i must be plotted on the same scales, linear or logarithmic. Fig. 18.14 shows P_f/P_i plotted against P_i on log scales for six successive years of cropping in small plots infested with *G. rostochiensis* Ro1 (Jones & Parrott, 1969). If a reasonable straight line can be plotted through the data, then for slopes less than −1, there will be a gradual approach to equilibrium without oscillations, for slopes between −1 and −2 equilibrium will be approached through damped oscillations and for slopes greater than −2 there will be increasing oscillations (see Williamson, 1972). For *Globodera rostochiensis* it is evident that the slope ($\tan \theta = -1{\cdot}75$) lies between −1 and −2 and that the slope would be greater if it were not for the unhatched eggs which are included in P_f.

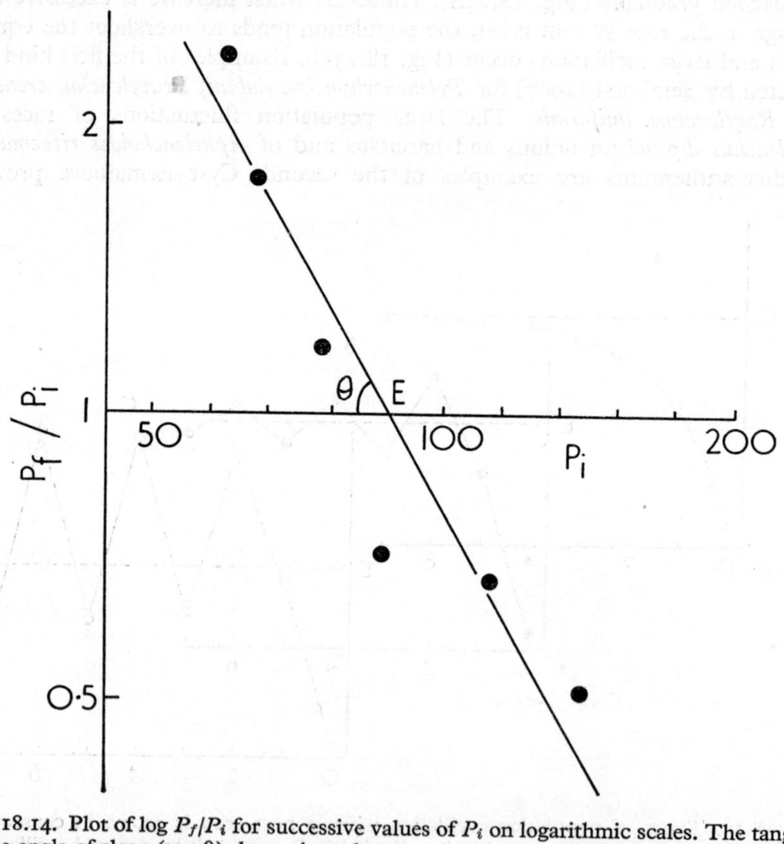

Fig. 18.14. Plot of log P_f/P_i for successive values of P_i on logarithmic scales. The tangent of the angle of slope (tan θ) determines the magnitude of oscillations about the equilibrium density. (Data from Jones & Parrott, 1969).

The behaviour of the population at the equilibrium may be investigated theoretically for the population equation (10). Differentiating with respect to P_i, the slope at the equilibrium,

$$\left[\frac{d(P_f - P_i)}{dP_i}\right]_{P_i = E} = (1 - C_p)\frac{a' - 1}{a'}(E \log z - 1),$$

and the nature of the equilibrium depends on whether the right-hand side of the equation is greater than −1 (a steady approach to E, no oscillations), less than −1 but greater than −2 (damped oscillations about E) or less than −2 (unstable equilibrium). The three regions are plotted on Fig. 18.15 for different values of C_p and the equilibrium ratio E/cE_l. While the multiplication rate a' is important in determining the position of and rate of approach to the equilibrium, it has little effect on the nature of the equilibrium for values of $a' > 10$ observed in practice.

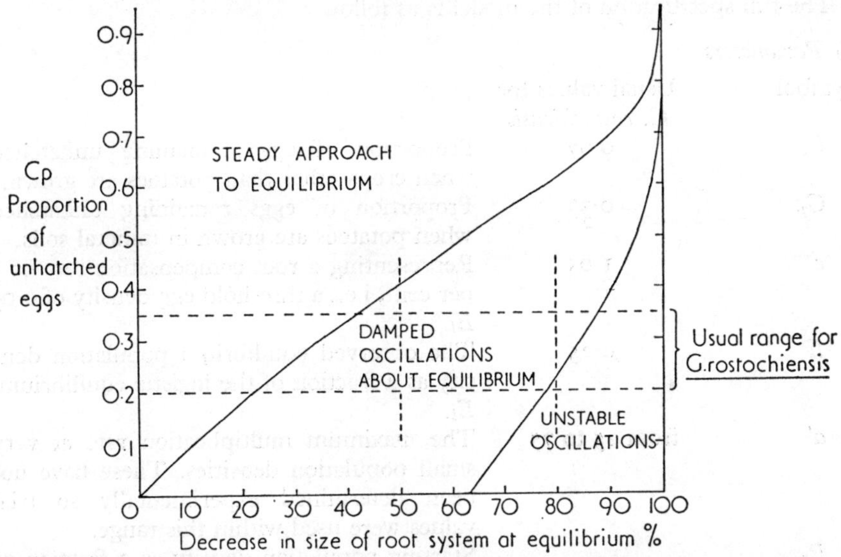

Fig. 18.15. The influence of the unhatched portion of the population, C_p, and the extent to which the root system is stunted at the equilibrium point (E expressed as a percentage of E_l) on the manner in which the population density fluctuates about the equilibrium point, E. At 0 per cent decrease in root size $E = E_l$ and the unmodified logistic equation applies.

Simulating Population Changes on a Computer

To model a system with any confidence we must be able to express its various components quantitatively. This requires good field data; indeed the modelling process is often useful in highlighting areas of work where new or better information is required.

Jones, Parrott & Ross (1967) attempted a model for *G. rostochiensis* Ro1. Besides trying to predict population changes under different rotations when susceptible potatoes, resistant potatoes or crops other than potatoes are grown, they were also interested in simulating the increase in frequency of a resistance-breaking gene in a nematode population when resistant potato cultivars are grown.

Their model did not allow for the effect of root damage on nematode density and used an alternative function for the sex ratio which was separate from the main specification of the model.

We have updated their model to include the effect of root damage, using equation (10) rather than the logistic equation (1). From the point of view of presentation of the computer output it is convenient to express all nematode densities as fractions of the logistic equilibrium density, E_l. Equation (10) then becomes

$$P_f = \frac{a'(1 - C_p)P_i}{1 + (a' - 1)\frac{P_i}{c} z_l{}^{-P_i}} + C_p\, P_i.$$

where $z_l = z^{El} = (E/c)^{1/E}$ (from equation 9), and the equilibrium density E is now expressed as a fraction of E_l, i.e., is the root size at equilibrium expressed as a fraction of maximum root size. If for example, $E = \frac{1}{4}$, then $z_l = {}^1/256$.

The full specification of the model is as follows:

(i) *Parameters*

Symbol	Usual values for *G. rostochiensis*	
C_o	0·67	Proportion of eggs remaining unhatched when crops other than potatoes are grown.
C_p	0·33	Proportion of eggs remaining unhatched when potatoes are grown in mineral soils.
c	1·05	Representing a root compensation rate of 5 per cent, i.e., a threshold egg density of 0·05 E_l.
E	0·25	The observed equilibrium population density as a fraction of the logistic equilibrium, E_l.
a'	from 25 to 75	The maximum multiplication rate at very small population densities. These have not been determined experimentally so trial values were used within this range.
P_o		Starting population density as a fraction of the equilibrium density.
q_o		The initial frequency of the recessive gene for hypotheses (b) and (c) in (iii).

(ii) *Linking equations*

Symbol	Equation	
O	$P_f = C_o P_i$	Density independent population change when crops other than potatoes are grown for one year.
R	$P_f = C_p P_i$	The same when resistant potatoes are grown for one year.
S	$P_f = m\, P_i$	Density dependent population change when susceptible potatoes are grown for one year. The multiplication factor m is given by equation (10).
N	$P_f = (1 - K)P_i$	The effect of a nematicide which kills a specified fraction K of the population.

(iii) *Hypotheses*

(a) *G. rostochiensis Ro1* is an interbreeding continuum, there are no fitness differences between gametes, and Hardy-Weinberg equilibrium between genotypes is achieved in one generation of random mating (Jones, 1975b).

(b) On the roots of hybrid potato plants possessing resistance gene H_1 (e.g., Maris Piper), males may have any genetic constitution (*AA*, *Aa* or *aa*) whereas females are double recessives (*aa*).

(c) The males are as in (b) but females must possess a dominant gene (*AA* or *Aa*).

(iv) *Rotations*

These are specified by arranging the crop symbols in order, e.g., OOOROOOS, RRRRR or NSOOONROOO, where N indicates a nematicide applied before the susceptible or the resistant potato crop.

(v) *Running the model*

The values of the model parameters and the rotation policy adopted are input to the computer. The computer reads each crop symbol in turn and uses the appropriate linking equation to predict final egg density after harvest from the initial pre-planting density. This final density then becomes the initial density for the next year, to be operated on by the appropriate linking equation for the next crop symbol in the rotation. Possible forms of computer output are shown in Tables 1 and 2.

(vi) *Fitting the model to observations*

We may compare observational data on changes in nematode density under a given rotation policy with changes predicted by the model and thus arrive at rough estimates of the model parameters. This was done for data from small plots where susceptible potatoes were grown continuously (Jones & Parrott, 1969). The model parameters were adjusted until the best fit between predicted and observed values was achieved. The fitted values mimicked well the damped oscillations about an equilibrium observed in practice (Fig. 18.16). The actual equilibrium value, E, was estimated at 87·3 eggs/g soil. Information on the proportion of eggs unhatched, C_p, and root damage, E/E_l, was confounded, similar patterns of oscillations being achieved by reducing C_p or E/E_l. Fixing C_p at 0·33, E/E_l, was estimated to be 0·19, close to what is often found in practice. The data provided practically no information for the estimation of the maximum multiplication rate, a', due to the lack of observations at small nematode densities.

Modifying the Model for HETERODERA AVENAE

The effects of the *Entomophthora*-like fungus that destroys females and eggs of *H. avenae* (Kerry, 1975) was probably responsible for the population decline observed by Gair, Mathias & Harvey (1969). Because of the activity of this and possibly of other fungal parasites, *H. avenae* populations do not behave as do most other *Heterodera* and *Globodera* populations. To take account of the

Table 1

Sample of Printout from Computer Model: for explanation of symbols and hypotheses see text

$a = 25$ $C_o = 0{\cdot}6667$ $C_p = 0{\cdot}3333$

		Hypothesis (b) $q_o{}^2 = 0{\cdot}001$		Hypothesis (c) $1 - q_o{}^2 = 0{\cdot}001$	
Year	Crop	Frequency of *aa*	Population size	Frequency of *AA* + *Aa*	Population size
0		0·0010	1·000	0·0010	1·000
1	R	0·0011	0·334	0·0020	0·334
2	R	0·0012	0·112	0·0074	0·113
3	R	0·0018	0·038	0·0526	0·041
4	R	0·0036	0·013	0·3120	0·032
5	R	0·0108	0·005	0·5295	0·105
6	R	0·0449	0·002	0·6290	0·298
7	R	0·1783	0·003	0·6988	0·483
8	R	0·4132	0·008	0·7492	0·608
9	R	0·5922	0·049	0·7875	0·689
10	R	0·7122	0·239	0·8178	0·746
11	R	0·7948	0·501	0·8421	0·786
12	R	0·8490	0·677	0·8621	0·818
0		0·0010	1·000	0·0010	1·000
1	R	0·0011	0·334	0·0020	0·334
2	O	0·0011	0·223	0·0020	0·223
3	O	0·0011	0·148	0·0020	0·148
4	S	0·0011	0·592	0·0020	0·592
5	O	0·0011	0·395	0·0020	0·395
6	O	0·0011	0·263	0·0020	0·263
7	R	0·0013	0·088	0·0087	0·089
8	O	0·0013	0·059	0·0087	0·059
9	O	0·0013	0·039	0·0087	0·039
10	S	0·0013	0·350	0·0087	0·351
11	O	0·0013	0·233	0·0087	0·234
12	O	0·0013	0·156	0·0087	0·156

Note: Nematode densities assumed to lie below threshold for crop damage.

effects of the fungi on the output of eggs by females an additional relationship signalled by the symbol M, is required after a susceptible cereal cultivar, e.g., SMOO ROO:

$$M \qquad P_f = (1 - D)P_i$$

where D is the fraction of eggs destroyed. Unfortunately this simple relationship is inadequate for several reasons. First, it is density independent whereas the effect of the fungi is almost certainly density dependent. Second, it relates to eggs when the effect of the most important fungus is on the females. More information is needed before a better quantitative relationship can be devised. Also, less is known about the parameters of *H. avenae* required for the model but 0·50 is a reasonable value for C_o and 0·20 for C_p.

Table 2 *Sample of Printout from Computer Model for Competition Between* Globodera rostochiensis Ro1 *and* G. pallida Pa3

P_i	Proportion of *G. pallida*	Kill by nematicide	$\acute{a}$ *G. rostoch.*	$\acute{a}$ *G. pall.*	C_p *G. rostoch.*	C_p *G. pall.*	C_o *G. rostoch.*	C_o *G. pall.*	E_{lp}/E_{lr}	v_r	v_p
0.10	0.01	0.00	40	50	0.33	0.70	0.67	0.80	1.00	0.0	0.0

Root damage

c	E_r	E_p
1.05	0.25	0.25

Crop rotation	Proportion of *G. rostochiensis*
-	0.990
R	0.727
R	0.133
R	0.032
R	0.010
R	0.003
R	0.001
R	0.000
R	0.000
R	0.000
R	0.000
-	0.990
S	0.920
O	0.906
O	0.890
R	0.299
O	0.263
O	0.230
O	0.539
O	0.495
O	0.451
O	0.173
O	0.149
O	0.128
S	0.475
O	0.431
O	0.389
R	0.138
O	0.119
O	0.101

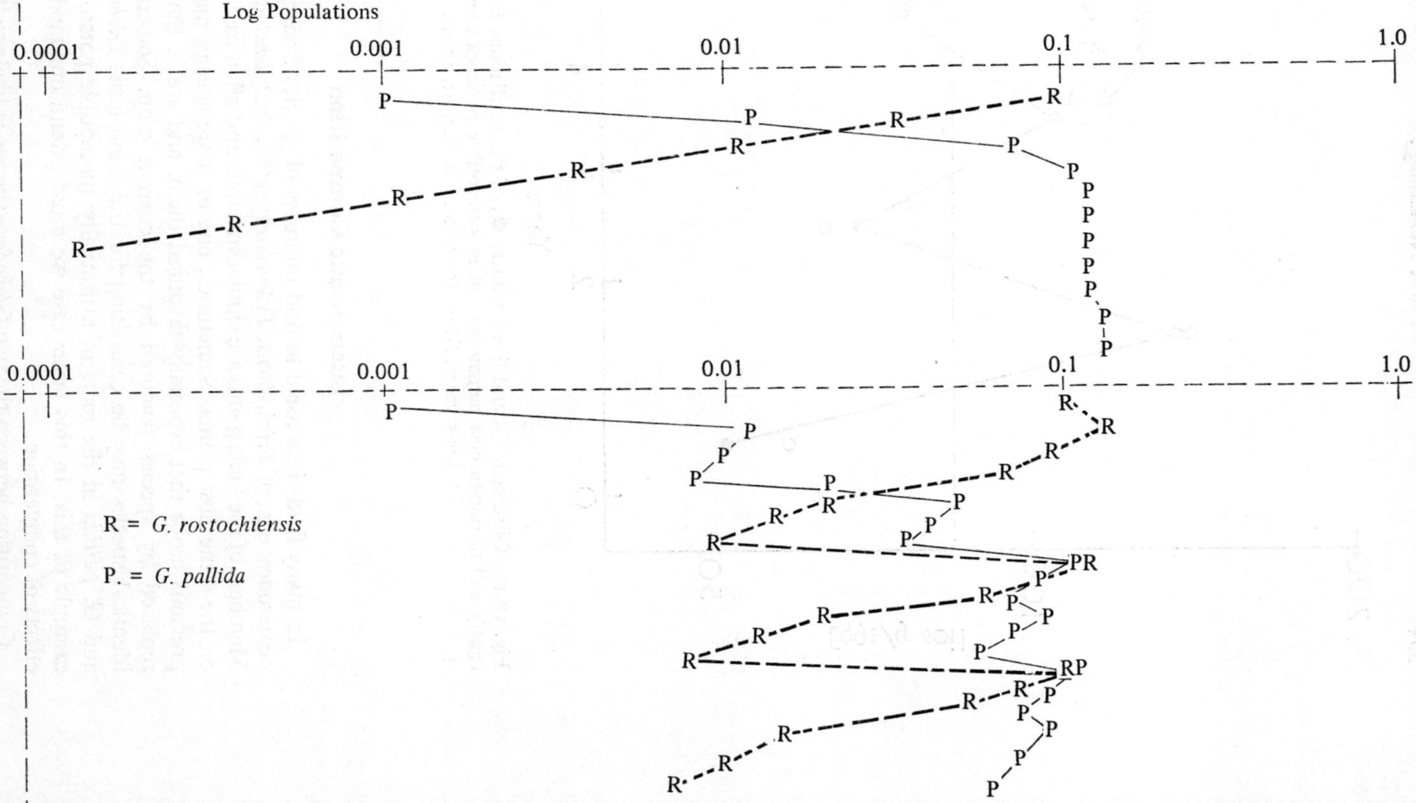

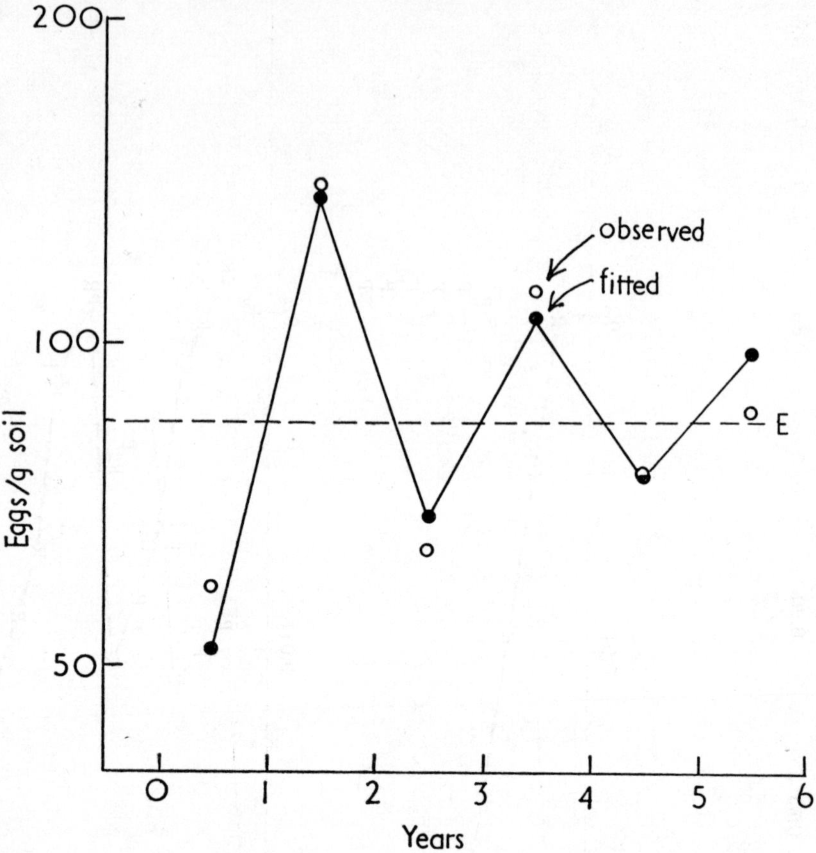

Fig. 18.16. Observed, ○, and fitted values, ●, of the oscillations in population density, eggs/g soil of potato cyst nematodes, when susceptible potatoes were grown every year for 5 years. (Data from Jones & Parrott, 1969).

Interspecific Competition

In many fields it is usual to find mixtures of species from the same genus of nematodes, e.g., of *Trichodorus*, *Tylenchorhynchus*, *Pratylenchus* and *Meloidogyne*. Members of the same genus are similar in habits and often feed almost identically on the same host plants. Sometimes, one or more species may be relicts from previous crops that especially favoured them, and are in the process of being replaced by species favoured by the current crop. Sometimes, two nearly identical species may be competing for the same host: *Globodera rostochiensis* and *G. pallida* in the roots of a mutually susceptible potato crop are a good example of this. In this latter case we must extend our model to include the effects of competition.

Competition between the two *Globodera* species is likely to be strongest in the search for root space to enable the larvae to become female and reproduce. The relative proportions of the two species becoming females will depend on relative hatching rates, the minimum necessary size of the female's territory in the rootlets, as well as other genetic differences. We extend the equations relating sex ratio to density to two species with densities, P_1, P_2 by writing

$$M_1/F_1 = b_1\,(P_1 + v_1P_2),$$
$$M_2/F_2 = b_2\,(P_2 + v_2P_1).$$

For this model, if the two species exploit totally different parts of the root system, then $v_1 = v_2 = 0$ and in the absence of root damage they will coexist independently. If their exploitation is identical then $v_1 = 1/v_2$, the species share the same ecological niche and coexistence is impossible.

Assuming densities are below the threshold at which root damage occurs, the difference equations relating final egg densities P_{f1}, P_{f2} to initial egg densities, P_{i1}, P_{i2} when a susceptible crop is grown are of the form:

$$P_{f1} = \frac{a'_1\,(1 - C_{p1})\,P_{i1}}{1 + (a'_1 - 1)\dfrac{(P_{i1} + v_1P_{i2})}{E_{l1}}} + C_{p1}\,P_{i1}$$

$$P_{f2} = \frac{a'_2\,(1 - C_{p2})\,P_{i2}}{1 + (a'_2 - 1)\dfrac{(P_{i2} + v_2P_{i1})}{E_{l2}}} + C_{p2}\,P_{i2}$$

where a'_1, a'_2 are the maximum multiplication rates; C_{p1}, C_{p2}, the proportions of eggs carried over unhatched each year; E_{l1}, E_{l2}, the egg densities at the logistic equilibrium when there is no competition.

The necessary and sufficient condition for the coexistence of both species in this system is that

$$\frac{E_{l2}}{v_1v_2} > \frac{E_{l1}}{v_1} > E_{l2}$$

There are in fact four possible outcomes for the system, as illustrated by Fig. 18.17 (cf. Williamson, 1972). Here the two axes represent the population densities of the two nematode species, and E_{l1}, E_{l2} are the equilibrium densities of each species in the absence of the other. The line running from E_{l1} and given by the equation, $N_1 + v_1N_2 = E_{l1}$, is the locus of the equilibria of species 1 in the presence of increasing numbers of species 2, where N_1 and N_2 are the numbers of species 1 and 2 respectively. The line running from E_{l2} represents the similar locus for species 2. The arrows show the direction of change from an initial set of population sizes P_{i1}, P_{i2} to a final set P_{f1}, P_{f2} after one generation. They are the resultants of the tendencies for the numbers of species 1 to move horizontally towards their equilibria points on the continuous line, and those of species 2 to move vertically towards their equilibria on the broken line.

The four possible outcomes from the model are as follows. In (1) movement is towards E_{l1} and species 2 will always eventually become extinct; in (2) the situation is the reverse and species 1 becomes extinct; in (3) the two equilibria loci cross and a stable equilibrium mixture is established with both species coexisting side by side; in (4) the equilibrium is unstable and either species may become extinct, depending on the initial population densities. In this last case stochastic variation may have an important effect on the outcome. Under this model the respective proportions of unhatched eggs C_{p1} and C_{p2} have no effect on the eventual outcome although of course they may considerably affect the rate of extinction of either species. Note that when a stable equilibrium exists

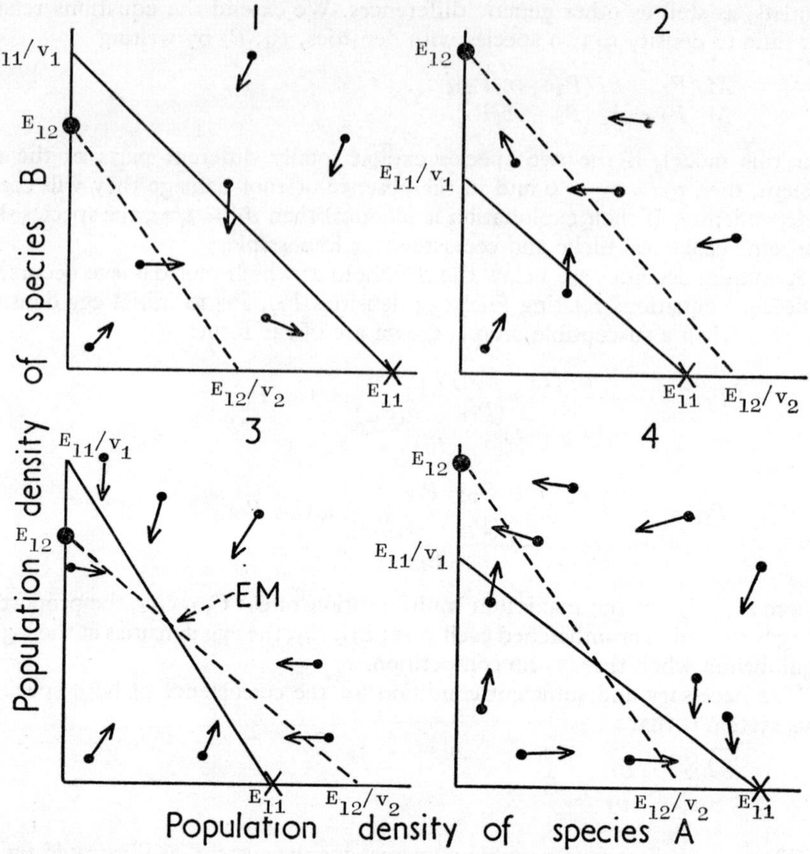

Fig. 18.17. Diagrams of the four simplest competitive situations. The numbers of species A are plotted against those of B: ———, is the locus of equilibrium points of species A in the presence of species B; X, E_{l1}, the equilibrium point of species A by itself; – – – –, the locus of equilibrium points of species B in the presence of species A; ●, E_{l2}, the equilibrium point of species B by itself. Arrows indicate the direction of change of the population mixtures from the composition indicated by the dot at the end of their tails. Which situation is appropriate is governed by the relative values for the equilibrium E_{l1}, E_{l2} and the species interaction parameters, v_1, v_2. For situation 1, $E_{l1}/v_1 > E_{l2}$ and $E_{l1} > E_{l2}/v_2$. In situation 2, the inequality signs are reversed. In situation 3, $E_{l1}/v_1 > E_{l2}$ and $E_{l2} > E_{l1}$; in situation 4 the inequality signs are reversed. For situation 3, when $v_1, v_2 > 0$, an equilibrium mixture EM is established if and only if $E_{l2}/v_1v_2 > E_{l1}/v_1 > E_{l2}$.

each species suppresses its own numbers more than those of its competitor, i.e., intraspecific competition is greater than interspecific. The joint equilibrium population is however greater than either single population.

For nematode densities above the threshold where root damage occurs, we may express the relative root size as a function of the densities of the two individual species by

$$y/y\text{max.} = ce^{-(k_1Pi_1+k_2Pi_2)} = cz_1^{Pi_1}z_2^{Pi_2}$$

When this factor is incorporated into the model the loci of equilibria for the individual species become curved. The criteria for determining the eventual outcome of the system are more complex, but one interesting possibility is that the joint equilibrium population and hence the crop damage for the two-species system may be smaller than when either species occurs separately. This requires further investigation.

Investigating the effects of competition experimentally requires a great range and variety of combinations of species population densities. These can be limited to manageable proportions only when there is some previous knowledge of likely equilibrium mixtures (Seinhorst, 1970). In the field the study of mixed populations is of more than academic importance especially when resistant varieties are available and the attacking nematodes are related species, as are *G. rostochiensis* and *G. pallida*. For races within species there may be competition between genotypes, as in hypotheses (b) and (c) of Jones *et al.*'s model (see also p. 351).

Predictions from the Models

Too much should not be expected of population models. They may provide qualitative information which is valuable in understanding the population dynamics of a species, but cannot seriously be used for prediction, particularly at the level of individual fields, where crop variety, local farm practices and other factors affect population development. Fortunately, soil climatic factors other than rainfall are relatively stable within an area as small as Great Britain. Farming practices for particular crops tend also to be stereotyped. Consequently generalizations for root endoparasitic nematodes have a wider validity than would otherwise be the case, but those for root ectoparasitic nematodes are more difficult (Jones, 1975a, 1975c). Models may be used to observe the character of a population fluctuating under different cropping regimes (Fig. 18.18). They also suggest that nematicides which kill some 70–90 per cent of cyst nematode

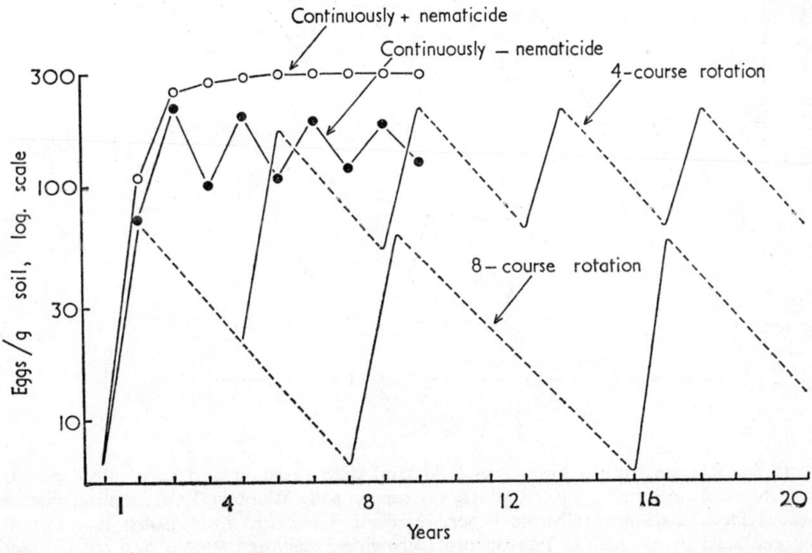

Fig. 18.18. Stylized population graphs drawn from the output of computer models: ———, susceptible potatoes grown; – – – –, other crops grown.

eggs, would, after harvest, tend to leave nematode numbers near the maximum of the population curve. This is confirmed by old field experiments (Peters & Fenwick, 1949) and by more recent ones, Fig. 18.19, which also suggest that such nematicides, rather than preventing the selection of a race or species able to reproduce on a resistant cultivar, would make selection more rapid. Other predictions are of the type shown in Table 3 which are in accord with experience. In fields on Long Island, New York, and in Britain, as yet there is no evidence that growing resistant potatoes with gene H_1 from *Solanum tuberosum* ssp. *andigena* has selected from *G. rostochiensis* Ro1 (genetic constitution *AA*) types equivalent to populations of race Ro2 such as occur in The Netherlands and Bolivia (Constitution *aa*; Jones, 1975b). Selection in the U.K. and in The Netherlands has usually resulted in the replacement of *G. rostochiensis* Ro1 by *G. pallida* Pa3 after growing up to five successive crops of a resistant potato cultivar bearing gene H_1. The observed data of Cole & Howard (1962b), and those of Huijsman (1961) to which Jones, Parrott & Ross's model was fitted, were inappropriate because it is now known that *G. pallida*, which does not interbreed with *G. rostochiensis*, was being selected, i.e., an interspecific competition model was required not an intraspecific one.

Table 2 shows an example of printout from the previously specified model based on interspecific competition. The relationship between P_i and P_f for *G. pallida* on potatoes susceptible to it differs in the values of a' (maximum observed multiplication rate), C_p (fraction that fails to hatch when potatoes are grown) and C_o (fraction that fails to hatch when other crops are grown). In this example,

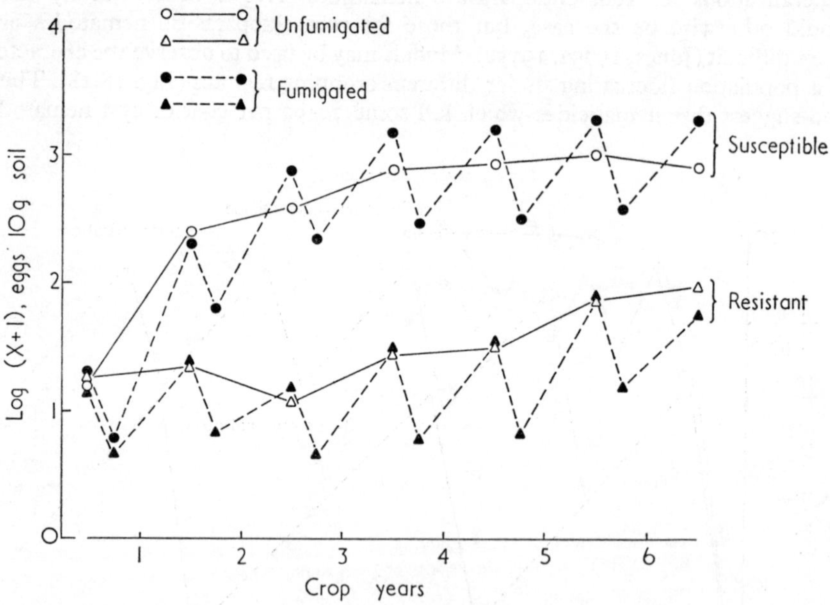

Fig. 18.19. Results from a long-term field trial with resistant and susceptible potatoes and fumigated and unfumigated plots, on sandy soil, Woburn Experimental Station. Pentland Dell, susceptible; Maris Piper, resistant. *Globodera rostochiensis* Ro1 initially, being replaced by *G. pallida* Pa3 on plots growing resistant potatoes and subsequently also on those growing the susceptible cultivar. (Years 1–3, D–D; years 4–6, Dazomet). (From Jones & Parrott, unpublished).

Table 3

Calculated Period in Years to Establishment of a Mutant or Immigrant Form of Potato Cyst Nematode able to Overcome Resistance

	Frequency of genotype initially					
	1 in 10 000		1 in 100 000		1 in 1 000 000	
Crop sequence	Dominant	Recessive	Dominant	Recessive	Dominant	Recessive
*RRRRRRR	9	> 40	11	> 40	> 40	> 40
ROOROO	34	> 60	> 40	> 60	> 60	> 60
ROOOROOO	53	> 60	> 60	> 60	> 60	> 60
ROOOOROOOO	> 60	> 60	> 60	> 60	> 60	> 60

*R, resistant potato cultivar grown; O, other crops grown.

the effect of interaction between *G. rostochiensis* and *G. pallida* larvae on their sex ratios was ignored, i.e., $v_r = v_p = 0$ and the logistic equilibria and root damage caused at equilibrium were assumed to be the same for both species. These simplifying assumptions will be dropped as soon as relevant information becomes available. The simple model suggests that rotations with frequent crops of potatoes bearing gene H_1 would eliminate *G. rostochiensis* whereas others containing potatoes with and without gene H_1 might establish an equilibrium mixture. However, the current distribution pattern of these two species in Britain suggests that they are mutually exclusive (p. 355). Either one species is displacing the other (competitive displacement; situation 2 or 3, Fig. 18.17), or whether a field is now dominated by one or other species may have been decided by which arrived first or in greatest numbers (Fig. 18.17, situation 4). Preliminary pot tests suggest that in all circumstances *G. pallida* replaces *G. rostochiensis* (Fig. 18.17, situation 2) because v_r (the effect of *pallida* on *rostochiensis*) is approximately 1 and v_p (the effect of *rostochiensis* on *pallida*) is negligibly small. If the pot tests are indicative of what happens in fields, it seems possible that *G. pallida* will ultimately replace *G. rostochiensis* in Britain whether potato cultivars with gene H_1 are grown or not (Jones, 1976). Evidently more work and better models are needed if we are to understand what is happening.

Integrating Control Measures

Integrated control originally meant the use of comprehensive measures which took account of all factors, including naturally occurring enemies and competitors, that tend to diminish pest numbers. As little is known of the effect of naturally occurring enemies of most plant-parasitic nematodes (see Chapter 13) except for those of *H. avenae* (Kerry, 1975), integrated control applies only to a rational combination of crop rotation, resistant cultivars and nematicides, and possibly also of fertilizers (Jones, 1977). Two criteria can be used to judge whether control measures are successful. The first is the yield increase from enhanced plant growth and the second the population density of the pest after harvest. As we have seen (p. 358, Figs. 18.18 & 18.19), a large yield increase may be accompanied by a much increased nematode population at the end of the year. How large a kill is needed to prevent such an increase depends on the multiplication rate of the nematode in question. Thus a kill of 75 per cent would leave 25 per cent as survivors and these would need to multiply only four times to replace the original population. Similarly a kill of 90 per cent would leave 10 per cent as survivors and require a multiplication rate of ten times to replace the original population. Table 4 illustrates the integration of control measures for *G. rostochiensis* Ro1.

Table 4

Integration of Control Methods for Globodera rostochiensis
(*Jones*, 1969*b*)

Control method(s)	Resulting population (% initial population)	Kill (%)	Population after growing and harvesting a susceptible cultivar, calculated at two assumed nematode multiplication rates* (% initial population) 30 ×	70 ×
1. 4 years without potatoes	3	97	90	> 100
2. 1 year with resistant potatoes	20	80	> 100	> 100
3. Nematicide treatment (fumigant)	25	75	> 100	> 100
4. 1 and 2	0·6	99·4	18	42
5. 1 and 3	0·75	99·25	22·5	52·5
6. 2 and 3	5	95·0	> 100	> 100
7. All three methods	0·15	99·85	4·5	10·5

*The observed maximum reproductive rate is thought to lie between 25 and 75 times but could be greater.

It is unrealistic to attempt to design methods that will eliminate nematodes. Practically, control is adequate if the population remaining after harvest is smaller than would cause harm if the crop were planted again. In many situations this may be asking too much and the degree of control is acceptable if the post-harvest population is equal to or slightly smaller than that at planting. Some granular nematicides (oxime carbamates and organophosphates) are sufficiently powerful to kill or inactivate enough cyst-nematode larvae to achieve this goal without being combined with other methods (Table 5). When combined with a short rotation or a resistant cultivar smaller rates of application might conceivably be used, but the rates recommended commercially are sometimes barely adequate when allowance is made for poor mixing of the granules with the surface soil. DD and other fumigants are less effective than the granular nematicides referred to above, but they are more than adequate for nematodes which multiply slowly (e.g., *Trichodorus, Paratrichodorus, Longidorus, Xiphinema*) and may take up to three years to replace their initial numbers. Single preplanting applications of the most efficient nematicides are barely adequate to control populations of those races of *Ditylenchus dipsaci* for which resistant cultivars are not available, or to control those of species like *Aphelenchoides ritzemabosi* which also have a large reproductive potential.

Table 5

Effect of Oxamyl and Aldicarb on Potato Cyst Nematode Increase in Ten Fields in Eastern England in 1973. (After Whitehead, 1975)

	Amount (kg a.i./ha)	Average nematode increase Peaty loams †(4)	Silty loams (4)	Sandy loams (2)
Untreated	0	× 41	× 14·5	× 13
Oxamyl	5·6	× 5	× 2	× 0·5
Aldicarb	5·6	× 4	× 1·5	× 0·3

† Number of trials.

To protect the potato starch industry in The Netherlands, integrated control is being applied on a large scale in the Groeningen area (Nollen & Mulder, 1969).

Concluding Remarks

The principles underlying population dynamics of plant-parasitic nematodes can only be studied in small field plots, microplots (Jones, 1956) or large pots, preferably the first. As a general rule the effects referred to by Seinhorst or Jones in their publications relate only to the plough layer of small plots, the minor variations in lateral or vertical distribution being ignored. However, full consideration of population dynamics would require the wider variations in fields, both vertically and horizontally, to be taken into account as well as the growth of roots (or other attacked plant parts) and the development of infestations in time. Seinhorst (1972, 1973c) has begun to consider some of these stochastic elements in nematode population dynamics and has listed the parameters of plant-nematode systems, the possibilities of measuring them and their importance in determining the amount of control needed. When some of these elements are included, the curves drawn to show mean effects may differ in form from those obtained by studying the populations of small plots.

For example, in Fig. 18.19 the mean curve for population trends derived from twelve unfumigated, large field plots suggests an approach to equilibrium density that would fit the logistic curve. This is because the oscillations of individual plots, and perhaps even parts of plots, are in different phases which, when summed, tend to cancel each other. The inherent tendency of populations to oscillate, which is sometimes accentuated by climatic and other factors, perhaps explains why nematode numbers are patchy in fields which have been treated uniformly and have reasonably uniform soil. It cannot be assumed that uniform conditions of food supply and environment would lead to uniform or stable populations. Foundation effects (Fig. 18.3) may also cause patchiness.

Bibliography

ABOUL-EID, H. Z. (1969). Electron microscope studies on the body wall and feeding apparatus of *Longidorus macrosoma*. *Nematologica* **15**, 451–463.

AIST, S. & RIGGS, R. D. (1969). Amino acids from *Heterodera glycines*. *J. Nematol.* **1**, 254–259.

AITKENHEAD, P. & JANSON, H. W. (Eds) (1970). Narcissus pests. *Bull. Minist. Agric. Fish. Fd* 51 (6th Edn), 40 pp. (H.M.S.O.).

ALCALA, J. V., TOBAR JIMENEZ, A. & MEDINA, J. M. M. (1970). Lesiones causadas y reacciones provocadas par algunos nematodes en las raices de ciertas plantas. *Revta iber. Parasit.* **30**, 547–566.

ALLEN, M. W. (1952). Taxonomic status of the bud and leaf nematodes related to *Aphelenchoides fragariae* (Ritzema Bos 1891). *Proc. helminth. Soc. Wash.* **19**, 108–120.

ALLEN, M. W. (1957). A review of the nematode genus *Trichodorus* with descriptions of ten new species. *Nematologica* **2**, 32–62.

ALLEN, M. W., HART, W. H. & BAGHOTT, K. (1970). Crop rotation controls barley root-knot nematode at Tulelake. *Calif. Agric.* **24**, 4–5.

ALLEN, M. W. & RASKI, D. J. (1950). The effect of soil type on the dispersion of soil fumigants. *Phytopathology* **40**, 1043–1053.

ALLEN, M. W. & SHER, S. A. (1967). Taxonomic problems concerning the phytoparasitic nematodes. *A. Rev. Phytopath.* **5**, 247–264.

ALPHEY, T. J. W. & BOAG, B. (1976). Distribution of trichodorid nematodes in Great Britain. *Ann. appl. Biol.* **84**, 371–381.

ANDERSEN, S. (1959). Resistance of barley to various populations of the cereal root eelworm (*Heterodera major*). *Nematologica* **4**, 91–98.

ANDERSEN, S. (1961). Resistens mod havreål. *Meddr K. Vet.- og Landbohøjsk. afd. Landbr. PlKult.* No. 68, 179 pp.

ANDERSEN, S. (1965). Heredity of race 1 and race 2 in *Heterodera avenae*. *Nematologica* **11**, 121–124.

ANDERSEN, S. & ANDERSEN, K. (1973). Linkage between marker genes on barley chromosome 2 and a gene for resistance to *Heterodera avenae*. *Hereditas* **73**, 271–276.

ANDERSON, R. C., CHABAUD, A. G. & WILMOTT, S. (Eds). (1974). *CIH Keys to the nematode parasites of vertebrates* No. 1, 17 pp. Farnham Royal, England, Commonw. Agric. Bur.

ANDERSSON, S. (1966). Angrepp av Stjälknematod (*Ditylenchus dipsaci* (Kühn) Filipjev) i havre. *Växtskyddsnotiser* **30**, 10–12.

ANDERSSON, S. (1973). En sannolikt ny cystnematod på stråsäd. *Växtskyddsnotiser* **37**, 74–76.

ANDRÁSSY, I. (1959a). Über das System der Mononchiden (Mononchidae Chitwood, 1937; Nematoda). *Annls. hist.-nat. Mus. natn. hung.* **9**, 151–171.

ANDRÁSSY, I. (1959b). Die Mundhöhlentypen der Mononchiden und der Schlüssel der *Mylonchulus*-Arten (Nematoda). *Opusc. zool., Bpest.* **3**, 3–12.

ANDRÁSSY, I. (1963). Nematologische Notizen, 12. *Annls Univ. Scient. bpest. Rolando Eötvös, Sect. biol.* **6**, 3–12.

ANDRÁSSY, I. (1968). Fauna Paraguayensis. 2. Nematoden aus den Galeriewäldern des Acaray-Flusses. *Opusc. zool., Bpest.* **8**, 167–315.

ANDRÁSSY, I. (1971). Zwei neue Arten der Familie Bunonematidae (Nematoda). *Zool. Anz.* **187**, 257–265.

ANDRÁSSY, I. (1976). *Evolution as a basis for the systematization of nematodes.* London, San Francisco & Melbourne, Pitman Publishing, 288 pp.

ANON. (1917). Lecture on narcissus disease. *Gdnrs' Chron.* **61**, 207, 217–218, 226–227.

ANON. (1954). International convention for the protection of plants and plant products. Rome, December 6, 1951. *Treaty Series* No. 16. London, H.M.S.O., 30 pp.

ANON. (1959). Soil sterilization. *Bull. Minist. Agric. Fish. Fd* 22 (5th Edn), 49 pp. (H.M.S.O.—out of print).

ANON. (1960). Mushroom growing. *Bull. Minist. Agric. Fish. Fd* 34 (7th Edn), 66 pp. (H.M.S.O.).

ANON. (1962). Outbreaks and new records. Canada. Occurrence of potato root eelworm. *Pl. Prot. Bull. F.A.O.* **10**, 115.

ANON. (1970). *Research report of the Research Branch for* 1969. Ottawa, Canada Department of Agriculture, 370 pp.

ANON. (1971). Rep. Rosewarne expl Hort. Stn (16th) 1970 (London, H.M.S.O.), 176 pp.

ANON. (1972a). Hot-water treatment of plant material. *Bull. Minist. Agric. Fish. Fd* 201 (2nd Edn), 46 pp. (H.M.S.O.).

ANON. (1972b). Plant parasitic nematode pathotypes. *Ann. appl. Biol.* **71**, 263–300.

ANON. (1975a). *A.D.A.S. Science Service Annual Report* 1974. London, H.M.S.O., 233 pp.

ANON. (1975b). *Rep. Rosewarne, Ellbridge & Isles of Scilly exp. Stns* 20th (1974), 206 pp.

ANON. (1976). Chrysanthemum eelworm. *Adv. Leafl. Minist. Agric. Fish. Fd* 339, 6 pp.

ANON. (1977). Cyst eelworms on potato. *Adv. Leafl. Minist. Agric. Fish. Fd.* 284, 8 pp.

ANSCOMBE, F. J. (1950). Soil sampling for potato root eelworm cysts. A report presented to the Conference of Advisory Entomologists. *Ann. appl. Biol.* **37**, 286–295.

APT, W. J. & GOULD, C. J. (1961). Control of root-lesion nematode, *Pratylenchus penetrans,* on narcissus. *Pl. Dis. Reptr* **45**, 290–295.

ARIAS, M. & NAVACERRADA, G. (1973). Geographical distribution of *Xiphinema* Cobb in Spanish vineyards. *Nematologia Mediterranea* **1**, 28–35.

ARROLD, N. P. & BLAKE, C. D. (1968). Some effects of the nematodes *Ditylenchus myceliophagus* and *Aphelenchoides composticola* on the yield of the cultivated mushroom. *Ann. appl. Biol.* **61**, 161–166.

ATANASOFF, D. (1925). The *Dilophospora* disease of cereals. *Phytopathology* **15**, 11–40.

ATKINSON, G. F. (1892). Some diseases of cotton. *Bull. Alabama agric. Exp. Stn* 41.

AYALA, A. & ALLEN, M. W. (1968). Transmission of the California tobacco rattle virus (CTRV) by three species of the nematode genus *Trichodorus. J. Agric. Univ. P. Rico* **52**, 101–125.

AYALA, A., ALLEN, M. W. & NOFFSINGER, E. M. (1970). Host range, biology, and factors affecting survival and reproduction of the stubby root nematode. *J. Agric. Univ. P. Rico* **54**, 341–369.

BAIN, D. C. (1959). Selection for resistance to root-knot of red and white clover. *Pl. Dis. Reptr* **43**, 318–322.

BAKER, A. D. (1962). *Check lists of the nematode superfamilies Dorylaimoidea, Rhabditoidea, Tylenchoidea, and Aphelenchoidea.* Leiden, E. J. Brill, 261 pp.

BAKER, A. D., BROWN, G. L. & JAMES, A. B. (1954). Relationships of fungi, mites and the potato-rot nematode. *Science* **119**, 92–93.

BAKER, C. R. B. (1972). An approach to determining potential pest distribution. *EPPO Bull.* No. 3, 5–22.

BALDWIN, J. G. & HIRSCHMANN, H. (1973). Fine structure of cephalic sense organs in *Meloidogyne incognita* males. *J. Nematol.* **5**, 285–302.

BALDWIN, J. G. & HIRSCHMANN, H. (1975a). Fine structure of cephalic sense organs in *Heterodera glycines* males. *J. Nematol.* **7**, 40–49.

BALDWIN, J. G. & HIRSCHMANN, H. (1975b). Body wall fine structure of the anterior region of *Meloidogyne incognita* and *Heterodera glycines* males. *J. Nematol.* **7**, 175–193.

BANYER, R. J. & FISHER, J. M. (1971a). Seasonal variation in hatching of eggs of *Heterodera avenae. Nematologica* **17**, 225–236.

BANYER, R. J. & FISHER, J. M. (1971b). Effect of temperature on hatching of eggs of *Heterodera avenae. Nematologica* **17**, 519–534.

BAQRI, Q. H. & JAIRAJPURI, M. S. (1968). On six new species of Dorylaimida (Nematoda). *J. Helminth.* **42**, 243–256.

BAQRI, S. Z. & JAIRAJPURI, M. S. (1974). Studies on Mononchida. V. The mononchs of El Salvador with descriptions of two new genera, *Actus* and *Paracrassibucca. Nematologica* **19**, 326–333.

BARKER, K. R. (1964). On the disease reduction and reproduction of the nematode *Aphelenchus avenae* on isolates of *Rhizoctonia solani. Pl. Dis. Reptr* **48**, 428–432.

BARKER, K. R. & DARLING, H. M. (1965). Reproduction of *Aphelenchus avenae* on plant tissues in culture. *Nematologica* **11**, 162–166.

BARKER, K. R., WORF, G. L. & EPSTEIN, A. H. (1965). Nematodes associated with the decline of azaleas in Wisconsin. *Pl. Dis. Reptr* **49**, 47–49.

BARRACLOUGH, R. & BLACKITH, R. E. (1962). Morphometric relationships in the genus *Ditylenchus. Nematologica* **8**, 51–58.

BARRONS, K. C. (1940). Root-knot resistance in beans. *J. Hered.* **31**, 35–38.

BAVER, L. D., GARDNER, W. H. & GARDNER, W. R. (1972). *Soil physics.* London, Wiley & Sons, 498 pp.

BAXTER, R. I. & BLAKE, C. D. (1969). Oxygen and the hatch of eggs and migration of larvae of *Meloidogyne javanica. Ann. appl. Biol.* **63**, 191–203.

BEDDING, R. A. (1973). Biology of *Deladenus siricidicola* (Neotylenchidae) an entomophagous-mycetophagous nematode parasitic in Siricid woodwasps. *Nematologica* **18** (Yr 1972), 482–493.

BEDDING, R. A. (1974). Five new species of *Deladenus* (Neotylenchidae), entomophagous-mycetophagous nematodes parasitic in Siricid woodwasps. *Nematologica* **20**, 204–225.

BEDI, A. S. (1968). A study of the variability in field populations of the potato cyst nematode (*Heterodera rostochiensis* Woll.) in relation to resistant potatoes. *Ph.D. Thesis, Edinburgh University.*

BELLO, A. (1971). El género *Tylenchus* (Nematoda, Tylenchina). *Boln R. Soc. esp. Hist. nat. (Biol.)* **69**, 279–301.

BELLONI, V. (1954). Comparsa in Italia di una anguillulosi die frumento e prove di lotta. *Notiz. Mal. Piante* **27**, 3–6.

BERGESON, G. B. (1955). The use of systemic phosphates for control of *Ditylenchus dipsaci* on alfalfa and daffodils. *Pl. Dis. Reptr* **39,** 705–709.

BERGESON, G. B. (1959). The influence of temperature on the survival of some species of the genus *Meloidogyne,* in the absence of a host. *Nematologica* **4,** 344–354.

BERGESON, G. B. (1966). Mobilization of minerals to the infection site of root knot nematodes. *Phytopathology* **56,** 1287–1289.

BERGESON, G. B. (1972). Concepts of nematode-fungus associations in plant disease complexes: a review. *Expl Parasit.* **32,** 301–314.

BERGESON, G. B., VAN GUNDY, S. D. & THOMASON, I. J. (1970). Effect of *Meloidogyne javanica* on rhizosphere microflora and Fusarium wilt of tomato. *Phytopathology* **60,** 1245–1249.

BERKELEY, M. J. (1885). *Gdnrs' Chron.* Yr. 1855, No. 14, 220.

BINDER, E. & HUTCHINSON, M. T. (1959). Further studies concerning the effect of the root-knot nematode *Meloidogyne incognita acrita* on the susceptibility of Chesapeake tomato to Fusarium wilt. *Pl. Dis. Reptr* **43,** 972–978.

BINGEFORS, S. (1967). International dispersal of nematodes. *Neth. J. Pl. Path.* **73** (Suppl. 1), 44–60.

BINGEFORS, S. (1969). The use of nematode-resistant varieties of grasses and legumes. *Herb. Abstr.* **39,** 107–111.

BINGEFORS, S. (1970). Resistance against stem nematodes *Ditylenchus dipsaci* (Kühn) Filipjev. *E.P.P.O. Public. Ser. A.* No. 54, 63–75.

BINGEFORS, S. (1971). Resistance to nematodes and the possible value of induced mutations. In: *Mutation breeding for disease resistance.* Vienna, International Atomic Energy Agency, 209–235.

BIRCHFIELD, W. & ANTONOPOULOS, A. A. (1976). Scanning electron microscopic observations of *Duboscqia penetrans* parasitizing root-knot larvae. *J. Nematol.* **8,** 272–273.

BIRD, A. F. (1959). Development of the root knot nematodes *Meloidogyne javanica* (Treub) and *Meloidogyne hapla* Chitwood in the tomato. *Nematologica* **4,** 31–42.

BIRD, A. F. (1960). Additional notes on the attractiveness of roots to plant parasitic nematodes. *Nematologica* **5,** 217.

BIRD, A. F. (1962). The inducement of giant cells by *Meloidogyne javanica. Nematologica* **8,** 1–10.

BIRD, A. F. (1967). Changes associated with parasitism in nematodes. I. Morphology and physiology of preparasitic and parasitic larvae of *Meloidogyne javanica. J. Parasit.* **53,** 768–776.

BIRD, A. F. (1968). Changes associated with parasitism in nematodes. IV. Cytochemical studies on the ampulla of the dorsal esophageal gland of *Meloidogyne javanica* and on exudations from the buccal stylet. *J. Parasit.* **54,** 879–890.

BIRD, A. F. (1969). The influence of tobacco ring spot virus and tobacco mosaic virus on the growth of *Meloidogyne javanica. Nematologica* **15,** 201–209.

BIRD, A. F. (1971). *The structure of nematodes.* London & New York, Academic Press, 318 pp.

BIRD, A. F. (1972a). Quantitative studies on the growth of syncytia induced in plants by root knot nematodes. *Internat. J. Parasit.* **2,** 157–170.

BIRD, A. F. (1972b). Cell wall breakdown during the formation of syncytia induced in plants by root knot nematodes. *Internat. J. Parasit.* **2,** 431–432.

BIRD, A. F. (1973). Observations on chromosomes and nucleoli in syncytia induced by *Meloidogyne javanica*. *Physiol. Pl. Path.* **3,** 387–391.

BIRD, A. F. (1974). Plant response to root-knot nematode. *A. Rev. Phytopath.* **12,** 69–85.

BIRD, A. F. & SOEFFKY, A. (1972). Changes in the ultrastructure of the gelatinous matrix of *Meloidogyne javanica* during dehydration. *J. Nematol.* **4,** 166–169.

BIRD, G. W. (1969). Depth of migration of *Meloidogyne incognita* (Nematodea) associated with greenhouse tomato and cucumber roots. *Can. J. Pl. Sci.* **49,** 90–92.

BIRD, G. W. (1971a). Digestive system of *Trichodorus porosus*. *J. Nematol.* **3,** 50–57.

BIRD, G. W. (1971b). Taxonomy: the science of classification. In: B. M. Zuckerman, W. F. Mai & R. A. Rohde (Eds), *Plant parasitic nematodes*. I. London & New York, Academic Press, 117–138.

BISHOP, D. (1955). The emergence of larvae of *Heterodera rostochiensis* under conditions of constant and alternating temperature. *Ann. appl. Biol.* **43,** 525–532.

BLAKE, C. D. (1961). Importance of osmotic potential as a component of the total potential of the soil water on the movement of nematodes. *Nature, Lond.* **192,** 144–145.

BLAKE, C. D. (1962a). The etiology of tulip-root disease in susceptible and in resistant varieties of oats infested by the stem nematode, *Ditylenchus dipsaci* (Kühn) Filipjev. I. Invasion of the host and reproduction by the nematode. *Ann. appl. Biol.* **50,** 703–712.

BLAKE, C. D. (1962b). The etiology of tulip-root disease in susceptible and resistant varieties of oats infested by the stem nematode, *Ditylenchus dipsaci* (Kühn) Filipjev. II. Histopathology of tulip-root and development of the nematode. *Ann. appl. Biol.* **50,** 713–722.

BOCK, K. R. (1966). Arabis mosaic and *Prunus* necrotic ringspot viruses in hop (*Humulus lupulus* L.). *Ann. appl. Biol.* **57,** 131–140.

BOUBAB, D., PISTRE, R. & DALMASSO, A. (1964). La lutte contre le court-noué de la vigne. Résultats obtenus à Frontignan en 1963. *Progrès agric. vitic.* **161,** 270–277, 293–304.

BOUWMAN, L. & ROSS, H. (1972). Differentiation between *Heterodera rostochiensis* and an undescribed allied species by female colour, morphometrics and pathogenicity. *Nematologica* **18,** 265–269.

BOWMAN, P. & BLOOM, J. R. (1966). Breaking the resistance of tomato varieties to Fusarium wilt by *Meloidogyne incognita*. (Abstr.). *Phytopathology* **56,** 871.

BRANDE, J. VAN DEN, KIPS, R. H. & D'HERDE, J. (1954a). Invloed van de vochtigheid bij de scheikundige bestrijding van het aardappelcystenaaltje *Heterodera rostochiensis* Woll. *Meded. LandbHogesch. OpzoekStns, Gent* **19,** 353–372.

BRANDE, J. VAN DEN, KIPS, R. H. & D'HERDE, J. (1954b). Veldproeven in verband met de invloed van het watergehalte van de bodem en van de bodemtemperatuur bij de scheikundige bestrijding van het aardappelcystenaaltje *Heterodera rostochiensis* Woll. *Meded. LandbHogesch. OpzoekStns, Gent* **19,** 765–776.

BRANDE, J. VAN DEN, KIPS, R. H. & D'HERDE, J. (1956). Survey of the results of four years experiments on the chemical control of the potato root eelworm. *Nematologica* **1,** 81–87.

Brathwaite, C. W. D. & Siddiqi, M. R. (1975). *Rhadinaphelenchus cocophilus. CIH Descriptions of Plant-parasitic Nematodes* Set 5, No. 72, 4 pp.

Braun, A. L. & Loof, P. A. A. (1966). *Pratylenchoides laticauda* n.sp., a new endoparasitic phytonematode. *Neth. J. Pl. Path.* **72,** 241–245.

Braun, A. L. & Lownsbery, B. F. (1975). The pin nematode, *Paratylenchus neoamblycephalus,* on Myrobalan plum and other hosts. *J. Nematol.* **7,** 336–343.

Breece, J. R. & Hart, W. H. (1959). A possible association of nematodes with the spread of peach yellow bud mosaic virus. *Pl. Dis. Reptr* **43,** 989–990.

Brendler, R. A., Isom, W. H., Radewald, J. D. & Shibuya, F. (1971). Oat variety testing for tolerance to nematode-caused 'tulip root'. *Calif. Agric.* **25** (7), 14–15.

Brewer, R. (1964). *Fabric and mineral analysis of soils.* London, Wiley & Sons, 470 pp.

Bridge, J. (1974). *Tylenchorhynchus dubius. CIH Descriptions of Plant-parasitic Nematodes* Set 4, No. 51, 4 pp.

Bromilow, R. H. (1973). Breakdown and fate of oximecarbamate nematicides in crops and soils. *Ann. appl. Biol.* **75,** 473–479.

Brown, E. B. (1955). Occurrence of the root-knot eelworm, *Meloidogyne hapla,* out of doors in Great Britain. *Nature, Lond.* **175,** 430.

Brown, E. B. (1956). A seed-borne attack of chrysanthemum eelworm (*Aphelenchoides ritzema-bosi*) on the annual aster (*Callistephus chinensis*). *J. Helminth.* **30,** 145–148.

Brown, E. B. (1957). Lucerne stem eelworm in Great Britain. *Nematologica* **2,** *Suppl.*, 369–375S.

Brown, E. B. (1958). Pea root eelworm in the eastern counties of England. *Nematologica* **3,** 257–268.

Brown, E. B. (1959). A broad red clover resistant to stem eelworm. *Pl. Path.* **8,** 124.

Brown, E. B. (1965). *Xiphinema diversicaudatum* and *Longidorus macrosoma* in rose glasshouses. *Pl. Path.* **14,** 45–46.

Brown, E. B. (1969). Assessment of the damage caused to potatoes by potato cyst eelworm, *Heterodera rostochiensis* Woll. *Ann. appl. Biol.* **63,** 493–502.

Brown, E. B. (1970). The behaviour of populations of potato cyst eelworm *Heterodera rostochiensis* Woll. towards resistant potato varieties derived from *Solanum tuberosum* ssp. *andigena* Juz & Buk. *Ann. appl. Biol.* **65,** 377–383.

Brown, E. B. & Franklin, M. T. (1953). Experiments on control of eelworm in black currants. *Pl. Path.* **2,** 101–102.

Brown, E. B. & Sykes, G. B. (1973). Control of tobacco rattle virus (spraing) in potatoes. *Ann. appl. Biol.* **75,** 462–464.

Brown, G., Callow, R. K., Green, C. D., Jones, F. G. W., Rayner, J. H., Shepherd, A. M. & Williams, T. D. (1971). The structure, composition and origin of the subcrystalline layer in some species of the genus *Heterodera. Nematologica* **17,** 591–599.

Brown, R. H. (1973). Chemical control of the cereal cyst nematode (*Heterodera avenae*)—a comparison of methods and rates of application of two systemic nematicides. *Aust. J. exp. Agric. Anim. Husb.* **13,** 587–592.

Brown, R. H. (1974). Further studies on the Victorian biotype of the cereal cyst nematode (*Heterodera avenae*). *Aust. J. exp. Agric. Anim. Husb.* **14,** 394–398.

Bryden, J. W. & Hodson, W. E. H. (1957). Control of chrysanthemum eelworm by parathion. *Pl. Path.* **6,** 20–24.

BRZESKI, M. (1962). Three new species of the genus *Acrobeloides* Cobb (Nematoda, Cephalobidae). *Bull. Acad. pol. Sci. Cl. II Sér. Sci. biol.* **10**, 335–339.

BRZESKI, M. W. (1970). Chemiczme zwalczanie węgorka miszczyka na cebuli. *Mater. IV Ogólnopolsk. Zjazdu Warzyw. Skierniew.*, 34–35.

BRZESKI, M. W. (1974). Taxonomy of Hemicycliophorinae (Nematoda, Tylenchida). *Zesz. Problem. Postęp. Nauk roln.* 154, 237–330.

BRZESKI, M. W. & HENDRICKS, E. K. (1971). The over-wintering and hatching of *Meloidogyne hapla* Chitw. *Zesz. probl. Postęp. Nauk roln.* 121, 235–244.

CADMAN, C. H. (1963). Biology of soil-borne viruses. *A. Rev. Phytopath.* **1**, 143–172.

CAIRASCHI, E. A. (1954). Observations sur une maladie vermiculaire du tabac en Alsace. *C.r. Acad. agric. Fr.* **40**, 75–77.

CALAM, C. T., RAISTRICK, H. & TODD, A. R. (1949). The potato eelworm hatching factor. I. The preparation of concentrates of the hatching factor and a method of bioassay. *Biochem. J.* **45**, 513–519.

CALAM, C. T., TODD, A. R. & WARING, W. S. (1949). The potato eelworm hatching factor. II. Purification of the factor by alkaloid salt fractionation. Anhydro-tetronic acid as an artificial hatching agent. *Biochem. J.* **45**, 520–524.

CALL, F. & HAGUE, N. G. M. (1962). The relationship between the concentration of ethylene dibromide and nematicidal effects in soil fumigation. *Nematologica* **7**, 186–192.

CAMERON, D. & SPEED, D. W. (1958). Resistance in oats to attack by the stem eelworm *Ditylenchus dipsaci* (Kühn). *Rep. Scott. Pl. Breed. Stn for* 1958, 66–74.

CANNING, E. U. (1973). Protozoal parasites as agents for biological control of plant-parasitic nematodes. *Nematologica* **19**, 342–348.

CARROLL, J., MCKAY, J. & HANNIGAN, M. (1957). Propagation of eelworm-free stocks of strawberry runners. *Econ. Proc. R. Dublin Soc.* **4**, 75–82.

CARROLL, J. & MCMAHON, E. (1937). Potato eelworm (*Heterodera schachtii*): further investigations. *J. Helminth.* **15**, 21–34.

CARROLL, J. & MCMAHON, E. (1939). Experiments on trap cropping with potatoes as a control measure against potato eelworm (*Heterodera schachtii*). *J. Helminth.* **17**, 101–112.

CAUBEL, G. (1973). Problèmes nématologiques du maïs en France: connaissances actuelles, possibilités d'intervention. *Phytiat.-Phytopharm.* **22**, 39–48.

CAVENESS, C. E. & CAVENESS, F. E. (1970). Hatching response of *Meloidogyne incognita acrita* to electric shock. *J. Nematol.* **2**, 294–297.

CAVENESS, F. E. & CAVENESS, C. E. (1970). Nematode electrocution. *J. Nematol.* **2**, 298–304.

CAYROL, J. C. (1964). Études préliminaires sur le cycle évolutif de *Ditylenchus myceliophagus* Goodey. *Nematologica* **10**, 361–368.

CAYROL, J. C. (1967). Étude du cycle évolutif d'*Aphelenchoides composticola*. *Nematologica* **13**, 23–32.

CAYROL, J. C. (1970). Contributions à l'étude de la biologie du nématode mycophage *Ditylenchus myceliophagus* J. B. Goodey, 1958 (Nematoda, Tylenchida). *Ph.D. Thesis Univ. de Nice*, 177 pp.

CAYROL, J. C. & RITTER, M. (1962). Recherches sur la maladie vermiculaire du muguet provoquée par *Pratylenchus convallariae* (Seinhorst, 1959). *Annls Épiphyt.* **13**, 301–328.

CHAMBERS, A. Y. & EPPS, J. M. (1967). Nature of soybean resistance to *Heterodera glycines*. *Pl. Dis. Reptr* **51**, 771–774.

CHANG, D. C. N., CAMPBELL, W. F. & GRIFFIN, G. D. (1973). Ultrastructure changes induced by stem nematodes in hypocotyl tissue of alfalfa. *J. Nematol.* **5**, 165–173.

CHANTANAO, A. & JENSEN, H. J. (1969). Transmission of an *Agrobacterium tumefaciens* phage by *Pristionchus lheritieri*. *J. Nematol.* **1**, 166–168.

CHEN, T. A. & WEN, G.-Y. (1972). Ultrastructure of the feeding apparatus of *Pratylenchus penetrans*. *J. Nematol.* **4**, 155–161.

CHIN, D. A. & ESTEY, R. H. (1966). Studies on the pathogenicity of *Aphelenchus avenae* Bastian, 1865. *Phytoprotection* **47**, 66–72.

CHITWOOD, B. G. (1949). Root-knot nematodes—Part I. A revision of the genus *Meloidogyne* Goeldi, 1887. *Proc. helminth. Soc. Wash.* **16**, 90–104.

CHITWOOD, B. G. (1958). The designation of official names for higher taxa of invertebrates. *Bull. zool. Nomencl.* **15**, 860–895.

CHITWOOD, B. G. & BLANTON, F. S. (1941). An evaluation of the results of treatments given narcissus bulbs for the control of the nematode *Ditylenchus dipsaci* (Kühn) Filipjev. *J. Wash. Acad. Sci.* **31**, 296–308.

CHITWOOD, B. G. & BUHRER, E. M. (1946). The life history of the golden nematode of potatoes, *Heterodera rostochiensis* Wollenweber, under Long Island, New York, conditions. *Phytopathology* **36**, 180–189.

CHITWOOD, B. G. & CHITWOOD, M. B. (1950). *An introduction to nematology*. Baltimore, Monumental Printing Co., 213 pp.

CHITWOOD, B. G., CLEMENT, R. L., MORGAN, R. & TANK, R. (1942). *Heterodera rostochiensis*, the golden nematode of potatoes, in New York state. *Pl. Dis. Reptr* **26**, 390–391.

CHITWOOD, B. G., HAASIS, F. A. & BLANTON, F. S. (1941). Hot-water-formalin treatment (at 110° to 111°F) of field-grown and of forced narcissus bulbs infected with the bulb or stem nematode, *Ditylenchus dipsaci*. *Proc. helminth. Soc. Wash.* **8**, 44–50.

CHITWOOD, B. G. & NEWHALL, A. G. (1942). The efficacy of certain nematocides in the control of onion bloat in muck soil. *Phytopathology* **32**, 254–258.

CHITWOOD, M. B. (1969). The systematics and biology of some parasitic nematodes. In: M. Florkin & B. T. Scheer (Eds), *Chemical Zoology* Vol. III. *Echinodermata, Nematoda and Acanthocephala*. London & New York, Academic Press, 223–244.

CHOI, Y. E. & GERAERT, E. (1974). Description of *Meloinema kerongense* n.g., n.sp. (Nematoda: Meloidogynidae) from Korea. *Nematologica* **19** (Yr 1973), 334–341.

CHOI, Y. E. & LOOF, P. A. A. (1974). Redescription of *Anguina moxae* Yokoo & Choi, 1968 (Tylenchina). *Nematologica* **19** (Yr 1973), 285–292.

CHRISTIE, J. R. & ARNDT, C. H. (1936). Feeding habits of the nematodes *Aphelenchoides parietinus* and *Aphelenchus avenae*. *Phytopathology* **26**, 698–701.

CHRISTIE, J. R. & PERRY, V. G. (1951). A root disease of plants caused by a nematode of the genus *Trichodorus*. *Science* **113**, 491–493.

CHRISTIE, J. R. & PERRY, V. G. (1959). Mechanism of nematode injury to plants. In: C. S. Holton *et al.* (Eds). *Plant pathology problems and progress* 1908–1958. Madison, Univ. Wisconsin Press, 419–426.

CHURCH, B. M., GOUGH, H. C. & SOUTHEY, J. F. (1959). Soil sampling procedures for potato root eelworm cysts. *Pl. Path.* **8**, 146–151.

CLAMOT, G. (1968). 'Greta' première variété belge d'avoine résistante au nématode de la tige. *Revue Agric., Brux.* 21e, 1205–1211.

CLARKE, A. J. (1968). The chemical composition of the cyst wall of the potato cyst-nematode, *Heterodera rostochiensis*. *Biochem. J.* **108,** 221–224.

CLARKE, A. J., COX, P. M. & SHEPHERD, A. M. (1967). The chemical composition of the egg shells of the potato cyst-nematode, *Heterodera rostochiensis* Woll. *Biochem. J.* **104,** 1056–1060.

CLARKE, A. J. & HENNESSY, J. A. (1975). Progress with work on hatching factors. In: Nematology Department. *Rep. Rothamsted exp. Stn for* 1974, *Part* 1, 181–182.

CLARKE, A. J. & HENNESSY, J. (1976). The distribution of carbohydrates in cysts of *Heterodera rostochiensis*. *Nematologica* **22,** 190–195.

CLARKE, A. J. & PERRY, R. N. (1977). Hatching of cyst-nematodes. *Nematologica* **23,** 350–368.

CLARKE, A. J. & SHEPHERD, A. M. (1964). Synthetic hatching agents for *Heterodera schachtii* Schm. and their mode of action. *Nematologica* **10,** 431–453.

CLARKE, A. J. & SHEPHERD, A. M. (1966). Inorganic ions and the hatching of *Heterodera* spp *Ann. appl. Biol.* **58,** 497–508.

CLARKE, A. J. & SHEPHERD, A. M. (1968). Hatching agents for the potato cyst-nematode *Heterodera rostochiensis* Woll. *Ann. appl. Biol.* **61,** 139–149.

CLARKE, A. J. & WIDDOWSON, E. (1966). The hatching factor of the potato-root eelworm. *Biochem. J.* **98,** 862–868.

COBB, N. A. (1914). Citrus-root nematode. *J. agric. Res.* **2,** 217–230.

COBB, N. A. (1917). The mononchs (*Mononchus* Bastian 1866) a genus of free-living predatory nematodes. *Contrib. Sci. Nematol.* 6, 129–187.

COBB, N. A. (1920). One hundred new nemas. (Type species of 100 new genera). *Contrib. Sci. Nematol.* 9, 217–343.

COHEN, M. (1971). Twenty years of E.P.P.O. *Agriculture, Lond.* **78,** 451–452.

COHN, E. & SHER, S. A. (1972). A contribution to the taxonomy of the genus *Xiphinema* Cobb, 1913. *J. Nematol.* **4,** 36–65.

COHN, E., TANNE, E. & NITZANY, F. E. (1970). *Xiphinema italiae,* a new vector of grapevine fanleaf virus. *Phytopathology* **60,** 181–182.

COLBRAN, R. C. (1971). Studies of plant and soil nematodes. 15. Eleven new species of *Radopholus* Thorne and a new species of *Radopholoides* de Guiran (Nematoda: Tylenchoidea) from Australia. *Qd. J. agric. anim. Sci.* **27,** 437–460.

COLE, C. S. & HOWARD, H. W. (1962a). The effect of growing resistant potatoes on a potato-root eelworm population—a microplot experiment. *Ann. appl Biol.* **50,** 121–127.

COLE, C. S. & HOWARD, H. W. (1962b). Further results from a field experiment on the effect of growing resistant potatoes on a potato root eelworm (*Heterodera rostochiensis*) population. *Nematologica* **7,** 57–61.

COLE, C. S. & HOWARD, H. W. (1966). The effects on a population of potato-root eelworm (*Heterodera rostochiensis*) of growing potatoes resistant to pathotype B. *Ann. appl. Biol.* **58,** 487–495.

COLLIS-GEORGE, N. & BLAKE, C. D. (1959). The influence of the soil moisture regime on the expulsion of the larval mass of the nematode *Anguina agrostis* from galls. *Aust. J. biol. Sci.* **12,** 247–256.

DE CONINCK, L. (1965). Systématique des nématodes. In: P.-P. Grassé, *Traité de zoologie* **4** (2), 586–681. Paris, Masson et Cie.

Conroy, J. J., Green, R. J., Jr. & Ferris, J. M. (1972). Interaction of *Verticillium albo-atrum* and the root lesion nematode, *Pratylenchus penetrans,* in tomato roots at controlled inoculum densities. *Phytopathology* **62,** 362–366.

Cook, R. (1972). Reaction of some oat cultivars to *Meloidogyne naasi. Pl. Path.* **21,** 41–43.

Cook, R. (1975a). Observations on the relationship between morphological variation and host range in populations of cereal cyst-nematode. *Ann. appl. Biol.* **81,** 199–205.

Cook, R. (1975b). Cereal cyst nematode (*Heterodera avenae*) resistance in oats. *Rep. Welsh Pl. Breed. Stn for* 1974, 34.

Cook, R. & Williams, T. D. (1972). Pathotypes of *Heterodera avenae. Ann. appl. Biol.* **71,** 267–271.

Cook, R., York, P. A. & Chungmanoch, Pontip (1974). Invasion of oats by cereal cyst nematode (*Heterodera avenae*). *Rep. Welsh Pl. Breed. Stn for* 1973, 62–63.

Cooke, D. A. & Draycott, A. P. (1969). The effect of soil fumigation and nitrogen on Docking disorder of sugar beet. *Proc. 5th Br. Insectic. Fungic. Conf. Vol.* 1, 260–264.

Cooke, D. A., Dunning, R. A. & Winder, G. H. (1974). The effect of nematicides applied to the seed rows in spring, on growth and yield of sugar beet in Docking-disorder-affected fields. *Ann. appl. Biol.* **76,** 289–298.

Cooke, R. C. (1961). Agar disk method for the direct observation of nematode-trapping fungi in the soil. *Nature, Lond.* **191,** 1411–1412.

Cooke, R. C. (1962). The ecology of nematode-trapping fungi in the soil. *Ann. appl. Biol.* **50,** 507–513.

Coolen, W. A., Hendrickx, G. & d'Herde, J. C. (1971). Nematologische problematiek in de kasplantenteelt en de bestrijding met systemische nematiciden van *Radopholus similis* en *Meloidogyne arenaria* in *Calathea makoyana* en *Monstera deliciosa. Rijksstation voor Nematologie en Entomologie, Wetteren, Publ. W7.*

Coolen, W. & d'Herde, J. (1970). Nematodes associated with glasshouse roses. *Zesz. probl. Postęp. Nauk roln.* No. 92, 259–266.

Coomans, A. (1965). Structure of the female gonads in members of the Dorylaimina. *Nematologica* **10** (Yr 1964), 601–622.

Coomans, A. (1971). Morphology and nematode systematics. *J. Parasitol.* **57** (4) (Sect. II, part 2). (2nd Int. Cong. Parasitol. part 5), 95–99.

Coomans, A. (1975). [Nematoda as a phylum] In: F. Lamberti, C. E. Taylor & J. W. Seinhorst (Eds), *Nematode vectors of plant viruses.* London & New York, Plenum Press, p. 14.

Coomans, A. & de Coninck, L. (1963). Observations on spear-formation in *Xiphinema. Nematologica* **9,** 85–96.

Coomans, A. & Loof, P. A. A. (1970). Morphology and taxonomy of Bathyodontina (Dorylaimida). *Nematologica* **16,** 180–196.

Cooper, B. A. (1953). Eelworm problems in north Fenland with special reference to crop rotation. *Rep. hort. Educ. Ass. for* 1953, 106–115.

Cooper, B. A. (1955). A preliminary key to British species of *Heterodera* for use in soil examination. In: D. K. McE. Kevan (Ed.), *Soil zoology.* London, Butterworths, 269–280.

Cooper, J. I. (1971). The distribution in Scotland of tobacco rattle virus and its nematode vectors in relation to soil type. *Pl. Path.* **20,** 51–58.

COOPER, J. I. & HARRISON, B. D. (1973). The role of weed hosts and the distribution and activity of vector nematodes in the ecology of tobacco rattle virus. *Ann. appl. Biol.* **73**, 53–66.

COOPER, J. I. & THOMAS, P. R. (1971). Chemical treatment of soil to prevent transmission of tobacco rattle virus to potatoes by *Trichodorus* spp. *Ann. appl. Biol.* **69**, 23–34.

CORBETT, D. C. M. (1969). *Pratylenchus pinguicaudatus* n.sp. (Pratylenchinae: Nematoda) with a key to the genus *Pratylenchus*. *Nematologica* **15**, 550–556.

CORBETT, D. C. M. (1972). The effect of *Pratylenchus fallax* on wheat, barley and sugar beet roots. *Nematologica* **18**, 303–308.

CORBETT, D. C. M. (1973). *Pratylenchus penetrans*. *CIH Descriptions of Plant-parasitic Nematodes* Set 2, No. 25, 4 pp.

CORBETT, D. C. M. (1974). *Pratylenchus vulnus*. *CIH Descriptions of Plant-parasitic Nematodes* Set 3, No. 37, 3 pp.

CORBETT, D. C. M. & HIDE, G. A. (1971). Interactions between *Heterodera rostochiensis* Woll. and *Verticillium dahliae* Kleb. on potatoes and the effect of CCC on both. *Ann. appl. Biol.* **68**, 71–80.

CORNU, M. (1879). Études sur le *Phylloxera vastatrix*. *Mém. Acad. Sci., Paris* **26**, 163–175, 328, 339–341.

COTTEN, J. (1960). Observations on the cytology of the potato-root eelworm, *Heterodera rostochiensis* Wollenweber. *Nematologica Suppl. II*, 123–126.

COTTEN, J. (1962). Effect of temperature on hatching in the cereal root eelworm. *Nature, Lond.* **195**, 308.

COTTEN, J. (1963). Resistance in barley and oats to the cereal root eelworm *Heterodera avenae* Wollenweber. *Nematologica* **9**, 81–84.

COTTEN, J. (1967). Cereal root eelworm pathotypes in England and Wales. *Pl. Path.* **16**, 54–59.

COTTEN, J. (1969). Cereal varieties resistant to *Heterodera avenae* and *Ditylenchus dipsaci*. *Proc. 5th Br. Insectic. Fungic. Conf. Vol.* 1, 164–168.

COTTEN, J. (1970a). Some aspects of breeding for cereal root eelworm (*Heterodera avenae* Woll.) resistance in barley. *Zesz. probl. Postęp. Nauk roln.* No. 92, 231–237.

COTTEN, J. (1970b). Field experiments with spring barley resistant to cereal cyst nematode, 1965–1968. *Ann. appl. Biol.* **65**, 163–168.

COTTEN, J. & HAYES, J. D. (1969). Genetic resistance to the cereal cyst nematode (*Heterodera avenae*). *Heredity, Lond.* **24**, 593–600.

COTTEN, J. & HAYES, J. D. (1972). Genetic studies of resistance to the cereal cyst nematode (*Heterodera avenae*) in oats (*Avena* spp.). *Euphytica* **21**, 538–542.

COURTNEY, W. D. (1960). Nematodes infecting narcissi and tulips in the Pacific Northwest. *Daffodil Tulip Yearb.* 1961 (26), 117–122.

COURTNEY, W. D. & BREAKEY, E. P. (1945). Forcing lily bulbs after treatment for control of nematodes. *Flor. Rev.* Sept. 27, 37–38.

COURTNEY, W. D. & GOULD, C. J. (1951). Tolerance of Wedgewood iris bulbs to a hot-water-formalin treatment. *Phytopathology* **41**, 40–45.

COURTNEY, W. D., POLLEY, D. & MILLER, V. L. (1955). TAF, an improved fixative in nematode technique. *Pl. Dis. Reptr* **39**, 570–571.

CREMER, M. C. & KOOISTRA, G. (1964). Investigations on notched leaf ('Kartelblad') of *Gladiolus* and its relation to rattle virus. *Nematologica* **10**, 69–70. (Abstr.).

CROLL, N. A. (1970). *The behaviour of nematodes*. London, Arnold, 117 pp.

CROLL, N. A. (Ed.) (1976). *The organization of nematodes*. London & New York, Academic Press, 428 pp.

CUANY, A. & DALMASSO, A. (1972). *Heterodera* spp. on carnations. *Abstracts, XIth International Symposium of Nematology, Reading, 3–8 September*, 1972 p. 15.

CUNNINGHAM, P. C. (1960). An investigation of winter dormancy in *Heterodera rostochiensis*. *Scient. Proc. R. Dublin Soc., Ser. B*, **1**, 1–4.

CURRIE, J. A. (1973). The seed-soil system. In: W. Heydecker (Ed.). *Seed ecology*. London, Butterworths, 463–480.

CURTIS, G. J. (1970). Resistance of sugar beet to the cyst-nematode *Heterodera schachtii* Schm. *Ann. appl. Biol.* **66**, 169–177.

DALMASSO, A. (1967). Connaissances actuelles sur les nématodes phytophages et leurs relations avec les maladies à virus. *Annls Épiphyt.* **18**, 249–272.

DALMASSO, A. (1969). Étude anatomique et taxonomique des genres *Xiphinema, Longidorus* et *Paralongidorus* (Nematoda: Dorylaimidae). *Mém. Mus. natn. Hist. nat., Paris, Sér. A. Zool.* **61**, 33–82.

DALMASSO, A. (1970). Influence directe de quelques facteurs écologiques sur l'activité biologique et la distribution des espèces françaises de la famille des Longidoridae (Nematoda-Dorylaimida). *Annls Zool. Écol. anim.* **2**, 163–200.

DALMASSO, A. (1973). La reproduction et la notion d'espèce chez le genre *Meloidogyne*. *EPPO Bull.* **3** (1), 67–73.

DANILOV, V. P. (1952). [The control of strawberry stem nematodes.] *Sad i Ogorod*, Yr. 1952, No. 7, 73–74 (in Russian). [*Helminth. Abstr.* **21**, 825a].

DAO D., F. (1970). Climatic influence on the distribution pattern of plant parasitic and soil inhabiting nematodes. *Meded. LandbHogesch. Wageningen* 70–72, 181 pp.

DARLING, H. M., FAULKNER, L. R. & WALLENDAL, P. (1957). Culturing the potato rot nematode. (Abstr.). *Phytopathology* **47**, 7.

DASGUPTA, D. R., RASKI, D. J. & SHER, S. A. (1968). A revision of the genus *Rotylenchulus* Linford and Oliveira, 1940 (Nematoda: Tylenchidae). *Proc. helminth. Soc. Wash.* **35**, 169–192.

DAULTON, R. A. C. & NUSBAUM, C. J. (1961). The effect of soil temperature on the survival of the root-knot nematodes *Meloidogyne javanica* and *M. hapla*. *Nematologica* **6**, 280–294.

DAULTON, R. A. C. & STOKES, W. M. (1952). The destruction or inhibition of root-knot nematodes by exposure to an electrostatic field. *Emp. J. exp. Agric.* **20**, 271–273.

DAVIDE, R. G. & TRIANTAPHYLLOU, A. C. (1967a). Influence of the environment on development and sex differentiation of root-knot nematodes. I. Effect of infection density, age of the host plant and soil temperature. *Nematologica* **13**, 102–110.

DAVIDE, R. G. & TRIANTAPHYLLOU, A. C. (1967b). Influence of the environment on development and sex differentiation of root-knot nematodes. II. Effect of host nutrition. *Nematologica* **13**, 111–117.

DAVIES, T. G. (1960). A spring oat variety resistant to the stem eelworm. *Nature, Lond.* **186**, 813.

DECKER, H. (1962). Zur Biologie und Ökologie von *Aphelenchus avenae* Bastian, 1865. *Wiss. Z. Univ. Rostock* **11**, 299–305.

DECRAEMER, W. (1974a). Scientific report on the Belgian expedition to the Great Barrier Reef in 1967. Nematodes IV: *Desmoscolex*-species (Nematoda: Desmoscolecida) from Yonge Reef and Nymphe Island with detailed study of *Desmoscolex longisetosus* Timm, 1970. *Biol. Jaarb.* **42**, 86–104.

DECRAEMER, W. (1974b). Scientific report on the Belgian expedition to the Great Barrier Reef in 1967. Nematodes II. *Desmoscolex* species (Nematoda, Desmoscolecida) from Yonge Reef, Lizard Island and Nymphe Island. *Zool. Scripta* **3**, 167–176.

DEKKER, W. H. (1972). 3-chloroallyl methyl sulfide, a product from the reaction of 1,3–dichloropropene and biological materials. *Meded. Fac. LandbWetensch. Rijksuniv. Gent* **37**, 865–868.

DERN, R. (1971). Kombination nematizider Granulate und Netzmittel. *Meded. Fac. LandbWetensch. Rijksuniv. Gent* **36**, 185–189.

DICKSON, D. W., HUISINGH, D. & SASSER, J. N. (1971). Dehydrogenases, acid and alkaline phosphatases and esterases for chemotaxonomy of selected *Meloidogyne, Ditylenchus, Heterodera* and *Aphelenchus* spp. *J. Nematol.* **3**, 1–16.

DICKSON, D. W., SASSER, J. N. & HUISINGH, D. (1970). Comparative disc-electrophoretic protein analyses of selected *Meloidogyne, Ditylenchus, Heterodera* and *Aphelenchus* spp. *J. Nematol.* **2**, 286–293.

DIERCKS, R. & KLEWITZ, R. (1962). Zur Samenübertragbarkeit einer Ackerbohnen vorkommenden Herkunf des Stengelälchens *Ditylenchus dipsaci* (Kühn). *Nematologica* **7**, 155–163.

DIJKSTRA, J. (1957). Symptoms of susceptibility and resistance in seedlings of red clover attacked by the stem nematode *Ditylenchus dipsaci* (Kühn) Filipjev. *Nematologica* **2**, 228–234.

DITTMANN, A. L. (1963). Ätiologische Zusammenhänge zwischen Nematoden und Pilzkrankheiten. Versuch einer Klassifierung ihre Beziehungen. *Zentbl. Bakt. ParasitKde, Abt.* 2 **116**, 716–749.

DIXON, G. M. (1963). The effect of spring rainfall on the host-parasite relationship between the cereal root eelworm (*Heterodera avenae* Woll.) and the oat plant (*Avena sativa* L.). *Nematologica* **9**, 521–526.

DIXON, G. M. (1969). The effect of cereal cyst eelworm on spring-sown cereals. *Pl. Path.* **18**, 109–112.

DIXON, G. M. & RYLAND, J. M. (1964). Morphometric relations in the genus *Heterodera. Nematologica* **10**, 180–181.

DOBBS, C. G. & HINSON, W. H. (1953). A widespread fungistasis in soils. *Nature, Lond.* **172**, 197–199.

DONCASTER, C. C. (1962). Nematode feeding mechanisms. I. Observations on *Rhabditis* and *Pelodera. Nematologica* **8**, 313–320.

DONCASTER, C. C. (1966). Nematode feeding mechanisms. 2. Observations on *Ditylenchus destructor* and *D. myceliophagus* feeding on *Botrytis cinerea. Nematologica* **12**, 417–427.

DONCASTER, C. C. (1971). Feeding in plant parasitic nematodes: mechanisms and behaviour. In: B. M. Zuckerman, W. F. Mai & R. A. Rohde (Eds), *Plant parasitic nematodes.* II. London & New York, Academic Press, 137–157.

DONCASTER, C. C. (1974). *Heterodera rostochiensis* (Nematoda). Egg-hatch. *Film* E2035/1974. *Inst. Wiss. Film. Göttingen.*

DONCASTER, C. C. & HOOPER, D. J. (1961). Nematodes attacked by Protozoa and tardigrades. *Nematologica* **6**, 333–335.

DONCASTER, C. C. & SHEPHERD, A. M. (1967). The behaviour of second-stage *Heterodera rostochiensis* larvae leading to their emergence from the egg. *Nematologica* **13**, 476–478.

DONEY, D. L., FIFE, J. M. & WHITNEY, E. D. (1970). The effect of the sugarbeet nematode *Heterodera schachtii* on the free amino acids in resistant and susceptible *Beta* species. *Phytopathology* **60**, 1727–1729.

DOUGHERTY, E. C. (1953). Problems of nomenclature for the growth of organisms of one species with and without associated organisms of other species. *Parasitology* **42**, 259–261.

DOUGHERTY, E. C. (1955). The genera and species of the subfamily Rhabditinae Micoletzky, 1922 (Nematoda): a nomenclatorial analysis—including an addendum on the composition of the family Rhabditidae Örley, 1880. *J. Helminth.* **29**, 105–152.

DRECHSLER, C. (1937). Some Hyphomycetes that prey on free-living terricolous nematodes. *Mycologia* **29**, 447–552.

DROPKIN, V. H. (1953). Studies on the variability of anal plate patterns in pure lines of *Meloidogyne* spp the root-knot nematode. *Proc. helminth. Soc. Wash.* **20**, 32–39.

DROPKIN, V. H. (1969a). Cellular responses of plants to nematode infections. *A. Rev. Phytopath.* **1**, 101–122.

DROPKIN, V. H. (1969b). The necrotic reaction of tomatoes and other hosts resistant to *Meloidogyne*: reversal by temperature. *Phytopathology*, **59**, 1632–1637.

DROPKIN, V. H. (1972). Pathology of *Meloidogyne*—galling, giant cell formation, effects on host physiology. *EPPO Bull.* No. 6, 23–32.

DROPKIN, V. H. & BOONE, W. R. (1966). Analysis of host-parasite relationships of root-knot nematodes by single-larva inoculations of excised tomato roots. *Nematologica* **12**, 225–236.

DROPKIN, V. H., HELGESON, J. P. & UPPER, C. D. (1969). The hypersensitivity reaction of tomatoes resistant to *Meloidogyne incognita*; reversal by cytokinins. *J. Nematol.* **1**, 55–61.

DROPKIN, V. H. & NELSON, P. E. (1960). The histopathology of root-knot nematode infections in soybeans. *Phytopathology* **50**, 442–447.

DROPKIN, V. H. & WEBB, R. E. (1967). Resistance of axenic tomato seedlings to *Meloidogyne incognita acrita* and to *M. hapla*. *Phytopathology* **57**, 584–587.

DUDDINGTON, C. L. (1954). Nematode-destroying fungi in agricultural soils. *Nature, Lond.* **173**, 500–501.

DUDDINGTON, C. L. & DUTHOIT, C. M. G. (1960). Green manuring and cereal root eelworm. *Pl. Path.* **9**, 7–9.

DUDDINGTON, C. L., EVERARD, C. O. R. & DUTHOIT, C. M. G. (1961). Effect of green manuring and a predacious fungus on cereal root eelworm in oats. *Pl. Path.* **10**, 108–109.

DUDDINGTON, C. L., JONES, F. G. W. & MORIARTY, F. (1956). The effect of predacious fungus and organic matter upon the soil population of beet eelworm, *Heterodera schachtii* Schm. *Nematologica* **1**, 344–348.

DUDDINGTON, C. L., JONES, F. G. W. & WILLIAMS, T. D. (1956). An experiment on the effect of a predacious fungus upon the soil population of potato root eelworm, *Heterodera rostochiensis* Woll. *Nematologica* **1**, 341–343.

DUGGAN, J. J. (1960). Effect of soil drying on the viability of *Heterodera major* cysts. *Nature, Lond.* **185**, 554–555.

DUGGAN, J. J. (1971). Control of stem nematode (*Ditylenchus dipsaci*) in onions. *Proc. 6th Br. Insectic. Fungic. Conf. Vol.* 1, 150–153.

DUGGAN, J. J. & MOORE, J. F. (1963). Observations on tuber-rot eelworm (*Ditylenchus destructor* Thorne 1945). *Ir. J. agric. Res.* **2**, 75–86.

DUNN, E. (1954). Factors influencing the emergence of *Heterodera rostochiensis* larvae. *Nature, Lond.* **173**, 780.

DUNN, E. (1962). Preconditioning of the cyst contents of potato root eelworm, *Heterodera rostochiensis* Woll., by temperature and its effect on the subsequent emergence of the larvae in water and root diffusate. *Nematologica* **7**, 177–185.

DUNN, E. & HUGHES, W. A. (1964). Interrelationship of the potato root eelworm, *Heterodera rostochiensis* Woll., *Rhizoctonia solani* Kühn and *Colletotrichum atramentarium* (B. & Br.) Toub., on the growth of the tomato plant. *Nature, Lond.* **201**, 413–414.

DUNNETT, J. M. (1957). Variation in pathogenicity of the potato root eelworm (*Heterodera rostochiensis* Woll.) and its significance in potato breeding. *Euphytica* **6**, 77–89.

EDMUNDS, J. E. & MAI, W. F. (1966). Effect of *Trichoderma viride, Fusarium oxysporum* and fungal enzymes upon the penetration of alfalfa roots by *Pratylenchus penetrans. Phytopathology* **56**, 1132–1135.

EDMUNDS, J. E. & MAI, W. F. (1967). Effect of *Fusarium oxysporum* on movement of *Pratylenchus penetrans* toward alfalfa roots. *Phytopathology* **57**, 468–471.

EDWARD, J. C. & MISRA, S. L. (1968). *Heterodera vigni* n.sp. and second stage larvae of *Heterodera* spp. in Uttar Pradesh, India. *Allahabad Farmer* **42**, 155–159.

EFREMOVA, T. G. (1966). [The damage caused by *Meloidogyne* and its control on vegetables grown by the hydroponic method.] In: [*Protection of vegetable crops from pests, diseases and weeds*]. Moscow, Izdatelstvo 'Kolos', 22–27 [In Russian]. (Abstract).

ELLENBY, C. (1945a). The influence of crucifers and mustard oil on the emergence of larvae of the potato-root eelworm, *Heterodera rostochiensis* Wollenweber. *Ann. appl. Biol.* **32**, 67–70.

ELLENBY, C. (1945b). Control of the potato-root eelworm, *Heterodera rostochiensis* Wollenweber, by allyl isothiocyanate, the mustard oil of *Brassica nigra* L. *Ann. appl. Biol.* **32**, 237–239.

ELLENBY, C. (1946a). Nature of the cyst wall of the potato-root eelworm *Heterodera rostochiensis,* Wollenweber, and its permeability to water. *Nature, Lond.* **157**, 302–303.

ELLENBY, C. (1946b). Ecology of the eelworm cyst. *Nature, Lond.* **157**, 451–452.

ELLENBY, C. (1951). Mustard oils and control of the potato-root eelworm, *Heterodera rostochiensis* Wollenweber: further field and laboratory experiments. *Ann. appl. Biol.* **38**, 859–875.

ELLENBY, C. (1954a). Environmental determination of the sex ratio of a plant parasitic nematode. *Nature, Lond.* **174**, 1016.

ELLENBY, C. (1954b). Tuber forming species and varieties of the genus *Solanum* tested for resistance to the potato root eelworm, *Heterodera rostochiensis* Wollenweber. *Euphytica* **3**, 195–202.

ELLENBY, C. (1955). The permeability to the hatching factor of the cyst wall of the potato-root eelworm, *Heterodera rostochiensis* Wollenweber. *Ann. appl. Biol.* **43**, 12–18.

ELLENBY, C. (1957). An investigation into the possibility of parthenogenesis in the potato-root eelworm, *Heterodera rostochiensis* Wollenweber. *Nematologica* **2**, 250–254.

ELLENBY, C. (1968). Determination of the water content of nematode worms by interference microscopy. *Experientia* **24**, 84–85.

ELLENBY, C. (1975). A simple method of comparing the rate of drying of nematode worms. *Nematologica* **21**, 115–116.

ELLENBY, C. & SMITH, L. (1968). Observations on the piezoelectric properties of nematode cuticle. *Comp. Biochem. Physiol.* **26,** 359–363.

ELLIS, P. R. (1970). The biology of potato cyst-eelworm *Heterodera rostochiensis* (Nematoda: Heteroderidae), especially in relation to the development of eelworm-resistant tomato cultivars. *Ph.D. Thesis, University of London.*

ELLIS, P. R. & HESLING, J. J. (1974). Morphology of cysts of isolates of *Heterodera rostochiensis* and *H. pallida. Nematologica* **20,** 226–241.

ELLIS, P. R. & SMITH, J. W. M. (1971). Inheritance of resistance to potato cyst-eelworm (*Heterodera rostochiensis* Woll.) in the genus *Lycopersicon. Euphytica* **20,** 93–101.

ELMILIGY, I. A. (1968). Three new species of the genus *Meloidogyne* Goeldi, 1887 (Nematoda: Heteroderidae). *Nematologica* **14,** 577–590.

ENDO, B. Y. (1964). Penetration and development of *Heterodera glycines* in soybean roots and related anatomical changes. *Phytopathology* **54,** 79–88.

ENDO, B. Y. (1965). Histological responses of resistant and susceptible soybean varieties, and backcross progeny to entry and development of *Heterodera glycines. Phytopathology* **55,** 375–381.

ENDO, B. Y. (1971). Nematode-induced syncytia (giant cells). Host-parasite relationships of Heteroderidae. In: B. M. Zuckerman, W. F. Mai & R. A. Rohde (Eds), *Plant parasitic nematodes.* II. London & New York, Academic Press, 91–117.

ERIKSSON, K. B. (1972). Nematode diseases of pasture legumes and turf grasses. In: J. M. Webster (Ed.), *Economic nematology.* London & New York, Academic Press, 66–96.

ERIKSSON, K. B. (1974). Intraspecific variation in *Ditylenchus dipsaci.* I. Compatibility tests with races. *Nematologica* **20,** 147–162.

ESSER, R. P. (1967). Foliar and other plant-parasitic nematodes associated with azalea in Florida. *Pl. Dis. Reptr* **51,** 46–49.

ESSER, R. P. (1974). A diagnostic species compendium of the genus *Xiphinema* Cobb 1913. *Proc. Soil Crop Sci. Soc. Fla* **33,** 88–92.

ESSER, R. P., PERRY, V. G. & TAYLOR, A. L. (1976). A diagnostic compendium of the genus *Meloidogyne* (Nematoda: Heteroderidae). *Proc. helminth. Soc. Wash.* **43,** 138–150.

ESSER, R. P. & SOBERS, E. K. (1964). Natural enemies of nematodes. *Proc. Soil Crop Sci. Soc. Fla* **24,** 326–353.

EVANS, A. A. F. (1971). Taxonomic value of gel electrophoresis of proteins from mycophagous and plant-parasitic nematodes. *Int. J. Biochem.* **2,** 72–79.

EVANS, A. A. F. & FISHER, J. M. (1970a). Some factors affecting the number and size of nematodes in populations of *Aphelenchus avenae. Nematologica* **16,** 295–304.

EVANS, A. A. F. & FISHER, J. M. (1970b). Ecological differences between four Australian isolates of *Aphelenchus avenae* Bastian. *Aust. J. biol. Sci.* **23,** 507–509.

EVANS, A. A. F. & FISHER, J. M. (1970c). The effect of environment on nematode morphometrics. Comparison of *Ditylenchus myceliophagus* and *D. destructor. Nematologica* **16,** 113–122.

EVANS, E. B. & WEBLEY, D. P. (1970). A guide to the morphological differences between pathotypes of *Heterodera rostochiensis* larvae. *Pl. Path.* **19,** 171–172.

EVANS, K. (1969). Changes in a *Heterodera rostochiensis* population through the growing season. *Ann. appl. Biol.* **64,** 31–41.

EVANS, K. (1970a). The effects of gamma radiation on *Heterodera rostochiensis*. *Nematologica* **16**, 284–294.

EVANS, K. (1970b). Longevity of males and fertilisation of females of *Heterodera rostochiensis*. *Nematologica* **16**, 369–374.

EVANS, K., FRANCO, J. & DE SCURRAH, M. M. (1975). Distribution of species of potato cyst-nematodes in South America. *Nematologica* **21**, 365–369.

EVANS, K. & PANDÉ, L. U. (1972). Nematodes in the potatoes of the ley-arable experiment at Woburn. In: Nematology department. *Rep. Rothamsted exp. Stn for* 1971, 175.

EVANS, K., PARKINSON, K. J. & TRUDGILL, D. L. (1975). Effects of potato cyst-nematodes on potato plants. III. Effects on the water relations and growth of a resistant and a susceptible variety. *Nematologica* **21**, 273–280.

EVANS, K. & TRUDGILL, D. L. (1971). Effects of amino acids on the reproduction of *Heterodera rostochiensis*. *Nematologica* **17**, 495–500.

EVANS, K. & TRUDGILL, D. L. (1978). Nematode pests of potatoes. In: P. M. Harris (Ed.), *The potato crop: the scientific basis for improvement*. London, Chapman & Hall (in the press).

EVANS, K., TRUDGILL, D. L. & BROWN, N. J. (1977). Effects of potato cyst-nematodes on potato plants V. Root system development in lightly- and heavily-infested susceptible and resistant varieties, and its importance in water and nutrient uptake. *Nematologica* **23**, 153–164.

FASSULIOTIS, G. (1958). Effects of ionizing radiations on the golden nematode *Heterodera rostochiensis*. *Radiation Res.* **9**, 112–113. [*Helminth. Abstr.* **29**, 925].

FASSULIOTIS, G. (1970). Resistance of *Cucumis* spp. to the root-knot nematode, *Meloidogyne incognita acrita*. *J. Nematol.* **2**, 174–178.

FASSULIOTIS, G. & SPARROW, A. H. (1955). Preliminary report of X-ray studies on the golden nematode. *Pl. Dis. Reptr* **39**, 572.

FAULKNER, L. R. & BOLANDER, W. J. (1966). Occurrence of large nematode populations in irrigation canals of south central Washington. *Nematologica* **12**, 591–600.

FAULKNER, L. R., BOLANDER, W. J. & SKOTLAND, C. B. (1970). Interaction of *Verticillium dahliae* and *Pratylenchus minyus* in Verticillium wilt of peppermint: influence of the nematode as determined by a double root technique. *Phytopathology* **60**, 100–103.

FAULKNER, L. R. & SKOTLAND, C. B. (1965). Interactions of *Verticillium dahliae* and *Pratylenchus minyus* in Verticillium wilt of peppermint. *Phytopathology* **55**, 583–586.

FEDER, W. A. (1952). Systox for nematode in daffodils. *Bull. N.Y. St. Flower Growers Inc.* No. 82, 3–4.

FENILI, G. A. & MARINARI, A. (1959). Prove di disinfestazione da nematodi galligeni con prodotto a base di 1,2-dibromo-3-chloropropano. *Redia* **44**, 219–227.

FENWICK, D. W. (1949). Investigations on the emergence of larvae from cysts of the potato-root eelworm *Heterodera rostochiensis*. I. Technique and variability. *J. Helminth.* **23**, 157–170.

FENWICK, D. W. (1956). The breakdown of potato-root diffusate in soil. *Nematologica* **1**, 290–302.

FENWICK, D. W. (1961). Estimation of field populations of cyst forming nematodes of the genus *Heterodera*. *J. Helminth., R. T. Leiper Suppl.*, 63–76.

FENWICK, D. W. & FRANKLIN, M. T. (1951). Further studies on the identification of *Heterodera* species by larval length. Estimation of the length parameters for eight species and varieties. *J. Helminth.* **25**, 57–76.

FENWICK, D. W., PETERS, B. G. & LIBBEY, R. P. (1953). Effects of repeated field injections of D-D mixture against potato-root eelworm. *Ann. appl. Biol.* **40**, 208–214.

FENWICK, D. W. & REID, E. (1953). Population studies on the potato root eelworm (*Heterodera rostochiensis* Woll.). *J. Helminth.* **27**, 119–128.

FERRIS, V. R. (1971). Taxonomy of the Dorylaimida. In: B. M. Zuckerman, W. F. Mai & R. A. Rohde (Eds), *Plant parasitic nematodes*. I. London & New York, Academic Press, 163–189.

FERRIS, V. R. & SIEGEL, B. M. (1957). Electron microscopy of golden nematode cyst wall. *Nematologica* **2**, 16–18.

FIDDIAN, W. E. H. & ALDRICH, D. T. A. (1964). The susceptibility of red clover varieties to clover stem eelworm. *Pl. Path.* **13**, 139–143.

FIELDING, M. J. (1951). Observations on the length of dormancy in certain plant infecting nematodes. *Proc. helminth. Soc. Wash.* **18**, 110–112.

FINNEY, D. J. (1962). *Probit analysis. A statistical treatment of the sigmoid response curve.* 2nd Edn. Cambridge Univ. Press, 318 pp. (see pp. 17–19).

FISHER, J. M. (1967). Effect of temperature and host on *Paratylenchus neoamblycephalus* and effect of the nematode on the host. *Aust. J. agric. Res.* **18**, 921–929.

FISHER, J. M. (1969). Investigations on fecundity of *Aphelenchus avenae*. *Nematologica* **15**, 22–28.

FISHER, J. M. (1970). Growth and development of *Aphelenchus avenae* Bastian. *Aust. J. biol. Sci.* **23**, 411–419.

FISHER, J. M. (1972). Observations on the effect of males on reproduction and fecundity of *Aphelenchus avenae*. *Nematologica* **18**, 179–189.

FISHER, J. M. & EVANS, A. A. F. (1967). Penetration and feeding by *Aphelenchus avenae*. *Nematologica* **13**, 425–428.

FISHER, R. A. (1938). *Statistical methods for research workers.* London, Oliver & Boyd, 356 pp.

FLEGG, J. J. M. (1966). The Z-organ in *Xiphinema diversicaudatum*. *Nematologica* **12**, 174.

FLEGG, J. J. M. (1967). A rapid identification method for British *Longidorus* species. *Pl. Path.* **16**, 167–169.

FLEGG, J. J. M. (1968). Life-cycle studies of some *Xiphinema* and *Longidorus* species in south eastern England. *Nematologica* **14**, 197–210.

FLEGG, J. J. M. (1969). Tests with potential nematode vectors of cherry leafroll virus. *Rep. E. Malling Res. Stn for* 1968, 155–157.

FORTUNER, R. (1970). On the morphology of *Aphelenchoides besseyi* Christie, 1942 and *A. siddiqii* n.sp (Nematoda, Aphelenchoidea). *J. Helminth.* **44**, 141–152.

FOX, J. A. (1967). Reproductive isolation in the genus *Heterodera*. *Nematologica* **13**, 143–144. (Abstr.).

FRANKLIN, M. T. (1937). The effect on the cyst contents of *Heterodera schachtii* of the cultivation of maize on potato sick land. *J. Helminth.* **15**, 61–68.

FRANKLIN, M. T. (1939). On the structure of the cyst wall of *Heterodera schachtii* (Schmidt). *J. Helminth.* **17**, 127–134.

FRANKLIN, M. T. (1951). *The cyst-forming species of* Heterodera. Farnham Royal, Commonw. Agric. Bur., 147 pp.

FRANKLIN, M. T. (1957). *Aphelenchoides composticola* n.sp. and *A. saprophilus* n.sp. from mushroom compost and rotting plant tissues. *Nematologica* **2**, 306–313.

FRANKLIN, M. T. (1961). A British root-knot nematode, *Meloidogyne artiellia* n.sp. *J. Helminth., R. T. Leiper Suppl.*, 85–92.

FRANKLIN, M. T. (1965). A root-knot nematode, *Meloidogyne naasi* n.sp., on field crops in England and Wales. *Nematologica* **11**, 79–86.

FRANKLIN, M. T. (1972). *Heterodera schachtii*. *CIH Descriptions of Plant-parasitic Nematodes* Set 1, No. 1, 4 pp.

FRANKLIN, M. T. (1973). *Meloidogyne naasi*. *CIH Descriptions of Plant-parasitic Nematodes* Set 2, No. 19, 4 pp.

FRANKLIN, M. T., CLARK, S. A. & COURSE, J. A. (1971). Population changes and development of *Meloidogyne naasi* in the field. *Nematologica* **17**, 575–590.

FRANKLIN, M. T. & SIDDIQI, M. R. (1972). *Aphelenchoides besseyi*. *CIH Descriptions of Plant-parasitic Nematodes* Set 1, No. 4, 3 pp.

FRANKLIN, M. T. & STONE, A. R. (1974). *Hemicycliophora arenaria*. *CIH Descriptions of Plant-parasitic Nematodes* Set 3, No. 43, 3 pp.

FRENCH, N. & BARRACLOUGH, R. M. (1962). Survival of *Aphelenchoides ritzemabosi* (Schwartz) in soil and dry leaves. *Nematologica* **7**, 309–316.

FRENCH, N. & CAIRNS, R. S. (1960). A new spring oat variety resistant to the stem and bulb eelworm. *Pl. Path.* **9**, 129.

FREUDENHAMMER, I. (1975). Desmoscolecida aus der Iberischen Tiefsee, zugleich eine Revision dieser Nematoden Ordnung. *'Meteor' Forsch-Ergebnisse,* Reihe D, No. 20, 65 pp.

FRITZSCHE, R. (1964). Untersuchungen über die Virusübertragung durch Nematoden. *Wiss. Z. Univ. Rostock. math.-naturw. Reihe* **13**, 343–347.

FRITZSCHE, R. (1970). Beiträge zum Übertragungsmechanismus pflanzenpathogener Viren durch Nematoden. *Zesz. probl. Postęp. Nauk roln.* No. 92, 293–300.

FRITZSCHE, R. & KEGLER, H. (1964). Die Übertragung des Blattrollvirus der Kirsche (cherry leaf-roll virus) durch Nematoden. *Naturwissenschaften* **12**, 299.

FRITZSCHE, R. & KEGLER, H. (1968). Nematoden als Vektoren von Viruskrankheiten der Obstgewächse. *TagBer. dt. Akad. LandwWiss. Berl.* No. 97, 289–295.

FRITZSCHE, R. & SCHMELZER, K. (1967). Übertragbarkeit des Nelkenringflecken-Virus durch Nematoden. *Naturwissenschaften* **54**, 498–499.

FRÖIER, K., HOFFMANN, W., SANDEGREN, E. & THUNAEUS, H. (1959). Gerste (*Hordeum vulgare* L.). In: H. Kappert & W. Rudorf (Eds), *Handbuch der Pflanzenzuchtung*. Vol. 2. Berlin, Parey, 276–426.

FULLER, J. M. & HOWARD, H. W. (1974). Breeding for resistance to the white potato cyst-nematode, *Heterodera pallida*. *Ann. appl. Biol.* **77**, 121–128.

FULTON, J. P. (1962). Transmission of tobacco ringspot virus by *Xiphinema americanum*. *Phytopathology* **52**, 375.

FUSHTEY, S. G. & JOHNSON, P. W. (1966). The biology of the oat cyst nematode, *Heterodera avenae* in Canada. I. The effect of temperature on the hatchability of cysts and emergence of larvae. *Nematologica* **12**, 313–320.

GAIR, R. (1965). Cereal root eelworm. In: J. F. Southey (Ed.), *Tech. Bull. Minist. Agric. Fish. Fd* 7 (2nd Edn), 199–211.

GAIR, R., MATHIAS, P. L. & HARVEY, P. N. (1969). Studies of cereal nematode populations and cereal yields under continuous or intensive culture. *Ann. appl. Biol.* **63**, 503–512.

GARRETT, S. D. (1956). *Biology of root-infecting fungi.* Cambridge Univ. Press, 293 pp.

GASSER, J. K. R. & PEACHEY, J. E. (1964). A note on the effects of some soil sterilants on the mineralisation and nitrification of soil-nitrogen. *J. Sci. Fd Agric.* **15**, 142–146.

GEIGER, R. (1965). *The climate near the ground.* Cambridge Mass., Harvard Univ. Press (Revised Edn), 611 pp.

GERAERT, E. (1968). Morphometric relations in nematodes. *Nematologica* **14**, 171–183.

GERAERT, E. (1971). Observations on the genera *Boleodorus* and *Boleodoroides* (Nematoda: Tylenchida). *Nematologica* **17**, 263–276.

GERAERT, E. (1972). A comparative study on the structure of the female gonads in plant-parasitic Tylenchida (Nematoda). *Annls Soc. r. zool. Belg.* **102**, 171–198.

GERAERT, E. & KHEIRI, A. (1970). The female gonads and the oesophageal structure in the genus *Pseudhalenchus* (Nematoda: Tylenchida). *Nematologica* **16**, 197–202.

GERLACH, S. A. & RIEMANN, F. (1973 & 1974). The Bremerhaven checklist of aquatic nematodes. *Veröff. Inst. Meeresforsch. Bremerh. Suppl.* 4 (1 & 2), 736 pp.

GIBBS, A. J. & HARRISON, B. D. (1964). A form of pea early-browning virus found in Britain. *Ann. appl. Biol.* **54**, 1–11.

GIEBEL, J. (1974). Biochemical mechanisms of plant resistance to nematodes: a review. *J. Nematol.* **6**, 175–184.

GIEBEL, J. & STOBIECKA, M. (1975). Role of amino acids in plant tissue response to *Heterodera rostochiensis.* I. Protein-proline and hydroxyproline content in roots of susceptible and resistant solanaceous plants. *Nematologica* **20**, 407–414.

GILLARD, A. (1961). Onderzoekingen ombrent de biologie, de verspreiding en de bestrijding van wortelknobbelaaltjes (*Meloidogyne* spp.). *Meded. Landb-Hogesch. OpzoekStns Gent* **26**, 515–646.

GILLARD, A. & VAN DEN BRANDE, J. (1959). Essais de lutte contre les nématodes des racines (*Meloidogyne* spp.) au moyen de chauffage électrique du sol. *Proc. IV internat. Congr. Crop Prot., Hamburg*, 1957, Vol. 1, 623–627.

GOELDI, E. A. (1887). Relatoria sôbre a molestia do cafeiro na provincia da Rio de Janeiro. (*Advance separate of*) *Archos Mus. nac., Rio de J.* **8**, 7–112 (1892).

GOFFART, H. (1941). Untersuchungen am Hafernematoden (*Heterodera schachtii*, Schmidt) unter besonderer Berücksichtigung der schleswigholsteinischen Verhaltnisse II. *Arb. biol. Reichsanst., Berl.* **23**, 141–161.

GOFFART, H. (1949). Nematodes, Fadenwurmer. In: P. Sorauer, *Handbuch der Pflanzenkrankheiten* **4**, (5e Aufl.) 4–95.

GOFFART, H. (1953). Beobachtungen an pflanzenschädlichen Nematoden I. *NachrBl. dt. PflSchutzdienst., Stuttg.* **5**, 150–153.

GOFFART, H. (1960a). Die Kutikularstruktur des Hinterendes bei der Gattung *Heterodera* als Artcharakteristikum. *Verh. dt. zool. Ges.* (*Zool. Anz.* **23** *Suppl.*) 387–393.

GOFFART, H. (1960b). Die taxonomische Bewertung morphologisch-anatomischer Merkmale bei den Zysten der Gattung *Heterodera* (Nematoda). *Mitt. biol. BundAnst. Ld- u. Forstw.* Nr. 99, 24–51.

GOFFART, H. & HEILING, A. (1962). Beobachtungen über die enzymatische Wirkung von Spiecheldrüsensekreten pflanzenparasitärer Nematoden. *Nematologica* **7**, 173–176.

GOLDEN, A. M. (1958). Interrelationships of certain *Beta* species and *Heterodera schachtii*, the sugar beet nematode. *Pl. Dis. Reptr* **42**, 1157–1162.

GOLDEN, A. M. (1971). Classification of the genera and higher categories of the order Tylenchida (Nematoda). In: B. M. Zuckerman, W. F. Mai & R. A. Rohde (Eds), *Plant parasitic nematodes* I. London & New York, Academic Press, 191–232

GOLDEN, A. M., EPPS, J. M., RIGGS, R. D., DUCLOS, L. A., FOX, J. A. & BERNARD, R. L. (1970). Terminology and identity of infraspecific forms of the soybean cyst nematode (*Heterodera glycines*). *Pl. Dis. Reptr* **54**, 544–546.

GOLDEN, A. M. & JENSEN, H. J. (1974). *Nacobbodera chitwoodi* n.gen., n.sp., (Nacobbidae: Nematoda) on Douglas fir in Oregon. *J. Nematol.* **6**, 30–37.

GOLDEN, A. M., RAU, G. J. & COBB, G. S. (1962). *Heterodera cyperi* (Heteroderidae), a new species of cyst-forming nematode. *Proc. helminth. Soc. Wash.* **29**, 168–173.

GOLDEN, A. M. & SHAFER, T. (1959). Susceptibility of tomato (*Lycopersicon esculentum*) to the sugar-beet nematode (*Heterodera schachtii*). *Pl. Dis. Reptr* **43**, 1196–1197.

GOLDEN, A. M. & TAYLOR, D. P. (1967). The barley root-knot nematode in Illinois. *Pl. Dis. Reptr* **51**, 974–975.

GOLDEN, J. K. & VAN GUNDY, S. D. (1972). Influence of *Meloidogyne incognita* on root rot development by *Rhizoctonia solani* and *Thielaviopsis basicola* in tomato. (Abstr.) *J. Nematol.* **4**, 225.

GOODEY, J. B. (1949). The control of *Anguillulina dipsaci* on the seed of teazle and red clover by fumigation with methyl bromide. *J. Helminth.* **23**, 171–174.

GOODEY, J. B. (1952a). Investigations into the host ranges of *Ditylenchus destructor* and *D. dipsaci*. *Ann. appl. Biol.* **39**, 221–228.

GOODEY, J. B. (1952b). The influence of the host on the dimensions of the plant parasitic nematode, *Ditylenchus destructor*. *Ann. appl. Biol.* **39**, 468–474.

GOODEY, J. B. (1958a). *Paraphelenchus myceliophthorus* n.sp. (Nematoda: Aphelenchidae). *Nematologica* **3**, 1–5.

GOODEY, J. B. (1958b). *Ditylenchus myceliophagus* n.sp. (Nematoda: Tylenchidae). *Nematologica* **3**, 91–96.

GOODEY, J. B. (1959). The excretory system of *Paraphelenchus* and the identity of the hemizonid. *Nematologica* **4**, 157–159.

GOODEY, J. B. (1960a). Gall-forming nematodes of grasses in Britain. *J. Sports Turf Res. Inst.* **10** (35), 54–60.

GOODEY, J. B. (1960b). Observations on the effects of the parasitic nematodes *Ditylenchus myceliophagus*, *Aphelenchoides composticola* and *Paraphelenchus myceliophthorus* on the growth and cropping of mushrooms. *Ann. appl. Biol.* **48**, 655–664.

GOODEY, J. B. (1965). *Ditylenchus* and *Anguina*. In: J. F. Southey (Ed.), Plant nematology. *Tech. Bull. Minist. Agric. Fish. Fd* 7 (2nd Edn), 47–58.

GOODEY, J. B., FRANKLIN, M. T. & HOOPER, D. J. (1965). *T. Goodey's The nematode parasites of plants catalogued under their hosts*. (3rd Edn). Farnham Royal, Commonwealth Agricultural Bureaux, 214 pp.

GOODEY, J. B. & HOOPER, D. J. (1958). Observations on the effects of *Ditylenchus dipsaci* and *Anguina tritici* on certain wheat and barley varieties. *Nematologica* **3**, 24–29.

GOODEY, J. B. & HOOPER, D. J. (1965). A neotype of *Aphelenchus avenae* Bastian, 1865 and the rejection of *Metaphelenchus* Steiner, 1943. *Nematologica* **11**, 55–65.

GOODEY, J. B., PEACOCK, F. C. & PITCHER, R. S. (1960). A redescription of *Xiphinema diversicaudatum* (Micoletzky, 1923 & 1927) Thorne, 1939 and observations on its larval stages. *Nematologica* **5**, 127–135.

GOODEY, T. (1932). Some observations on the biology of the root-gall nematode, *Anguillulina radicicola* (Greeff, 1872). *J. Helminth.* **10**, 33–44.

GOODEY, T. (1933a). *Anguillulina graminophila* n.sp., a nematode causing galls on the leaves of fine bent-grass. *J. Helminth.* **9**, 45–56.

GOODEY, T. (1933b). *Plant parasitic nematodes.* London, Methuen, 306 pp.

GOODEY, T. (1937). Observations on the susceptibility of certain varieties of oats to tulip-root caused by the stem eelworm, *Anguillulina dipsaci. J. Helminth.* **15**, 203–214.

GOODEY, T. (1938). Observations on *Anguillulina millefolii* (Löw, 1874) Goodey, 1932, from galls on the leaves of yarrow, *Achillea millefolium* L. *J. Helminth.* **16**, 93–108.

GOODEY, T. (1940). On *Anguillulina multicincta* (Cobb) and other species of *Anguillulina* associated with the roots of plants. *J. Helminth.* **18**, 21–38.

GOODEY, T. (1941). Observations on a giant race of stem eelworm, *Anguillulina dipsaci,* attacking broad beans, *Vicia faba* L. *J. Helminth.* **19**, 114–122.

GOODEY, T. (1945a). *Anguillulina dipsaci* on onion seed and its control by fumigation with methyl bromide. *J. Helminth.* **21**, 45–59.

GOODEY, T. (1945b). *Anguillulina brenani* n.sp., a nematode causing galls on the moss, *Pottia bryoides* Mitt. *J. Helminth.* **21**, 105–110.

GOODEY, T. (1963). *Soil and freshwater nematodes.* 2nd Edn revised and re-written by J. B. Goodey. London, Methuen, 544 pp.

GOPLEN, B. P. & STANFORD, E. H. (1959). Studies on the nature of resistance in alfalfa to two species of root-knot nematodes. *Agron. J.* **51**, 486–488.

GOPLEN, B. P., STANFORD, E. H. & ALLEN, M. W. (1959). Demonstration of physiological races within three root-knot nematode species attacking alfalfa. *Phytopathology* **49**, 653–656.

GRAF, A., JOSEPH, E., KELLER, E., LIECHTI, H. & SAVARY, A. (1964). Essais de lutte contre *Ditylenchus dipsaci* Kühn nématode agent de la pourriture du collet des betteraves. *Agriculture romande, Série A,* **3**, 4–7.

GRAHAM, C. W. & STONE, L. E. W. (1975). Field experiments on the cereal cyst-nematode (*Heterodera avenae*) in south-east England, 1967–72. *Ann. appl. Biol.* **80**, 61–73.

GRAHAM, T. W. (1958). Root-knot and other nematodes in relation to the development of tobacco black shank. *Phytopathology* **48**, 343.

GRAHAM, T. W. (1969). A new pathogenic race of *Meloidogyne incognita* on flue-cured tobacco. *Tobacco* **168**, 28–29.

GRAINGER, J. (1951). The golden eelworm. Studies on the ecology and control of the potato root eelworm, *Heterodera rostochiensis. Res. Bull. W. Scotl. agric. Coll.* 10, 72 pp.

GRANEK, I. (1955). Additional morphological differences between the cysts of *Heterodera rostochiensis* and *Heterodera tabacum. Pl. Dis. Reptr* **39**, 716–718.

GRANEK, I. (1968). Keys to the *Heterodera* 'groups' and species of the *Heterodera rostochiensis* 'group'. *Pl. Dis. Reptr* **52**, 827–828.

GRASSÉ, P. P. (Ed.) (1965). *Traité de zoologie. Anatomie, systématique, biologie* **4** (2), 1–731; **4** (3), 732–1497. Paris, Masson et Cie.

GREEN, C. D. (1963). Killing of bulb eelworm by hot-water treatment. *Nature, Lond.* **198**, 303.

GREEN, C. D. (1971). The morphology of the terminal area of the round-cyst nematodes, s.g. *Heterodera rostochiensis* and allied species. *Nematologica* **17**, 34–46.

GREEN, C. D. (1972). The structure of the females of the round-cyst nematodes. *Ann. appl. Biol.* **71**, 286–289.

GREEN, C. D. & GREET, D. N. (1972). The location of the secretions that attract male *Heterodera schachtii* and *H. rostochiensis* to their females. *Nematologica* **18**, 347–352.

GREEN, C. D., GREET, D. N. & JONES, F. G. W. (1970). The influence of multiple mating on the reproduction and genetics of *Heterodera rostochiensis* and *H. schachtii*. *Nematologica* **16**, 309–326.

GREEN, C. D. & MILLER, L. I. (1969). Attraction of males to females. In: Nematology department. *Rep. Rothamsted exp. Stn for* 1968, Part 1, 153–155.

GREEN, C. D. & PLUMB, S. (1967). The effect of ultra-violet radiation on the invasion, survival and fertility of larvae of *Heterodera rostochiensis*. *Nematologica* **13**, 186–190.

GREEN, C. D. & PLUMB, S. C. (1970). The interrelationships of some *Heterodera* spp indicated by the specificity of the male attractants emitted by their females. *Nematologica* **16**, 39–46.

GREEN, C. D. & WEBSTER, J. M. (1966). The effect of ultraviolet radiation on the stem and bulb nematode *Ditylenchus dipsaci* (Kühn). *Nematologica* **11** (Yr 1965), 638–642.

GREENWOOD, D. J. (1968). Root growth and oxygen distribution in soil. *9th Int. Congr. Soil Sci. Adelaide Vol.* 1, 823–832.

GREET, D. N. (1972). Electrophoresis and morphometrics of the round-cyst nematodes. *Ann. appl. Biol.* **71**, 283–286.

GREET, D. N. (1975). The response of five round-cyst nematodes (Heteroderidae) to five artificial hatching agents. *Nematologica* **20**, 363–364.

GRIFFIN, G. D., ANDERSON, J. L. & JORGENSON, E. C. (1968). Interaction of *Meloidogyne hapla* and *Agrobacterium tumefaciens* in relation to raspberry cultivars. *Pl. Dis. Reptr* **52**, 492–493.

GRIFFIN, G. D. & DARLING, H. M. (1964). An ecological study of *Xiphinema americanum* Cobb in an ornamental spruce nursery. *Nematologica* **10**, 471–479.

GRIFFITHS, D. J., HOLDEN, J. W. H. & JONES, J. M. (1957). Investigations on resistance of oats to stem eelworm, *Ditylenchus dipsaci* Kühn. *Ann. appl. Biol.* **45**, 709–720.

DE GRISSE, A. T., LIPPENS, P. L. & COOMANS, A. (1974). The cephalic sensory system of *Rotylenchus robustus* and a comparison with some other Tylenchids. *Nematologica* **20**, 88–95.

DE GRISSE, A. & LOOF, P. A. A. (1965). Revision of the genus *Criconemoides* (Nematoda). *Meded. LandbHogesch. OpzoekStns Gent* **30**, 577–603.

GRUJIČIĆ, G. (1962). Opojavi stabljikine nematode (*Ditylenchus dipsaci* Kühn) na pšenici u Jugoslaviji. *Savremen. Poljopr., Novi Sad* **10**, 86–94.

GRUJIČIĆ, G. (1967). Korenova nematoda (*Meloidogyne naasi* Franklin) in Srbiji. *Zašt. Bilja* **18**, 193–197.

GRUJIČIĆ, G. (1974). Rasprostanjenost štetnih nematoda na duvanu na proizvodnom području duvanske industrije Vranje u 1971 godiny. *Tutun* **23**, 5–6, 163–170.

GRUJIČIĆ, G. & PAUNOVIĆ, M. (1971). Prilog proučavanju korenove nematode (*Meloidogyne hapla* Chitwood). *Zašt. Bilja* **22**, 147–152.

GUILE, C. T. (1966). Cyst chromogenesis in potato cyst eelworm pathotypes. *Pl. Path.* **15**, 125–128.

GUILE, C. T. (1967). On cyst colour changes, bionomics and distribution of potato cyst-eelworm (*Heterodera rostochiensis* Woll.) pathotypes in the East Midlands. *Ann. appl. Biol.* **60**, 411–419.

GUILE, C. T. (1970). Further observations on cyst colour changes in potato cyst eelworm pathotypes. *Pl. Path.* **19**, 1–6.

DE GUIRAN, G. (1960). Étude comparative de la pénétration des larves de *Meloidogyne javanica* (Treub, 1885) Chitwood, 1949 et de *Meloidogyne incognita acrita* Chitwood, 1949 dans les racines des plantes hôtes et non hôtes. Résultats préliminaires. *Meded. LandbHogesch. OpzoekStns Gent* **25**, 1047–1056.

DE GUIRAN, G. (1974). Partial diapause within the egg masses of *Meloidogyne incognita. XII Simposio Internacional de Nematologia, Granada* 1 to 7 Sept. 1974. *Abstracts*, 37–39.

GÜNTHER, B. (1972). Untersuchungen zum Kutikula-aufbau und zum Häutungsverlauf bei einigen Nematodenarten. *Nematologica* **18**, 275–287.

GÜNTHER, B. & KÄMPFE, L. (1967). Bau und Veränderung des Integumentes im Entwicklungszyklus cystenbildender Nematoden. *Verh. dt. zool. Ges.*, Yr 1966 (*Zool. Anz.* **30** *Suppl.*) 152–166.

GUPTA, P. & SWARUP, G. (1968). On the ear-cockle and yellow rot disease of wheat. I. Symptoms and histopathology. *Indian Phytopath.* **21**, 318–323.

GUPTA, P. & SWARUP, G. (1972). Ear-cockle and yellow ear-rot diseases of wheat: II. Nematode bacterial association. *Nematologica* **18**, 320–324.

HAGLUND, W. A. & KING, T. H. (1961). Effect of parasitic nematodes on the severity of common root rot of canning peas. *Nematologica* **6**, 311–314.

HAGUE, N. G. M. (1963). Fumigation of Agricultural Products. XVIII.—Effect of methyl bromide on the bent grass nematode, *Anguina agrostis* (Steinbuch 1799) Filipjev 1936, and on the germination of bent grass, *Agrostis tenuis*. *J. Sci. Fd Agric.* **14**, 577–579.

HAGUE, N. G. M. (1966). Chemical control of eelworm in daffodils and tulips. *Commercial Grower* August 1966, 260–261.

HAGUE, N. G. M. (1968). Fumigation of shallot seed with methyl bromide to control stem eelworm, *Ditylenchus dipsaci*. *Pl. Path.* **17**, 127–128.

HAGUE, N. G. M. (1972). Nematode diseases of flower bulbs, glasshouse crops and ornamentals. In: J. M. Webster (Ed.), *Economic nematology*. London & New York, Academic Press, 409–434.

HAGUE, N. G. M., LUBATTI, O. F. & PAGE, A. B. P. (1964). Methyl bromide fumigation of nematodes in soil under gas-proof sheets. *Hort. Res., Edinb.* **3**, 84–101.

HAGUE, N. G. M. & SOOD, U. (1963). Soil sterilization with methyl bromide to control soil nematodes. *Pl. Path.* **12**, 88–90.

HAMER, M., HUFFMAN, C. W. & RISTICH, S. S. (1970). Synthesis and nematocidal activity of 2-pyridyl alkanesulphonates. *J. agric. Fd Chem., Ser. B*, **18**, 57–59.

HAMLEN, R. A., BLOOM, J. R. & LUKEZIC, F. L. (1973). Hatching of *Meloidogyne incognita* eggs in the neutral carbohydrate fraction of root exudates of gnotobiotically grown alfalfa. *J. Nematol.* **5**, 142–146.

HAMS, A. F. & WILKIN, G. D. (1961). Observations on the use of predacious fungi for the control of *Heterodera* spp. *Ann. appl. Biol.* **49**, 515–523.

HANNA, M. R. & HAWN, E. J. (1965). Seedling inoculation studies with alfalfa stem nematode. *Can. J. Pl. Sci.* **45**, 357–363.

HANSEN, E. L., BUECHER, E. J. & YARWOOD, E. A. (1972). Sex differentiation of *Aphelenchus avenae* in axenic culture. *Nematologica* **18**, 253–260.

HARE, W. W. (1965). The inheritance of resistance of plants to nematodes. *Phytopathology* **55**, 1162–1167.

HARLAN, D. P. & JENKINS, L. (1967). Elimination of root-knot nematodes from plants by chemical bare-root dips or soil drenches. *Pl. Dis. Reptr* **51**, 103–107.

HARRANGER, J. (1956). Le salsifis de Lunéville, son avenir est-il menacé par les anguillules? *Phytoma* **8**, 7–10.

HARRISON, B. D. (1960). The biology of soil-borne plant viruses. *Adv. Virus Res.* **7**, 131–161.

HARRISON, B. D. (1961). Soil-borne viruses. Tobacco rattle virus. *Rep. Rothamsted exp. Stn for* 1960, p. 118.

HARRISON, B. D. (1964a). The transmission of plant viruses in soil. In: M. K. Corbett & H. D. Sisler (Eds). *Plant virology*. Gainesville, Florida University Press, 118–147.

HARRISON, B. D. (1964b). Specific nematode vectors for serologically distinctive forms of raspberry ringspot and tomato black ring viruses. *Virology* **22**, 544–550.

HARRISON, B. D. (1966). Further studies on a British form of pea early-browning virus. *Ann. appl. Biol.* **57**, 121–129.

HARRISON, B. D. (1967). Pea early-browning virus (PEBV). *Rep. Rothamsted exp. Stn for* 1966, p. 115.

HARRISON, B. D. (1968). Reactions of some old and new British potato cultivars to tobacco rattle virus. *Eur. Potato J.* **11**, 165–176.

HARRISON, B. D. & CADMAN, C. H. (1959). Role of a dagger nematode (*Xiphinema* sp.) in outbreaks of plant diseases caused by arabis mosaic virus. *Nature, Lond.* **184**, 1624–1626.

HARRISON, B. D., MOWAT, W. P. & TAYLOR, C. E. (1961). Transmission of a strain of tomato black ring virus by *Longidorus elongatus* (Nematoda). *Virology* **14**, 480–485.

HARRISON, B. D., PEACHEY, J. E. & WINSLOW, R. D. (1963). The use of nematicides to control the spread of arabis mosaic virus by *Xiphinema diversicaudatum* (Micol.). *Ann. appl. Biol.* **52**, 243–255.

HASTINGS, R. J. (1933). Treatment of narcissus bulbs with hot water. *Gdnrs' Chron.* **94**, 313–314.

HASTINGS, R. J. & NEWTON, W. (1934). The effect of temperature upon the pre-adult larvae of the bulb nematode *Anguillulina dipsaci* (Kühn, 1858) Gerv. and v. Ben., 1859, in relation to time and moisture. *Can. J. Res.* **10**, 793–797.

HAWN, E. J. (1963). Transmission of bacterial wilt of alfalfa by *Ditylenchus dipsaci* (Kühn). *Nematologica* **9**, 65–68.

HAYES, J. D. (1970). Arable crop breeding (iv) Cereal cyst nematode (*Heterodera avenae*). *Jubilee Rep. Welsh Pl. Breed. Stn* 1919–1969, 128–130.

HAYES, J. D. & COTTEN, J. (1971). Breeding for nematode resistance with particular reference to *Heterodera avenae* Woll. In: R. A. Nilan (Ed.), *Barley Genetics II*, 527–534.

HEALY, W. B., WIDDOWSON, J. P. & YEATES, G. W. (1973). Effect of root nematodes on the growth of seedlings and established white clover on a yellow-brown loam. *N.Z. Jl agric. Res.* **16**, 70–76.

HECHLER, H. C. (1962). The development of *Aphelenchus avenae* Bastian, 1865 in fungus culture. *Proc. helminth. Soc. Wash.* **29**, 162–167.

HEELEY, W. (1951). D-D—a soil fumigant against *Heterodera marioni* in the United Kingdom. *Proc. 2nd int. Congr. Crop Prot.* 1949, 193–202.

D'HERDE, J. (1965). Een nieuw wortelknobbelaaltje, parasiet van de bieteteelt. *Meded. LandbHogesch. OpzoekStns Gent* **30,** 1429–1436.

D'HERDE, J. & VAN DEN BRANDE, J. (1964). Verspreiding en bestrijding van *Xiphinema* en *Longidorus* spp in aardbeien in Belgie. *Meded. LandbHogesch. OpzoekStns Gent* **29,** 788–796.

D'HERDE, J., VAN DEN BRANDE, J. & GILLARD, A. (1960). Control of *Pratylenchus penetrans* (Cobb), causal agent of root rot in lilies. *Nematologica Suppl. II,* 64–67.

D'HERDE, J., DE CLERCQ, R. & COOLEN, W. A. (1969). Over het gevaar en de gevolgen van het overbrengen van tropische en subtropische parasieten met geimporteerd plantenmaterial. *Publ. W4/W5, Rijksstation voor Nematologie en Entomologie, Wetteren,* 14 pp.

HESLING, J. J. (1956). Some observations on *Heterodera major. Nematologica* **1,** 56–63.

HESLING, J. J. (1957). The hatching response of *Heterodera major* (O. Schmidt) to certain root diffusates. *Nematologica* **2,** 123–125.

HESLING, J. J. (1958a). *Heterodera major* O. Schmidt, 1930—population changes in the field and in pots of fallow soil. *Nematologica* **3,** 274–282.

HESLING, J. J. (1958b). The efficiency of certain grasses as hosts of cereal root eelworm. *Pl. Path.* **7,** 141–143.

HESLING, J. J. (1961a). *Heterodera rostochiensis* (Woll. 1923) on *Solanum demissum*—a population study. *Ann. appl. Biol.* **49,** 350–359.

HESLING, J. J. (1961b). Problems in the routine hot-water treatment of chrysanthemum stools. *Pl. Path.* **10,** 139–141.

HESLING, J. J. (1965). *Heterodera*: morphology and identification. In: J. F. Southey (Ed.), Plant nematology. *Tech. Bull. Minist. Agric. Fish. Fd* No. 7 (2nd edn), 103–130.

HESLING, J. J. (1966). Preliminary experiments on the control of mycophagous eelworms in mushroom beds, with a note on their swarming. *Pl. Path.* **15,** 163–166.

HESLING, J. J. (1972). Nematode pests of mushrooms. In: J. M. Webster (Ed.), *Economic nematology.* London & New York, Academic Press, 435–468.

HESLING, J. J. (1973). The estimation of Granek's ratio in round-cyst *Heteroderas. Nematologica* **19,** 119–120.

HESLING, J. J. (1974). *Ditylenchus myceliophagus. CIH Descriptions of Plant-parasitic Nematodes* Set 3, No. 36, 4 pp.

HESLING, J. J. & ELLIS, P. R. (1974). The fenestralia of *Heterodera rostochiensis* and *H. pallida. Nematologica* **20,** 43–51.

HESLING, J. J. & KEMPTON, R. J. (1969). The control of nematodes in mushroom compost. *Proc. 5th Br. Insectic. Fungic. Conf.* Vol. 1, 185–188.

HESLING, J. J., PAWELSKA, K. & SHEPHERD, A. M. (1961). The response of potato root eelworm, *Heterodera rostochiensis* Wollenweber and beet eelworm, *H. schachtii* Schmidt to root diffusates of some grasses, cereals and of *Tagetes minuta. Nematologica* **6,** 207–213.

HESLING, J. J. & WALLACE, H. R. (1961a). Observations on the biology of chrysanthemum eelworm *Aphelenchoides ritzema-bosi* (Schwartz) Steiner in florists' chrysanthemum. I. Spread of eelworm infestation. *Ann. appl. Biol.* **49,** 195–203.

HESLING, J. J. & WALLACE, H. R. (1961b). Susceptibility of varieties of chrysanthemum to infestation by *Aphelenchoides ritzema-bosi* (Schwartz). *Nematologica* **5**, 297–302.

HEWITT, T. R. (1914). Eelworms in narcissus bulbs. *J. Dep. Agric. Ire.* **14**, 345–353.

HEWITT, W. B., RASKI, D. J. & GOHEEN, A. C. (1958). Nematode vector of soil-borne fanleaf virus of grapevines. *Phytopathology* **48**, 586–595.

HEYNS, J. (1968). A monographic study of the nematode families Nygolaimidae and Nygolaimellidae. *Entomology Mem. Dep. agric. tech. Serv. Repub. S. Afr.* No. 19, 144 pp.

HEYNS, J. (1971). *A guide to the plant and soil nematodes of South Africa.* Cape Town, A. A. Balkema, 233 pp.

HIDE, G. A. & CORBETT, D. C. M. (1973). Controlling early death of potatoes caused by *Heterodera rostochiensis* and *Verticillium dahliae*. *Ann. appl. Biol.* **75**, 461–462.

HIJINK, M. J. (1972). An integrated approach in the control of potato root eelworm. *EPPO Bull.* No. 7, 41–48.

HIJINK, M. J. & VAN ROSSEN, H. (1968). Inoculaties met *Pratylenchoides laticauda* Braun & Loof bij *Monarda* Hybride 'Croftway Pink' en *Mentha piperita* L. *Meded. Rijksfak. LandbWetensch. Gent* **33**, 729–737.

HIJINK, M. J. & WINOTO SUATMADJI, R. (1967). Influence of different Compositae on population density of *Pratylenchus penetrans* and some other root-infesting nematodes. *Neth. J. Pl. Path.* **73**, 71–82.

HIJNER, J. A., OOSTENBRINK, M. & DEN OUDEN, H. (1953). Morfologische verschillen tussen de belangrijkste *Heterodera*-sorten in Nederland. *Tijdschr. PlZiekt.* **59**, 245–251.

HIRLING, W. (1971). Zur Technik der Untersuchung von Erdbeerpflanzen und Chrysanthemenblättern auf Blattälchen (*Aphelenchoides fragariae* und *A. ritzemabosi*). I, II. *Anz. Schädlingsk.* **44**, 171–174, 182–185.

HIRLING, W. (1972). Zur Technik der Untersuchung von Erdbeerpflanzen und Chrysanthemenblättern auf Blattälchen (*Aphelenchoides fragariae* und *A. ritzemabosi*). III. *Anz. Schädlingsk.* **45**, 6–10.

HIRLING, W. (1974). Schädliche Nematoden an Mais in Baden-Württemberg I. Die Umfallkrankheit beim Mais durch *Ditylenchus dipsaci*. *Anz. Schädlingsk.* **47**, 33–39.

HIRSCHMANN, H. (1956). Comparative morphological studies on the soybean cyst nematode, *Heterodera glycines* and the clover cyst nematode, *H. trifolii* (Nematoda: Heteroderidae). *Proc. helminth. Soc. Wash.* **23**, 140–151.

HIRSCHMANN, H. (1971). Comparative morphology and anatomy. In: B. M. Zuckerman, W. F. Mai & R. A. Rohde (Eds), *Plant parasitic nematodes* I. London & New York, Academic Press, 11–63.

HIRUMI, H., CHEN, T. A., LEE, K. J. & MARAMOROSCH, K. (1968). Ultrastructure of the feeding apparatus of the nematode *Trichodorus christiei*. *J. Ultrastruct. Res.* **24**, 434–453.

HIRUMI, H. & HUNG, C. L. (1969). Ultrastructure of intestinal epithelium in a plant-parasitic nematode, *Trichodorus christiei*. *J. Nematol.* **1**, 292.

HODSON, W. E. H. (1933). The control of chrysanthemum eelworm. *H.E.A. Yearb.* **2**, 85–89.

HOLDEMAN, Q. L. (1956). The effect of the tobacco stunt nematode on the incidence of Fusarium wilt in flue-cured tobacco. *Phytopathology* **46**, 129.

HOLLIS, J. P. & RODRIGUEZ-KABANA, R. (1966). Rapid kill of nematodes in flooded soil. *Phytopathology* **56**, 1015–1019.

HOLTZMANN, O. V. (1965). Effect of soil temperature on resistance of tomato to root-knot nematode (*Meloidogyne incognita*). *Phytopathology* **55**, 990–992.

VAN HOOF, H. A. (1962). *Trichodorus pachydermus* and *T. teres*, vectors of the early browning virus of peas. *Tijdschr. PlZiekt.* **68**, 391–396.

VAN HOOF, H. A. (1968). Transmission of tobacco rattle virus by *Trichodorus* species. *Nematologica* **14**, 20–24.

VAN HOOF, H. A. (1971). Viruses transmitted by *Xiphinema* species in the Netherlands. *Neth. J. Pl. Path.* **77**, 30–31.

HOOPER, D. J. (1961). A redescription of *Longidorus elongatus* (de Man, 1876) Thorne & Swanger, 1936, (Nematoda, Dorylaimidae) and descriptions of five new species of *Longidorus* from Great Britain. *Nematologica* **6**, 237–257.

HOOPER, D. J. (1962). Effects of a nematode on the growth of mushroom mycelium. *Nature, Lond.* **193**, 496–497.

HOOPER, D. J. (1969a). Extraction and handling of plant and soil nematodes. In: J. E. Peachey (Ed.), Nematodes of tropical crops. *Tech. Commun. Commonw. Bur. Helminth.* 40, 20–36.

HOOPER, D. J. (1969b). Some problems in the systematics of soil nematodes. In: J. G. Sheals (Ed.), The soil ecosystem. *Systematics Assn Publ.* No. 8, 131–141.

HOOPER, D. J. (1969c). Identification of plant and soil nematodes. In: J. E. Peachey (Ed.), Nematodes of tropical crops. *Tech. Commun. Commonw. Bur. Helminth.* 40, 37–66.

HOOPER, D. J. (1971). Stem eelworm (*Ditylenchus dipsaci*), a seed and soil-borne pathogen of field beans (*Vicia faba*). *Pl. Path.* **20**, 25–27.

HOOPER, D. J. (1972). *Ditylenchus dipsaci. CIH Descriptions of Plant-parasitic Nematodes* Set 1, No. 14, 4 pp.

HOOPER, D. J. (1973a). Identification of *Longidorus* and *Paralongidorus* species found in the British Isles. In: *The Longidoridae.* Harpenden, Rothamsted Experimental Station, 11–36.

HOOPER, D. J. (1973b). *Ditylenchus destructor. CIH Descriptions of Plant-parasitic Nematodes* Set 2, No. 21, 4 pp.

HOOPER, D. J. (1973c). Nematodes. In: A. J. Gibbs (Ed.), *Viruses and invertebrates.* Amsterdam & London, North Holland, 212–228.

HOOPER, D. J. (1974a). *Aphelenchus avenae. CIH Descriptions of Plant-parasitic Nematodes* Set 4, No. 50, 4 pp.

HOOPER, D. J. (1974b). *Cephalenchus emarginatus. CIH Descriptions of Plant-parasitic Nematodes* Set 3, No. 35, 2 pp.

HOOPER, D. J. (1975). Morphology of Trichodorid nematodes. In: F. Lamberti, C. E. Taylor & J. W. Seinhorst (Eds), *Nematode vectors of plant viruses.* London & New York, Plenum Press, 91–101.

HOOPER, D. J. (1976). Stem eelworm *Ditylenchus dipsaci* on field beans (*Vicia faba*). *Rep. Rothamsted exp. Stn for* 1975, *Part* 1, 202–203.

HOOPER, D. J. & DONCASTER, C. C. (1972). Stem and inflorescence galls on yarrow (*Achillea millefolium* L.) caused by the nematode *Anguina millefolii* (Löw, 1874) Filipjev, 1936. *Pl. Path.* **21**, 46.

HOOPER, D. J. & SOUTHEY, J. F. (1973). The Longidoridae. In: *The Longidoridae.* Harpenden, Rothamsted Experimental Station, 3–10.

HOPE, W. D. & MURPHY, D. G. (1972). A taxonomic hierarchy and checklist of the genera and higher taxa of marine nematodes. *Smithson. Contr. Zool.* 137, 101 pp.

HOPPER, B. E. & CAIRNS, E. J. (1959). *Taxonomic keys to plant, soil and aquatic nematodes.* Auburn, Alabama Polytechnic Institute, 176 pp.

HOWARD, H. W. (1970). *Genetics of the potato,* Solanum tuberosum. London, Logos Press.

HOWARD, H. W. (1972a). Pathotypes of potato cyst nematode. *Ann. appl. Biol.* **71,** 263–266.

HOWARD, H. W. (1972b). Breeding potatoes for resistance to cyst nematodes. *A.D.A.S. q. Rev.* No. 7, 132–138.

HOWARD, H. W., COLE, C. S. & FULLER, J. M. (1970). Further sources of resistance to *Heterodera rostochiensis* Woll. in the Andigena potatoes. *Euphytica* **19,** 210–216.

HOWARD, H. W. & FULLER, J. M. (1971). Resistance to the cream and white potato cyst nematodes. *Pl. Path.* **20,** 32–35.

HSU, D. & HENDRIX, F. F., Jr. (1973). Influence of *Criconemoides quadricornis* on pecan feeder root necrosis caused by *Pythium irregulare* and *Fusarium solani* at different temperatures. *Can. J. Bot.* **51,** 1421–1424.

HUIJSMAN, C. A. (1961). The influence of resistant potato varieties on the soil population of *Heterodera rostochiensis* Woll. *Nematologica* **6,** 177–180.

HUNGER, F. W. T. (1901). Een bacterie-ziekte den tomaat. *Meded. PlTuin. Batavia* **48,** 4–57.

HUNT, O. J., FAULKNER, L. R. & PEADEN, R. N. (1972). Breeding for nematode resistance. In: C. H. Hanson (Ed.), *Alfalfa science and technology.* Madison, American Society of Agronomy, Inc., 812 pp.

HUSAIN, I. (1976). Phylogeny and interrelationship of the superfamily Heteroderoidea (Skarbilovich, 1947) Golden, 1971. *Geobios* **3,** 9–12.

HUSAIN, S. I. & KHAN, A. M. (1967). On the status of the genera of the superfamily Aphelenchoidea (Fuchs, 1937) Thorne, 1949 with the descriptions of six new species of nematodes from India. *Proc. helminth. Soc. Wash.* **34,** 167–174.

HUSSEY, N. W. (1967). Considerations in the use of pesticides in mushroom culture. *Proc. 4th Br. Insectic. Fungic. Conf.* Vol. 1, 76–81.

HUSSEY, N. W., READ, W. H. & HESLING, J. J. (1969). *The pests of protected cultivation.* London, Arnold, 404 pp.

HUTCHINSON, M. T. & STREU, H. T. (1960). Tardigrades attacking nematodes. *Nematologica* **5,** 149.

HYMAN, L. H. (1951). *The Invertebrates V. Smaller coelomate groups.* New York, McGraw-Hill.

ICHIKAWA, S. T., GILPATRICK, J. D. & MCBETH, C. W. (1955). Soil diffusion pattern of 1,2–dibromo–3–chloropropane. *Phytopathology* **45,** 576–578.

ICHINOHE, M. (1952). On the soybean nematode, *Heterodera glycines* n.sp., from Japan. *Ôyô-Dôbutsugaku-Zasshi* **17,** 1–4.

ICHINOHE, M. (1961). Studies on the soybean cyst nematode, *Heterodera glycines. Rep. Hokkaido natn. agric. Exp. Stn* No. 56, 80 pp.

INAGAKI, H. & KEGASAWA, K. (1973). Discovery of the potato cyst nematode, *Heterodera rostochiensis* Wollenweber, 1923, (Tylenchida: Heteroderidae) from Peru guano. *Appl. Ent. Zool., Tokyo* **8,** 97–102.

INAGAKI, H. & POWELL, N. T. (1969). Influence of the root-lesion nematode on black shank symptom development in flue-cured tobacco. *Phytopathology* **59,** 1350–1355.

IRIZARRI, H., JENKINS, W. R. & CHILDERS, N. F. (1971). Interaction of soil temperature and *Meloidogyne* spp. on resistance of the common bean, *Phaseolus vulgaris* L., to the root-knot disease. *Nematropica* **1,** 41–42.

ISHIBASHI, N. (1970). Variations of the electrophoretic protein patterns of Heteroderidae (Nematodea: Tylenchida) depending on the developmental stages of the nematode and the growing conditions of the host plants. *Appl. Ent. Zool., Tokyo* **5,** 23–32.

ITOH, Y., OHSHIMA, Y. & ICHINOHE, M. (1969). A root-knot nematode, *Meloidogyne mali* n.sp. on apple-tree from Japan (Tylenchida: Heteroderidae). *Appl. Ent. Zool., Tokyo* **4,** 194–202.

IVANOVA, T. S. (1972). [Ectoparasitic nematodes in Tadzhikistan.] (In Russian). *Zashch. Rast. Vredit. Bolez.* **7,** 21.

IWAKI, M. & KOMURO, Y. (1971). [Viruses isolated from narcissus (*Narcissus* spp.) in Japan. II. Tomato ringspot virus and its transmission by *Xiphinema americanum.*] *Ann. phytopath. Soc. Japan* **37,** 108–116.

S'JACOB, J. J. (1960). Der Einfluss einiger Gewächse auf die Population von *Meloidogyne hapla. Nematologica Suppl.* II, 141–143.

S'JACOB, J. J. (1962). Beobachtungen an *Ditylenchus radicicola* (Greeff). *Nematologica* **7,** 231–234.

JAIRAJPURI, M. S. (1969). Studies on Mononchida of India. I. The genera *Hadronchus, Iotonchus* and *Miconchus* and a revised classification of Mononchida, new order. *Nematologica* **15,** 557–581.

JAIRAJPURI, M. S. (1970a). Studies on Mononchida of India. II. The genera *Mononchus, Clarkus* n.gen. and *Prionchulus* (Family Mononchidae Chitwood, 1937). *Nematologica* **16,** 213–221.

JAIRAJPURI, M. S. (1970b). Studies on Mononchida of India. III. The genus *Mylonchulus* (Family Mylonchulidae Jairajpuri, 1969). *Nematologica* **16,** 434–456.

JAIRAJPURI, M. S. (1971). Studies on Mononchida of India. IV. The genera *Sporonchulus, Bathyodontus* and *Oionchus* (Nematoda). *Nematologica* **17,** 407–412.

JAIRAJPURI, M. S. & SIDDIQI, M. R. (1969). *Paurodontoides* n.gen. (Paurodontidae) with an outline classification of Neotylenchoidea n.rank. *Nematologica* **15,** 287–288.

JAKOBSEN, J. (1973). New pests in 1972. In: Plantesygdomme i Danmark 1972. 89. *Årsoversigt samlet ved Statens plantepatologiske Forsøg, Lyngby* 53 pp. (see p. 52).

JAMES, G. L. (1966). Effect of the causal fungus of brown root rot of tomatoes on the hatch of the potato root eelworm, *Heterodera rostochiensis* Woll. *Nature, Lond.* **212,** 1466.

JAMES, G. L. (1968). The interrelationships of the causal fungus of brown root rot of tomatoes and potato root eelworm, *Heterodera rostochiensis* Woll. *Ann. appl. Biol.* **61,** 503–510.

JARNEVIC, N. B. & COFFEE, E. G. (1965). Control of root knot nematodes with organophosphate insecticides. *Pl. Dis. Reptr* **49,** 603–604.

JEFFREY, C. (1973). *Biological nomenclature.* London, Arnold, 69 pp.

JENKINS, W. R. & COURSEN, B. W. (1957). The effect of root-knot nematodes, *Meloidogyne incognita acrita* and *M. hapla,* on Fusarium wilt of tomato. *Pl. Dis. Reptr* **41,** 182–186.

JENSEN, H. J. & ALLEN, T. C., Jr. (1964). Transmission of tobacco rattle virus by a stubby-root nematode, *Trichodorus allius. Pl. Dis. Reptr* **48,** 333–334.

JHA, A. & POSNETTE, A. F. (1959). Transmission of a virus to strawberry plants by a nematode (*Xiphinema* sp.). *Nature, Lond.* **184,** 962–963.

JOBERT, C. (1878). Sur une maladie du cafeier observée au Bresil. *C.r.hebd. Séanc. Acad. Sci., Paris* **87,** 941–943.

JOHNSON, A. W. & GILL, D. L. (1972). Control of the root-knot nematode, *Meloidogyne incognita,* on mimosa (*Albizzia julibrissin*) by chemical dips. *J. Nematol.* **4,** 68–69.

JOHNSON, A. W., RATCLIFFE, T. J. & FREEMAN, G. C. (1970). Control of *Meloidogyne incognita* on dogwood seedlings by chemical dips. *Pl. Dis. Reptr* **54,** 952–955.

JOHNSON, H. A. & POWELL, N. T. (1969). Influence of root knot nematodes on bacterial wilt development in flue-cured tobacco. *Phytopathology* **59,** 486–491.

JOHNSON, P. W., VAN GUNDY, S. D. & THOMSON, W. W. (1970a). Cuticle ultrastructure of *Hemicycliophora arenaria, Aphelenchus avenae, Hirschmanniella gracilis* and *Hirschmanniella belli. J. Nematol.* **2,** 42–58.

JOHNSON, P. W., VAN GUNDY, S. D. & THOMSON, W. W. (1970b). Cuticle formation in *Hemicycliophora arenaria, Aphelenchus avenae* and *Hirschmanniella gracilis. J. Nematol.* **2,** 59–79.

JOHNSON, R. N. & VIGLIERCHIO, D. R. (1961). The accumulation of plant parasitic nematode larvae around carbon dioxide and oxygen. *Proc. helminth. Soc. Wash.* **28,** 171–174.

JOHNSON, R. N. & VIGLIERCHIO, D. R. (1969). Sugar beet nematode (*Heterodera schachtii*) reared on axenic *Beta vulgaris* root explants. II. Selected environmental and nutritional factors affecting development and sex-ratio. *Nematologica* **15,** 144–152.

JONES, F. G. W. (1945). Soil populations of beet eelworm (*Heterodera schachtii* Schm.) in relation to cropping. *Ann. appl. Biol.* **32,** 351–380.

JONES, F. G. W. (1950). Observations on the beet eelworm and other cyst-forming species of *Heterodera. Ann. appl. Biol.* **37,** 407–440.

JONES, F. G. W. (1951a). Further observations on the distribution of the beet eelworm. *J. Helminth.* **25,** 223–230.

JONES, F. G. W. (1951b). The Sugar Beet Eelworm Order 1943. *Ann. appl. Biol.* **38,** 535–537.

JONES, F. G. W. (1952). Differentiation of species of *Heterodera* (II). *Proc. Int. Nem. Symposium,* Harpenden, 1951. Rome, F.A.O., 52–57.

JONES, F. G. W. (1954). First steps in breeding for resistance to potato-root eelworm. *Ann. appl. Biol.* **41,** 348–353.

JONES, F. G. W. (1955). Quantitative methods in nematology. *Ann. appl. Biol.* **42,** 372–381.

JONES, F. G. W. (1956). Soil populations of beet eelworm (*Heterodera schachtii* Schm.) in relation to cropping. II. Microplot and field plot results. *Ann. appl. Biol.* **44,** 25–56.

JONES, F. G. W. (1957). Resistance-breaking biotypes of the potato root eelworm (*Heterodera rostochiensis* Woll.). *Nematologica* **2,** 185–192.

JONES, F. G. W. (1958). Resistance-breaking populations of potato root eelworm. *Pl. Path.* **7,** 24–25.

JONES, F. G. W. (1960). Some observations and reflections on host finding by plant nematodes. *Meded. LandbHogesch. OpzoekStns Gent* **25,** 1009–1024.

JONES, F. G. W. (1961). The potato root-eelworm *Heterodera rostochiensis* Woll. in India. *Curr. Sci.* **30,** 187.

JONES, F. G. W. (1962). A note on Granek's ratio for the separation of *Heterodera rostochiensis* W. from *H. tabacum* L. & L. *Nematologica* **7,** 256–258.

JONES, F. G. W. (1965). Structure and classification of nematodes. In: J. F. Southey (Ed.), Plant nematology. *Tech. Bull. Minist. Agric. Fish. Fd* 7 (2nd Edn), 3–29.

JONES, F. G. W. (1966). The population dynamics and population genetics of the potato cyst-nematode *Heterodera rostochiensis* Woll. on susceptible and resistant potatoes. *Rep. Rothamsted exp. Stn for* 1965, 301–316.

JONES, F. G. W. (1969a). Some reflections on quarantine, distribution and control of plant nematodes. In: J. E. Peachey (Ed.), Nematodes of tropical crops. *Tech. Commun. Commonw. Bur. Helminth.* 40, 67–80.

JONES, F. G. W. (1969b). Integrated control of the potato cyst-nematode. *Proc. 5th Br. Insectic. Fungic. Conf.* Vol. 3, 646–656.

JONES, F. G. W. (1970). The control of the potato cyst-nematode. *Jl R. Soc. Arts* **118**, 179–199.

JONES, F. G. W. (1972a). Management of nematode populations in Great Britain. *Proc. ann. Tall Timbers Conf. on ecological Animal Control by Habitat Management* 1972, 81–107.

JONES, F. G. W. (1972b). Nematodes and cereals. *A.D.A.S. q. Rev.* No. 6, 20–36.

JONES, F. G. W. (1972c). Pathotypes in perspective. *Ann. appl. Biol.* **71**, 296–300.

JONES, F. G. W. (1974). Control of nematode pests, background and outlook for biological control. In: D. Price Jones & M. E. Solomon (Eds), *Biology in pest and disease control.* Oxford, London, Edinburgh, Melbourne; Blackwell Scientific Publications, 249–268.

JONES, F. G. W. (1975a). The soil as an environment for plant parasitic nematodes. *Ann. appl. Biol.* **79**, 113–139.

JONES, F. G. W. (1975b). Host parasite relationships of potato cyst-nematodes: a speculation arising from the gene for gene hypothesis. *Nematologica* **20** (Yr 1974), 437–443.

JONES, F. G. W. (1975c). Accumulated temperature and rainfall as measures of nematode development and activity. *Nematologica* **21**, 62–70.

JONES, F. G. W. (1976). Nematology department. *Rep. Rothamsted exp. Stn for* 1975 Part 1, 191–212.

JONES, F. G. W. (1977). Pests, resistance and fertilizers. *Proc. 12th Colloquium Int. Potash Inst. Izmir*, 1976, 233–258.

JONES, F. G. W. & DUNNING, R. A. (1972). Sugar beet pests. *Bull. Minist. Agric. Fish. Fd* 162 (3rd Edn), 114 pp. (H.M.S.O.).

JONES, F. G. W. & JONES, M. G. (1974). *Pests of field crops.* (2nd Edn). London, Edward Arnold, 448 pp.

JONES, F. G. W., LARBEY, D. W. & PARROTT, D. M. (1969). The influence of soil structure and moisture on nematodes, especially *Xiphinema, Longidorus, Trichodorus* and *Heterodera* spp. *Soil Biol. Biochem.* **1**, 153–165.

JONES, F. G. W. & MORIARTY, F. (1956). Further observations on the effects of peas, beans and vetch upon soil population levels of pea root eelworm, *Heterodera göttingiana* Liebscher. *Nematologica* **1**, 268–273.

JONES, F. G. W. & PARROTT, D. M. (1965). The genetic relationships of pathotypes of *Heterodera rostochiensis* Woll. which reproduce on hybrid potatoes with genes for resistance. *Ann. appl. Biol.* **56**, 27–36.

JONES, F. G. W. & PARROTT, D. M. (1968). Potato production using resistant varieties on land infested with potato cyst-eelworm, *Heterodera rostochiensis* Woll. *Outlook on Agriculture* **5**, 215–222.

JONES, F. G. W. & PARROTT, D. M. (1969). Population fluctuations of *Heterodera rostochiensis* Woll. when susceptible potato varieties are grown continuously. *Ann. appl. Biol.* **63,** 175–181.
JONES, F. G. W., PARROTT, D. M. & ROSS, G. J. S. (1967). The population genetics of the potato cyst-nematode, *Heterodera rostochiensis*: mathematical models to simulate the effects of growing eelworm-resistant potatoes bred from *Solanum tuberosum* ssp. *andigena*. *Ann. appl. Biol.* **60,** 151–171.
JONES, F. G. W. & THOMASSON, A. J. (1976). Bulk density as an indicator of pore-space in soils usable by nematodes. *Nematologica* **22,** 133–137.
JONES, J. M., GRIFFITHS, D. J. & HOLDEN, J. W. H. (1955). Varietal resistance in oats to attacks by the stem and bulb eelworm. *Pl. Path.* **4,** 35–43.
JONES, M. G. K. & DROPKIN, V. H. (1975). Scanning electron microscopy of syncytial transfer cells induced in roots by cyst-nematodes. *Physiol. Pl. Path.* **7,** 259–263.
JONES, M. G. K. & NORTHCOTE, D. H. (1972a). Nematode-induced syncytium—a multinucleate transfer cell. *J. Cell Sci.* **10,** 789–809.
JONES, M. G. K. & NORTHCOTE, D. H. (1972b). Multinucleate transfer cells induced in coleus roots by the root-knot nematode, *Meloidogyne arenaria*. *Protoplasma* **75,** 381–395.
JORGENSON, E. C. (1970). Antagonistic interaction of *Heterodera schachtii* Schmidt and *Fusarium oxysporum* (Woll.) on sugarbeets. *J. Nematol.* **2,** 393–398.
KAAI, C. & WINDRICH, W. A. (1971). Bestrijding stengelaaltjes. *Jversl. Inst. plziektenk. Onderz.* 1970, 113.
KÄMPFE, L. (1962a). Vergleichende Untersuchungen zur Autökologie von *Heterodera rostochiensis* Wollenweber und *Heterodera schachtii* Schmidt sowie einiger anderer Nematodenarten abweichender Lebensstätten. *Parasit. SchrReihe* No. 14, 205 pp. (Jena, Fischer).
KÄMPFE, L. (1962b). Zur Wirkung von Ultraschall auf cystenbildende und freilebende Nematoden. *Nematologica* **7,** 164–172.
KANTZES, J. G., JENKINS, W. R. & DAVIS, R. A. (1959). Control of root-knot nematodes on vegetables on the eastern shore of Maryland with 1,2–dibromo–3–chloropropane-fertilizer mixtures. *Pl. Dis. Reptr* **43,** 1231–1235.
KANTZES, J. G. & MORGAN, O. D. (1962). Comparison of nematocides for control of root-knot nematodes on strawberries. *Phytopathology* **52,** 164 (Abstr.).
KEEN, B. A. (1931). *The physical properties of the soil.* London, Longmans Green, 380 pp.
KEHR, A. E. (1966). Current status and opportunities for the control of nematodes by plant breeding. In: Pest control by chemical, biological, genetic and physical means. *Agricultural Research Service, United States Department of Agriculture. ARS* 33–110, pp. 126–133.
KEMPER, W. D. (1960). Water and ion movement in thin films as influenced by the electrostatic charge and diffuse layer of cations associated with clay mineral surfaces. *Proc. Soil Sci. Soc. Am.* **25,** 10–16.
KEMPTON, R. J. & MAW, G. A. (1972). Soil fumigation with methyl bromide: bromide accumulation by lettuce plants. *Ann. appl. Biol.* **72,** 71–79.
KENNETH, J. H. (1963). *Henderson's dictionary of biological terms.* (8th Edn). Edinburgh & London, Oliver & Boyd.
KERRY, B. R. (1974). A fungus associated with young females of the cereal cyst-nematode, *Heterodera avenae*. *Nematologica* **20,** 259–260.
KERRY, B. R. (1975). Fungi and the decrease of cereal cyst-nematode populations in cereal monoculture. *EPPO Bull.* **5,** 353–361.

KERSTAN, U. (1969). Die Beeinflussung des Geschlechterverhältnisses in der Gattung *Heterodera*. II. Minimallebensraum—selektive Absterberate der Geschlechter-Geschlechterverhältnis (*Heterodera schachtii*). *Nematologica* **15**, 210–228.

KEYWORTH, W. G. & DAVIES, D. L. G. (1946). Nettlehead disease of the hop (*Humulus lupulus*). *J. Pomol.* **22**, 134–139.

KEYWORTH, W. G. & HITCHCOCK, M. M. (1948). Aerial surveys of the incidence of nettlehead disease of the hop on former hedgerow and pasture sites. *Rep. E. Malling Res. Stn for* 1947, 153–156.

KHOURY, F. Y. & ALCORN, S. M. (1973). Effect of *Meloidogyne incognita acrita* on the susceptibility of cotton plants to *Verticillium albo-atrum*. *Phytopathology* **63**, 485–490.

KHURANA, S. M. P., GOSWAMI, B. K. & RAYCHAUDHURI, S. P. (1970). Interaction of maize mosaic with root-knot nematode *Meloidogyne incognita* (Kofoid & White) Chitwood in maize (*Zea mays* L.). *Phytopath. Z.* **69**, 267–272.

KIRBY, M. F., DICKSON, D. W. & SMART, G. C., Jr. (1975). Physiological variation within species of *Meloidogyne* occurring in Florida. *Pl. Dis. Reptr* **59**, 353–356.

KIRKPATRICK, J. D., MAI, W. F., FISHER, E. G. & PARKER, K. G. (1959a). Population levels of *Pratylenchus penetrans* and *Xiphinema americanum* in relation to potassium fertilization of Montmorency sour cherries on Mazzard root stock. *Phytopathology* **49**, 543 (Abstr.).

KIRKPATRICK, J. D., MAI, W. F., FISHER, E. G. & PARKER, K. G. (1959b). Relation of nematode populations to nutrition of sour cherries. *Phytopathology* **49**, 543 (Abstr.).

KIR'YANOVA, E. S. (1955). [Quackgrass nematode—*Paranguina agropyri* Kiryanova n.g., n.sp. (Nematoda).] (In Russian). *Trudy zool. Inst. Akad. Nauk SSSR* **18**, 42–52.

KIR'YANOVA, E. S. (1959). [On the systematics of the nematodes belonging to the genus *Heterodera* Schmidt, 1871.] (In Russian). *Zool. Zh.* **38**, 1620–1626.

KIR'YANOVA, E. S. (1970). The structure of the sub-crystalline layer in cyst-forming nematodes of the genus *Heterodera* (Nematoda: Heteroderidae). *Zesz. probl. Postęp Nauk roln.* No. 92, 63–68.

KIR'YANOVA, E. S. & IVANOVA, T. S. (1968). [New species of the genus *Paranguina* Kirjanova, 1955 (Nematoda: Tylenchidae) from Tadzhikistan.] (In Russian). In: M. N. Narzikulov (Ed.), *Ushchel'e Kondara*, Vol. 2, Dushanbe, Akad. Nauk Tadzhik. SSR, 200–217.

KIR'YANOVA, E. S. & KRALL, E. L. (1969). [*Parasitic nematodes of plants and methods for controlling them.*] (In Russian). Vol. 1. Leningrad, Izdatel'stvo 'Nauka', 447 pp.

KLINGLER, J. & LENGWEILER-REY, V. (1969). Die Wärmeempfindlichkeit von aktiven und trockenstarren *Ditylenchus myceliophagus, D. dipsaci* und *Anguina tritici* bei verschiedenen Feuchtigkeitsbedingungen. *Z. PflKrankh. PflPath. PflSchutz* **76**, 193–208.

KLINK, J. W. & BARKER, K. R. (1968). Effect of *Aphelenchus avenae* on the survival and pathogenic activity of root-rotting fungi. *Phytopathology* **58**, 228–232.

KLINKENBERG, C. H. & SEINHORST, J. W. (1956). De nematicide werking van Na N-methyl dithiocarbaminaat (Vapam) bij toepassing in de herfst. *Meded. LandbHogesch. OpzoekStns Gent* **21**, 397–400.

KOCHBA, J. & SAMISH, R. M. (1971). Effect of kinetin and 1–napthylacetic acid on root-knot nematodes in resistant and susceptible peach rootstocks. *J. Am. Soc. hort. Sci.* **96**, 458–461.

KOLIOPANOS, C. N. & TRIANTAPHYLLOU, A. C. (1972). Effect of infection density on sex ratio of *Heterodera glycines*. *Nematologica* **18**, 131–137.

KOMURO, Y., YOSHINO, M. & ICHINOHE, M. (1970). [Tobacco rattle virus isolated from aster showing ringspot syndrome and its transmission by *Trichodorus minor* Colbran.] (In Japanese). *Ann. phytopath. Soc. Japan* **36**, 17–26.

KORT, J. (1962). Effect of population density on cyst production in *Heterodera rostochiensis* Woll. *Nematologica* **7**, 305–308.

KORT, J. (1972). Nematode diseases of cereals of temperate climates. In: J. M. Webster (Ed.), *Economic nematology*. London & New York, Academic Press, 97–126.

KORT, J., DANTUMA, G. & VAN ESSEN, A. (1964). On biotypes of the cereal root eelworm (*Heterodera avenae*) and resistance in oats and barley. *Neth. J. Pl. Path.* **70**, 9–17.

KORT, J. & JASPERS, C. P. (1974). Shift of pathotypes of *Heterodera rostochiensis* under susceptible potato cultivars. *Nematologica* **19** (Yr 1973), 538–545.

KORT, J., JASPERS, C. P. & DIJKSTRA, D. L. (1972). Testing for resistance to pathotype C of *Heterodera rostochiensis* and the practical application of *Solanum vernei*-hybrids in The Netherlands. *Ann. appl. Biol.* **71**, 289–294.

KORT, J., ROSS, H., RUMPENHORST, H. J. & STONE, A. R. (1977). An international scheme for identifying and classifying pathotypes of potato cyst-nematodes. *Globodera rostochiensis* and *G. pallida*. *Nematologica* **23**, 333–339.

KOSTINA, K. & ZHOLUDEVA, Z. (1974). [Promising potato varieties.] (In Russian). *Kartofel' i Ovoshchi* No. 3, 20–21.

KRALL, E. (1970). [Study of foliar plant nematodes of the genus *Anguina* in the Estonian S.S.R.] (In Russian). In: *Problemy Parazitologii v Pribaltike*. Riga, Izdatel'stvo 'Zinatne', 79–81.

KRALL, E. & KRALL, H. (1970a). [Evolution of the mutual parasitic relations between phytohelminths of the family Heteroderidae and their host plants.] (In Russian). *Sb. nauch. Trud. estonsk. sel'skokhoz. Akad.* **70**, 152–154.

KRALL, E. & KRALL, H. (1970b). [A new biological race of *Anguina radicicola* parasitizing fodder grasses in Estonia.] (In Russian). In: *Materialy 7go Pribaltiyskogo Soveshchaniya po Zashchite Rasteniy, Part I*. Elgava, U.S.S.R.; Ministerstvo Sel'skogo Khozyaystva Latviyskoi SSR, pp. 57–60.

KRNJAIĆ, D. (1973). Inokulaciona sposobnost larvi žitne nematode (*Anguina tritici* Steinbuch 1799, Filipjev, 1936) ozračenih gama zracima (Co^{60}). *Zaštita Bilja* **24**, 61–65.

KRUSBERG, L. R. (1963). Host response to nematode infection. *A. Rev. Phytopath.* **1**, 219–240.

KRUSBERG, L. R. (1971). Control of *Pratylenchus penetrans* in heavy greenhouse soil. *Pl. Dis. Reptr* **55**, 276–279.

KÜHN, J. (1891). Neuere Versuche zur Bekämpfung der Rübennematoden. *Z. PflKrankh.* **1**, 85–86.

KUIPER, K. (1960). Resistance of white clover varieties to the clover cyst-eelworm, *Heterodera trifolii* Goffart. *Nematologica Suppl.* II, 95–96.

KÜTHE, K. & DERN, R. (1970). Erfahrungen bei der Untersuchung von *Ditylenchus*-Befall an Mais (*Zea mays*) in Hessen. *Gesunde Pfl.* **22**, 101–104.

LAAN, P. A. VAN DER (1956). The influence of organic manuring on the development of the potato root eelworm, *Heterodera rostochiensis*. *Nematologica* **1**, 112–125.

Labruyère, R. E., den Ouden, H. & Seinhorst, J. W. (1959). Experiments on the interaction of *Hoplolaimus uniformis* and *Fusarium oxysporum* f. *pisi* race 3 and its importance in 'early yellowing' of peas. *Nematologica* **4**, 336–343.

Ladygina, N. M. (1976). [The genetic and physiological compatibility of various forms of stem nematode.] (In Russian). *Parazitologiya* **10**, 40–47.

Lamberti, F. (1975). Taxonomy of *Longidorus* (Micoletzky) Filipjev and *Paralongidorus* Siddiqi, Hooper and Khan. In: F. Lamberti, C. E. Taylor & J. W. Seinhorst (Eds), *Nematode vectors of plant viruses*. London & New York, Plenum Press, 71–90.

Lamberti, F. & Martelli, G. P. (1971). Notes on *Xiphinema mediterraneum* (Nematoda: Longidoridae). *Nematologica* **17**, 75–81.

Lamberti, F., Roca, F. & Bleve-Zacheo, T. (1972). Geographical distribution of species of *Xiphinema* Cobb in some Italian regions. *XIth International Symposium of Nematology, Reading*, 1972. *Abstracts* p. 40.

Lamberti, F., Taylor, C. E. & Seinhorst, J. W. (Eds) (1975). *Nematode vectors of plant viruses*. London & New York, Plenum Press, 460 pp.

Lapp, N. A. & Triantaphyllou, A. C. (1969). Relative DNA content of some members of Heteroderidae. *J. Nematol.* **1**, 296.

Latta, R. (1932). The vapor-heat treatment as applied to the control of narcissus pests. *J. econ. Ent.* **25**, 1020–1026.

Laughlin, C. W., Lugthart, J. A. & Vargas, J. M., Jr. (1974). Observations on the emergence of *Heterodera iri* from the egg. *J. Nematol.* **6**, 100–101.

Lear, B. & Jacob, F. C. (1955). Results of laboratory experiments with high-voltage, non-thermal electrical treatments for control of root-knot nematodes. *Pl. Dis. Reptr* **39**, 397–399.

Lear, B. & Raski, D. J. (1956). Survival of root-knot nematodes in grape and tomato roots recovered from soils fumigated with Nemagon. *Phytopathology* **46**, 18.

Lear, B. & Raski, D. J. (1962). Survival of root-knot nematodes in excised grape roots in moist soil fumigated with ethylene dibromide. *Phytopathology* **52**, 1309–1310.

Lee, D. L. & Atkinson, H. J. (1976). *The physiology of nematodes*. London, Macmillan, 215 pp.

Leistra, M. (1972). Diffusion and adsorption of the nematicide 1,3–dichloropropene in soil. *Doctoral thesis, Wageningen*, ISBN 90/220 0379/5, 105 pp. Wageningen, Centre for Agricultural Publishing and Documentation.

Lewis, G. D. & Mai, W. F. (1958). Chemical control of *Ditylenchus dipsaci* (Kühn) Filipjev in organic soils of southern New York. *Pl. Dis. Reptr* **42**, 1360–1363.

Lewis, G. D. & Mai, W. F. (1960). Overwintering and migration of *Ditylenchus dipsaci* in organic soils of southern New York. *Phytopathology* **50**, 341–343.

Lewis, G. D. & Mai, W. F. (1963). Temperature and moisture in relation to the toxicity of 1,3–dichloropropene to *Ditylenchus dipsaci* (Kühn) Filipjev in organic soil. *Pl. Dis. Reptr* **47**, 1097–1101.

Lewis, S. & Webley, D. (1966). Observations on two nematodes infesting grasses. *Pl. Path.* **15**, 184–186.

Libman, G., Leach, J. G. & Adams, R. E. (1964). Role of certain plant-parasitic nematodes in infection of tomatoes by *Pseudomonas solanacearum*. *Phytopathology* **54**, 151–153.

Licopoli, G. (1875). Sopra alcuni tubercoli radicellari continente anguillole. *Rc. Accad. sci. fiz. Napoli* **14**, 41–42.

LIMBER, D. P. (1973). Notes on the longevity of *Anguina tritici* (Steinbuch, 1799) Filipjev, 1936, and its ability to invade wheat seedlings after thirty-two years of dormancy. *Proc. helminth. Soc. Wash.* **40**, 272–274.

VAN DER LINDE, W. J. (1956). The *Meloidogyne* problem in South Africa. *Nematologica* **1**, 177–183.

LINFORD, M. B. (1937a). Notes on the feeding of *Ditylenchus dipsaci* (Nematoda: Tylenchidae). *Proc. helminth. Soc. Wash.* **4**, 46–47.

LINFORD, M. B. (1937b). Stimulated activity of natural enemies of nematodes. *Science* **85**, 123–124.

LINFORD, M. B. & YAP, F. (1939). Root-knot nematode injury restricted by a fungus. *Phytopathology* **29**, 596–609.

LISTER, R. M. (1964). Strawberry latent ringspot: a new nematode-borne virus. *Ann. appl. Biol.* **54**, 167–176.

LOEWENBERG, J. R., SULLIVAN, T. & SCHUSTER, M. L. (1959). A virus disease of *Meloidogyne incognita incognita*, the southern root knot nematode. *Nature, Lond.* **184**, 1896.

LOOF, P. A. A. (1961). The nematode collection of Dr. J. G. de Man. *Beaufortia* **8**, 169–254.

LOOF, P. A. A. (1965). *Trichodorus anemones* n.sp. with a note on *T. teres* Hooper, 1962 (Nematoda: Enoplida). *Versl. Meded. Plziektenk. Dienst Wageningen* (*Jaarboek*, 1964) 142, 132–136.

LOOF, P. A. A. (1968). Taxonomy of *Hemicycliophora* species from west and central Europe (Nematoda: Criconematoidea). *Meded. LandbHogesch. Wageningen* 68–14, 43 pp.

LOOF, P. A. A. (1974a). *Tylenchorhynchus claytoni*. *CIH Descriptions of Plant-parasitic Nematodes* Set 3, No. 39. 2 pp.

LOOF, P. A. A. (1974b). *Macroposthonia curvata*. *CIH Descriptions of Plant-parasitic Nematodes* Set 4, No. 58, 3 pp.

LOOF, P. A. A. (1975a). Taxonomy of Trichodoridae. In: F. Lamberti, C. E. Taylor & J. W. Seinhorst (Eds), *Nematode vectors of plant viruses*. London & New York, Plenum Press, 103–127.

LOOF, P. A. A. (1975b). *Paratylenchus projectus*. *CIH Descriptions of Plant-parasitic Nematodes* Set 5, No. 71. 3 pp.

LOOF, P. A. A. & COOMANS, A. (1972). The oesophageal gland nuclei of Longidoridae (Dorylaimida). *Nematologica* **18**, 213–233.

LOOF, P. A. A. & DE GRISSE, A. (1973). Interrelationships of the genera of Criconematidae (Nematoda: Tylenchida). *Meded. Fak. LandbWetensch. Gent* **38**, 1303–1328.

LOOF, P. A. A. & JAIRAJPURI, M. S. (1968). Taxonomic studies on the genus *Tylencholaimus* de Man, 1876 (Dorylaimoidea) with a key to the species. *Nematologica* **14**, 317–350.

LOOF, P. A. A. & MAAS, P. W. TH. (1972). The genus *Xiphinema* (Dorylaimida) in Surinam. *Nematologica* **18**, 92–119.

LÓPEZ-ABELLA, D., JIMÉNEZ-MILLÁN, F. & GARCIA-HIDALGO, F. (1967). Electron microscope studies of some cephalic structures of *Xiphinema americanum*. *Nematologica* **13**, 283–286.

LORDELLO, L. G. E. (1972). Nematode pests of coffee. In: J. M. Webster (Ed.), *Economic nematology*. London & New York, Academic Press, 268–284.

LOWNSBERY, B. F. & VIGLIERCHIO, D. R. (1961). Importance of response of *Meloidogyne hapla* to an agent from germinating tomato seeds. *Phytopathology* **51**, 219–221.

LUC, M. (1961). Structure de la gonade femelle chez quelques espèces du genre *Xiphinema* Cobb, 1913 (Nematoda—Dorylaimoidea). *Nematologica* **6**, 144–154.

LUC, M. (1970). Contribution à l'étude du genre *Criconemoides* Taylor, 1936 (Nematoda: Criconematidae). *Cah. ORSTOM, Sér. Biol.* **11**, 69–131.

LUC, M. & DALMASSO, A. (1963). Trois nouveaux *Xiphinema* associés à la vigne (Nematoda—Dorylaimidae). *Nematologica* **9**, 531–541.

LUC, M. & DALMASSO, A. (1975). Considerations on the genus *Xiphinema* Cobb, 1913 (Nematoda: Longidoridae) and a 'lattice' for the identification of species. *Cah. ORSTOM, Sér. Biol.* **10**, 303–327.

LUC, M., GERMANI, G. & NETSCHER, C. (1973). Description de *Sarisodera africana* n.sp. et considérations sur les relations entre les genres *Sarisodera* Wouts & Sher, 1971 et *Heterodera* A. Schmidt, 1871 (Nematoda: Tylenchida). *Cah. ORSTOM, Sér. Biol.* No. 21, 35–43.

LUC, M., LIMA, M. B., WEISCHER, B. & FLEGG, J. J. M. (1964). *Xiphinema vuittenezi* n.sp. (Nematoda: Dorylaimidae). *Nematologica* **10**, 151–163.

LUCAS, G. B. & KRUSBERG, L. R. (1956). The relationship of the stunt nematode to Granville wilt resistance in tobacco. *Pl. Dis. Reptr* **40**, 150–152.

LUCAS, G. B., SASSER, J. N. & KELMAN, A. (1955). The relationship of root-knot nematodes to Granville wilt resistance in tobacco. *Phytopathology* **45**, 537–540.

LÜCKE, E. (1969). Untersuchungen zum Hafernematodenproblem. *Mitt. biol. BundAnst. Ld- u. Forstw.* **136**, 41–49.

MABBOTT, T. W. (1956). Potato root eelworm. A report on an experiment to free seed potatoes from adhering soil and cysts. *Scot. Agric.* **36**, 73–74.

MABBOTT, T. W. (1960). Observations on the development of potato root eelworm, *Heterodera rostochiensis* Woll., on the potato tuber and the importance of such development in the spread of this nematode on washed tubers. *Eur. Potato J.* **3**, 236–244.

MABBOTT, T. W. (1968). Serological techniques as an aid to potato-root eelworm strain identification. *Reports 8th International Symposium of Nematology, Antibes*, 1965, p. 46.

MABBOTT, T. W. (1974). Distribution of potato cyst eelworm in Scotland. *EPPO Bull.* **4**, 481–493.

MACKINTOSH, G. M. (1960). The morphology of the brassica root eelworm *Heterodera cruciferae* Franklin, 1945. *Nematologica* **5**, 158–165.

MAGGENTI, A. R. (1961a). Morphology and biology of the genus *Plectus* (Nematoda: Plectidae). *Proc. helminth. Soc. Wash.* **28**, 118–130.

MAGGENTI, A. R. (1961b). Revision of the genus *Plectus* (Nematoda: Plectidae). *Proc. helminth. Soc. Wash.* **28**, 139–166.

MAGGENTI, A. R. (1962). The production of the gelatinous matrix and its taxonomic significance in *Tylenchulus* (Nematoda: Tylenchulinae). *Proc. helminth. Soc. Wash.* **29**, 139–144.

MAGGENTI, A. R. (1971). Nemic relationships and the origins of plant parasitic nematodes. In: B. M. Zuckerman, W. F. Mai & R. A. Rohde (Eds), *Plant parasitic nematodes* I. London & New York, Academic Press, 65–81.

MAGGENTI, A. R. (1976). Taxonomic position of Nematoda among the pseudocoelomate bilateria. In: N. A. Croll (Ed.), *The organization of nematodes*. London & New York, Academic Press, 428 pp.

MAGGENTI, A. R. & ALLEN, M. W. (1960). The origin of the gelatinous matrix in *Meloidogyne*. *Proc. helminth Soc. Wash.* **27**, 4–10.

MAGGENTI, A. R. & HART, W. H. (1975a). Carbamate and phosphate nematicidal granules, drenches and dips for the control of *Ditylenchus destructor* on bulbous iris, variety Wedgewood. *Pl. Dis. Reptr* **59**, 233–235.

MAGGENTI, A. R. & HART, W. H. (1975b). Postplanting control of *Ditylenchus dipsaci* in narcissus with granular organic phosphates and carbamates. *Pl. Dis. Reptr* **59**, 281–282.

MAGGENTI, A. R., HART, W. H. & PAXMAN, G. A. (1974). A new genus and species of gall forming nematode from *Danthonia californica*, with a discussion of its life history. *Nematologica* **19** (Yr 1973), 491–497.

MAI, W. F. (1958). Effectiveness of di-electric heat in killing encysted golden nematode larvae. *Pl. Dis. Reptr* **42**, 449–450.

MALEK, R. B. (1974). Control of *Meloidogyne hapla* on peony by chemical bare-root dip. *Pl. Dis. Reptr* **58**, 997–999.

MALO, S. E. (1964). A review of plant breeding for nematode resistance. *Proc. Soil Crop Sci. Soc. Fla* **24**, 345–365.

MANKAU, R. (1961). The use of nematode-trapping fungi to control root-knot nematodes. *Nematologica* **6**, 326–332.

MANKAU, R. (1962). Soil fungistasis and nematophagus fungi. *Phytopathology* **52**, 611–615.

MANKAU, R. (1975). *Bacillus penetrans* n.comb. causing a virulent disease of plant-parasitic nematodes. *J. Invert. Path.* **26**, 333–339.

MANKAU, R. & LINFORD, M. B. (1960). Host-parasite relationships of clover cyst nematode, *Heterodera trifolii* Goffart. *Bull. Ill. agric. Exp. Stn* 667, 50 pp.

MANKAU, R. & MANKAU, S. K. (1963). The role of mycophagous nematodes in the soil. I. The relationships of *Aphelenchus avenae* to phytopathogenic fungi. In: J. Doeksen & J. van der Drift (Eds). *Soil organisms. Proceedings of the colloquium on soil fauna, soil microflora and their relationships. Oosterbeek, The Netherlands*, Sept. 10–16, 1962, 271–280.

MARCINOWSKI, [K.] (1909). Untersuchungen über Nematoden. *Mitt. biol. ZentAnst. Berl.* No. 8, 41–46.

MARIËN, A. (1956). Bestrijdingsproef op het schorsenerenaaltje. *Tuinbouwberichten* **20**, 89–91.

MARSHALL, G. M. (1960). The incidence of certain seed-borne diseases in commercial seed samples. IV. Bunt of wheat, *Tilletia caries* (DC.) Tul. V. Earcockles of wheat, *Anguina tritici* (Stein.) Filipjev. *Ann. appl. Biol.* **48**, 34–38.

MARTELLI, G. P. & LAMBERTI, F. (1967). Le specie di *Xiphinema* Cobb, 1913 trovate in Italia e commenti sulla presenza di *Xiphinema americanum* Cobb (Nematoda, Dorylaimoidea). *Phytopath. medit.* **6**, 65–85.

MARTIN, G. C. (1947). 'DD' as a means of controlling *Heterodera rostochiensis* (Woll.). *Nature, Lond.* **160**, 720.

MARTIN, W. J. (1962). Elimination of root-knot nematodes from infested sweet-potato roots and plants. *Pl. Dis. Reptr* **46**, 21–23.

MARTIN, W. J., NEWSOM, L. D. & JONES, J. E. (1956). Relationship of nematodes to the development of Fusarium wilt in cotton. *Phytopathology* **46**, 285–289.

MASSEY, C. L. (1974). Biology and taxonomy of nematode parasites and associates of bark beetles in the United States. *Agric. Handb. U.S.D.A.* No. 446, 233 pp.

MATHUR, R. S. & MISRA, M. P. (1961). Simultaneous occurrence of *Tilletia foetida* (Wallr.) Liro and *Anguillulina tritici* (S.) G. Ben. in the same ear and grains of wheat in Pauri-Garhwal, Uttar Pradesh. *Curr. Sci.* **30**, 307.

MAYOL, P. S. & BERGESON, G. B. (1970). The role of secondary invaders in *Meloidogyne incognita* infection. *J. Nematol.* **2**, 80–83.

MAYR, E. (1969). *Principles of systematic zoology*. New York, McGraw Hill, 428 pp.

MCBETH, C. W. & TAYLOR, A. L. (1944). Immune and resistant cover crops valuable in root-knot infested peach orchards. *Proc. Am. Soc. hort. Sci.* **45**, 158–166.

MCBETH, C. W., WHITE, L. V. & ICHIKAWA, S. T. (1964). A new residual nematocide: 2,4-dichlorophenyl methanesulphonate. *Pl. Dis. Reptr* **48**, 634–635.

MCCLURE, M. A., MISAGHI, I. & NIGH, E. L., Jr. (1973). Shared antigens of parasitic nematodes and host plants. *Nature, Lond.* **244**, 306.

MCCUBBIN, W. A. (1954). The plant quarantine problem. *Ann. cryptog. phytop., Waltham, Mass.* **11** (Copenhagen, Ejnar Munksgaard), 255 pp.

MCELROY, F. D. (1970). *Xiphinema bakeri*, a nematode pest of raspberry. (Abstr.). *J. Parasit.* **56** (4, Sect. 2) 448.

MCGUIRE, J. M., KIM, K. S. & DOUTHIT, L. B. (1970). Tobacco ringspot virus in the nematode *Xiphinema americanum*. *Virology* **42**, 212–216.

MCKEEN, C. D. & MOUNTAIN, W. B. (1960). Synergism between *Pratylenchus penetrans* (Cobb) and *Verticillium albo-atrum* R. & B. in eggplant wilt. *Can. J. Bot.* **38**, 789–794.

MCKEEN, C. D. & SAYRE, R. M. (1964). Control of *Verticillium dahliae* and *Meloidogyne incognita* in greenhouse tomatoes by soil fumigation. *Can. J. Pl. Sci.* **44**, 466–470.

MCLEOD, K., GOLIGHTLY, W. H. & PRICE, C. D. (1954). 'Tulip root' of oats in Durham. *Agriculture, Lond.* **61**, 229–231.

MCLEOD, R. W. (1971). Graminaceous hosts of root knot nematodes from New South Wales. *J. Aust. Inst. agric. Sci.* **37**, 240–241.

MCLEOD, R. W. (1973). Suppression of *Aphelenchoides composticola* and *Ditylenchus myceliophagus* on *Agaricus bisporus* by thiabendazole and benomyl. *Nematologica* **19**, 236–241.

MEAGHER, J. W. (1970). Seasonal fluctuations in numbers of larvae of the cereal cyst nematode (*Heterodera avenae*) and of *Pratylenchus minyus* and *Tylenchorhynchus brevidens* in soil. *Nematologica* **16**, 333–347.

MEAGHER, J. W. (1975). Cryptobiosis of the cereal cyst nematode (*Heterodera avenae*), and effects of temperature and relative humidity on survival of eggs in storage. *Nematologica* **20** (Yr 1974), 323–336.

MEIJNEKE, C. A. R. & OOSTENBRINK, M. (1958). *Tagetes* ter bestrijding van aaltjesaantastingen. *Meded. Dir. Tuinb.* **21**, 283–290.

MEIJNEKE, C. A. R. & SCHEELE, J. (1970). Plantuitjes stengelaaltjes en warmwaterbehandelingen. *Zaadbelangen* **24**, 248–250. [*Hort. Abstr.* **42**, 1365].

MELENDÉZ, P. L. & POWELL, N. T. (1967). Histological aspects of the Fusarium wilt - root knot complex in flue-cured tobacco. *Phytopathology* **57**, 286–292.

MELENDÉZ, P. L. & POWELL, N. T. (1969). The influence of *Meloidogyne* on root decay in tobacco caused by *Pythium* and *Trichoderma*. *Phytopathology* **59**, 1348. (Abstr.)

MELENDÉZ, P. L. & POWELL, N. T. (1970). The *Pythium* - root knot nematode complex in flue-cured tobacco. *Phytopathology* **60**, 1303. (Abstr.)

METCALFE, G. (1940). *Bacterium rhaptonticum* (Millard) Dowson, a cause of crown-rot disease of rhubarb. *Ann. appl. Biol.* **27**, 502–508.

METLITZKY, O. Z. (1969). [The influence of different host-plants on size and proportions of the body of *Ditylenchus dipsaci*]. (In Russian). *Parazitologiya* **3,** 266–272.

MEYL, A. H. (1961). Die freilebenden Erd- und Süsswassernematoden (Fadenwurmer). In: *Die Tierwelt Mitteleuropas*. Leipzig, Quelle & Meyer, 164 pp.

MICHELL, R. E., MALEK, R. B., TAYLOR, D. P. & EDWARDS, D. I. (1973a). Races of barley root-knot nematode, *Meloidogyne naasi*. I. Characterization by host preference. *J. Nematol.* **5,** 41–44.

MICHELL, R. E., MALEK, R. B., TAYLOR, D. P. & EDWARDS, D. I. (1973b). Races of barley root-knot nematode, *Meloidogyne naasi*. II. Developmental rates. *J. Nematol.* **5,** 44–46.

MICHELL, R. E., MALEK, R. B., TAYLOR, D. P. & EDWARDS, D. I. (1973c). Races of barley root-knot nematode, *Meloidogyne naasi*. III. Reproduction and pathogenicity on creeping bentgrass. *J. Nematol.* **5,** 47–49.

MILLER, H. N. & PERRY, V. G. (1965). Elimination of nematodes from nursery plants by chemical bare-root dips. *Pl. Dis. Reptr* **49,** 51–53.

MILLER, L. I. (1966). Variation in development of two morphologically different isolates of *Heterodera glycines* obtained from the same field. *Phytopathology* **56,** 585. (Abstr.).

MILLER, L. I. (1969). Physiologic variation of six isolates of *Heterodera glycines*. *Phytopathology* **59,** 1558. (Abstr.).

MILLER, L. I. (1971). Physiologic variation within the Virginia-2 population of *Heterodera glycines*. *J. Nematol.* **3,** 318.

MILLER, L. I. & GRAY, B. J. (1968). Horsenettle cyst nematode, *Heterodera virginiae* n.sp., a parasite of solanaceous plants. *Nematologica* **14,** 535–543.

MILLER, L. I. & GRAY, B. J. (1972). *Heterodera solanacearum* n.sp., a parasite of solanaceous plants. *Nematologica* **18,** 404–413.

MILLER, P. M. & HAWKINS, A. (1969). Long term effects of preplant fumigation of potato fields. *Am. Potato J.* **46,** 387–397.

MILLER, P. M. & KRING, J. B. (1970). Reduction of nematode and insect damage to potatoes by band application of systemic insecticides and soil fumigation. *J. econ. Ent.* **63,** 186–189.

MILLIKAN, C. R. (1938). Eelworm (*Heterodera schachtii* Schmidt) disease of cereals. *J. Dep. Agric. Vict.* **36,** 452–468, 509–520.

MOJTAHEDI, H. & LOWNSBERY, B. F. (1975). Pathogenicity of *Criconemoides xenoplax* to prune and plum rootstocks. *J. Nematol.* **7,** 114–119.

MONOSON, H. L. (1971). Effect of nematode-trapping fungi, media, and temperature on the morphometrics of *Aphelenchus avenae*. *Nematologica* **17,** 219–224.

MOORE, J. F. (1971). Potato varieties susceptible to *Ditylenchus destructor*, the potato tuber nematode. *Ir. J. agric. Res.* **10,** 239–240.

MOORE, W. C. (1953). International trade in plants and the need for healthy planting material. *Rep. 13th int. hort. Congress,* 1952 Vol. 1, 552–558.

MORETON, B. D. & JOHN, M. E. (1962). Peak heat: is yours what you think it is? *M.G.A. Bull.* No. 147, 91–95.

MORGAN, D. O. (1925). Investigations on eelworm in potatoes in south Lincolnshire. *J. Helminth.* **3,** 185–192.

MORGAN, G. T. & MCALLAN, J. W. (1962). Hydrolytic enzymes in plant-parasitic nematodes. *Nematologica* **8,** 209–215.

MORIARTY, F. (1961). The effects of red beet and of *Hesperis matronalis* L. on a population of *Heterodera schachtii* Schm. *Nematologica* **6,** 214–221.

MORIARTY, F. (1963). A population experiment with *Heterodera goettingiana* Lieb. and a relatively inefficient host, broad bean. *Nematologica* **9**, 152–156.

MORSCHEL, J. R. (1971). *Introduction to plant quarantine*. Canberra, Australian Government Publishing Service, 71 pp.

MOSS, S. R., CRUMP, D. & WHITEHEAD, A. G. (1975). Control of potato cyst-nematodes, *Heterodera rostochiensis* and *H. pallida*, in sandy, peaty and silt loam soils by oximecarbamate and organophosphate nematicides. *Ann. appl. Biol.* **81**, 359–365.

MOUNTAIN, W. B. (1954). Studies of nematodes in relation to brown root rot of tobacco in Ontario. *Can. J. Bot.* **32**, 737–759.

MOUNTAIN, W. B. (1960). Acceptable standards of proof and approaches for evaluating plant-nematode relationships. In: J. N. Sasser & W. R. Jenkins (Eds), *Nematology*. Chapel Hill, Univ. N. Carolina Press, 422–425.

MOUNTAIN, W. B. (1961). Studies on the pathogenicity of *Pratylenchus*. In: *Recent advances in botany*. Toronto Univ. Press, Vol. 1, 414–417.

MOUNTAIN, W. B. (1965). Pathogenesis by soil nematodes. In: K. F. Baker & W. C. Snyder (Eds), *Ecology of Soil-borne plant pathogens*. Berkeley, Univ. Calif. Press, 285–301.

MOUNTAIN, W. B. & MCKEEN, C. D. (1962). Effect of *Verticillium dahliae* on the population of *Pratylenchus penetrans*. *Nematologica* **7**, 261–266.

MOUNTAIN, W. B. & MCKEEN, C. D. (1965). Effects of transplant injury and nematodes on incidence of Verticillium wilt of eggplant. *Can. J. Bot.* **43**, 619–624.

MOUNTAIN, W. B. & PATRICK, Z. A. (1959). The peach replant problem in Ontario. VII. The pathogenicity of *Pratylenchus penetrans* (Cobb, 1917) Filip. & Stek. 1941. *Can. J. Bot.* **37**, 459–470.

MUKHOPADHAYA, M. C., DALAL, M. R., SARAN, S. & KHARUB, S. S. (1973). Studies on the 'molya' disease of wheat and barley. *Indian J. Nematol.* **2** (Yr 1972), 11–20.

MÜLLER, G. (1958). Morphologische Untersuchungen zur Variabilität des Kartoffelnematoden *Heterodera rostochiensis* W. *Biol. Zbl.* **77**, 673–714.

MULVEY, R. H. (1959a). Investigations on the clover cyst nematode, *Heterodera trifolii* (Nematoda: Heteroderidae). *Nematologica* **4**, 147–156.

MULVEY, R. H. (1959b). Giant eggs of the clover cyst nematode, *Heterodera trifolii* Goffart, 1932. *Nature, Lond.* **184**, 1662–1663.

MULVEY, R. H. (1960a). The value of cone top and underbridge structures in the separation of some cyst-forming nematodes. In: J. N. Sasser & W. R. Jenkins (Eds), *Nematology*. Chapel Hill, Univ. N. Carolina Press, 212–215.

MULVEY, R. H. (1960b). Oögenesis in some species of *Heterodera* and *Meloidogyne* (Nematoda: Heteroderidae). In: J. N. Sasser & W. R. Jenkins (Eds), *Nematology*. Chapel Hill, Univ. N. Carolina Press, 323–330.

MULVEY, R. H. (1960c). Abnormalities in second-stage larvae of *Heterodera trifolii* Goffart, 1932 (Nematoda: Heteroderidae). *Can. J. Zool.* **38**, 777–779.

MULVEY, R. H. (1972). Identification of *Heterodera* cysts by terminal and cone top structures. *Can. J. Zool.* **50**, 1277–1292.

MULVEY, R. H. (1973). Morphology of the terminal areas of white females and cysts of the genus *Heterodera* (s.g. *Globodera*). *J. Nematol.* **5**, 303–311.

MULVEY, R. H. & STONE, A. R. (1976). Description of *Punctodera matadorensis* n.gen., n.sp. (Nematoda: Heteroderidae) from Saskatchewan with lists of species and generic diagnoses of *Globodera* (n. rank), *Heterodera*, and *Sarisodera*. *Can. J. Zool.* **54**, 772–785.

MULVEY, R. H., TOWNSHEND, J. L. & POTTER, J. W. (1975). *Meloidogyne microtyla* sp. nov. from southwestern Ontario, Canada. *Can. J. Zool.* **53,** 1528–1536.

MURANT, A. F. & LISTER, R. M. (1967). Seed-transmission in the ecology of nematode-borne viruses. *Ann. appl. Biol.* **59,** 63–76.

MURANT, A. F. & TAYLOR, C. E. (1965). Treatment of soil with chemicals to prevent transmission of tomato black ring and raspberry ringspot viruses by *Longidorus elongatus* (de Man). *Ann. appl. Biol.* **55,** 227–237.

MURDOCH, G. & MURDOCH, S. M. (1963). Soil sterilants in the control of potato root eelworm on tomatoes. *Expl Hort.* No. 8, 27–37.

MURPHY, P. W. & DONCASTER, C. C. (1957). A culture method for soil meiofauna and its application to the study of nematode predators. *Nematologica* **2,** 202–214.

MYERS, R. F. & DROPKIN, V. H. (1959). Impracticability of control of plant parasitic nematodes with ionizing radiations. *Pl. Dis. Reptr* **43,** 311–313.

MYUGE, S. G. (1961). On the interrelations of some phytonematodes with its plant host. *Növényvédel. Tudom. Tanácsk.* (*Conf. on Scientific Problems of Plant Protection, Budapest, July* 19–22, 1960) Vol. 2, 333–338.

NAIR, P. (1973a). The genus *Axonchium* (Nematoda: Belondiridae). II. Typical species. *Nematologica* **18** (Yr 1972), 514–554.

NAIR, P. (1973b). The genus *Axonchium* (Nematoda: Belondiridae). III. Species with vaginal sclerotization (Part 1). *Nematologica* **19,** 169–189.

NAIR, P. & COOMANS, A. (1973). The genus *Axonchium* (Nematoda: Belondiridae). I. Morphology, juvenile stages, diagnosis and list of species. *Nematologica* **18** (Yr 1972), 494–513.

NAIR, P. & COOMANS, A. (1974). The genus *Axonchium* (Nematoda: Belondiridae). IV. Species with vaginal sclerotization (Part 2). *Nematologica* **20,** 69–87.

NARANG, H. K. (1972). The excretory system of nematodes: structure and ultrastructure of the excretory system of *Panagrellus redivivus, Ditylenchus myceliophagus* with some observations on *D. dipsaci* and *Heterodera rostochiensis*. *Parasitology* **64,** 253–268.

NEEDHAM, J. T. (1743). Concerning certain chalky tubulous concretions, called malm; with some microscopical observations on the farina of the red lilly, and of worms discovered in smutty corn. *Phil. Trans.* **42,** 634. Abridged edition of 1809, **8,** 729–732.

NELMES, A. J. (1971). Trends in plant chemotherapy for the control of nematodes. *Pest Articles and News Summaries* **17,** 248–249.

NELMES, A. J., TRUDGILL, D. L. & CORBETT, D. C. M. (1973). Chemotherapy in the study of plant parasitic nematodes. In: A. E. R. Taylor & R. Muller (Eds), *Chemotherapeutic agents in the study of parasites*. Oxford, Blackwell, 95–112.

NETSCHER, C. & PERNES, J. (1971). Étude concernant l'influence de la constitution génétique sur la longeur des larves d'*Heterodera oryzae*. *Nematologica* **17,** 336–346.

NEUBERT, E. (1967). Über das Vorkommen von Biotypen des Haferzystenälchens (*Heterodera avenae* Wollenweber, 1924) im Norden der DDR. *NachrBl. dt. PflSchutzdienst, Berl.* **21,** 66–68.

NEWMAN, E. I. (1974). Root and soil water relations. In: E. W. Carson (Ed.), *The plant root and its environment*. Charlottesville, University Press of Virginia, 363–440.

NEWTON, H. C. F. & DUTHOIT, C. M. G. (1954). Stem and bulb eelworm on potatoes. *Pl. Path.* **3**, 139–140.

NEWTON, W., HASTINGS, R. J. & BOSHER, J. E. (1933). Sterilization of narcissus bulbs by immersion in silver nitrate - potassium cyanide solution in vacuo. *Can. J. Res.* **9**, 31–36.

NICHOLAS, W. L. (1975). *The biology of free-living nematodes.* Oxford, Clarendon Press, 219 pp.

NICKLE, W. R. (1967). On the classification of the insect parasitic nematodes of the Sphaerulariidae Lubbock, 1861 (Tylenchoidea: Nematoda). *Proc. helminth. Soc. Wash.* **34**, 72–94.

NICKLE, W. R. (1970). A taxonomic review of the genera of the Aphelenchoidea (Fuchs, 1937) Thorne, 1949 (Nematoda: Tylenchida). *J. Nematol.* **2**, 375–392.

NIELSEN, C. HOLM (1967). Untersuchungen über die Vererbung der Resistenz gegen den Getreidenematoden (*Heterodera avenae*) beim Weizen. *Nematologica* **12** (Yr 1966), 575–578.

NILSSON-EHLE, H. (1903). Nematoder, en hotende fare for vor sädesodling. *Sver. Utsädesför. Tidskr.* **13**, 34–68.

NILSSON-EHLE, H. (1920). Uber Resistenz gegen *Heterodera schachtii* bei gewissen Gerstensorten ihre Vererbungsweise und Bedeutung für die Praxis. *Hereditas* **1**, 1–34.

NIRULA, K. K., KHUSHU, C. L. & RAJ, B. T. (1969). Resistance in tuber-bearing *Solanum* species to root knot nematode, *Meloidogyne incognita. Am. Potato J.* **46**, 251–253.

NOLLEN, H. M. & MULDER, A. (1969). A practical method for economic control of potato cyst nematode (*Heterodera rostochiensis*). *Proc. 5th Br. Insectic. Fungic. Conf. Brighton,* Vol. 3, 671–674.

NONAKA, F. (1959). [On the relation between the stem rot caused by *Leptosphaeria salvinii* and the rice nematode disease 'white tip', and on the changes of the respiration rate and activities of respiratory enzymes of the 'white tip' plants]. (In Japanese; Engl. Summary). *Sci. Bull. Fac. Agric. Kyushu Univ.* **17**, 1–8.

NORTON, D. C. (1958). The association of *Pratylenchus hexincisus* with charcoal rot of sorghum. *Phytopathology* **48**, 355–358.

NORTON, D. C. (1963). Population fluctuations of *Xiphinema americanum* in Iowa. *Phytopathology* **53**, 66–68.

NUSBAUM, C. J. & FERRIS, H. (1973). The role of cropping systems in nematode population management. *A. Rev. Phytopath.* **11**, 423–440.

O'BANNON, J. H. & ESSER, R. P. (1970). Control of *Heterodera cacti* infecting *Zygocactus truncatus. Pl. Dis. Reptr* **54**, 692–694.

O'BANNON, J. H. & TAYLOR, A. L. (1967). Control of nematodes on citrus seedlings by chemical bare-root dip. *Pl. Dis. Reptr* **51**, 995–998.

O'BRIEN, D. G. & PRENTICE, E. G. (1932). A nematode disease of potatoes caused by *Heterodera schachtii* (Schmidt). *Res. Bull. W. Scotl. agric. Coll.* 2, 63 pp.

OCHSE, J. J. & BREWTON, W. S. (1954). Preliminary report on *Crotalaria* versus nematodes. *Proc. Fla St. hort. Soc.* **67**, 218–219.

OGBUJI, R. O. & JENSEN, H. J. (1972). Pacific northwest biotypes of *Meloidogyne hapla. Pl. Dis. Reptr* **56**, 520–523.

OLIFF, K. E. (1965). Laboratory control of the nematode *Ditylenchus myceliophagus* living on *Agaricus campestris. Hort. Res., Edinb.* **5**, 36–38.

OLIFF, K. E. (1966a). The control of plant parasitic nematodes by water dispersed nematicides. III. Foliar application of thionazin to narcissus and tulip to control the stem eelworm, *Ditylenchus dipsaci* (Kühn) Filipjev. *Hort. Res., Edinb.* **6,** 79–84.

OLIFF, K. E. (1966b). The control of plant parasitic nematodes by water dispersed nematicides. IV. The effect of dipping narcissus bulbs infested with *Ditylenchus dipsaci* (Kühn) Filipjev in thionazin. *Hort. Res., Edinb.* **6,** 85–90.

OLTHOF, T. H. A. & POTTER, J. W. (1972). Relationship between population densities of *Meloidogyne hapla* and crop losses in summer-maturing vegetables in Ontario. *Phytopathology* **62,** 981–986.

OMIDVAR, A. M. (1961). On the effects of root diffusates from *Tagetes* spp. on *Heterodera rostochiensis* Woll. *Nematologica* **6,** 123–129.

OMIDVAR, A. M. (1962). The nematicidal effects of *Tagetes* spp. on the final population of *Heterodera rostochiensis* Woll. *Nematologica* **7,** 62–64.

ONIONS, T. G. (1953). Giant larvae of the potato root eelworm *Heterodera rostochiensis* Wollenweber. *Nature, Lond.* **172,** 249–250.

OOSTENBRINK, M. (1950). Het aardappelaaltje (*Heterodera rostochiensis* Wollenweber), een gevaarlijke parasiet voor de eenzijdige aardappelcultuur. *Versl. Meded. plziektenk. Dienst Wageningen* 115, 230 pp.

OOSTENBRINK, M. (1954). Over de betekenis van vrijlevende wortelaaltjes in land- en tuinbouw. *Versl. Meded. plziektenk. Dienst Wageningen* 124, 196–223.

OOSTENBRINK, M. (1955). Over de waardplanten van het bietencystenaaltje *Heterodera schachtii* Schmidt. *Versl. Meded. plziektenk. Dienst Wageningen* 127, 186–193.

OOSTENBRINK, M. (1956). Over de invloed van verschillende gewassen op de vermeerdering van en de schade door *Pratylenchus pratensis* en *Pratylenchus penetrans* (Vermes, Nematoda), met vermelding van een afwijkend moeheidsverschijnsel bij houtige gewassen. *Tijdschr. PlZiekt.* **62,** 189–203.

OOSTENBRINK, M. (1960a). Estimating nematode populations by some selected methods. In: J. N. Sasser & W. R. Jenkins (Eds), *Nematology*. Chapel Hill, Univ. N. Carolina Press, 85–102.

OOSTENBRINK, M. (1960b). The genus *Heterodera*. In: J. N. Sasser & W. R. Jenkins (Eds), *Nematology*. Chapel Hill, Univ. N. Carolina Press, 206–211.

OOSTENBRINK, M. (1961). Nematodes in relation to plant growth. III. *Pratylenchus penetrans* (Cobb) in tree crops, potatoes and red clover. *Neth. J. agric. Sci.* **9,** 188–209.

OOSTENBRINK, M. (1966). Major characteristics of the relation between nematodes and plants. *Meded. LandbHogesch. Wageningen* **66,** 3–46.

OOSTENBRINK, M., S'JACOB, J. J. & KUIPER, K. (1956). An interpretation of some crop rotation experiences based on nematode surveys and population studies. *Nematologica* **1,** 202–215.

OOSTENBRINK, M., KUIPER, K. & S'JACOB, J. J. (1957). *Tagetes* als Feindpflanzen von *Pratylenchus*-Arten. *Nematologica* **2** *Suppl.* 424–433S.

ORCHARD, W. R. (1965). Occurrence of the golden nematode on Vancouver Island, British Columbia. *Can. Pl. Dis. Surv.* **45,** 89.

ORION, D. & MINZ, G. (1971). The influence of morphactin on the root-knot nematode, *Meloidogyne javanica*, and its galls. *Nematologica* **17,** 107–112.

ORTON WILLIAMS, K. J. (1972a). *Meloidogyne javanica. CIH Descriptions of Plant-parasitic Nematodes* Set 1, No. 3, 4 pp.

ORTON WILLIAMS, K. J. (1972b). *Macroposthonia xenoplax. CIH Descriptions of Plant-parasitic Nematodes* Set 1, No. 12, 2 pp.

ORTON WILLIAMS, K. J. (1973). *Meloidogyne incognita. CIH Descriptions of Plant-parasitic Nematodes* Set 2, No. 18, 4 pp.

ORTON WILLIAMS, K. J. (1974). *Meloidogyne hapla. CIH Descriptions of Plant-parasitic Nematodes* Set 3, No. 31, 4 pp.

ORTON WILLIAMS, K. J. (1975). *Meloidogyne arenaria. CIH Descriptions of Plant-parasitic Nematodes* Set 5, No. 62, 4 pp.

ORTON WILLIAMS, K. J. & SIDDIQI, M. R. (1973). *Radopholus similis. CIH Descriptions of Plant-parasitic Nematodes* Set 2, No. 27, 4 pp.

OSCHE, G. (1952). Systematik und Phylogenie der Gattung *Rhabditis* (Nematoda). *Zool. Jb.* **81**, 190–280.

DEN OUDEN, H. (1960a). Periodicity in spontaneous hatching of *Heterodera rostochiensis* in the soil. *Nematologica Suppl.* II, 101–105.

DEN OUDEN, H. (1960b). A note on parthenogenesis and sex determination in *Heterodera rostochiensis* Woll. *Nematologica* **5**, 215–216.

DEN OUDEN, H. (1967). The influence of volunteer potato plants in oats on the population density of *Heterodera rostochiensis. Nematologica* **13**, 325–335.

OVERMAN, A. J. (1970). Gladiolus corm dips for root-knot nematode control. *Proc. Fla St. hort. Soc.* **82**, 362–366.

OWENS, R. G. & SPECHT, H. N. (1966). Biochemical alterations induced in host tissues by root-knot nematodes. *Contr. Boyce Thompson Inst. Pl. Res.* **23**, 181–198.

ØYDVIN, J. (1974). The usefulness of some larval dimensions to distinguish *Heterodera rostochiensis* and *H. pallida. Nematologica* **19** (Yr 1973), 435–442.

OYEKAN, P. O. & MITCHELL, J. E. (1971). Effect of *Pratylenchus penetrans* on the resistance of a pea variety to Fusarium wilt. *Pl. Dis. Reptr* **55**, 1032–1035.

PAGE, A. B. P., HAGUE, N. G. M., JAKABSONS, V. & GOLDSMITH, R. E. (1959). Fumigation of lucerne seed with methyl bromide for the control of the stem eelworm (*Ditylenchus* (*Anguillulina*) *dipsaci*). *J. Sci. Fd Agric.* **10**, 461–467.

PANAYI, C. & ESTEY, R. H. (1971). Control of the root-knot nematode *Meloidogyne hapla* with the insecticide - nematicide Lannate. *Can. J. Pl. Sci.* **51**, 405–408.

PARAMONOV, A. A. (1962). [*Principles of phytonematology. Vol.* 1. *The origin of nematodes. Ecological and morphological characteristics of phytonematodes. General principles of taxonomy*]. (In Russian), Moscow, Izdatelstvo Akad. Nauk SSSR, 480 pp. (See also: *Plant parasitic nematodes. Vol.* 1. Translation from Russian by Israel Program for Sci. Transl., Jerusalem, 1968, 390 pp).

PARAMONOV, A. A. (1970). [*Principles of phytonematology. Vol.* 3. *Taxonomy of nematodes of the superfamily Tylenchoidea*]. (In Russian), Moscow, Izdatelstvo 'Nauka', 253 pp. (See also: Translation from Russian by Israel Program for Sci., Transl., Washington, U.S.D.A. & Natn Sci. Foundation, 1972, 200 pp).

PARROTT, D. M. (1972). Mating of *Heterodera rostochiensis* pathotypes. *Ann. appl. Biol.* **71**, 271–273.

PARROTT, D. M. & TRUDGILL, D. L. (1972). The resistance of hybrids of *Solanum tuberosum* ssp. *andigena* and *S. multidissectum* to *Heterodera rostochiensis* pathotype E. *Pl. Path.* **21**, 86–88.

PAULSON, R. E. & WEBSTER, J. M. (1970). Giant cell formation in tomato roots caused by *Meloidogyne incognita* and *Meloidogyne hapla* (Nematoda) infection. A light and electron microscope study. *Can. J. Bot.* **48**, 271–276.

PEACHEY, J. E., RAO, G. N. & CHAPMAN, M. R. (1963). Field tests of experimental and commercial soil sterilants against the potato root eelworm, *Heterodera rostochiensis* Woll. *Ann. appl. Biol.* **52**, 19–31.

PEACHEY, J. E. & WINSLOW, R. D. (1962). Effects of soil treatments on populations of soil nematodes and on carrot crops grown for two years after treatment. *Nematologica* **8,** 75–79.

PEACOCK, F. C. (1966). Nematode control by plant chemotherapy. *Nematologica* **12,** 70–86.

PERRY, D. A. (1963). Interaction of root knot and Fusarium wilt of cotton. *Emp. Cott. Grow. Rev.* **40,** 41–47.

PERRY, R. N. (1977). Desiccation survival of larval and adult stages of the plant parasitic nematodes, *Ditylenchus dipsaci* and *D. myceliophagus*. *Parasitology* **74,** 139–148.

PERRY, V. G. & MILLER, H. N. (1964). Recommendations for control of nematodes infecting ornamental foliage plants in Florida. *Proc. Fla St. hort. Soc.* **76,** 449–454.

PERSSON, S. (1968). Nematoder av släktet *Trichodorus* i sydsvenska åkerjordar och deras förmåga att överföra rattlevirus. *Meddn St. VäxtskAnst.* **14,** 167–199.

PETERS, B. G. (1951). The term 'eelworm-free soil' in plant quarantine regulations. *Nature, Lond.* **167,** 368.

PETERS, B. G. (1954). Nematology department. *Rep. Rothamsted exp. Stn for* 1953, 97–102.

PETERS, B. G. (1955). The combined use of nematicidal soil fumigants and solubilized chemicals. *J. Helminth.* **29,** 81–86.

PETERS, B. G. & FENWICK, D. W. (1949). Field trials with D–D mixture against potato-root eelworm. *Ann. appl. Biol.* **36,** 364–382.

PETHERBRIDGE, F. R. & JONES, F. G. W. (1944). Beet eelworm (*Heterodera schachtii* Schm.) in East Anglia, 1934–1943. *Ann. appl. Biol.* **31,** 320–332.

PETROVA, Z. I. (1961). [Observations on the biology of the onion stem nematode *Ditylenchus allii* Beijerinck, 1883.] (In Russian). *Trud. vsesoyuz. Inst. Zashch. Rast.* No. 16, 174–198.

PITCHER, R. S. (1961). Nematodes as primary incitants of root rot disease. In: *Recent advances in botany*. Vol. 1. Toronto Univ. Press, 477–481.

PITCHER, R. S. (1963). Symposium on interrelationships between nematodes and other agents causing plant diseases. Role of plant-parasitic nematodes in bacterial diseases. *Phytopathology* **53,** 35–39.

PITCHER, R. S. (1965). Interrelationships of nematodes and other pathogens of plants. *Helminth. Abstr.* **34,** 1–17.

PITCHER, R. S. (1975). Nettlehead and related hop diseases associated with the hop strain of arabis mosaic virus and its vector *Xiphinema diversicaudatum*. In: F. Lamberti, C. E. Taylor & J. W. Seinhorst (Eds), *Nematode vectors of plant viruses*. London & New York, Plenum Press, p. 281.

PITCHER, R. S. & CROSSE, J. E. (1958). Studies in the relationship of eelworms and bacteria to certain plant diseases. II. Further analysis of the strawberry cauliflower disease complex. *Nematologica* **3,** 244–256.

PITCHER, R. S. & FLEGG, J. J. M. (1965). Observation of root-feeding by the nematode *Trichodorus viruliferus* Hooper. *Nature, Lond.* **207,** 317.

PITCHER, R. S. & MCNAMARA, D. G. (1970). The effect of nutrition and season of year on the reproduction of *Trichodorus viruliferus*. *Nematologica* **16,** 99–106.

PITCHER, R. S. & MCNAMARA, D. G. (1973). The control of *Xiphinema diversicaudatum*, the vector of arabis mosaic virus in hops. *Ann. appl. Biol.* **75,** 468–469.

PITCHER, R. S., PATRICK, Z. A. & MOUNTAIN, W. B. (1960). Studies on the host-parasite relations of *Pratylenchus penetrans* (Cobb) to apple seedlings. I. Pathogenicity under sterile conditions. *Nematologica* **5**, 309–314.

PITCHER, R. S., SIDDIQI, M. R. & BROWN, D. J. F. (1974). *Xiphinema diversicaudatum. CIH Descriptions of Plant-parasitic Nematodes* Set 4, No. 60, 4 pp.

POGOSYAN, E. E. (1962). [The incidence of nematodes of the family Heteroderidae (Nematoda) in Armenian SSR. In: Nematodes harmful to agriculture and their control. *Proc. 5th all-union Conference Phytohelminthologists, Samarkand,* 1960]. pp. 228–250. (In Russian, English summary).

POINAR, G. (1975). *Entomogenous nematodes. A manual and host list of insect-nematode associations.* Leiden, E. J. Brill, 317 pp.

POLYCHRONOPOULOS, A. G., HOUSTON, B. R. & LOWNSBERY, B. F. (1969). Penetration and development of *Rhizoctonia solani* in sugar beet seedlings infected with *Heterodera schachtii. Phytopathology* **59**, 482–485.

PORTER, D. M. & POWELL, N. T. (1967). Influence of certain *Meloidogyne* species on Fusarium wilt development in flue-cured tobacco. *Phytopathology* **57**, 282–285.

POTTER, H. S. & MORGAN, O. D. (1956). Nemagon control of root-knot nematode on strawberries. *Pl. Dis. Reptr* **40**, 187–189.

POTTER, J. W. & FOX, J. A. (1965). Hybridization of *Heterodera schachtii* and *H. glycines. Phytopathology* **55**, 800–801.

POTTER, J. W. & TOWNSHEND, J. L. (1973). Distribution of plant-parasitic nematodes in field crop soils of southwestern and central Ontario. *Can. Pl. Dis. Surv.* **53**, 39–48.

POWELL, D. F. (1974). Fumigation of field beans against *Ditylenchus dipsaci. Pl. Path.* **23**, 110–113.

POWELL, N. T. (1963). The role of plant-parasitic nematodes in fungus diseases. *Phytopathology* **53**, 28–35.

POWELL, N. T. & BATTEN, C. K. (1969). Complexes in tobacco involving *Meloidogyne incognita, Fusarium oxysporum* f.sp. *nicotianae,* and *Alternaria tenuis. Phytopathology* **59**, 1044. (Abstr.)

POWELL, N. T., MELENDÉZ, P. L. & BATTEN, C. K. (1971). Disease complexes in tobacco involving *Meloidogyne incognita* and certain soil-borne fungi. *Phytopathology* **61**, 1332–1337.

POWELL, N. T. & NUSBAUM, C. J. (1960). The black shank-root-knot complex in flue-cured tobacco. *Phytopathology* **50**, 899–906.

PRASAD, S. K. & WEBSTER, J. M. (1967). The effect of amino acid antimetabolites on four nematode species and their host plants. *Nematologica* **13**, 318–320.

PROESELER, G. (1964). Die Übertragung pflanzlicher Viren durch Nematoden. *NachrBl. dt. PflSchutzdienst, Berl.* **18**, 31–35.

PURNELL, R. E. (1964). The control of plant parasitic nematodes by water dispersed nematicides. I. Laboratory methods with the stem nematode of narcissus, *Ditylenchus dipsaci* (Kühn) Filipjev. *Hort. Res., Edinb.* **4**, 42–48.

PURNELL, R. E. & HAGUE, N. G. M. (1965a). Fumigation of agricultural products. XIX—Methyl bromide fumigation of narcissus bulbs infested with stem nematode, *Ditylenchus dipsaci* (Kühn). *J. Sci. Fd Agric.* **16**, 421–424.

PURNELL, R. E. & HAGUE, N. G. M. (1965b). The control of plant parasitic nematodes by water-dispersed nematicides. II. Field experiments on the control of *Ditylenchus dipsaci* (Kühn) Filipjev in narcissus bulbs. *Hort. Res., Edinb.* **5**, 65–75.

PUTZ, C. & STOCKY, G. (1970). Premières observations sur une souche de strawberry latent ringspot virus transmise par *Xiphinema coxi* Tarjan et associée à une maladie du framboisier en Alsace. *Annls Phytopath.* **2**, 329–347.

RAMSBOTTOM, J. K. (1918). Experiments on the control of eelworm disease of narcissus. *J. R. hort. Soc.* **43**, 65–78.

RANA, G. L. & ROCA, F. (1973). Trasmissione con nematodi del virus latente italiano del carciofo (AILV). *Proceedings of the 2nd International Meeting on Globe Artichoke, Bari, November*, 1973, p. 139.

RASKI, D. J. (1950). The life history and morphology of the sugar beet nematode, *Heterodera schachtii* Schmidt. *Phytopathology* **40**, 135–152.

RASKI, D. J. (1975a). Revision of the genus *Paratylenchus* Micoletzky, 1922 and descriptions of new species. Part I of 3 parts. *J. Nematol.* **7**, 15–34.

RASKI, D. J. (1975b). Revision of the genus *Paratylenchus* Micoletzky, 1922 and descriptions of new species. Part II of 3 parts. *J. Nematol.* **7**, 274–295.

RASKI, D. J. & GOLDEN, A. M. (1966). Studies on the genus *Criconemoides* Taylor, 1936 with descriptions of eleven new species and *Bakernema variabile* n.sp. (Criconematidae: Nematoda). *Nematologica* **11** (Yr 1965), 501–565.

RASKI, D. J. & HEWITT, W. B. (1963). Plant-parasitic nematodes as vectors of plant viruses. *Phytopathology* **53**, 39–47.

RASKI, D. J., HEWITT, W. B., GOHEEN, A. C., TAYLOR, C. E. & TAYLOR, R. H. (1965). Survival of *Xiphinema index* and reservoirs of fanleaf virus in fallowed vineyard soil. *Nematologica* **11**, 349–352.

RASKI, D. J., HEWITT, W. & SCHMITT, R. V. (1971). Controlling fanleaf virus - dagger nematode disease complex in vineyards by soil fumigation. *Calif. Agric.* **25**, 11–14.

RASKI, D. J., MAGGENTI, A. R. & JONES, N. O. (1973). Location of grapevine fanleaf and yellow mosaic virus particles in *Xiphinema index*. *J. Nematol.* **5**, 208–211.

RASKI, D. J. & SIDDIQUI, I. A. (1975). *Tylenchocriconema alleni* n.g. n.sp. from Guatemala (Tylenchocriconematidae n.fam.; Tylenchocriconematoidea n.superfam.; Nematoda). *J. Nematol.* **7**, 247–251.

RAZAK, A. R. BIN A. (1975). Host parasite relationship of *Rotylenchulus reniformis* on cowpea and cotton. *Ph.D. Thesis, London University*, 176 pp.

RAZJIVIN, A. A., O'RELLY, J. P. & MILIAN, J. R. P. (1973). Nuevas especies de nemátodos (Nematoda: Dorylaimidae y Hoplolaimidae) parásitos de la caña de azúcar en Cuba. *Poeyana* No. 108, 1–12.

REEVE, M. J., SMITH, P. D. & THOMASSON, A. J. (1973). The effect of density on water retention properties of field soils. *J. Soil Sci.* **24**, 355–367.

REYNOLDS, H. W., CARTER, W. W. & O'BANNON, J. H. (1970). Symptomless resistance of alfalfa to *Meloidogyne incognita acrita*. *J. Nematol.* **2**, 131–134.

RHOADES, H. L. (1968). Pathogenicity and control of *Pratylenchus penetrans* on leatherleaf fern. *Pl. Dis. Reptr* **52**, 383–385.

RHOADES, H. L. (1970). A comparison of chemical treatments with hot water for control of root-knot nematodes in *Caladium* tubers. *Pl. Dis. Reptr* **54**, 411–413.

RHOADES, H. L. & LINFORD, M. B. (1959a). Molting of preadult nematodes of the genus *Paratylenchus* stimulated by root diffusates. *Science* **130**, 1476–1477.

RHOADES, H. L. & LINFORD, M. B. (1959b). Control of *Pythium* root rot by the nematode *Aphelenchus avenae*. *Pl. Dis. Reptr* **43**, 323–328.

RIEDEL, R. M. & LARSEN, P. O. (1974). Interrelationship of *Aphelenchoides fragariae* and *Xanthomonas begoniae* on Rieger begonia. *J. Nematol.* **6**, 215–216.

RIFFLE, J. W. (1963). *Meloidogyne ovalis* (Nematoda: Heteroderidae), a new species of root-knot nematode. *Proc. helminth. Soc. Wash.* **30,** 287–292.

RIFFLE, J. W. (1969). Effect of nematodes on root-inhabiting fungi. [Proceedings of the 1st North American Conference on Mycorrhizae, April, 1969]. *Misc. Publs. U.S. Dep. Agric. For. Serv.* No. 1189, 97–113.

RIGGS, R. D., SLACK, D. A. & HAMBLEN, M. B. (1968). New biotype of soybean cyst nematode. *Arkansas Farm Research* **17** (5), 11.

RIGGS, R. D. & WINSTEAD, N. N. (1959). Studies on resistance in tomato to root-knot nematodes and on the occurrence of pathogenic biotypes. *Phytopathology* **49,** 716–724.

RILEY, R. & CHAPMAN, V. (1957). Chromosomes of the potato root eelworm. *Nature, Lond.* **180,** 662.

RITTER, M. (1960). Les nématodes nuisibles de l'oeillet *C.r. Journées florales, Nice-Antibes,* 16–17 *May* 1960 (Reprint 14 pp).

RITTER, M (1972) Rôle économique et importance des *Meloidogyne* en Europe et dans le bassin Méditerranéen. *EPPO Bull.* No. 6, 17–22.

ROBERTSON, W. M. & TAYLOR, C. E. (1975). The ultrastructure of the feeding apparatus in Trichodorids. In: F. Lamberti, C. E. Taylor & J. W. Seinhorst (Eds), *Nematode vectors of plant viruses.* London & New York, Plenum Press, 195–201.

ROCA, F., MARTELLI, G. P. & RANA, G. L. (1975). Distribution of artichoke Italian latent virus and its nematode vector in Apulia. In: F. Lamberti, C. E. Taylor & J. W. Seinhorst (Eds), *Nematode vectors of plant viruses.* London & New York, Plenum Press, 279–280.

RODRIGUEZ-KABANA, R., JORDAN, J. W. & HOLLIS, J. P. (1965). Nematodes: biological control in rice fields: role of hydrogen sulfide. *Science* **148,** 524–526.

ROGGEN, D. R., RASKI, D. J. & JONES, N. O. (1967). Further electron microscope observations of *Xiphinema index. Nematologica* **13,** 1–16.

ROHDE, R. A. (1965). The nature of resistance in plants to nematodes. *Phytopathology* **55,** 1159–1162.

ROHDE, R. A. (1972). Expression of resistance in plants to nematodes. *A. Rev. Phytopath.* **10,** 233–252.

ROHDE, R. A. & JENKINS, W. R. (1957a). Host range of a species of *Trichodorus* and its host-parasite relationships on tomato. *Phytopathology* **47,** 295–298.

ROHDE, R. A. & JENKINS, W. R. (1957b). Effect of temperature on the life cycle of stubby-root nematodes. *Phytopathology* **47,** 29. (Abstr.)

ROHDE, R. A. & JENKINS, W. R. (1958). The chemical basis of resistance of asparagus to the nematode *Trichodorus christiei. Phytopathology* **48,** 463. (Abstr.)

ROMANENKO, N. D. (1971). [Distribution of nematodes—vectors of viruses on fruit and berry crops]. (In Russian). *Plodovodye i Yagodov Nechernoz Polosy* **3,** 383–386.

ROMASÇU, E. (1969). Cercetări privind rezistenta unor soiuri grĭu de toamnă și de primăvară la atacul nematodului *Anguina* (*Tylenchus*) *tritici* Steinb. *Comunicări Zool., Bucuresti, Year* 1969, 91–98.

ROSS, J. P. (1958). Host-parasite relationship of the soybean cyst nematode in resistant soybean roots. *Phytopathology* **48,** 578–579.

ROSS, J. P. (1960). *Heterodera trifolii,* a foliage pathogen of white clover. *Phytopathology* **50,** 866–867.

ROSS, J. P. (1962). Physiological strains of *Heterodera glycines. Pl. Dis. Reptr* **46,** 766–769.

ROSS, J. P. (1965). Predisposition of soybeans to Fusarium wilt by *Heterodera glycines* and *Meloidogyne incognita*. *Phytopathology* **55**, 361–364.

ROSS, J. P. & BRIM, C. A. (1957). Resistance of soybeans to the soybean cyst nematode as determined by a double-row method. *Pl. Dis. Reptr* **41**, 923–924.

ROVIRA, A. D. & DAVY, C. B. (1974). Biology of the rhizosphere. In: E. W. Carson (Ed.), *The plant root and its environment*. Charlottesville, University Press of Virginia, 153–204.

ROY, A. K. (1968). The formation of giant cells in tomato roots infected by the potato cyst nematode, *Heterodera rostochiensis* and the grey sterile fungus, G.S.F. *Nematologica* **14**, 313–314.

SAFYANOV, S. P. (1965). [Nutritional specialization of the potato stem nematode]. (In Russian). *Vest. sel'.-khoz. Nauki, Alma-Ata* No. 8, 121–126.

SANDNER, H. (1962). Morphological differences between *Heterodera galeopsidis* (Goff.) and *Heterodera trifolii* (Goff.). *Nematologica* **7**, 17.

SANGER, H. L. (1961). Untersuchungen über schwer übertragbare Formen des Rattlevirus. *Proc. 4th Conf. Potato Virus Diseases, Brunswick* 1960, 22–28.

SANTOS, M. S. N. de A. (1968). *Meloidogyne ardenensis* n.sp. (Nematoda: Heteroderidae), a new British species of root-knot nematode. *Nematologica* **13** (Yr 1967), 593–598.

SANWAL, K. C. (1971). *Geraldius* n. gen., Macrolaiminae n.subfam., with a revision of the subfamilies and genera of Chambersiellidae (Nematoda). *Can. J. Zool.* **49**, 965–967.

SASSER, J. N. (1972). Physiological variation in the genus *Meloidogyne* as determined by differential hosts. *EPPO Bull.* No. 6, 41–48.

SASSER, J. N., LUCAS, G. B. & POWERS, H. R., Jr. (1955). The relationship of root-knot nematodes to black-shank resistance in tobacco. *Phytopathology* **45**, 459–461.

SAVITSKY, H. (1975). Hybridization between *Beta vulgaris* and *B. procumbens* and transmission of nematode (*Heterodera schachtii*) resistance to sugarbeet. *Can. J. Genet. Cytol.* **17**, 197–209.

SAYNOR, M. (1975). The distribution of pathotypes of the cereal cyst-eelworm *Heterodera avenae* in England and Wales. *Ann. appl. Biol.* **81**, 215–218.

SAYRE, R. M. (1963). Winter survival of root-knot nematodes in southwestern Ontario. *Can. J. Pl. Sci.* **43**, 361–364.

SAYRE, R. M. (1971). Biotic influences in soil environment. In: B. M. Zuckerman, W. F. Mai & R. A. Rohde (Eds), *Plant parasitic nematodes*. I. London & New York, Academic Press, 235–256.

SCHAERFFENBERG, B. (1950). Untersuchungen über die Bedeutung der Enchytraeiden als Humusbildner und Nematodenfeinde. *Z. PflKrankh. PflPath. PflSchutz* **57**, 183–191.

SCHINDLER, A. F. (1957). Parasitism and pathogenicity of *Xiphinema diversicaudatum* an ectoparasitic nematode. *Nematologica* **2**, 25–31.

SCHINDLER, A. F. & STEWART, R. N. (1956). Fusarium wilt of carnations retarded by fungus-eating nematodes, *Ditylenchus* spp. *Phytopathology* **46**, 469. (Abstr.)

SCHINDLER, A. F., STEWART, R. N. & SEMENIUK, P. (1961). A synergistic Fusarium-nematode interaction in carnations. *Phytopathology* **51**, 143–146.

SCHMIDT, H. B., FRITZSCHE, R. & LEHMANN, W. (1963). Die Übertragung des Weidelgrasmosaik-Virus durch Nematoden. *Naturwissenschaften* **50**, 386.

SCHMIDT, J. (1954). Der gegenwärtige Stand der Kartoffelnematodenforschung. *Wiss. Z. Univ. Rostock, math.-nat.* **3**, 371–377.

SCHNEIDER, J. (1967). Un nouveau nématode du genre *Meloidogyne* parasite des céréales en France. *Phytoma* No. 185, 21–25.

SCHULDT, P. H. & BLUESTONE, H. (1957). Nematocidal properties of 3,4-dichlorotetrahydrothiophene-1,1-dioxide (PRD). *Contr. Boyce Thompson Inst. Pl. Res.* **19,** 63–75.

SCHUSTER, M. L. & WAGNER, L. J. (1972). Control of bacterial canker and root knot of hydroponic tomatoes. *Pl. Dis. Reptr* **56,** 139–140.

SCOTT, H. A. & RIGGS, R. D. (1971) Immunoelectrophoretic comparisons of three plant-parasitic nematodes. *Phytopathology* **61,** 751–752.

SCOTTO LA MASSESE, C. (1967). Découverte d'un nématode phytophage non encore signale en Europe. *Phytoma* **19** (192), 29–33.

SCOTTO LA MASSESE, C. (1972). The use of 'isopathological curves' for estimation of parasitic pathogenicity, exemplified by *Tylenchulus semipenetrans* on citrus. *XIth International Symposium of Nematology, Reading,* 1972. Abstracts, 61–62.

SEINHORST, J. W. (1956). Population studies on stem eelworms (*Ditylenchus dipsaci*). *Nematologica* **1,** 159–164.

SEINHORST, J. W. (1957). *Phytonematology in Western Europe.* Tech. Committee, Southern Regional Nematology Project (S–19), Alabama. 39 pp.

SEINHORST, J. W. (1961). Plant-nematode inter-relationships. *A. Rev. Microbiol.* **15,** 177–196.

SEINHORST, J. W. (1963). A redescription of the male of *Trichodorus primitivus* (de Man) and the description of a new species *T. similis. Nematologica* **9,** 125–130.

SEINHORST, J. W. (1965). The relation between nematode density and damage to plants. *Nematologica* **11,** 137–154.

SEINHORST, J. W. (1966a). The relationships between population increase and population density in plant parasitic nematodes. *Nematologica* **12,** 157–169.

SEINHORST, J. W. (1966b). Killing nematodes for taxonomic study with hot f.a. 4: 1. *Nematologica* **12,** 178.

SEINHORST, J. W. (1967). The relationship between population increase and population density in plant parasitic nematodes. III. Definition of the terms host, host status and resistance. IV. The influence of external conditions on the regulation of population density. *Nematologica* **13,** 429–442.

SEINHORST, J. W. (1968a). The relationships between population increase and population density in plant parasitic nematodes. V. Influence of damage to the host on multiplication. *Nematologica* **13** (Yr 1967), 481–492.

SEINHORST, J. W. (1968b). Underpopulation in plant parasitic nematodes. *Nematologica* **14,** 549–553.

SEINHORST, J. W. (1970). Dynamics of populations of plant parasitic nematodes. *A. Rev. Phytopath.* **8,** 131–156.

SEINHORST, J. W. (1972). Principles and possibilities efficient economic nematode control. *I.P.O. Wageningen. Cyclostyled paper distr. at 11th internat. Nematol. Symposium, Reading,* 1972, 19 pp.

SEINHORST, J. W. (1973a). Dosage of nematicidal fumigants and mortality of nematodes. *Neth. J. Pl. Path.* **79,** 180–188.

SEINHORST, J. W. (1973b). The combined effects of dichloropropane-dichloropropene mixture injected into the soil and of dazomet applied to the soil surface. *Neth. J. Pl. Path.* **79,** 194–206.

SEINHORST, J. W. (1973c). Principles and possibilities of determining degrees of nematode control leading to maximum returns. I. Protection of one crop sown or planted soon after treatment. *Nematologia Mediterranea* **1,** 93–105.

SEINHORST, J. W. & KOERT, J. L. (1969). Stengelaaltjes op uienzaad. *Gewasbescherming* **2,** 25–31.

SEINHORST, J. W. & KUNIYASU, K. (1971). Interaction of *Pratylenchus penetrans* and *Fusarium oxysporum* forma *pisi* race 2 and of *Rotylenchus uniformis* and *F. oxysporum* f. *pisi* race 1 on peas. *Nematologica* **17,** 444–452.

SEINHORST, J. W. & DEN OUDEN, H. (1971). The relation between density of *Heterodera rostochiensis* and growth and yield of two potato varieties. *Nematologica* **17,** 347–369.

SEMBDNER, G. (1963). Anatomie der *Heterodera rostochiensis* - Gallen an Tomatenwurzeln. *Nematologica* **9,** 55–64.

SESHADRI, A. R. & DASGUPTA, D. R. (1975). *Ditylenchus angustus. CIH Descriptions of Plant-parasitic Nematodes* Set 5, No. 64, 3 pp.

SEWELL, R. (1967). Some records of plant-parasitic nematodes encountered in Canada in 1966. *Can. Pl. Dis. Surv.* **47,** 23–24.

SEYMOUR, M. K. & DONCASTER, C. C. (1973). Defaecation behaviour of *Aphelenchoides blastophthorus. Nematologica* **18** (Yr 1972), 463–468.

SEYMOUR, M. K. & SHEPHERD, A. M. (1974). Cell junctions acting as intestinal valves in nematodes. *J. Zool., Lond.* **173,** 517–523.

SHARMA, R. D. (1971). Studies on the plant parasitic nematode *Tylenchorhynchus dubius. Meded. LandbHogesch. Wageningen* 71–1, 154 pp.

SHEPHERD, A. M. (1958). Experimental methods in testing for resistance to beet eelworm, *Heterodera schachtii* Schmidt. *Nematologica* **3,** 127–135.

SHEPHERD, A. M. (1959). The invasion and development of some species of *Heterodera* in plants of different host status. *Nematologica* **4,** 253–267.

SHEPHERD, A. M. (1962a). Dyes as artificial hatching agents for beet eelworm, *Heterodera schachtii* Schm. *Nature, Lond.* **196,** 391–392.

SHEPHERD, A. M. (1962b). *The emergence of larvae from cysts in the genus* Heterodera. Farnham Royal, England; Commonwealth Agricultural Bureaux, 90 pp.

SHEPHERD, A. M. (1962c). New Blue R, a stain that differentiates between living and dead nematodes. *Nematologica* **8,** 201–208.

SHEPHERD, A. M. (1963). The emergence of larvae of *Heterodera goettingiana* Liebs. *in vitro* and a comparison between field populations of *H. goettingiana* and *H. rostochiensis* Woll. *Nematologica* **9,** 143–151.

SHEPHERD, A. M. & CLARK, S. A. (1976). Structure of the anterior alimentary tract of the passively feeding nematode *Hexatylus viviparus* (Neotylenchidae: Tylenchida). *Nematologica* **22,** 332–342.

SHEPHERD, A. M., CLARK, S. A. & DART, P. J. (1972). Cuticle structure in the genus *Heterodera. Nematologica* **18,** 1–17.

SHEPHERD, A. M., CLARK, S. A. & KEMPTON, A. (1974). Spermatogenesis and sperm ultrastructure in some cyst nematodes, *Heterodera* spp. *Nematologica* **19** (Yr 1973), 551–560.

SHEPHERD, A. M. & CLARKE, A. J. (1971). Molting and hatching stimuli. In: B. M. Zuckerman, W. F. Mai & R. A. Rohde (Eds), *Plant parasitic nematodes* Vol. 2. London & New York, Academic Press, 267–287.

SHEPHERD, A. M. & COX, P. M. (1967). Observations on periodicity of hatching of eggs of the potato cyst-nematode, *Heterodera rostochiensis* Woll. *Ann. appl. Biol.* **60,** 143–150.

SHEPHERD, C. J. (1952). Tainting of tobacco by a dichloropropene-dichloropropane soil fumigant. *Nature, Lond.* **170,** 1073–1074.

SHER, S. A. (1965). Revision of the Hoplolaiminae (Nematoda). V. *Rotylenchus* Filipjev, 1936. *Nematologica* **11**, 173–198.

SHER, S. A. (1966). Revision of the Hoplolaiminae (Nematoda) VI. *Helicotylenchus* Steiner, 1945. *Nematologica* **12**, 1–56.

SHER, S. A. (1970). Revision of the genus *Pratylenchoides* Winslow, 1958 (Nematoda: Tylenchoidea). *Proc. helminth. Soc. Wash.* **37**, 154–166.

SHER, S. A. & BELL, A. H. (1975). Scanning electron micrographs of the anterior region of some species of Tylenchoidea (Tylenchida: Nematoda). *J. Nematol.* **7**, 69–83.

SHERMAN, G. W. (1936). The control of the root-knot nematode, *Heterodera marioni* (Cornu) (Anguillulinidae), on tuberoses by hot water and vapor heat. *Proc. helminth. Soc. Wash.* **2**, 111.

SIDDIQI, M. R. (1971). Structure of the oesophagus in the classification of the superfamily Tylenchoidea (Nematoda). *Indian J. Nematol.* **1**, 25–43.

SIDDIQI, M. R. (1972a). *Tylenchorhynchus cylindricus. CIH Descriptions of Plant-parasitic Nematodes* Set 1, No. 7, 2 pp.

SIDDIQI, M. R. (1972b). *Helicotylenchus dihystera. CIH Descriptions of Plant-parasitic Nematodes* Set 1, No. 9, 3 pp.

SIDDIQI, M. R. (1972c). *Rotylenchus robustus. CIH Descriptions of Plant-parasitic Nematodes* Set 1, No. 11, 3 pp.

SIDDIQI, M. R. (1973). *Helicotylenchus multicinctus. CIH Descriptions of Plant-parasitic Nematodes* Set 2, No. 23, 3 pp.

SIDDIQI, M. R. (1974a). Systematics of the genus *Trichodorus* Cobb, 1913 (Nematoda: Dorylaimida), with descriptions of three new species. *Nematologica* **19** (Yr 1973), 259–278.

SIDDIQI, M. R. (1974b). *Aphelenchoides ritzemabosi. CIH Descriptions of Plant-parasitic Nematodes* Set 3, No. 32, 4 pp.

SIDDIQI, M. R. (1974c). *Pratylenchoides crenicauda. CIH Descriptions of Plant-parasitic Nematodes* Set 3, No. 38, 2 pp.

SIDDIQI, M. R. (1974d). *Rotylenchus buxophilus. CIH Descriptions of Plant-parasitic Nematodes* Set 4, No. 55, 3 pp.

SIDDIQI, M. R. (1975). *Aphelenchoides fragariae. CIH Descriptions of Plant-parasitic Nematodes* Set 5, No. 74, 4 pp.

SIDDIQI, M. R. (1976). *Tylenchorhynchus annulatus. CIH Descriptions of Plant-parasitic Nematodes* Set 6, No. 85, 4 pp.

SIDDIQI, M. R. & HUSAIN, Z. (1967). Studies on the genus *Alaimus* de Man, 1880 with description of six new species from India. *Proc. helminth. Soc. Wash.* **34**, 158–167.

SIDDIQUI, I. A. & TAYLOR, D. P. (1970a). The biology of *Meloidogyne naasi. Nematologica* **16**, 133–143.

SIDDIQUI, I. A. & TAYLOR, D. P. (1970b). Symptoms and varietal reaction of oats to the Illinois isolate of the barley root-knot nematode, *Meloidogyne naasi. Pl. Dis. Reptr* **54**, 972–975.

SIDHU, G. S. (1975). Gene-for-gene relationships in plant parasitic systems. *Sci. Prog., Oxf.* **62**, 467–485.

SKARBILOVICH, T. S. (1959). On the structure of systematics of nematodes Order Tylenchida Thorne, 1949. *Acta parasit. pol.* **7**, 117–132.

SKARBILOVICH, T. S. (1972). [*Ditylenchus humuli* on hops]. (In Russian). *Problemy parazitologii. Trudy VII Nauchnoi Konferentsii Parazitologov SSSR. Part II. Kiev, USSR; Izdatel'stvo 'Naukova Dumka'*, 258–259.

SLACK, D. A., RIGGS, R. D. & HAMBLEN, M. L. (1972). The effect of temperature and moisture on the survival of *Heterodera glycines* in the absence of a host. *J. Nematol.* **4**, 263–266.

VAN SLOGTEREN, E. (1920). De nematoden-bestrijding in de bloembollenstreek. III. Bestrijding der ziekte. *Tijdschr. PlZiekt.* **26**, 161–171, 177–188.

VAN SLOGTEREN, E. (1931). Les helminthiases des plantes. Traitement. (*Rapp.*) *IIe Congr. int. Path. comp.* 432–447.

SLOOTMAKER, L. A. J., LANGE, W., JOCHEMSEN, G. & SCHEPERS, J. (1974). Monosomic analysis in bread wheat of resistance to cereal root eelworm. *Euphytica* **23**, 497–503.

SLOOTWEG, A. F. G. (1956). Root rot of bulbs caused by *Pratylenchus* and *Hoplolaimus* spp. *Nematologica* **1**, 192–201.

SLOOTWEG, A. F. G. (1963). Control of nematodes in flowerbulbs with organic phosphates. *Meded. LandbHogesch. OpzoekStns Gent* **28**, 623–629.

SMART, G. C., Jr. (1964). Physiological strains and one additional host of the soybean cyst nematode, *Heterodera glycines. Pl. Dis. Reptr* **48**, 542–543.

SMITH, D. H. (1963). Effect of fumigants on the soil status and plant uptake of certain elements. *Proc. Soil Sci. Soc. Am.* **27**, 538–541.

SMITH, J. J. & MAI, W. F. (1965). Host-parasite relationships of *Allium cepa* and *Meloidogyne hapla. Phytopathology* **55**, 693–697.

SMITH, L. & ELLENBY, C. (1967). Some biochemical changes in the contents of the cyst of *Heterodera rostochiensis* during its maturation. *Nematologica* **13**, 395–405.

SOL, H. H., VAN HEUVEN, J. C. & SEINHORST, J. W. (1960). Transmission of rattle virus and *Atropa belladonna* mosaic virus by nematodes. *Tijdschr. PlZiekt.* **66**, 228–231.

SOUTHARDS, C. J. & PRIEST, M. F. (1973). Variation in pathogenicity of seventeen isolates of *Meloidogyne incognita. J. Nematol.* **5**, 63–67.

SOUTHEY, J. F. (1952). Unusual chrysanthemum eelworm symptoms. *Pl. Path.* **1**, 48–49.

SOUTHEY, J. F. (1956). National survey work for cereal root eelworm (*Heterodera major* (O. Schmidt) Franklin). *Nematologica* **1**, 64–71.

SOUTHEY, J. F. (1957a). Observations on *Heterodera cacti* Filipjev et Sch. Stekhoven and *Meloidogyne* spp. on imported cactus plants with a list of new host records. *Nematologica* **2**, 1–6.

SOUTHEY, J. F. (1957b). Observations on races of *Ditylenchus dipsaci* infesting bulbs. *J. Helminth.* **31**, 39–46.

SOUTHEY, J. F. (1965). The incidence and location of stem eelworms on onion seed. *Pl. Path.* **14**, 55–59.

SOUTHEY, J. F. (1969). A gall-forming nematode (*Anguina* sp.) parasitic on cocksfoot grass. *Pl. Path.* **18**, 164–166.

SOUTHEY, J. F. (Ed.) (1970). Laboratory methods for work with plant and soil nematodes. *Tech. Bull. Minist. Agric. Fish. Fd* 2 (5th edn), 148 pp.

SOUTHEY, J. F. (1972). *Anguina tritici. CIH Descriptions of Plant-parasitic Nematodes* Set 1, No. 13, 4 pp.

SOUTHEY, J. F. (1973a). *Anguina agrostis. CIH Descriptions of Plant-parasitic Nematodes* Set 2, No. 20, 3 pp.

SOUTHEY, J. F. (1973b). Identification of *Xiphinema* species. In: *The Longidoridae*. Harpenden, Rothamsted Experimental Station, 37–58.

SOUTHEY, J. F. (1974a). Methods for detection of potato cyst nematodes. *EPPO Bull.* **4** (4), 463–473.

SOUTHEY, J. F. (1974b). *Anguina graminis. CIH Descriptions of Plant-parasitic Nematodes* Set 4, No. 53, 3 pp.

SOUTHEY, J. F. & JONES, A. G. (1963). Hot-water treatment of onion sets. *Exptl Hort.* No. 9, 57–63.

SOUTHEY, J. F. & LUC, M. (1974). Redefinition of *Xiphinema ensiculiferum* (Cobb, 1893) Thorne, 1937, and description of *Xiphinema loosi* n.sp. and *Xiphinema hygrophilum* n.sp. (Nematoda: Dorylaimoidea). *Nematologica* **19**, 293–307.

SPEARS, J. F. (1968). The golden nematode handbook. Survey, laboratory, control and quarantine procedures. *Agric. Handb. agric. Res. Serv. U.S.* No. 353, 81 pp.

SPEARS, J. F. (1969). Golden nematode found in Delaware. *Pl. Dis. Reptr* **53**, 243.

STANILAND, L. N. (1933). The treatment of narcissus bulbs with hot water. *J. Minist. Agric.* **40**, 343–355.

[STANILAND, L. N.] (1936). Hot water treatment of grape hyacinths. *Rep. Dep. Pl. Path. Seale-Hayne agric. Coll.* 12th, p. 5.

[STANILAND, L. N.] (1937). Hot water treatment of gladiolus corms. *Rep. Dep. Pl. Path. Seale-Hayne agric. Coll.* 13th, p. 7.

[STANILAND, L. N.] (1939). Hot water treatment of potatoes. *Rep. Dep. Pl. Path. Seale-Hayne agric. Coll.* 15th, 18–20.

STANILAND, L. N. (1950a). Experiments on the control of chrysanthemum eelworm (*Aphelenchoides ritzemabosi*, Schwartz) by hot water treatment. *Ann. appl. Biol.* **37**, 11–18.

STANILAND, L. N. (1950b). Notes on the use of iodine and chlorphenol against certain plant nematodes. *J. Helminth.* **24**, 91–99.

STANILAND, L. N. (1953). Hot-water treatment of strawberry runners. *Pl. Path.* **2**, 44–48.

STANILAND, L. N. (1957). The swarming of rhabditid eelworms in mushroom houses. *Pl. Path.* **6**, 61–62.

STANILAND, L. N. & BARBER, D. R. (1937). The efficiency of baths used for the hot-water treatment of narcissus bulbs. *Bull. Minist. Agric. Fish. Fd* 105, 29 pp. (H.M.S.O., out of print).

STANILAND, L. N. & GOODEY, T. (1934). A dwarfing disease of cultivated violets associated with the eelworm, *Aphelenchoides olesistus*. *J. Helminth.* **12**, 13–22.

STARK, F. L., Jr. (1948). Investigations of chloropicrin as a soil fumigant. *Mem. Cornell Univ. agric. Exp. Stn* No. 278, 61 pp.

STARK, F. L., Jr. & LEAR, B. (1947). Miscellaneous greenhouse tests with various soil fumigants for the control of fungi and nematodes. *Phytopathology* **37**, 698–711.

STEELE, A. E. (1965). The host range of the sugar beet nematode, *Heterodera schachtii* Schmidt. *J. Am. Soc. Sug. Beet Technol.* **13**, 573–603.

STEELE, A. E. (1971a). Orientation and development of *Heterodera schachtii* larvae on tomato and sugar beet roots. *J. Nematol.* **3**, 424–426.

STEELE, A. E. (1971b). Invasion of non-host plant roots by larvae of the sugar beet nematode, *Heterodera schachtii*. *J. Am. Soc. Sug. Beet Technol.* **16**, 457–460.

STEGARESCU, O. P. (1966). [The application of the polytomous principle to the identification of species of the genus *Xiphinema* Cobb]. (In Russian). In: *Politomicheskiy printsip opredeleniya zhivotnykhi rasteniy*. Kishinev, 53–59.

STEIN, W. (1965). Untersuchungen über die Ausbreitung von *Meloidogyne hapla* Chitwood unter Freilandbedingungen. *Nematologica* **11**, 291–296.

STEINER, G. (1936a). The status of the nematode *Aphelenchus avenae* Bastian, 1865, as a plant parasite. *Phytopathology* **26**, 294–295.

STEINER, G. (1936b). *Anguillulina askenasyi* (Bütschli, 1873), a gall forming nematode parasite of the common fern moss, *Thuidium delicatulum* (L.) Hedw. *J. Wash. Acad. Sci.* **26**, 410–414.

STELTER, H. (1958). Untersuchungen über den Kartoffelnematoden, *Heterodera rostochiensis* Wollenweber. IV. Der Einfluss von Mais (*Zea mays*) auf den Kartoffelnematoden. *NachrBl. dt. PflSchutzdienst Berl.* **12**, 133–135.

STETTMEIER, W. (1957). Bekämpfung des Wurzelgallenälchens im Gemüsebau mittels Natrium-methyldithiocarbamat. *Pflanzenschutz, Munich* **9**, 183–184.

STEUDEL, W. (1969). Untersuchungen zur Wirkung einiger Nematizide auf *Trichodorus* spp *Mitt. biol. BundAnst. Ld- u. Forstw.* 136, 77–86.

STEWART, R. N. & SCHINDLER, A. F. (1956). The effect of some ectoparasitic and endoparasitic nematodes on the expression of bacterial wilt in carnations. *Phytopathology* **46**, 219–222.

STØEN, M. (1971). *Heterodera avenae,* rase- og resistensundesokelser. *N.J.F. - Kongressen Uppsala, Seksion IV,* 101–102.

STOKES, W. M. & MARTIN, G. C. (1954). Failure of an attempt to control root-knot nematode by electrical means. *Rhod. Tobacco* Yr 1954, No. 6, 30–32. [*Helminth. Abstr.* **23**, 315a.]

STONE, A. R. (1971). Effect of processing on measurements of *Heterodera rostochiensis* larvae. *Nematologica* **17**, 167–171.

STONE, A. R. (1973). *Heterodera pallida* n.sp. (Nematoda: Heteroderidae), a second species of potato cyst nematode. *Nematologica* **18** (Yr 1972), 591–606.

STONE, A. R. (1975a). Head morphology of second-stage juveniles of some Heteroderidae (Nematoda: Tylenchoidea). *Nematologica* **21**, 81–88.

STONE, A. R. (1975b). Taxonomy of potato cyst-nematodes. *EPPO Bull.* **5**, 79–86.

STONE, A. R. & COURSE, J. A. (1974). *Heterodera goettingiana. CIH Descriptions of Plant-parasitic Nematodes* Set 4, No. 47, 4 pp.

STONE, A. R. & WILLIAMS, T. D. (1974). The morphology and soluble protein electrophoresis of *Heterodera avenae* pathotypes. *Ann. appl. Biol.* **76**, 231–236.

STONE, L. E. W. (1956). On the control of potato-root eelworm (*Heterodera rostochiensis* Wollenw.) by chlorinated phenol and *p–m*–cresol in solubilized form. *Ann. appl. Biol.* **44**, 292–306.

STONE, L. E. W. (1957). Observations on the control of potato-root eelworm (*Heterodera rostochiensis* Wollenw.) under glass by D–D and solubilized *p–m* cresol. *Ann. appl. Biol.* **45**, 256–260.

STONE, L. E. W. (1960). Field experiments on cereal root eelworm. *Ann. appl. Biol.* **48**, 681–686.

STONE, L. E. W. (1961). Oats as a trap-crop for cereal root eelworm. *Pl. Path.* **10**, 164.

STONE, L. E. W. (1968). Cereal cyst eelworm in spring barley damage assessment experiments 1963–4. *Pl. Path.* **17**, 148–150.

STOYANOV, D. (1969). [Results from investigations concerning the identification of cyst-forming nematodes in Bulgaria. 2. *Heterodera avenae*]. (In Bulgarian, Engl. Summary). *Rasten. Nauki, Sof.* **6**, 169–178.

STOYANOV, D. (1974). [Two species of *Xiphinema* parasitic on vines in Bulgaria]. *Rastitelna Zashchita* **22**, 35–38.

STRÜMPEL, H. (1967). Beobachtungen zur Lebensweise von *Aphelenchoides fragariae* in Lorraine-Begonien. *Nematologica* **13**, 67–72.

STUBBS, L. L. (1971). Plant Pathology in Australia. *Rev. Pl. Path.* **50,** 461–478.

STURHAN, D. (1966). Wirtspflanzenuntersuchungen an Bastardpopulation von *Ditylenchus dipsaci*-Rassen. *Z. PflKrankh. PflPath. PflSchutz.* **73,** 168–174.

STURHAN, D. (1969). Das Rassenproblem bei *Ditylenchus dipsaci. Mitt. biol. BundAnst. Ld- u. Forstw.* 136, 87–98.

STURHAN, D. & FRIEDMAN, W. (1965). *Ditylenchus convallariae* n.sp. (Nematoda: Tylenchida). *Nematologica* **11,** 219–223.

SUTHERLAND, J. R. & DUNN, T. G. (1970). Nematodes in coastal British Columbia forest nurseries and association of *Xiphinema bakeri* with a root disease of Douglas-fir seedlings. *Pl. Dis. Reptr* **54,** 165–168.

SUTHERLAND, J. R., DUNN, T. G. & COUSENS, N. B. F. (1970). Morphometric and morphological characteristics for identifying the life stages of *Xiphinema bakeri* Williams, 1961 (Nematoda: Longidorinae). *Can. J. Zool.* **48,** 771–773.

SUTHERLAND, J. R. & FORTIN, J. A. (1968). Effect of the nematode *Aphelenchus avenae* on some ectotrophic, mycorrhizal fungi and on a red pine mycorrhizal relationship. *Phytopathology* **58,** 519–523.

SVESHNIKOVA, N. M. (1956). A review of the study of nematodes in the families Heteroderidae and Tylenchidae, causing crop diseases in the U.S.S.R. *Nematologica* **1,** 151–158.

SVESHNIKOVA, N. M. (Ed.) (1967). [*Nematode diseases of crops*]. (In Russian). Moscow Izdatelstvo 'Kolos', 231 pp.

SWINK, J. F. (1954). Breeding for resistance to the sugar beet nematode. *Proc. Am. Soc. Sug. Beet Technol.* **8,** 109–111.

SZCZYGIEL, A. (1970). Distribution of leaf and bud nematodes (*Aphelenchoides* spp.) and stem nematode (*Ditylenchus dipsaci*) in strawberry fields in Poland. *Zesz. probl. Postęp. Nauk roln.* No. 92, 321–329.

SZCZYGIEL, A. & GIEBEL, J. (1970). Phenols in the leaf buds of two strawberry varieties, resistant and susceptible to *Aphelenchoides fragariae* (Ritzema Bos) Christie. *Zesz. probl. Postęp. Nauk roln.* No. 92, 247–253.

SZCZYGIEL, A. & HASIOR, H. (1971). Possibility of persistence of leaf and bud nematodes (*Aphelenchoides fragariae*) on strawberry plants and in the soil. *Zesz. probl. Postęp. Nauk roln.* No. 121, 101–106.

TACONIS, P. J. & KUIPER, K. (1964). Overdracht van het *Nicotiana* virus 5 door aaltjes van het geslacht *Trichodorus* in zaailingen van 5 gewassen. *Versl. Meded. plziektenk. Dienst Wageningen* No. 141, 177–178.

TARJAN, A. C. (1952). Comparative studies of some root-knot nematodes infecting the common snapdragon, *Antirrhinum majus* L. *Phytopathology* **42,** 641–644.

TARJAN, A. C. (1967). [Variability of diagnostic characters among some plant and soil nematodes]. (In Polish—English summary). *Pr. nauk. Inst. Ochr. Rośl.* **9,** 105–116.

TARJAN, A. C. (1969). Variation within the *Xiphinema americanum* group (Nematoda: Longidoridae). *Nematologica* **15,** 241–252.

TARJAN, A. C. (1973). A synopsis of the genera and species in the Tylenchorhynchinae (Tylenchoidea, Nematoda). *Proc. helminth. Soc. Wash.* **40,** 123–144.

TAYLOR, A. L. (1957). *Heterodera* taxonomy. *Proc. S–19 Workshop in Phytonematology,* 1957. Univ. Tennessee, pp. (*H.* Taxon:) 1–12.

TAYLOR, A. L. (1971). *Introduction to research on plant nematology.* (Revised edn). Rome, F.A.O. 133 pp.

TAYLOR, A. L., DROPKIN, V. H. & MARTIN, G. C. (1955). Perineal patterns of root-knot nematodes. *Phytopathology* **45**, 26–34.

TAYLOR, C. E. (1962). Transmission of raspberry ringspot virus by *Longidorus elongatus* (de Man) (Nematoda: Dorylaimidae). *Virology* **17**, 493–494.

TAYLOR, C. E. (1967). The multiplication of *Longidorus elongatus* (de Man) on different host plants with reference to virus transmission. *Ann. appl. Biol.* **59**, 275–281.

TAYLOR, C. E. (1971). Nematodes as vectors of plant viruses. In: B. M. Zuckerman, W. F. Mai & R. A. Rohde (Eds), *Plant parasitic nematodes*. Vol. 2. London & New York, Academic Press, 185–211.

TAYLOR, C. E. (1972a). Transmission of viruses by nematodes. In: C. I. Kado & H. O. Agrawal (Eds), *Principles and techniques in plant virology*. New York, Van Nostrand Reinhold, 226–247.

TAYLOR, C. E. (1972b). Nematode transmission of plant viruses. *Pest Articles & News Summaries* **18**, 269–282.

TAYLOR, C. E. & BROWN, D. J. F. (1976). The geographical distribution of *Xiphinema* and *Longidorus* nematodes in the British Isles and Ireland. *Ann. appl. Biol.* **84**, 383–402.

TAYLOR, C. E. & CADMAN, C. H. (1969). Nematode vectors. In: K. Maramorosch (Ed.), *Viruses, vectors and vegetation*. New York, Interscience, 55–94.

TAYLOR, C. E. & MURANT, A. F. (1967). Nematicidal activity of aqueous extracts from raspberry canes and roots. *Nematologica* **12**, 488–494.

TAYLOR, C. E. & MURANT, A. F. (1968). Chemical control of raspberry ringspot and tomato black ring viruses in strawberry. *Pl. Path.* **17**, 171–178.

TAYLOR, C. E. & MURANT, A. F. (1969). Transmission of strains of raspberry ringspot and tomato black ring viruses by *Longidorus elongatus* (de Man). *Ann. appl. Biol.* **64**, 43–48.

TAYLOR, C. E. & ROBERTSON, W. M. (1969). The location of raspberry ringspot and tomato black ring viruses in the nematode vector, *Longidorus elongatus* (de Man). *Ann. appl. Biol.* **64**, 233–237.

TAYLOR, C. E. & ROBERTSON, W. M. (1970a). Location of tobacco rattle virus in the nematode vector, *Trichodorus pachydermus* Seinhorst. *J. gen. Virol.* **6**, 179–182.

TAYLOR, C. E. & ROBERTSON, W. M. (1970b). Sites of virus retention in the alimentary tract of the nematode vectors, *Xiphinema diversicaudatum* (Micol.) and *X. index* Thorne and Allen. *Ann. appl. Biol.* **66**, 375–380.

TAYLOR, C. E. & ROBERTSON, W. M. (1973). Nematology. Electron microscopy. *Rep. Scott. hort. Res. Inst.* 19th (1972), p. 77.

TAYLOR, C. E. & THOMAS, P. R. (1968). The association of *Xiphinema diversicaudatum* (Micoletzky) with strawberry latent ringspot and arabis mosaic viruses in a raspberry plantation. *Ann. appl. Biol.* **62**, 147–157.

TAYLOR, C. E., THOMAS, P. R., ROBERTSON, W. M. & ROBERTS, I. M. (1970). An electron microscope study of the oesophageal region of *Longidorus elongatus*. *Nematologica* **16**, 6–12.

TAYLOR, D. P., NGUNDO, B. W. & OTHIENO, S. M. (1972). Nematology. *Rec. Res. E. Afr. Agric. For. Res. Org.* 1971, 177–185.

TAYLOR, D. P. & PILLAI, J. K. (1967). *Paraphelenchus acontioides* n.sp. (Nematoda: Paraphelenchidae), a mycophagous nematode from Illinois, with observations on its feeding habits and a key to the species of *Paraphelenchus*. *Proelc. hminth. Soc. Wash.* **34**, 51–54.

TAYLOR, D. P. & WYLIE, T. D. (1959). Interrelationship of root knot nematodes and *Rhizoctonia solani* on soybean emergence. *Phytopathology* **49,** 552. (Abstr.)

TAYLOR, J. B., CANTER-VISSCHER, T. W., NEWHOOK, F. J. & GRANDISON, G. S. (1970). Prolonged control of *Verticillium* wilt and root-lesion nematodes by soil fumigation. *N.Z. Jl Sci.* **13,** 591–602.

TERENT'EVA, T. G. (1967). [The use of statistics for testing the variability of taxonomic characters of species of the genus *Meloidogyne*]. In: N. M. Sveshnikova (Ed.), [*Nematode diseases of agricultural crops.*] Moscow, Izdatelstvo 'Kolos', 223–227. (In Russian).

TEŠIĆ, T. (1969). Prilog prončavanju otpornosti sorti pšenic prema pšeničnoj nematodi. *Savremen. Poljopr., Novi Sad* **17,** 541–543.

THOMAS, H. E. (1939). The stem nematode disease of oats and peas. *Phytopathology* **29,** 827.

THOMAS, P. R. (1965a). Studies on the biology, morphology and taxonomy of the nematode genus *Acrobeles* von Listow, 1877. *PhD Thesis, Univ. Calif.* (*Diss. Abstr.* **26,** 1583).

THOMAS, P. R. (1965b). Methyl bromide fumigation of narcissus bulbs infested with *Ditylenchus dipsaci* (Kühn) Filipjev. *Hort. Res., Edinb.* **5,** 76–80.

THOMAS, P. R. (1969). Population development of *Longidorus elongatus* on strawberry in Scotland with observations on *Xiphinema diversicaudatum* on raspberry. *Nematologica* **15,** 582–590.

THOMAS, P. R., WIGGELL, P. & MOYSE, W. J. (1958). Observations on eelworm control in narcissus in the Isles of Scilly. *Pl. Path.* **7,** 49–50.

THOMAS, W. & PROCTER, C. H. (1972). Arabis mosaic virus in *Cyphomandra betaceae* Sendt. *N.Z. Jl agric. Res.* **15,** 395–404.

THOMASON, I. J. (1962). Reaction of cereals and Sudan grass to *Meloidogyne* spp. and the relation of soil temperature to *M. javanica* populations. *Phytopathology* **52,** 787–791.

THOMASON, I. J., VAN GUNDY, S. D. & MCKINNEY, H. E. (1960). Thermotherapy for root-knot nematodes, *Meloidogyne* spp., of sweetpotato and tarragon propagating stocks. *Pl. Dis. Reptr* **44,** 354–358.

THOMASON, I. J. & LEAR, B. (1961). Rate of reproduction of *Meloidogyne* spp. as influenced by soil temperature. *Phytopathology* **51,** 520–524.

THOMPSON, H. W. (1949). The potato root eelworm (*Heterodera rostochiensis* Woll.) in the United Kingdom. *Emp. J. exp. Agric.* **17,** 60–71.

THOMPSON, L. S. & WILLIS, C. B. (1970). Effect of nematicides on root lesion nematodes and forage legume yields. *Can. J. Pl. Sci.* **50,** 577–581.

THORNE, G. (1927). The life history, habits and economic importance of some Mononchs. *J. agric. Res.* **34,** 265–286.

THORNE, G. (1928). Nematodes inhabiting the cysts of sugar-beet nematode (*Heterodera schachtii* Schmidt). *J. agric. Res.* **37,** 571–575.

THORNE, G. (1934). Some plant-parasitic nemas, with descriptions of three new species. *J. agric. Res.* **49,** 755–763.

THORNE, G. (1945). *Ditylenchus destructor,* n.sp., the potato rot nematode and *Ditylenchus dipsaci* (Kühn, 1857) Filipjev, 1936, the teasel nematode (Nematoda: Tylenchidae). *Proc. helminth. Soc. Wash.* **12,** 27–34.

THORNE, G. (1955). Fifteen new species of the genus *Hemicycliophora* with an emended description of *H. typica* de Man (Tylenchida Criconematidae). *Proc. helminth. Soc. Wash.* **22,** 1–16.

THORNE, G. (1961). *Principles of Nematology.* New York, McGraw-Hill, 553 pp.

THORNE, G. (1969). *Hypsoperine ottersoni* sp.n. (Nemata, Heteroderidae) infesting canary grass, *Phalaris arundinacea* (L.) Reed in Wisconsin. *Proc. helminth. Soc. Wash.* **36,** 98–102.

THORNLEY, J. H. M. & HESLING, J. J. (1973a). Models describing sex ratio data in *Heterodera rostochiensis.* I. Environmentally determined sex. *Nematologica* **18,** 563–570.

THORNLEY, J. H. M. & HESLING, J. J. (1973b). Models describing sex ratio data in *Heterodera rostochiensis.* II. Genetically determined sex. *Nematologica* **18,** 571–579.

THRESH, J. M. & PITCHER, R. S. (1971). The spread and control of nettlehead and other diseases of hop associated with arabis mosaic virus. *Proc. 6th Br. Insectic. Fungic. Conf.* Vol. 1, 314–318.

TIMM, R. W. (1969). The genus *Isolamium* Cobb, 1920 (Order Isolaimida, Isolaimidae new family). *J. Nematol.* **1,** 97–106.

TIMM, R. W. (1970). A revision of the nematode order Desmoscolecida Filipjev, 1929. *Univ. Calif. Publs Zool.* **93,** 1–115.

TJEPKEMA, J. P., FERRIS, V. R. & FERRIS, J. M. (1971). Review of the genus *Aporcelaimellus* Heyns, 1965 and six species groups of the genus *Eudorylaimus* Andrássy, 1959 (Nematoda: Dorylaimida). *Res. Bull. Purdue Univ.* 882, 52 pp.

TOBAR JIMENEZ, A. (1973). Nematodes de los 'Secanos' de la comarca de Alhama, I. Niveles de pobación y cultivos hospedadores. *Revta iber. Parasit.* **33,** 525–556.

TORGESON, D. C. & HAYNES, H. L. (1961). Nematocidal properties of diphenyl sulfide. *Contr. Boyce Thompson Inst. Pl. Res.* **21,** 11–19.

TOWNSHEND, J. L. (1968). Gamma irradiation of *Heterodera schachtii. Nematologica* **13** (Yr 1967), 586–592.

TOWNSHEND, J. L. (1972). Fungal diet and morphometric relationships in *Aphelenchus avenae.* In: *International Symp. Nematol.* (11*th*) *Reading, UK, 3–8 Sept.,* 1972. *Abstracts,* 76–77.

TRIANTAPHYLLOU, A. C. (1960). Sex determination in *Meloidogyne incognita* Chitwood, 1949 and intersexuality in *M. javanica* (Treub, 1885) Chitwood, 1949. *Ann. Inst. phytopath. Benaki, N.S.* **3,** 12–31.

TRIANTAPHYLLOU, A. C. (1962). Oogenesis in the root-knot nematode *Meloidogyne javanica. Nematologica* **7,** 105–113.

TRIANTAPHYLLOU, A. C. (1970). Cytogenetic aspects of evolution of the family Heteroderidae. *J. Nematol.* **2,** 26–32.

TRIANTAPHYLLOU, A. C. (1971). Genetics and cytology. In: B. M. Zuckerman, W. F. Mai & R. A. Rohde (Eds), *Plant parasitic nematodes.* Vol. 2. Academic Press, London & New York, 1–34.

TRIANTAPHYLLOU, A. C. (1973). Environmental sex differentiation of nematodes in relation to pest management. *A. Rev. Phytopath.* **11,** 441–462.

TRIANTAPHYLLOU, A. C. (1975). Oogenesis and the chromosomes of twelve bisexual species of *Heterodera* (Nematoda: Heteroderidae). *J. Nematol.* **7,** 34–40.

TRIANTAPHYLLOU, A. C. & HIRSCHMANN, H. (1960). Post-infection development of *Meloidogyne incognita* Chitwood, 1949 (Nematoda: Heteroderidae). *Ann. Inst. phytopath. Benaki, N.S.* **3,** 3–11.

TRIANTAPHYLLOU, A. C. & HUSSEY, R. S. (1973). Modern approaches in the study of relationships in the genus *Meloidogyne. EPPO Bull.* No. 9, 61–66.

TRIANTAPHYLLOU, A. C. & SASSER, J. N. (1960). Variation in perineal patterns and host specificity of *Meloidogyne incognita. Phytopathology* **50,** 724–735.

TRIFFITT, M. J. (1929a). Preliminary researches on mustard as a factor inhibiting cyst-formation in *Heterodera schachtii*. *J. Helminth.* 7, 81–92.

TRIFFITT, M. J. (1929b). On the occurrence and significance of *Heterodera schachtii* infesting certain weeds. *J. Helminth.* 7, 215–222.

TRIFFITT, M. J. (1930). On the bionomics of *Heterodera schachtii* on potatoes, with special reference to the influence of mustard on the escape of the larvae from the cysts. *J. Helminth.* **8**, 19–48.

TRIFFITT, M. J. (1931). On the occurrence of *Heterodera radicicola* associated with *Heterodera schachtii* as a field parasite in Britain. *J. Helminth.* **9**, 205–208.

TRIFFITT, M. J. (1934). Experiments with root excretions of grasses as a possible means of eliminating *Heterodera schachtii* from infected soil. *J. Helminth.* **12**, 1–12.

TRUDGILL, D. L. (1967). The effect of environment on sex determination in *Heterodera rostochiensis*. *Nematologica* **13**, 263–272.

TRUDGILL, D. L. (1974). The influence of DL-tyrosine on the sex of *Heterodera rostochiensis*. *Nematologica* **20**, 34–38.

TRUDGILL, D. L. (1976). Observations on the feeding of *Pungentus thornei* on the roots of *Petunia hybrida*. *Pl. Path.* **25**, 117–118.

TRUDGILL, D. L. & CARPENTER, J. M. (1971). Disk electrophoresis of proteins of *Heterodera* species and pathotypes of *Heterodera rostochiensis*. *Ann. appl. Biol.* **69**, 35–41.

TRUDGILL, D. L., EVANS, K. & PARROTT, D. M. (1975a). Effects of potato cyst nematodes on potato plants. I. Effects in a trial with irrigation and fumigation on the growth and nitrogen and potassium contents of a resistant and a susceptible variety. *Nematologica* **21**, 169–182.

TRUDGILL, D. L. EVANS, K. & PARROTT, D. M. (1975b). Effects of potato cyst nematodes on potato plants. II. Effects on haulm size, concentration of nutrients in haulm tissue and tuber yield of a nematode resistant and a nematode susceptible potato variety. *Nematologica* **21**, 183–191.

TRUDGILL, D. L. & PARROTT, D. M. (1972). Disc electrophoresis and larval dimensions of British, Dutch and other populations of *Heterodera rostochiensis*, as evidence of the existence of two species, each with pathotypes. *Nematologica* **18**, 141–148.

TRUDGILL, D. L. & PARROTT, D. M. (1973). Effects of growing resistant potatoes with gene H_1 from *Solanum tuberosum* ssp. *andigena* on populations of *Heterodera rostochiensis* British pathotype A. *Ann. appl. Biol.* **73**, 67–75.

TSALOLIKHIN, S. YA. (1974). [New species of nematodes—commensals of Baikal sponges]. (In Russian, Engl. Summary). *Zool. Zh.* **53**, 1081–1085.

TURLIGINA, E. S. & VERSHINSKIY, N. V. (1957). [Application of an electric current for the destruction of nematodes in the soil]. *Priroda, Moskva*, Yr 1957, No. 8, 97–98. (In Russian). [*Biol. Abstr.* **35**, 25280].

UHLENBROEK, J. H. & BIJLOO, J. D. (1958). Investigations on nematicides. I. Isolation and structure of a nematicidal principle occurring in *Tagetes* roots. *Recl Trav. chim. Pays-Bas Belg.* 77, 1004–1009.

VALDEZ, R. B., MCNAMARA, D. G., ORMEROD, P. J., PITCHER, R. S. & THRESH, J. M. (1974). Transmission of the hop strain of arabis mosaic virus by *Xiphinema diversicaudatum*. *Ann. appl. Biol.* **76**, 113–122.

VAN GUNDY, S. D., BIRD, A. F. & WALLACE, H. R. (1967). Aging and starvation in larvae of *Meloidogyne javanica* and *Tylenchulus semipenetrans*. *Phytopathology* **57**, 559–571.

Van Gundy, S. D., Stolzy, L. H., Szuszkiewicz, T. E. & Rackham, R. L. (1962). Influence of oxygen supply on survival of plant parasitic nematodes in soil. *Phytopathology* **52**, 628–632.

Videgård, G. (1973). Rasinventering av havrecystnematod. *Växtskyddsnotiser* **37**, 77–79.

Viglierchio, D. R. (1971a). Nematodes and other pathogens in auxin-related plant-growth disorders. *Bot. Rev.* **37**, 1–21.

Viglierchio, D. R. (1971b). Race genesis in *Ditylenchus dipsaci*. *Nematologica* **17**, 386–392.

Viglierchio, D. R. & Lownsbery, B. F. (1960). The hatching response of *Meloidogyne* species to the emanations from the roots of germinating tomatoes. *Nematologica* **5**, 153–157.

Walkinshaw, C. H., Griffin, G. D. & Larson, R. H. (1961). *Trichodorus christiei* as a vector of potato corky ringspot (tobacco rattle) virus. *Phytopathology* **51**, 806–808.

Wallace, E. R. & Wood, J. (1943). Stem eelworm in onion bulbs, probably seed-borne in origin. *J. Helminth.* **21**, 33–36.

Wallace, H. R. (1954). Hydrostatic pressure-deficiency and the emergence of larvae from cysts of the beet eelworm. *Nature, Lond.* **173**, 502–503.

Wallace, H. R. (1955a). The influence of soil moisture on the emergence of larvae from cysts of the beet eelworm, *Heterodera schachtii* Schmidt. *Ann. appl. Biol.* **43**, 477–484.

Wallace, H. R. (1955b). Factors influencing the emergence of larvae from cysts of the beet eelworm, *Heterodera schachtii* Schmidt. *J. Helminth.* **29**, 3–16.

Wallace, H. R. (1956). Soil aeration and the emergence of larvae from cysts of the beet eelworm, *Heterodera schachtii* Schm. *Ann. appl. Biol.* **44**, 57–66.

Wallace, H. R. (1958a). Movement of eelworms. I. The influence of pore size and moisture content of the soil on the migration of larvae of the beet eelworm, *Heterodera schachtii* Schmidt. *Ann. appl. Biol.* **46**, 74–85.

Wallace, H. R. (1958b). Movement of eelworms. III. The relationship between eelworm length, activity and mobility. *Ann. appl. Biol.* **46**, 662–668.

Wallace, H. R. (1958c). Observations on the emergence from cysts and the orientation of larvae of three species of the genus *Heterodera* in the presence of host plant roots. *Nematologica* **3**, 236–243.

Wallace, H. R. (1959). The movement of eelworms in water films. *Ann. appl. Biol.* **47**, 366–370.

Wallace, H. R. (1960). Movement of eelworms. VI. The influences of soil type, moisture gradients and host plant roots on the migration of the potato-root eelworm *Heterodera rostochiensis* Wollenweber. *Ann. appl. Biol.* **48**, 107–120.

Wallace, H. R. (1961). The nature of resistance in chrysanthemum varieties to *Aphelenchoides ritzemabosi*. *Nematologica* **6**, 49–58.

Wallace, H. R. (1963). *The biology of plant parasitic nematodes*. London, Edward Arnold, 280 pp.

Wallace, H. R. (1968). The dynamics of nematode movement. *A. Rev. Phytopath.* **6**, 91–114.

Wålstedt, I. (1967). Havrenematodraser och sortresistens. *Sver. Utsädesför. Tidskr.* **77**, 177–241.

Walter, J. M. & Kelsheimer, E. G. (1949). In-the-row application of soil fumigants. *Market Growers Journal* **78**, 12.

Weaving, G. S. (1960). Efficient hot water treatment plant. *Commercial Grower* No. 3349, 619–625.

WEBER, A. PH., ZWILLENBERG, L. O. & VAN DER LAAN, P. A. (1952). A predacious amoeboid organism destroying larvae of the potato root eelworm and other nematodes. *Nature, Lond.* **169,** 834–835.

WEBLEY, D. (1970). A morphometric study of the three pathotypes of the potato cyst eelworm (*Heterodera rostochiensis*) recognised in Great Britain. *Nematologica* **16,** 107–112.

WEBSTER, J. M. (1964). Biological races in species of plant parasitic nematodes. *Parasitology* **54,** 8 P. (Abstr.)

WEBSTER, J. M. (1967). The significance of biological races of *Ditylenchus dipsaci* and their hybrids. *Ann. appl. Biol.* **59,** 77–83.

WEBSTER, J. M. (1969). The host-parasite relationships of plant parasitic nematodes. *Adv. Parasit.* **7,** 1–40.

WEBSTER, J. M. (Ed.) (1972). *Economic nematology.* London & New York, Academic Press, 563 pp.

WEBSTER, J. M. & HOOPER, D. J. (1968). Serological and morphological studies on the inter- and intraspecific differences of the plant-parasitic nematodes *Heterodera* and *Ditylenchus. Parasitology* **58,** 879–891.

WEHUNT, E. J. & EDWARDS, D. I. (1968). *Radopholus similis* and other nematode species on banana. In: G. C. Smart & V. G. Perry (Eds), *Tropical nematology* Gainesville, Univ. Florida Press, 1–19.

WEISCHER, B. (1957). Die Wirkung ionisierender Strahlen auf die Entwicklung von *Heterodera rostochiensis* und *H. schachtii. Nematologica* **2,** 300–305.

WEISCHER, B. (1959). Experimentelle Untersuchungen über die Wanderung von Nematoden. *Nematologica* **4,** 172–186.

WEISCHER, B. (1968a). Wechselwirkungen zwischen Nematoden und anderen Schaderregern an Nutzpflanzen. *C.R.* 8me *Symp. internat. Nématol. Antibes* 1965, 91–107.

WEISCHER, B. (1968b). Virusübertragung durch Nematoden. *Mitt. biol. Bund-Anst. Ld- u. Forstw.* No. 128, 24–38.

WEISCHER, B. (1969a). Vermehrung und Schadwirkung von *Aphelenchoides ritzemabosi* und *Ditylenchus dipsaci* in virusfreiem und in TMV-infiziertem Tabak. *Nematologica* **15,** 334–336.

WEISCHER, B. (1969b). [German translation from Stegarescu (1966)]. *Nematology News* No. 16, 20–23.

WEISCHER, B. (1975). Ecology of *Xiphinema* and *Longidorus.* In: F. Lamberti, C. E. Taylor & J. W. Seinhorst (Eds), *Nematode vectors of plant viruses.* London & New York, Plenum Press, 291–307.

WHITE, J. H. (1953). Wind-borne dispersal of potato root eelworm. *Nature, Lond.* **172,** 686–687.

WHITEHEAD, A. G. (1959). *Nothanguina cecidoplastes* n.comb. syn. *Anguina cecidoplastes* (Goodey, 1934) Filipjev, 1936 (Nothotylenchinae: Tylenchida) *Nematologica* **4,** 70–75.

WHITEHEAD, A. G. (1968). Taxonomy of *Meloidogyne* (Nematodea: Heteroderidae) with descriptions of four new species. *Trans. zool. Soc. Lond.* **31,** 263–401.

WHITEHEAD, A. G. (1969). The distribution of root-knot nematodes (*Meloidogyne* spp) in tropical Africa. *Nematologica* **15,** 315–333.

WHITEHEAD, A. G. (1972). Nematicides for field crops. *Proc. 6th Br. Insectic. Fungic. Conf.,* 1971, Vol. 3, 662–672.

WHITEHEAD, A. G. (1973). Control of cyst-nematodes (*Heterodera* spp.) by organophosphates, oximecarbamates and soil fumigants. *Ann. appl. Biol.* **75**, 439–453.

WHITEHEAD, A. G. (1974). Nematicides for temperate crops. *Proc. 7th Br. Insectic. Fungic. Conf.*, 1973, Vol. 3, 955–962.

WHITEHEAD, A. G. (1975). Chemical control of potato cyst-nematode. *ARC Res. Rev.* **1**, 17–23.

WHITEHEAD, A. G., BROMILOW, R. H., LORD, K. A., MOSS, S. R. & SMITH, J. (1975). Incorporating granular nematicides in soil. *Proc. 8th Br. Insectic. Fungic. Conf.* Vol. 1, 133–144.

WHITEHEAD, A. G., DUNNING, R. A. & COOKE, D. A. (1971). Docking disorder and root ectoparasitic nematodes of sugar beet. *Rep. Rothamsted exp. Stn for* 1970, *Part* 2, 219–236.

WHITEHEAD, A. G., FRASER, J. E., FRENCH, E. M. & WRIGHT, S. M. (1975). Chemical control of potato cyst-nematode, *Heterodera pallida,* on tomato grown under glass. *Ann. appl. Biol.* **80**, 75–84.

WHITEHEAD, A. G., FRASER, J. E. & GREET, D. N. (1970). The effect of D–D, chloropicrin and previous crops on numbers of migratory root-parasitic nematodes and on the growth of sugar beet and barley. *Ann. appl. Biol.* **65**, 351–359.

WHITEHEAD, A. G., FRASER, J. E. & STOREY, G. (1972). Chemical control of potato cyst-nematode in sandy clay soil. *Ann. appl. Biol.* **72**, 81–88.

WHITEHEAD, A. G. & HOOPER, D. J. (1970). Needle nematodes (*Longidorus* spp) and stubby-root nematodes (*Trichodorus* spp.) harmful to sugar beet and other field crops in England. *Ann. appl. Biol.* **65**, 339–350.

WHITEHEAD, A. G., TITE, D. J., FINCH, P., FRASER, J. E., FRENCH, E. M. & WRIGHT, S. M. (1974). [Trials with nematicides]. In: Nematology department. *Rep. Rothamsted exp. Stn for* 1973, *Part* 1, 160–163.

WHITEHEAD, A. G., TITE, D. J. & FRASER, J. E. (1970). The effect of small doses of nematicides on migratory root-parasitic nematodes and on the growth of sugar beet and barley in sandy soils. *Ann. appl. Biol.* **65**, 361–375.

WHITEHEAD, A. G., TITE, D. J. & FRASER, J. E. (1973a). Chemical control of stem nematode. In: Nematology Department. *Rep. Rothamsted exp. Stn for* 1972, *Part* 1, 166–167.

WHITEHEAD, A. G., TITE, D. J. & FRASER, J. E. (1973b). Control of potato cyst nematode, *Heterodera rostochiensis,* in sandy loam, by Du Pont 1410 (*S*–methyl 1–(dimethylcarbamoyl)–*N*–[(methylcarbamoyl)oxy] thioformimidate) applied to the soil at planting time. *Ann. appl. Biol.* **73**, 325–328.

WHITEHEAD, A. G., TITE, D. J., FRASER, J. E. & FRENCH, E. M. (1972). Control of potato cyst-nematode, *Heterodera rostochiensis,* in peaty loam soil by D–D, aldicarb and a resistant variety of potato. *Ann. appl. Biol.* **72**, 307–312.

WHITEHEAD, A. G., TITE, D. J., FRASER, J. E. & FRENCH, E. M. (1973a). Control of potato cyst-nematode, *Heterodera rostochiensis,* in silt and peat loams by ten pesticides applied to the soil at planting time. *Ann. appl. Biol.* **73**, 197–201.

WHITEHEAD, A. G., TITE, D. J., FRASER, J. E. & FRENCH, E. M. (1973b). Treating potato ridges in spring with aldicarb, D–D or dazomet to control potato cyst-nematode, *Heterodera rostochiensis,* in sandy clay and peat loam soils. *Ann. appl. Biol.* **73**, 203–210.

WHITEHEAD, A. G., TITE, D. J., FRASER, J. E. & FRENCH, E. M. (1973c). Effects of D–D, Telone or dazomet applied to potato ridges in spring on potato cyst-nematode, *Heterodera rostochiensis,* in sandy loam and silt loam soils. *Ann. appl. Biol.* **74**, 105–111.

WHITEHEAD, A. G., TITE, D. J., FRASER, J. E. & FRENCH, E. M. (1973d). Control of potato cyst-nematode, *Heterodera rostochiensis,* in three soils by small amounts of aldicarb, Du Pont 1410 or Nemacur applied to the soil at planting time. *Ann. appl. Biol.* **74,** 113–118.

WHITEHEAD, A. G., TITE, D. J., FRASER, J. E. & FRENCH, E. M. (1973e). Control of potato cyst-nematode, *Heterodera rostochiensis,* in sandy, peaty and silt loam soils by dazomet and Telone applied in different ways. *Ann. appl. Biol.* **75,** 257–265.

WHITEHEAD, A. G., TITE, D. J., FRASER, J. E. & FRENCH, E. M. (1974). Control of pea cyst-nematode, *Heterodera goettingiana,* by small amounts of aldicarb, Du Pont 1410, Ciba-Geigy 10576 or Dowco 275 applied to the soil at planting time. *Ann. appl. Biol.* **78,** 331–335.

WHITEHEAD, A. G., TITE, D. J., FRASER, J. E., FRENCH, E. M., BROMILOW, R. H., LORD, K. A. & SMITH, J. (1974). Incorporation of nematicide granules in soil by rotavation. In: Nematology department. *Rep. Rothamsted exp. Stn for* 1973, *Part* 2, 164–165.

WHITEHEAD, A. G., TITE, D. J., FRASER, J. E., FRENCH, E. M. & SMITH, J. (1975). Incorporating granular nematicides in soil to control potato cyst-nematode, *Heterodera rostochiensis. Ann. appl. Biol.* **80,** 85–92.

WIDDOWSON, E., DONCASTER, C. C. & FENWICK, D. W. (1958). Observations on the development of *Heterodera rostochiensis* Woll. in sterile root cultures. *Nematologica* **3,** 308–314.

WIDDOWSON, J. P., YEATES, G. W. & HEALY, W. B. (1973). The effect of root nematodes on the utilization of phosphorus by white clover on a yellow-brown loam. *N.Z. Jl agric. Res.* **16,** 77–80.

WIESER, W. (1953). On the structure of the cyst wall in four species of *Heterodera* Schmidt. *Meddn St. VäxtskAnst* No. 65, 15 pp.

WILLIAMS, R. J. B. (1971). Relationships between the composition of soils and physical measurements made on them. *Rep. Rothamsted exp. Stn for* 1970, *Part* 2, 5–35.

WILLIAMS, T. D. (1956). The resistance of potatoes to root eelworm. *Nematologica* **1,** 88–93.

WILLIAMS, T. D. (1957). Development of isolated female larvae of the potato-root eelworm (*Heterodera rostochiensis* Woll.). *Nature, Lond.* **180,** 1000.

WILLIAMS, T. D. (1958). Potatoes resistant to root eelworm. *Proc. Linn. Soc. Lond.* **169,** 93–104.

WILLIAMS, T. D. (1961). *Xiphinema bakeri* n.sp. (Nematoda: Longidorinae) from the Fraser River Valley, British Columbia, Canada. *Can. J. Zool.* **39,** 407–412.

WILLIAMS, T. D. (1969). The effects of formalin, nabam, irrigation and nitrogen on *Heterodera avenae* Woll., *Ophiobolus graminis* Sacc. and the growth of spring wheat. *Ann. appl. Biol.* **64,** 325–334.

WILLIAMS, T. D. (1970). Barley segregates resistant and susceptible to the cereal cyst-nematode (*Heterodera avenae* Woll.). *Ann. appl. Biol.* **66,** 339–346.

WILLIAMS, T. D. & BEANE, J. (1972). Effect of temperature and root exudates on *H. avenae* hatch. In: Nematology department. *Rep. Rothamsted exp. Stn for* 1971, *Part* 1, 171–172.

WILLIAMS, T. D. & SALT, G. A. (1970). The effects of soil sterilants on the cereal cyst-nematode (*Heterodera avenae* Woll.) take-all (*Ophiobolus graminis* Sacc.) and yields of spring wheat and barley. *Ann. appl. Biol.* **66,** 329–338.

WILLIAMS, T. D. & SIDDIQI, M. R. (1972). *Heterodera avenae. CIH Descriptions of Plant-parasitic Nematodes* Set 1, No. 2, 4 pp.

WILLIAMS, W. M. (1972). Laboratory screening of white clover for resistance to stem nematode. *N.Z. Jl agric. Res.* **15**, 363–370.

WILLIAMSON, M. (1972). *The analysis of biological populations.* London, Edward Arnold, 180 pp.

WILSKI, A. & GIEBEL, J. (1966). β-Glucosidase in *Heterodera rostochiensis* and its significance in resistance of potato to this nematode. *Nematologica* **12**, 219–224.

WILSON, E. M. (1968). Vulval bodies in certain species of *Heterodera. Nature, Lond.* **217**, 879.

WILSON, E. M. (1971). The changes in cyst form during the development of the cyst of *Heterodera rostochiensis* and *H. schachtii. J. Helminth.* **45**, 211–212.

WILSON, J. D. & HEDDEN, O. (1964). Effect of repeated applications of nematocides on vegetable yields and nematode populations in muck soils. *Pl. Dis. Reptr* **48**, 698–702.

WILSON, W. R. & FRENCH, N. (1975). Population studies of stem eelworm *Ditylenchus dipsaci* in microplots, field plots and fields in the north of England under different crop rotations. *Pl. Path.* **24**, 176–183.

WINDRICH, W. A. (1970). Control of *Ditylenchus dipsaci* (Kuehn) in tulip bulbs by immersion in thionazin. *Hort. Res., Edinb.* **10**, 83–87.

WINFIELD, A. L. (1965). Microplot experiments with pea cyst nematode. *Pl. Path.* **14**, 158–161.

WINFIELD, A. L. (1969). Control of stem nematode (*Ditylenchus dipsaci*) in flower bulbs with hot water or thionazin. *Proc. 5th Br. Insectic. Fungic. Conf.* Vol. 1, 274–283.

WINFIELD, A. L. (1970a). Control [of stem eelworm] by chemical dips. In: Narcissus Pests. *Bull. Minist. Agric. Fish. Fd* 51 (6th Edn) 20–21. (H.M.S.O.).

WINFIELD, A. L. (1970b). Factors affecting the control by hot water treatment of stem nematode *Ditylenchus dipsaci* (Kühn) Filipjev in narcissus bulbs. *J. hort. Sci.* **45**, 447–456.

WINFIELD, A. L. (1971). Control of bulb scale mite and stem nematode of narcissus, and reclaiming forced bulbs. *Pl. Path.* **20**, 10–13.

WINFIELD, A. L. (1972). Observations on the control of the stem nematode *Ditylenchus dipsaci* of tulips with hot water or thionazin treatment. *J. hort. Sci.* **47**, 357–364.

WINFIELD, A. L. (1973). Chemical control of *Ditylenchus dipsaci* in flower bulbs and onions. *Ann. appl. Biol.* **75**, 454–460.

WINFIELD, A. L. (1974). Observations on the occurrence, pathogenicity and control of *Pratylenchus vulnus, P. thornei* and *Xiphinema diversicaudatum* associated with glasshouse roses. *Ann. appl. Biol.* **77**, 297–307.

WINFIELD, A. L., MURDOCH, G. & JOHN, M. E. (1971). Chemical control of stem nematodes (*Ditylenchus dipsaci*) of bulb onions. *Proc. 6th Br. Insectic. Fungic. Conf.* Vol. 1, 141–149.

WINNER, C. (1957). Über die activierende Wirkung von Aminoacridinen auf *Heterodera schachtii. Nematologica* **2**, 126–130.

WINSLOW, R. D. (1954). Provisional lists of host plants of some root eelworms (*Heterodera* spp.). *Ann. appl. Biol.* **41**, 591–605.

WINSLOW, R. D. (1955a). The hatching responses of some root eelworms of the genus *Heterodera. Ann. appl. Biol.* **43**, 19–36.

WINSLOW, R. D. (1955b). Experiments on the control of the potato root eelworm by trap cropping with black nightshade. *Pl. Path.* **4,** 139–140.

WINSLOW, R. D. (1956). Seasonal variations in the hatching responses of the potato-root eelworm, *Heterodera rostochiensis* Wollenweber, and related species. *J. Helminth.* **30,** 157–164.

WINSLOW, R. D. & WILLIAMS, T. D. (1957). Amoeboid organisms attacking larvae of the potato root eelworm (*Heterodera rostochiensis* Woll.) in England and the beet eelworm (*H. schachtii* Schm.) in Canada. *Tijdschr. PlZiekt.* **63,** 242–243.

WONG, C. L. & WILLETTS, H. J. (1969). Gall formation in aerial parts of plants inoculated with *Meloidogyne javanica. Nematologica* **15,** 425–428.

WONG, T. K. & MAI, W. F. (1973a). Pathogenicity of *Meloidogyne hapla* to lettuce as affected by inoculum level, plant age at inoculation and temperature. *J. Nematol.* **5,** 126–129.

WONG, T. K. & MAI, W. F. (1973b). *Meloidogyne hapla* in organic soil: effects of environment on hatch, movement and root invasion. *J. Nematol.* **5,** 130–138.

WONG, T. K. & MAI, W. F. (1973c). Effect of temperature on growth, development and reproduction of *Meloidogyne hapla* in lettuce. *J. Nematol.* **5,** 139–142.

WOOD, F. C. & GOODEY, J. B. (1957). Effects of gamma-ray irradiation on nematodes infesting cultivated mushroom beds. *Nature, Lond.* **180,** 760–761.

WOOD, F. H. (1973). Nematode feeding relationships. Feeding relationships of soil-dwelling nematodes. *Soil Biol. Biochem.* **5,** 593–601.

WOOD, F. H. & FOOT, M. A. (1975). Treatment of potato tubers to destroy cysts of potato cyst nematode: a note. *N.Z. Jl exp. Agric.* **3,** 349–350.

WOODVILLE, H. C. (1964). Lethal times and temperatures for two species of eelworm. *Expl Hort.* No. 10, 90–95.

WOODVILLE, H. C. & MORGAN, H. G. (1961). Lethal times and temperatures for bulb eelworm (*Ditylenchus dipsaci*). *Expl Hort.* No. 5, 19–22.

WOUTS, W. M. (1973a). A revision of the family Heteroderidae (Nematoda: Tylenchoidea). I. The family Heteroderidae and its subfamilies. *Nematologica* **18** (Yr 1972), 439–446.

WOUTS, W. M. (1973b). A revision of the family Heteroderidae (Nematoda: Tylenchoidea). II. The subfamily Meloidoderinae. *Nematologica* **19,** 218–235.

WOUTS, W. M. (1974). A revision of the family Heteroderidae (Nematoda: Tylenchoidea). III. The subfamily Ataloderinae. *Nematologica* **19** (Yr 1973), 279–284.

WOUTS, W. M. & SHER, S. A. (1971). The genera of the subfamily Heteroderinae (Nematoda: Tylenchoidea) with a description of two new genera. *J. Nematol.* **3,** 129–144.

WRIGHT, K. A. (1965). The histology of the oesophageal region of *Xiphinema index* Thorne & Allen, 1950, as seen with the electron microscope. *Can. J. Zool.* **43,** 689–700.

WU L.-Y. (1968). Morphological study of *Anguina calamagrostis* Wu, 1967 (Tylenchidae: Nematoda) with the electron microscope. *Can. J. Zool.* **46,** 467–468.

WUYTS, F. (1971). Chemische grondontsmetting in de schorsenerenteelt. *Tuinbouwberichten* **35,** 143–144.

WYSS, U. (1969). Untersuchungen über das schadauftreten wandernder Wurzelnematoden an Erdbeerkulturen in Niedersachsen. *Mitt. biol. BundAnst. Ld- u. Forstw.* No. 136, 110–126.

WYSS, U. (1971). Der Mechanismus der Nahrungsaufnahme bei *Trichodorus similis*. *Nematologica* **17**, 508–518.

WYSS, U. (1973). Feeding of *Tylenchorhynchus dubius*. *Nematologica* **19**, 125–136.

YADAV, B. S. & VERMA, A. C. (1971). Cyst forming nematode (*Heterodera* sp.) attacking *Sesamum orientale* in India. *Curr. Sci.* **40**, 612.

YAGITA, H. & KOMURO, Y. (1972). [Transmission of mulberry ringspot virus by *Longidorus martini* Merny.] *Ann. phytopath. Soc. Japan* **38**, 275–283.

YASSIN, A. M. (1969). Glasshouse and laboratory studies on the biology of the needle nematode, *Longidorus elongatus*. *Nematologica* **15**, 169–178.

YEATES, G. W. (1970). Failure of *Heterodera carotae*, *H. cruciferae* and *H. goettingiana* to interbreed *in vitro*. *Nematologica* **16**, 153–154.

YEATES, G. W., HEALY, W. B. & WIDDOWSON, J. P. (1973). Screening of legume varieties for resistance to the root nematodes *Heterodera trifolii* and *Meloidogyne hapla*. *N.Z. Jl agric. Res.* **16**, 81–86.

YOKOO, T. (1964). On a new species of *Aphelenchoides* (Aphelenchidae: Nematoda), parasite of bulb of lily, from Japan. *Agric. Bull. Saga Univ.* **20**, 67–69.

YOUNGSON, C. R. & GORING, C. A. I. (1962). Diffusion and nematode control by 1,2–dibromoethane, 1,3–dichloropropene, and 1,2–dibromo–3–chloropropane in soil. *Soil Sci.* **93**, 306–316.

YOUNGSON, C. R. & GORING, C. A. I. (1963). Nellite nematocide for root-knot nematode control. Laboratory and greenhouse studies. *Biokemia* **1**, 2–8.

YUEN, P.-H. (1967). Electron microscopical studies on *Ditylenchus dipsaci* (Kühn) I. Stomatal region. *Can. J. Zool.* **45**, 1019–1033.

YUKSEL, H. S. (1960). Observations on the life cycle of *Ditylenchus dipsaci* on onion seedlings. *Nematologica* **5**, 289–296.

ZOPF, W. (1888). Zur Kenntnis der Infektions-Krankheiten niederer Thiere und Pflanzen. *Nova Acta Leop. Carol.* **52**, 314–376.

ZUCKERMAN, B. M. (1969). Gnotobiotic culture and its significance to plant nematology. *Helminth. Abstr.* **38**, 1–15.

ZUCKERMAN, B. M., MAI, W. F. & ROHDE, R. A. (Eds). *Plant parasitic nematodes*. Vol. 1. *Morphology, anatomy, taxonomy, and ecology*. 345 pp. Vol. 2. *Cytogenetics, host-parasite interactions, and physiology*. 347 pp. London & New York, Academic Press.

Read and follow carefully the instructions on all pesticide labels

Pesticides Safety Precautions Scheme

Users should consult the 'Recommendations for Safe Use of Chemical Compounds used in Agriculture and Food Storage' published by the Ministry of Agriculture, Fisheries and Food. Official recommendations have been published for many of the pesticides mentioned in this book.

There are statutory obligations affecting employers and workers who use certain poisonous substances; users are strongly advised to read the leaflet, 'The Safe Use of Poisonous Chemicals on the Farm' obtainable from Health and Safety Executive, Baynards House, Chepstow Place, Westbourne Grove, London W2 4TY.

Agricultural Chemicals Approval Scheme

AGRICULTURAL CHEMICALS APPROVAL SCHEME

Proprietary products based on chemicals used for pest, disease and weed control can be officially approved under the Agricultural Chemicals Approval Scheme. It is strongly recommended that approved products should be used. Approval is indicated on the containers by the mark shown above. A List of Approved Products is published in February each year and may be bought from Government Bookshops, through booksellers or from the Ministry of Agriculture, Fisheries and Food (Publications), Block C, Tolcarne Drive, Pinner, Middlesex HA5 2DT.

Index

Printed in Scotland by Her Majesty's Stationery Office at HMSO Press, Edinburgh
Dd 586984 K10 8/78 (14582)